Ranaviruses

Matthew J. Gray • V. Gregory Chinchar
Editors

Ranaviruses

Emerging Pathogens of Ectothermic Vertebrates

Second Edition

 Springer

Editors
Matthew J. Gray
School of Natural Resources
University of Tennessee at Knoxville
Knoxville, TN, USA

V. Gregory Chinchar
Department of Cell and Molecular Biology
University of Mississippi Medical Center
Jackson, MS, USA

ISBN 978-3-031-64975-2   ISBN 978-3-031-64973-8   (eBook)
https://doi.org/10.1007/978-3-031-64973-8

This Springer imprint is published by the registered company Springer Nature Switzerland AG
The registered company address is: Gewerbestrasse 11, 6330 Cham, Switzerland

If disposing of this product, please recycle the paper.

Acknowledgements

We thank the following organizations for providing funds to support Open Access publishing of the second edition: University of Tennessee's (UT) Open Publishing Support Fund, UT School of Natural Resources, UT Center for Wildlife Health, Washington State University Libraries. University of Mississippi Medical Center, Gordon State College, Association of Reptile and Amphibian Veterinarians, and the Global Ranavirus Consortium.

Contents

Immune Defenses Against Ranavirus Infections 83

Leon Grayfer, Eva-Stina Edholm, V. Gregory Chinchar, Yongming Sang,
and Jacques Robert

**Characterization, Pathogenesis, and Immuno-Biological Control
of Singapore Grouper Iridovirus (SGIV)** 121

Youhua Huang, Shaowen Wang, Xiaohong Huang, Jingguang Wei,
and Qiwei Qin

Ranavirus Distribution and Host Range. 155

Rachel E. Marschang, Jonathan I. Meddings, Thomas B. Waltzek,
Paul Hick, Matthew C. Allender, Wytamma Wirth,
and Amanda L. J. Duffus

Contributors

Matthew C. Allender Wildlife Epidemiology Lab, University of Illinois, Urbana, IL, USA

Brookfield Zoo Chicago, Brookfield, IL, USA

Ellen Ariel College of Public Health, Medical and Veterinary Sciences, James Cook University, Townsville, QLD, Australia

Jesse L. Brunner School of Biological Sciences, Washington State University, Pullman, WA, USA

V. Gregory Chinchar Department of Cell and Molecular Biology, University of Mississippi Medical Center, Jackson, MS, USA

Amanda L. J. Duffus Department of Natural Sciences, Gordon State College, Barnesville, GA, USA

Julia E. Earl School of Biological Sciences, Louisiana Tech University, Ruston, LA, USA

Eva-Stina Edholm The Arctic University of Norway, Norwegian College of Fishery Science, Tomsø, Norway

María J. Forzán Department of Veterinary Sciences, University of Wyoming, Laramie, WY, USA

Trenton W. J. Garner Institute of Zoology, Zoological Society of London, London, UK

Matthew J. Gray Center for Wildlife Health, School of Natural Resources, University of Tennessee, Knoxville, TN, USA

Leon Grayfer Department of Biological Sciences, George Washington University, Washington, DC, USA

Paul Hick Elizabeth Macarthur Agricultural Institute, Department of Primary Industries, Menangle, NSW, Australia

Jason T. Hoverman Department of Forestry and Natural Resources, Purdue University, West Lafayette, IN, USA

Xiaohong Huang College of Marine Sciences, South China Agricultural University, Guangzhou, China

Youhua Huang College of Marine Sciences, South China Agricultural University, Guangzhou, China

James K. Jancovich Department of Biological Sciences, California State University, San Marcos, CA, USA

Rachel E. Marschang Laboklin GmbH & Co. KG, Bad Kissingen, Germany

Jonathan I. Meddings Faculty of Medicine, Nursing and Health Sciences, Eastern Health Clinical School, Monash University, Melbourne, Victoria, Australia

Debra L. Miller One Health Initiative, Center for Wildlife Health and Department of Biomedical and Diagnostic Sciences, University of Tennessee, Knoxville, TN, USA

Angela Peace Department of Mathematics and Statistics, Texas Tech University, Lubbock, TX, USA

Allan P. Pessier College of Veterinary Medicine, Washington State University, Pullman, WA, USA

Qiwei Qin College of Marine Sciences, South China Agricultural University, Guangzhou, China

Jacques Robert University of Rochester Medical Center, Department of Microbiology and Immunology, and Environmental Medicine, Rochester, Rochester, NY, USA

Emily H. Le Sage Biology Department, Skidmore College, Saratoga Springs, NY, USA

Yongming Sang Department of Agricultural and Environmental Sciences, College of Agriculture, Tennessee State University, Nashville, TN, USA

Andrew Storfer School of Biological Sciences, Washington State University, Pullman, WA, USA

Kuttichantran Subramaniam Department of Infectious Diseases & Immunology, College of Veterinary Medicine, University of Florida, Gainesville, FL, USA

Thomas B. Waltzek College of Veterinary Medicine, Washington State University, Pullman, WA, USA

Shaowen Wang College of Marine Sciences, South China Agricultural University, Guangzhou, China

Jingguang Wei College of Marine Sciences, South China Agricultural University, Guangzhou, China

Richard J. Whittington Sydney School of Veterinary Science, Faculty of Science, University of Sydney, Sydney, NSW, Australia

Wytamma Wirth Peter Doherty Institute for Infection and Immunity, University of Melbourne, Parkville, VIC, Australia

Qi-Ya Zhang Institute of Hydrobiology, Chinese Academy of Sciences, Wuhan, Hubei, China

Introduction to Ranaviruses: Past, Present, and Future

V. Gregory Chinchar and Matthew J. Gray

Since the first edition of *Ranaviruses: Lethal Pathogens of Ectothermic Vertebrates* in 2015, an increasingly large number of studies focused on the biology and ecology of ranaviruses and other related genera within the family *Iridoviridae* have been conducted. Indicative of continuing interest in this virus family, a PubMed search, covering the 8 years from 2015 to 2023 and using "iridovirus, ranavirus, and megalocytivirus" as keywords, identified about as many papers published in those 8 years (916) as in the preceding 60-year period (1187). In view of this growing interest, this updated and expanded edition will focus primarily on findings reported since the first edition.

Of the three genera within the family *Iridoviridae* that infect cold-blooded vertebrates, *Lymphocystivirus*, *Megalocytivirus*, and *Ranavirus*, the last has received the most attention for a variety of reasons. Unlike the first two genera whose members infect only fish, ranaviruses, despite their eponymous designation, display a broad host range that encompasses fish and reptiles in addition to amphibians. Moreover, they are propagated readily in cell culture allowing facile in vitro analysis of various aspects of viral biogenesis. Finally, ranaviruses are pathogens of both farmed amphibians and fish, as well as ecologically important wild species. Among commercially important species (e.g., grouper and red seabream), ranaviruses and megalocytiviruses are responsible for considerable economic losses, whereas among native amphibians, ranavirus infections have led to localized die-offs that threaten the survival and population persistence of some species. In addition to their study as pathogens, ranaviruses have proven to be a very useful tool with which to probe the

V. G. Chinchar
Department of Cell and Molecular Biology, University of Mississippi Medical Center, Jackson, MS, USA

M. J. Gray (✉)
Center for Wildlife Health, School of Natural Resources, University of Tennessee at Knoxville, Knoxville, TN, USA
e-mail: mgray11@utk.edu

M. J. Gray, V. G. Chinchar (eds.), *Ranaviruses*,
https://doi.org/10.1007/978-3-031-64973-8_1

1

immune system of lower vertebrates. Studies using frog virus 3 (FV3), the best studied member of the genus *Ranavirus*, and the African clawed toad, *Xenopus laevis*, whose immune system has been the subject of extensive research, provide a powerful model for examining innate and acquired anti-viral immunity in an ectothermic vertebrate and for determining the specific role that host anti-viral immune proteins and cells play in protection from severe disease and death.

The discovery of the first ranavirus was serendipitous and a classic example of chance favoring the prepared mind. Allan Granoff, chair of the Division of Virology at St. Jude Children's Research Hospital in Memphis, TN, was interested in tumor viruses, specifically Lucke tumor virus (LTV), a herpesvirus shown to be the causative agent of renal carcinomas in leopard frogs. Earlier work showed that exposure of tadpoles to LTV led to the development of tumors during warm summer months. Interestingly, the tumors did not contain virus particles. However, with the onset of cold weather in the fall and continuing during winter hibernation, the tumors regressed and viral replication was initiated leading to the release of virus and the subsequent infection of susceptible young the following spring. Thus, although viral DNA was present in summer tumors, the only way to initiate viral replication and obtain virus particles was to expose tumor bearing frogs to low temperature (e.g., 8–9 C for 8–16 weeks). Under those conditions, virions were formed and released in the urine. As this method of virion production was awkward, Granoff, seeking a better approach, speculated that frog kidney cultures might prove a suitable in vitro system with which to propagate LTV. Primary cell cultures were established from a number of ostensibly healthy and tumor-bearing leopard frogs (*Lithobates* (formerly *Rana*) *pipiens*). After several days, spontaneous cytopathic effect was seen in two cultures prepared from healthy frogs and one from a tumor-bearing animal; frog viruses 1 and 2 (FV1, FV2) were isolated from the healthy animals, whereas FV3 was obtained from the tumor-bearing animal. While it is probable that all three viruses were the same viral species, FV3 was chosen for further study because of its association with a tumor-bearing animal. Although subsequent study showed that FV3 was not linked to tumor development, its study continued and allowed Granoff and others to characterize, in ways not possible with previously identified lymphocystiviruses and invertebrate iridescent viruses (IIV), the life cycle of viruses with the family *Iridoviridae*. Although early work by Granoff and others focused primarily on molecular aspects of FV3 replication, it became apparent by the mid-1980s that ranaviruses were linked to disease outbreaks among various fish and amphibian species. The growing impact of ranavirus infections on a large number of ectothermic vertebrates is currently reflected in the increased publication of reports dealing with various issues related to viral ecology and diagnostics.

The first chapters of this monograph focus on taxonomic, genetic, molecular, and immunological aspects of ranaviruses and their replication. Included in the first section is one new chapter covering *Singapore grouper iridovirus* (SGIV), the most phylogenetically diverse member of the genus *Ranavirus*. Because of its adverse impact on commercially important fish species primarily in Asia, SGIV has been the focus of considerable molecular and ecological studies. Furthermore, marked gene conservation within the family *Iridoviridae* suggests that insights gained in the study of SGIV will inform ongoing research into the biology and ecology of "amphibian ranaviruses" such as frog virus 3 (FV3), common midwife toad virus (CMTV), Ambystoma tigrinum virus (ATV), and others. The latter chapters deal with viral ecology and discuss ranavirus geographic distribution and host range, pathology, diagnosis, and the design and analysis of ranavirus surveillance studies. Below, we summarize chapter contents and highlight key additions to each.

Chapter "Introduction to Ranaviruses: Past, Present, and Future".

Chapter "Ranavirus Taxonomy and Phylogeny". Here, Waltzek and collaborators provide an update on viral taxonomy at both the genus and the family level. Since the first edition in 2015 of this work, two new genera within the family have been identified among invertebrates as well as additional species within the genus *Ranavirus*. In addition, the expansion of viral taxonomy to include higher taxa, i.e., realm, kingdom, phylum, class, and order as well as recent changes in viral nomenclature, will be discussed.

Chapter "Ranavirus Replication: New Studies Provide Answers to Old Questions". The ranavirus life cycle will be reviewed and new findings presented including an expanded section on the determination of viral gene function. Although the outlines of the ranavirus life cycle have been known since the mid-1980s, specific details are lacking. Ranaviruses encode between 100 and 162 putative genes, yet the function of many is only inferred by homology to genes in other systems. Here, Jancovich and co-workers discuss a variety of approaches to unravel gene function and detail how those studies inform our understanding of ranavirus biogenesis. Focus is on the role of specific viral genes from various ranavirus species in viral entry, DNA and RNA synthesis, and virion morphogenesis, as well as viral accessory/efficiency and immune evasion genes. While the latter two gene categories may not be required for replication in vitro, they enhance replication in vivo and in some in vitro systems by maintaining dNTP pools, providing functions that are limiting in some cells or tissues and by blocking various effectors of innate and acquired immunity. Identification and characterization of the latter two classes of genes may inform attempts to develop attenuated vaccine strains that may protect commercially important or ecologically endangered animals.

Chapter "Immune Defenses Against Ranavirus Infections". As more is learned about the amphibian immune system, interactions between host and virus will elucidate key aspects of the innate and acquired immune systems that are required to prevent severe disease and clear viral infections. Here, Grayfer and co-authors discuss host immunity at the molecular, cellular, and organismal level. These studies provide important information about the immunopathogenesis of ranaviral disease and provide insights into the role of various innate and acquired elements of the amphibian immune system and their role in combatting viral infections. Furthermore, this work will advance the development of vaccines designed to protect valuable commercial, and, perhaps endangered, species.

Chapter "Characterization, Pathogenesis, and Immuno-Biological Control of Singapore Grouper Iridovirus (SGIV)". Singapore grouper iridovirus (SGIV) is a major threat to commercially-farmed fish in Asia and for that reason has been the focus of much recent attention. Here, Qin and co-workers discuss molecular, pathogenic, and immune aspects of SGIV infection and the development of vaccines targeting SGIV. While the life cycle of SGIV is considered to be essentially the same as that of FV3, recent studies using SGIV have determined the function of multiple viral genes and provided a markedly more detailed understanding of virion structure at the near-atomic level.

Chapter "Ranavirus Distribution and Host Range". The discovery of FV3 began the systematic investigation of a little studied virus family previously found only in fish and insects. Initially, there was little effort to link FV3 with amphibian disease as several isolates were derived from ostensibly health animals. At most, FV3 and several similar viruses were viewed as minor pathogens that, although capable of causing disease and death in some cases, did not adversely impact overall populations. However, as research, driven by expanded studies and aided by molecular tools that permitted rapid and specific diagnosis, continued ranaviruses came increasingly to be linked to morbidity and mortality in a number of animal systems. In this chapter, Marschang and colleagues build on their previous work and summarize the expanding world of ranavirus-susceptible vertebrates.

Chapter "Ranavirus Ecology: From Individual Infections to Population Epidemiology and Community Impacts". In this comprehensive chapter, Brunner and others update our understanding of ranavirus ecology including the impact of seasonality, susceptibility among various vertebrate taxa, the impacts of developmental stage, ambient temperature, predation, and stress on susceptibility, viral persistence in the environment, the impact of co-infections on amphibian survival, and the ability of ranavirus infection to trigger transient declines or population extinctions.

Chapter "Pathology and Diagnostics". Our understanding of the impact of ranaviruses on ectothermic vertebrates has advanced hand-in-hand with our ability to accurately and rapidly diagnose ranavirus infections. Here Miller and co-workers discuss the pathology of ranavirus infections and the various diagnostic approaches used to identify and confirm these infections.

Chapter "Design and Analysis of Ranavirus Studies: Insights into Planning Surveillance, Modeling Host-Pathogen Dynamics, and Performing Risk Analyses". In the first part of this chapter, Gray, Brunner, Earl, and Peace discuss contemporary approaches for designing ranavirus surveillance studies, analysis of different types of surveillance data, and use of surveillance and other data to model pathogen dynamics within and between populations. New topic additions include discussing the usefulness of dose-response experiments and contemporary approaches to estimating R_0. In addition to discussing commonly used approaches, R code is provided to facilitate practical application of concepts. The last quarter of the chapter is dedicated to risk analyses, which is covered by Australian colleagues, Wirth and Ariel.

Collectively, these revised and updated chapters highlight advances in our understanding of ranavirus biology and ecology and provide directions for further research. Given the impact of ranavirus disease on ecologically important cold-blooded vertebrates and commercially important fish and amphibians, it may also provide a theoretical basis for protecting these important species through management techniques and vaccination.

Acknowledgments We thank the following organizations for providing funds to support Open Access publishing of the second edition: University of Tennessee's (UT) Open Publishing Support Fund, UT School of Natural Resources, UT Center for Wildlife Health, Washington State University Libraries, University of Mississippi Medical Center, Gordon State College, Association of Reptile and Amphibian Veterinarians, and the Global Ranavirus Consortium.

Ranavirus Taxonomy and Phylogeny

Thomas B. Waltzek, Kuttichantran Subramaniam, and James K. Jancovich

1 Introduction

The phylum *Nucleocytoviricota*, known informally as the Nucleocytoplasmic Large DNA Viruses (NCLDV), is a monophyletic assemblage of viruses that infect eukaryotes, ranging from single-celled organisms to humans, worldwide. The NCLDV phylum encompasses two classes (*Megaviricetes*, *Pokkesviricetes*), five orders (*Pimascovirales*, *Imitervirales*, *Algavirales*, *Asfuvirales*, *Chitovirales*), and 11 families, including the family *Iridoviridae* (https://ictv.global/). Under this new format, the *Ranavirus* genus (family *Iridoviridae*) is placed within class *Megaviricetes*, order *Pimascovirales*. Members of the NCLDV group have some of the largest known viral genomes. For example, members of the family *Mimiviridae* have genomes that are ~1.2 million base pairs (bp) in size and encode more than 1000 viral genes (Raoult et al. 2004). Members of the phylum replicate within the cytoplasm of infected cells, although some members (e.g., family *Iridoviridae*) also include a nuclear stage during their replication cycle. As a result, NCLDV members encode many of the genes necessary for replication within the cytoplasm but still rely completely on the host translational machinery along with a number of other host encoded proteins. Comparative analysis of NCLDV genomes reveals a core set of five viral genes that are conserved among the NCLDV (Colson et al. 2013), supporting the hypothesis that this large assemblage of viruses originated from a common ancestor. Although the best-characterized family within the NCLDV is the

T. B. Waltzek (✉)
College of Veterinary Medicine, Washington State University, Pullman, WA, USA
e-mail: thomas.waltzek@wsu.edu

K. Subramaniam
Department of Infectious Diseases & Immunology, College of Veterinary Medicine, University of Florida, Gainesville, FL, USA

J. K. Jancovich
Department of Biological Sciences, California State University, San Marcos, CA, USA

© The Author(s) 2025
M. J. Gray, V. G. Chinchar (eds.), *Ranaviruses*,
https://doi.org/10.1007/978-3-031-64973-8_2

Poxviridae, which includes a major human pathogen (smallpox virus), our understanding of the molecular biology, ecology, and infection dynamics of other families within the NCLDV, particularly members of the family *Iridoviridae*, has increased significantly in recent decades.

The family *Iridoviridae* is composed of two subfamilies and seven genera: The *Iridovirus*, *Chloriridovirus*, *Daphniairidovirus*, and *Decapodiridovirus* genera whose members infect invertebrate hosts (subfamily *Betairidovirinae*) and the *Megalocytivirus*, *Lymphocystivirus*, and *Ranavirus* genera that infect cold-blooded vertebrates (subfamily *Alphairidovirinae*) (Chinchar et al. 2017). Iridoviruses have linear dsDNA genomes that are circularly permutated and terminally redundant (Goorha and Murti 1982). Genome size, which includes both unique and redundant regions, is highly variable within the family and ranges from 140 to 303 kbp. However, because genomes are terminally redundant, unit-length genome sizes (i.e., the sum of the size of only the unique genes) are smaller and range from 103 to 220 kbp (Jancovich et al. 2012; Chinchar et al. 2017). Although earlier studies reported 26 genes common to all members of the family (Eaton et al. 2007), the absence of one or more of these in various genera has lowered the number of core genes to 21 (Table 1). This cluster of core genes includes viral structural proteins as well as proteins involved in the regulation of gene expression, virus replication, and virulence (Jancovich et al. 2015; Grayfer et al. 2015). Sequence analysis of the 21 core genes has been used to generate high-resolution phylogenies (Fig. 1) for members of the family *Iridoviridae* as well as members of the genus *Ranavirus* (Jancovich et al. 2012).

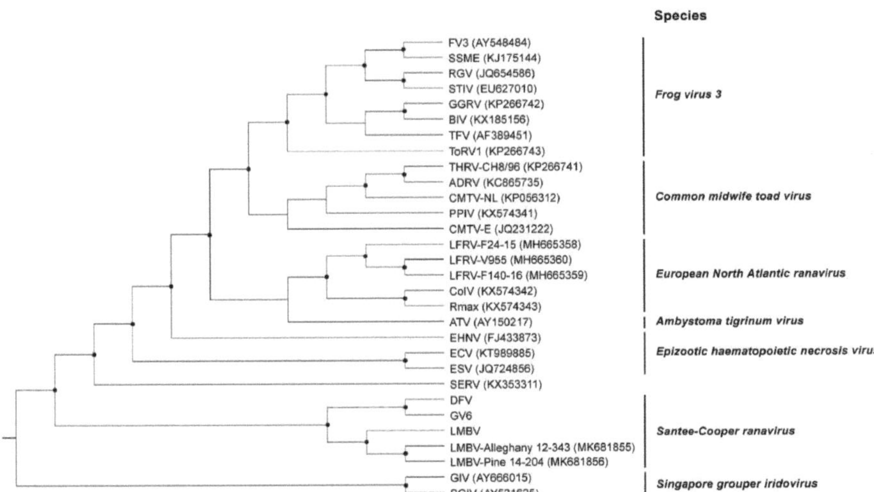

Fig. 1 Cladogram depicting the evolutionary relationships among 30 fully sequenced ranaviruses, based on the aligned deduced amino acid (AA) sequences of the concatenated 21 conserved iridovirus genes (Table 1). The data set contained 11,835 aligned AA positions. The maximum likelihood analysis was conducted in IQ-TREE (Nguyen et al. 2015). Nodes with black circles are supported by bootstrap values ≥80% (1000 replicates). See Table 2 for taxa abbreviations

Table 1 Twenty-one core genes present in all members of the family *Iridoviridae*

ORF[a]	Putative function
1R	Putative replication factor and/or DNA binding/packaging protein
2L	Myristylated membrane protein
8R	DNA-dependent, RNA polymerase II, α subunit
9L	NTPase/helicase
12L	Unknown function
15R	AAA-ATPase, similar to poxvirus A32, required for DNA packaging
21L	Helicase family protein
22R	D5 family NTPase involved in DNA replication
37R	NIF-NLI interacting factor
41R	Vaccinia virus early transcription factor
53R	Myristylated membrane protein
57R	Serine/threonine protein kinase
60R	DNA polymerase family B exonuclease
62L	DNA-dependent, RNA polymerase II, β subunit
80L	RNase III
81R	Transcription elongation factor TFIIS
88R	Erv/Alr family of thiol oxidoreductases
90R	Major capsid protein (MCP)
91R	Immediate early protein ICP-46
94L	Uvr/REP helicase
95R	Putative RAD2-type nuclease

[a]ORF designations are based on their positions within the genome of frog virus 3 (AY548484)

2 Ranavirus Taxonomy

Members of the genus *Ranavirus* are a promiscuous group of viruses capable of infecting a wide variety of cold-blooded vertebrate hosts including fish, amphibians, and reptiles (Marschang 2011; Miller et al. 2011; Whittington et al. 2010). In addition, it has been hypothesized that ranaviruses have jumped recently in their evolutionary history from fish to amphibians and reptiles (Jancovich et al. 2010; Mavian et al. 2012a). This wide host range has been the focus of much ranavirus research, as investigators seek to understand how ranaviruses are able to infect such a wide variety of hosts (Brenes et al. 2014), when in evolutionary history jumps from fish to other cold-blooded vertebrates occur (Chen et al. 2013; Jancovich et al. 2010; Mavian et al. 2012a), and what genetic elements contribute to ranavirus host range and pathogenesis (Jancovich et al. 2015).

There are currently seven species recognized by the ICTV within the genus *Ranavirus* (Jancovich et al. 2012). These species include *Frog virus 3* (FV3), the best-characterized member of the family *Iridoviridae*; *Ambystoma tigrinum virus* (ATV); *Epizootic hematopoietic necrosis virus* (EHNV); *Santee-Cooper ranavirus* (SCRV); *Singapore grouper iridovirus* (SGIV); *Common midwife toad virus*

Table 2 Sequenced genomes from members of the genus *Ranavirus*

Virus species	Isolate name (abbreviation)	Host	Genome size (bp)	GC%	Predicted ORFs	GenBank accession number	Reference
Santee-Cooper ranavirus	Largemouth bass virus (LMBV)	Fish	99,025	52	N/A	N/A	Unpublished
	Largemouth bass virus isolate Alleghany 12-343 (LMBV Alleghany 12-343)	Fish	99,827	52	86	MK681855	Unpublished
	Largemouth bass virus isolate Pine 14-204 (LMBV Pine 14-204)	Fish	99,290	52	87	MK681856	Unpublished
	Doctorfish virus (DFV)	Fish	99,810	52	N/A	N/A	Unpublished
	Guppy virus 6 (GV6)	Fish	106,323	52	N/A	N/A	Unpublished
Singapore grouper iridovirus	Singapore grouper iridovirus (SGIV)	Fish	140,131	48	162	AY521625	Song et al. (2004)
	Grouper iridovirus (GIV)	Fish	139,793	49	139	AY666015	Tsai et al. (2005)
Frog virus 3	Frog virus 3	Frog	105,903	55	98	AY548484	Tan et al. (2004)
	Frog virus 3 isolate SSME (SSME)	Salamander	105,070	55	95	KJ175144	Morrison et al. (2014)
	Soft-shelled turtle iridovirus (STIV)	Turtle	105,890	55	105	EU627010	Huang et al. (2009)
	Rana grylio iridovirus (RGV)	Frog	105,791	55	106	JQ654586	Lei et al. (2012)
	Tiger frog virus (TFV)	Frog	105,057	55	105	AF389451	He et al. (2002)
	Bohle iridovirus (BIV)	Frog	103,531	55	100	KX185156	Hick et al. (2016)
	German gecko ranavirus (GGRV)	Gecko	103,681	55	73	KP266742	Stöhr et al. (2015)
Ambystoma tigrinum virus	Ambystoma tigrinum virus (ATV)	Salamander	106,332	54	96	AY150217	Jancovich et al. (2003)
Epizootic haematopoietic necrosis virus	Epizootic haematopoietic necrosis virus (EHNV)	Fish	127,011	54	100	FJ433873	Jancovich et al. (2010)
	European catfish virus (ECV)	Fish	127,549	54	136	KT989885	Fehér et al. (2016)
	European sheatfish virus (ESV)	Fish	127,732	54	136	JQ724856	Mavian et al. (2012b)

Common midwife toad virus	Common midwife toad virus (CMTV-ES)	Frog	106,878	55	104	JQ231222	Mavian et al. (2012a)
	Common midwife toad virus (CMTV-NL)	Frog	107,772	55	104	KP056312	van Beurden et al. (2014)
	Testudo hermanni ranavirus (THRV-CH8/96)	Tortoise	105,811	55	75	KP266741	Stöhr et al. (2015)
	Tortoise ranavirus 1 isolate (ToRV1)	Tortoise	103,876	55	76	KP266743	Stöhr et al. (2015)
	Andrias davidianus ranavirus (ADRV)	Salamander	106,734	55	101	KC865735	Chen et al. (2013)
	Pike-perch iridovirus (PPIV)	Fish	108,041	57	109	KX574341	Holopainen et al. (2016)
European North Atlantic ranavirus	Lumpfish ranavirus isolate F24-15 (LMRV-F24-15)	Fish	115,616	54	97	MH665358	Stagg et al. (2020)
	Lumpfish ranavirus isolate F140-16 (LMRV-F140-16)	Fish	115,877	54	97	MH665359	Stagg et al. (2020)
	Lumpfish ranavirus isolate V4955 (LMRV-V4955)	Fish	115,971	55	97	MH665360	Stagg et al. (2020)
	Ranavirus maximus (Rmax)	Fish	115,510	55	100	KX574343	Ariel et al. (2016)
	Cod iridovirus (CoIV)	Fish	114,865	57	98	KX574342	Ariel et al. (2016)
Unclassified	Short-finned eel ranavirus (SERV)	Fish	126,965	57	111	KX353311	Subramaniam et al. (2016)

(CMTV); and the *European North Atlantic ranavirus* (ENARV) (Chinchar et al. 2017). Moreover, there are other genetically distant ranaviruses that have not yet been recognized as species by the ICTV Iridoviridae Study Group due to a paucity of characterization data. One such example is the short-finned eel ranavirus (SERV; Subramaniam et al. 2016).

Multiple criteria are used to delineate members within the genus *Ranavirus* including restriction endonuclease fragment length polymorphism (RFLP) profiles of genomic DNA, virus protein profiles, DNA sequence analysis, and host specificity (Jancovich et al. 2012). In addition to these criteria, dot-plot analysis using complete genomic sequence information as well as phylogenetic analyses of individual and concatenated gene sequences or locally colinear gene block sequences have provided insights into the taxonomy of the ranaviruses (Eaton et al. 2007; Jancovich et al. 2010; Mavian et al. 2012a; Tan et al. 2004; Wang et al. 2014; Claytor et al. 2017). Dot-plot analyses offer a general overview of ranavirus genomic organization and a visual way to identify insertions, deletions, and inversions within viral genomes. Dot-plot studies clearly indicate that although ranaviruses share the majority of their genes, gene order is not conserved and may serve as a way to distinguish evolutionarily related isolates or species. For example, gene order is conserved among FV3 strains (e.g., tiger frog virus [TFV] and soft-shelled turtle virus [STIV]) and distinct from that seen with ATV and EHNV strains (Jancovich et al. 2015).

Phylogenetic analysis using the 21 core genes from sequenced ranaviruses supports at least 7 ranavirus species (Fig. 1; Table 1): FV3 strains including TFV, BIV, and STIV; CMTV strains including Andrias davidianus ranavirus (ADRV) and Rana esculentus virus (REV); ATV strains, EHNV strains including European catfish virus (ECV), SGIV strains including grouper iridovirus (GIV), SCRV strains including largemouth bass virus (LMBV), guppy virus 6 (GV6), and doctorfish virus (DFV); and ENARV strains including the lumpfish ranavirus (LRV), ranavirus maximus (Rmax, cod iridovirus (CoIV). Some ranavirus species infect fish (e.g., EHNV, SGIV, SCRV, ENARV) or amphibians (ATV); whereas, other species infect fish, amphibians, and reptiles (e.g., FV3, CMTV). As suggested by core gene analysis, SERV likely constitutes a new ranavirus species. Therefore, phylogenetic analyses will enable investigators to identify and classify newly discovered ranaviruses.

Novel ranaviruses are characterized as viral species following sequencing of one or more viral genes, such as the major capsid protein (MCP). For example, phylogenetic analysis and taxonomic classification of newly isolated ranaviruses have focused on a single, highly conserved gene (e.g., MCP gene; Allender et al. 2013; Duffus and Andrews 2013; Geng et al. 2011; George et al. 2014; Kolby et al. 2014; Marsh et al. 2002; Waltzek et al. 2014), or on a concatenated set composed of multiple viral genes (Holopainen et al. 2009; Iwanowicz et al. 2013). While analysis of the MCP gene is convenient, the highly conserved nature of this protein may mask differences between virus isolates. Collectively, either approach provides a useful starting point to characterize and classify ranavirus isolates. However, having complete genomic sequence information available from a variety of ranavirus isolates will help in developing more rapid, sensitive, and universal approaches for the detection and classification of new ranaviruses. For example, identifying primers that flank hypervariable regions within the genome may allow viral isolates to be more readily distinguished.

Multiple strains of some ranavirus species have been sequenced (Table 2). For example, FV3 (Morrison et al. 2014) and closely related viruses (He et al. 2002; Huang et al. 2009; Lei et al. 2012) have been sequenced and shown to be members of the same species. Comparative dot-plot analysis of completely sequenced ranavirus genomes will be discussed in detail in another chapter of this book (Jancovich et al. 2024). That said, there are currently five unique genomic organizations identified among ranavirus genomes (Chen et al. 2013; Eaton et al. 2007; Jancovich et al. 2003, 2010; Mavian et al. 2012a, b; Song et al. 2004; Tan et al. 2004; Tsai et al. 2005), and additional ranavirus genomic organizations may yet to be discovered. Interestingly, whole genome dot-plot analyses show that ranaviruses with a similar genomic organization cluster together upon phylogenetic analysis using the 21 core genes. Therefore, there appears to be a direct correlation between ranavirus genomic organization and the 21-gene based phylogenies.

It is unclear why there is such diversity in overall genomic architecture among ranaviruses. To that end, no other member of the NCLDV has such a diverse genomic organization. For example, all poxviruses possess genomes that display a conserved central core and variable, inverted terminal repeat regions. The core contains replicative genes common to all poxviruses, whereas the terminal repeat regions encode genes that influence host specificity and pathogenesis (Gubser et al. 2004; Upton et al. 2003). In contrast, although most genes are conserved, gene order differs among the aforementioned ranavirus lineages. Perhaps the diverse genomic organization reflects their inherently high recombination frequency (Chinchar and Granoff 1986) that leads to marked rearrangement of the viral genome. Therefore, if recombination of the viral genome increases over time, then ranaviruses showing greater sequence divergence may also show lower sequence collinearity. Moreover, rearrangements in gene order may be possible because elements regulating temporal gene expression may be located immediately upstream of each gene. Thus rearrangements may not adversely impact either gene expression or temporal appearance. In view of this, future work should focus on understanding this genomic variability and diversity among ranaviruses and its relationship to viral ecology, host range, and pathogenesis.

Comparative genomic sequence analysis has provided valuable insights into ranavirus taxonomy and the evolutionary relationships among viral species. Phylodynamics, or phylogenomics, the process of examining viral gene phylogenies in the context of epidemiology, immunology, genomic organization, and changes in viral spatial dynamics (Volz et al. 2013), are helping advance our understanding of the relationships among, and the origins of, ranaviruses. Phylodynamic analyses have been used to understand taxonomy and evolutionary processes in other viral systems (Grenfell et al. 2004; Diaz-Canova et al. 2022; Guinat et al. 2023), and there are studies that have been performed that will help future work understand relationships among ranavirus species. Claytor et al. (2017) examined the phylogenomics of a CMTV strain, a virus previously known only from China and Europe, isolated during a disease outbreak in bullfrogs at a North American farm. The authors also determined that a virus responsible for another disease outbreak on the same farm some years later was a chimeric CMTV ranavirus with a FV3 backbone. Their findings suggest that the international trade in ranavirus host species has played a role in this outbreak and the generation of this highly pathogenic chimeric ranavirus isolate. Genomic sequence analysis of Canadian FV3 strains determined that FV3-CMTV chimeric viruses may be widespread (Vilaca et al. 2019). They suggest that these viruses were introduced in the past 70–100 years, a timeframe linked to human trading of amphibians. In addition, Epstein and Storfer (2016) examined the genomes of 15 ATV isolates collected across North America while incorporating geographic information to generate estimations of viral emergence. They concluded that the ATV ranaviruses are a monophyletic group in the genus that emerged several hundred years ago before the bait trade was established. Their work also suggests the bait trade impacts ranavirus dissemination to new populations of susceptible host species. FV3 phylodynamic analysis suggests that this virus may have originated in the 1500s from Asia or North America and then spread throughout the world because of human translocation of the pathogen in ectothermic vertebrate hosts (Price et al. 2016; Vilaca et al. 2019; Owen 2022). While the precise origin of FV3 has yet to be determined, these analyses have provided important insight into FV3 evolutionary biology and subsequent taxonomy of this viral strain. Therefore, future phylodynamic and phylogenomic analyses of ranavirus genomes will help our understanding of the relationships between and among ranavirus isolates while further shaping our understanding of ranavirus taxonomy, especially as new viral genome sequences become available.

3 Current Ranavirus Taxonomy

The ICTV defines a species as "a monophyletic group of viruses whose properties can be distinguished from those of other species by multiple criteria" (Adams et al. 2013). The criteria for defining a viral species is determined by individual ICTV study groups and may include "natural and experimental host range, cell and tissue tropism, pathogenicity, vector specificity, antigenicity, and the degree of relatedness of their genomes or genes" (Adams et al. 2013). However, the critical component is that a viral species must be defined by "multiple" criteria, not a single distinguishing criterion. In addition, the ICTV recognizes a genus as "a group of species that share certain common criteria" (Adams et al. 2013).

As discussed above, ranavirus taxonomy has been based on RFLP profiles of genomic DNA, virus protein profiles, DNA sequence analysis, and host specificity (Jancovich et al. 2012). Unfortunately, these criteria do not allow us to quantify and differentiate between intraspecific and interspecific diversity in order to delineate one species from another. However, our understanding of ranavirus diversity has significantly increased in recent years through sequence analysis of individual rana-virus genes and sequencing of complete viral genomes. As a result, the Iridoviridae Study Group will soon need to reassess the criteria used to assign species and genera within the family. For example, the grouper iridoviruses, GIV and SGIV, appear to be the most distantly related viruses among the current isolates of the genus *Ranavirus* (Fig. 1). Whole genome dot-plot analysis shows collinearity between the genomes of GIV and SGIV. However, grouper iridoviruses possess few regions of collinearity with other ranaviruses (Jancovich et al. 2015). In addition, GIV and SGIV are unusual in that they lack the DNA methyltransferase gene seen among other ranaviruses and, as a result, do not have a methylated genome (Song et al. 2004; Tsai et al. 2005). Therefore, GIV/SGIV may need to be considered as a new genus. Similarly, strains of SCRV (e.g., LMBV, DFV, GV6) may also need to be considered as a new genus in the family. Finally, the SERV should receive consideration as a possible new species in the genus *Ranavirus* (Subramaniam et al. 2016). To that end, the Iridoviridae Study Group will need to assess the impact of these features in construction of a consistent taxonomy.

4 Ranavirus Taxonomy: Adoption of a bionomial nomenclature

The ICTV has mandated that all viral species be renamed using a binomial format. Historically, species in the family *Iridoviridae* have not been named using a single set of unifying rules. For example, *Santee-Cooper ranavirus* (subfamily *Alphairidovirinae*, genus *Ranavirus*) was named after the site, the Santee-Cooper Reservoir (USA), from which infected largemouth bass (*Micropterus salmoides*) were sampled leading to the isolation of this virus. In contrast, *Infectious spleen and kidney necrosis virus* and *Scale drop disease virus* (subfamily *Alphairidovirinae*, genus *Megalocytivirus*) were named based on the pathology observed in the respective hosts. The names of other species, such as *Lymphocystivirus disease virus 1–4* (subfamily *Alphairidovirinae*, genus *Lymphocystivirus*), reflect the recognized disease and a numeral as a unique identifier. Lastly, many viruses were named based on the host from which they were first isolated (e.g., *Frog virus 3*, *Ambystoma tigrinum virus*).

The ICTV Iridoviridae Study Group has developed a binomial format proposal in which the first word of the species name is the genus to which the viral species belongs. The second word (the species epithet) consists of two elements. The first element is the name of the genus of the host from which the virus was isolated, written in lowercase letters. For iridoviruses that infect multiple host genera, the name of a prominent host genus will be incorporated with preference given to the genus in which the virus was initially characterized. A taxonomic change to the name of a host genus that occurs after the associated iridovirus was initially characterized will not alter the iridovirus species epithet. The second element is a numeral to serve as a unique identifier. For example, largemouth bass (genus *Micropterus*) is a prominent host for largemouth bass virus of the current species *Santee-Cooper ranavirus*. Thus, under the new system, *Santee-Cooper ranavirus* would become *Ranavirus micropterus1*. When reporting on research methods, the correct approach identifies both the viral strain and the viral species. For example, "Largemouth bass virus (*Ranavirus micropterus1*) has been isolated from the tissues of a variety of a fish species."

The proposed binomial species nomenclature for the family *Iridoviridae* should be easily implemented across viruses from both vertebrate and invertebrate hosts. Species names will convey important information, such as a prominent host (i.e., genus) of the virus and a number to differentiate it from other viruses infecting the same host genus but belonging to separate species. In Table 3, the proposed binomial designations of the current 22 species within the family *Iridoviridae*, including the 7 ranavirus species, are presented. The ICTV Iridoviridae Study Group's binomial proposal was submitted to the ICTV and approved in April 2024. The previous and newly approved species names are shown in Table 3.

5   Final Thoughts

The taxonomy of ranaviruses is continually evolving, especially as new isolates are discovered worldwide. Taxonomic classification of newly discovered ranavirus isolates has been based on single and multiple viral genes as well as host, protein, serological, and morphological characteristics; however, single-gene taxonomic analysis is unlikely to be as robust as whole genome analysis or phylogenetic comparisons using the 21 core ranavirus genes. Many complete genomic sequences have become available, leading to much improved understanding of the diversity and complexity of ranavirus taxonomy. Finally, the ICTV Iridoviridae Study Group has made a proposal to the ICTV to rename the 22 species in the family *Iridoviridae* using the mandated binomial format, although the common virus names will remain unchanged. Newly discovered ranavirus species will be named following the binomial rules outlined above.

Table 3 Proposed binomial name changes in the family *Iridoviridae*

Genus	Current species	Exemplar virus (abbreviation)	Host of exemplar virus	Proposed species name (binomial system)	GenBank Number
Ranavirus	*Frog virus 3*	Frog virus 3 (FV3)	*Rana pipiens*	*Ranavirus rana1*	AY548484
	Ambystoma tigrinum virus	Ambystoma tigrinum virus (ATV)	*Ambystoma tigrinum*	*Ranavirus ambystoma1*	AY150217
	Epizootic haematopoietic necrosis virus	Epizootic haematopoietic necrosis virus (EHNV)	*Perca fluviatilis*	*Ranavirus perca1*	FJ433873
	European North Atlantic ranavirus	Cod iridovirus (CoIV)	*Gadus morhua*	*Ranavirus gadus1*	KX574343
	Common midwife toad virus	Common midwife toad virus-E (CMTV-E)	*Alytes obstetricans*	*Ranavirus alytes1*	FM165473
	Santee-Cooper ranavirus	Largemouth bass virus (LMBV)	*Micropterus salmoides*	*Ranavirus micropterus1*	AF080250
	Singapore grouper iridovirus	Singapore grouper iridovirus (SGIV)	*Epinephelus tauvina*	*Ranavirus epinephelus1*	AY521625
Lymphocystivirus	*Lymphocystis disease virus 1*	Lymphocystis disease virus 1 (LCDV-1)	*Platichthys flesus*	*Lymphocystivirus platichthys1*	L63545
	Lymphocystis disease virus 2	Lymphocystis disease virus China (LCDV-C)	*Paralichthys olivaceus*	*Lymphocystivirus paralichthys1*	AY380826
	Lymphocystis disease virus 3	Lymphocystis disease virus-Sparus aurata (LCDV-Sa)	*Sparus aurata*	*Lymphocystivirus sparus1*	KX643370
	Lymphocystis disease virus 4	Lymphocystis disease virus-white croaker (LCDV-WC)	*Micropogonias furnieri*	*Lymphocystivirus micropogonias1*	MN803438
Megalocytivirus	*Infectious spleen and kidney necrosis virus*	Red seabream iridovirus (RSIV)	*Pagrus major*	*Megalocytivirus pagrus1*	AB104413
	Scale drop disease virus	Scale drop disease virus (SDDV)	*Lates calcarifer*	*Megalocytivirus lates1*	KR139659

(continued)

Table 3 (continued)

Genus	Current species	Exemplar virus (abbreviation)	Host of exemplar virus	Proposed species name (binomial system)	GenBank Number
Iridovirus	*Invertebrate iridescent virus 6*	Invertebrate iridescent virus 6 (IIV6)	*Chilo suppressalis*	*Iridovirus chilo1*	AF303741
	Invertebrate iridescent virus 31	Invertebrate iridescent virus 31 (IIV31)	*Armadillidium vulgare*	*Iridovirus armadillidium1*	HF920637
Chloriridovirus	*Invertebrate iridescent virus 3*	Invertebrate iridescent virus 3 (IIV3)	*Aedes taeniorhynchus*	*Chloriridovirus aedes1*	DQ643392
	Invertebrate iridescent virus 9	Invertebrate iridescent virus 9 (IIV9)	*Wiseana cervinata*	*Chloriridovirus wiseana1*	GQ918152
	Invertebrate iridescent virus 22	Invertebrate iridescent virus 22 (IIV22)	*Simulium* sp.	*Chloriridovirus simulium1*	HF920633
	Invertebrate iridescent virus 25	Invertebrate iridescent virus 25 (IIV25)	*Simulium* sp.	*Chloriridovirus simulium2*	HF920635
	Anopheles minimus iridovirus	Anopheles minimus iridovirus (AMIV)	*Anopheles minimus*	*Chloriridovirus anopheles1*	KF938901
Decapodiridovirus	*Decapod iridescent virus 1*	Shrimp hemocyte iridescent virus (SHIV)	*Litopenaeus vannamei*	*Decapodiridovirus litopenaeus1*	MF599468
Daphniairidovirus	*Daphniairidovirus tvaerminne*	Daphnia iridescent virus 1 (DIV-1)	*Daphnia magna*	*Daphniairidovirus daphnia1*	LS484712

Acknowledgments We would like to thank Trevor Williams for a critical review of this manuscript. This work was partially funded by the California State University San Marcos (J.K.J.); the University of Florida (K.S.); and Washington State University (T.B.W.). We thank the following organizations for providing funds to support Open Access publishing of the second edition: University of Tennessee's (UT) Open Publishing Support Fund, UT School of Natural Resources, UT Center for Wildlife Health, Washington State University Libraries, University of Mississippi Medical Center, Gordon State College, Association of Reptile and Amphibian Veterinarians, and the Global Ranavirus Consortium.

References

Adams MJ, Lefkowitz EJ, King AMQ, Carstens EB (2013) Recently agreed changes to the International Code of Virus Classification and Nomenclature. Arch Virol 158:2633–2639

Allender MC, Bunick D, Mitchell MA (2013) Development and validation of TaqMan quantitative PCR for detection of frog virus 3-like virus in eastern box turtles (*Terrapene carolina carolina*). J Virol Methods 188:121–125

Ariel E, Steckler NK, Subramaniam K, Olesen NJ, Waltzek TB (2016) Genomic sequencing of ranaviruses isolated from turbot (*Scophthalmus maximus*) and Atlantic cod (*Gadus morhua*). Genome Announc 4(6):e01393–e01316. https://doi.org/10.1128/genomeA.01393-16. PMID: 27979944; PMCID: PMC5159577

Brenes R, Gray MJ, Waltzek TB, Wilkes RP, Miller DL (2014) Transmission of ranavirus between ectothermic vertebrate hosts. PLoS One 9:e92476

Chen ZY, Gui JF, Gao XC, Pei C, Hong YJ, Zhang QY (2013) Genome architecture changes and major gene variations of Andrias davidianus ranavirus (ADRV). Vet Res 44:101–114

Chinchar VG, Granoff A (1986) Temperature-sensitive mutants of frog virus 3: biochemical and genetic characterization. J Virol 58:192–202

Chinchar VG, Hick P, Ince IA et al (2017) ICTV virus taxonomy profile: Iridoviridae. J Gen Virol 98(5):890–891

Claytor SC, Subramaniam K, Landrau-Giovannetti N, Chinchar VG, Gray MJ, Miller DL, Mavian C, Salemi M, Wisely S, Waltzek TB (2017) Ranavirus phylogenomics: signatures of recombination and inversions among bullfrog ranaculture isolates. Virology 511:330–343

Colson P, De Lamballerie X, Yutin N, Asgari S, Bigot Y, Bideshi DK, Cheng XW, Federici BA, Van Etten JL, Koonin EV, La Scola B, Raoult D (2013) "Megavirales", a proposed new order for eukaryotic nucleocytoplasmic large DNA viruses. Arch Virol 158:2517–2521

Diaz-Canova D, Mavian C, Brinkmann A, Nitsche A, Moens U, Okeke MI (2022) Genomic sequencing and phylogenomics of cowpox virus. Viruses 14(10):2134

Duffus AL, Andrews AM (2013) Phylogenetic analysis of a frog virus 3-like ranavirus found at a site with recurrent mortality and morbidity events in southeastern Ontario, Canada: partial major capsid protein sequence alone is not sufficient for fine-scale differentiation. J Wildl Dis 49:464–467

Eaton HE, Metcalf J, Penny E, Tcherepanov V, Upton C, Brunetti CR (2007) Comparative genomic analysis of the family *Iridoviridae*: re-annotating and defining the core set of iridovirus genes. Virol J 4:11–28

Epstein B, Storfer A (2016) Comparative genomics of an emerging amphibian virus. G3 6(1):15–27

Fehér E, Doszpoly A, Horváth B, Marton S, Forró B, Farkas SL et al. (2016) Whole genome sequencing and phylogenetic characterization of brown bullhead (*Ameiurus nebulosus*) origin ranavirus strains from independent disease outbreaks. Infect Genet Evol 45:402–407

Geng Y, Wang KY, Zhou ZY, Li CW, Wang J, He M, Yin ZQ, Lai WM (2011) First report of a ranavirus associated with morbidity and mortality in farmed Chinese giant salamanders (*Andrias davidianus*). J Comp Pathol 145:95–102

George MR, John KR, Mansoor MM, Saravanakumar R, Sundar P, Pradeep V (2014) Isolation and characterization of a ranavirus from koi, *Cyprinus carpio* L., experiencing mass mortalities in India. J Fish Dis. https://doi.org/10.1111/jfd.12246

Goorha R, Murti KG (1982) The genome of frog virus-3, an animal DNA virus, is circularly permuted and terminally redundant. Proc Natl Acad Sci USA 79:248–252

Grayfer L, Edholm E-S, De Jesús Andino F, Chinchar VG, Robert J (2015) Ranavirus host immunity and immune evasion. In: Gray MJ, Chinchar VG (eds) Ranaviruses: lethal pathogens of ectothermic vertebrates. Springer, New York

Grenfell BT, Pybus OG, Gog JR, Wood JLN, Daly JM, Holmes EC (2004) Unifying the epidemiological and evolutionary dynamics of pathogens. Science 303:327–332

Gubser C, Hue S, Kellam P, Smith GL (2004) Poxvirus genomes: a phylogenetic analysis. J Gen Virol 85:105–117

Guinat C, Tang H, Yang Q (2023) Bayesian phylodynamics reveals the transmission dynamics of avian influenza A(H7N9) virus at the human-live bird market interface in China. Proc Natl Acad Sci USA 120(17):e2215610120

He JG, Lu L, Deng M, He HH, Weng SP, Wang XH, Zhou SY, Long QX, Wang XZ, Chan SM (2002) Sequence analysis of the complete genome of an iridovirus isolated from the tiger frog. Virology 292:185–197

Hick PM, Subramaniam K, Thompson P, Whittington RJ, Waltzek TB (2016) Complete genome sequence of a Bohle iridovirus isolate from ornate burrowing frogs (*Limnodynastes ornatus*) in Australia. Genome Announc 4(4):e00632–e00616. https://doi.org/10.1128/genomeA.00632-16. PMID: 27540051; PMCID: PMC4991696

Holopainen R, Ohlemeyer S, Schutze H, Bergmann SM, Tapiovaara H (2009) Ranavirus phylogeny and differentiation based on major capsid protein, DNA polymerase and neurofilament triplet H1-like protein genes. Dis Aquat Org 85:81–91

Holopainen R, Subramaniam K, Steckler NK, Claytor SC, Ariel E, Waltzek TB (2016) Genome Sequence of a Ranavirus Isolated from Pike-Perch *Sander lucioperca*. Genome Announc 4(6):e01295–e01216. https://doi.org/10.1128/genomeA.01295-16. PMID: 27856591; PMCID: PMC5114383

Huang YH, Huang XH, Liu H, Gong J, Ouyang ZL, Cui HC, Cao JH, Zhao YT, Wang XJ, Jiang YL, Qin QW (2009) Complete sequence determination of a novel reptile iridovirus isolated from soft-shelled turtle and evolutionary analysis of Iridoviridae. BMC Genomics 10:224–238

Iwanowicz L, Densmore C, Hahn C, McAllister P, Odenkirk J (2013) Identification of largemouth bass virus in the introduced Northern Snakehead inhabiting the Chesapeake Bay watershed. J Aquat Anim Health 25:191–196

Jancovich JK, Mao J, Chinchar VG, Wyatt C, Case ST, Kumar S, Valente G, Subramanian S, Davidson EW, Collins JP, Jacobs BL (2003) Genomic sequence of a ranavirus (family *Iridoviridae*) associated with salamander mortalities in North America. Virology 316:90–103

Jancovich JK, Bremont M, Touchman JW, Jacobs BL (2010) Evidence for multiple recent host species shifts among the ranaviruses (family *Iridoviridae*). J Virol 84:2636–2647

Jancovich JK, Chinchar VG, Hyatt A, Myazaki T, Williams T, Zhang QY (2012) Family *Iridoviridae*. In: King AMQ (ed) Ninth report of the International Committee on Taxonomy of Viruses. Elsevier, San Diego

Jancovich JK, Qin Q, Zhang Q-Y, Chinchar VG (2015) Ranavirus replication: molecular, cellular, and immunological events. In: Gray MJ, Chinchar VG (eds) Ranaviruses: lethal pathogens of ectothermic vertebrates. Springer, New York

Jancovich JK, Zhang, Q-Y, and Chinchar VG. Ranavirus replication: New Studies provide answers to old questions. IN: Ranaviruses: Emerging pathogens of ectothermic vertebrates, MJ Gray and VG Chinchar, eds., . Springer OPEN, 2024.

Kolby JE, Smith KM, Berger L, Karesh WB, Preston A, Pessier AP, Skerratt LF (2014) First evidence of amphibian chytrid fungus (*Batrachochytrium dendrobatidis*) and ranavirus in Hong Kong amphibian trade. PLoS One 9:e90750

Lei XY, Ou T, Zhu RL, Zhang QY (2012) Sequencing and analysis of the complete genome of Rana grylio virus (RGV). Arch Virol 157:1559–1564

Marschang RE (2011) Viruses infecting reptiles. Viruses 3:2087–2126

Marsh IB, Whittington RJ, O'Rourke B, Hyatt AD, Chisholm O (2002) Rapid differentiation of Australian, European and American ranaviruses based on variation in major capsid protein gene sequence. Mol Cell Probes 16:137–151

Mavian C, Lopez-Bueno A, Balseiro A, Casais R, Alcami A, Alejo A (2012a) The genome sequence of the emerging common midwife toad virus identifies an evolutionary intermediate within ranaviruses. J Virol 86:3617–3625

Mavian C, Lopez-Bueno A, Fernandez Somalo MP, Alcami A, Alejo A (2012b) Complete genome sequence of the European sheatfish virus. J Virol 86:6365–6366

Miller D, Gray M, Storfer A (2011) Ecopathology of ranaviruses infecting amphibians. Viruses 3:2351–2373

Morrison EA, Garner S, Echaubard P, Lesbarreres D, Kyle CJ, Brunetti CR (2014) Complete genome analysis of a frog virus 3 (FV3) isolate and sequence comparison with isolates of differing levels of virulence. Virol J 11:46–59

Nguyen L-T, Schmidt HA, von Haeseler A, Minh BQ (2015) IQ-TREE: a fast and effective stochastic algorithm for estimating maximum-likelihood phylogenies. Mol Biol Evol 32:268–274

Owen CJ (2022) Global genomic diversity of a major wildlife pathogen: ranavirus, past and present. Doctoral thesis, University College of London

Price SJ, Garner TWJ, Cunningham AA, Langton TES, Nichols RA (2016) Reconstructing the emergence of a lethal infectious disease of wildlife supports a key role for spread through translocations by humans. Proc R Soc B 283:20160952

Raoult D, Audic S, Robert C, Abergel C, Renesto P, Ogata H, La Scola B, Suzan M, Claverie JM (2004) The 1.2-megabase genome sequence of Mimivirus. Science 306:1344–1350

Song WJ, Qin QW, Qiu J, Huang CH, Wang F, Hew CL (2004) Functional genomics analysis of Singapore grouper iridovirus: complete sequence determination and proteomic analysis. J Virol 73:12576–12590

Stagg HEB, Guðmundsdóttir S, Vendramin N, Ruane NM, Sigurðardóttir H, Christiansen DH, Cuenca A, Petersen PE, Munro ES, Popov VL, Subramaniam K, Imnoi K, Waltzek TB, Olesen NJ (2020) Characterization of ranaviruses isolated from lumpfish *Cyclopterus lumpus* L. in the North Atlantic area: proposal for a new ranavirus species (European North Atlantic Ranavirus). J Gen Virol 101(2):198–207. https://doi.org/10.1099/jgv.0.001377. Epub 2019 Dec 20

Stöhr AC, López-Bueno A, Blahak S, Caeiro MF, Rosa GM, Alves de Matos AP, Martel A, Alejo A, Marschang RE (2015) Phylogeny and differentiation of reptilian and amphibian ranaviruses detected in Europe. PLoS One 10(2):e0118633. https://doi.org/10.1371/journal.pone.0118633. PMID: 25706285; PMCID: PMC4338083

Subramaniam K, Toffan A, Cappellozza E, Steckler NK, Olesen NJ, Ariel E, Waltzek TB (2016) Genomic sequence of a ranavirus isolated from short-finned eel (*Anguilla australis*). Genome Announc 4(4):e00843–e00816. https://doi.org/10.1128/genomeA.00843-16. PMID: 27540067; PMCID: PMC4991712

Tan WGH, Barkman TJ, Chinchar VG, Essani K (2004) Comparative genomic analyses of frog virus 3, type species of the genus *Ranavirus* (family *Iridoviridae*). Virology 323:70–84

Tsai CT, Ting JW, Wu MH, Wu MF, Guo IC, Chang CY (2005) Complete genome sequence of the grouper iridovirus and comparison of genomic organization with those of other iridoviruses. J Virol 79:2010–2023

Upton C, Slack S, Hunter AL, Ehlers A, Roper RL (2003) Poxvirus orthologous clusters: toward defining the minimum essential poxvirus genome. J Virol 77:7590–7600

van Beurden SJ, Hughes J, Saucedo B, Rijks J, Kik M, Haenen OL, Engelsma MY, Gröne A, Verheije MH, Wilkie G (2014) Complete genome sequence of a common midwife toad virus-like ranavirus associated with mass mortalities in wild amphibians in the Netherlands. Genome Announc 2(6):e01293–e01214. https://doi.org/10.1128/genomeA.01293-14

Vilaca ST, Bienentreu JF, Brunetti CR, Lesbarreres D, Murray DL, Kyle CJ (2019) Frog virus 3 genomes reveal prevalent recombination between ranavirus lineages and their origins in Canada. J Virol 93(20):e00765–e00719

Volz EM, Koelle K, Bedford T (2013) Viral phylodynamics. PLoS Comput Biol 9(3):e1002947

Waltzek TB, Miller DL, Gray MJ, Drecktrah B, Briggler JT, MacConnell B, Hudson K, Hopper L, Friary J, Yun SC, Malm KV, Weber ES, Hedrick RP (2014) New disease records for hatchery-reared sturgeon. I. Expansion of the host range of frog virus 3 into hatchery-reared pallid sturgeon *Scaphirhynchus albus*. Dis Aquat Organ 111:219–227

Wang N, Zhang M, Zhang L, Jing H, Jiang Y, Wu S, Lin X (2014) Complete genome sequence of a ranavirus isolated from Chinese giant salamander (*Andrias davidianus*). Genome Announc 2:e01032–e01013

Whittington RJ, Becker JA, Dennis MM (2010) Iridovirus infections in finfish—critical review with emphasis on ranaviruses. J Fish Dis 33:95–122

Ranavirus Replication: New Studies Provide Answers to Old Questions

James K. Jancovich, Qi-Ya Zhang, and V. Gregory Chinchar

1   Ranavirus Taxonomy and Physical Characteristics

Even within established taxa, viral taxonomy is constantly changing as new isolates are identified, sequenced, and analyzed. As with other virus families, taxonomy within the family *Iridoviridae* has undergone several changes in recent years including division of the family into two subfamilies, the *Alphairidovirinae* and *Betairidovirinae*, infecting ectothermic vertebrates and invertebrates respectively, the establishment of two additional genera, *Decapodiridovirus* and *Daphniairidovirus*, within the subfamily *Betairidovirinae*, and the recognition of multiple new viral species (Chinchar et al. 2017a; Lefkowitz et al. 2018; Toenshoff et al. 2018). Furthermore, the International Committee on Taxonomy of Viruses (ICTV) has mandated that by the end of 2023 new and existing virus species are to be designated using a "binomial nomenclature." Accordingly for the family *Iridoviridae*, the first word in the new format will be the name of the viral genus to which the species is assigned and the second word is an epithet, reflecting the genus of the principal host or the host from which the virus was first isolated (Zerbini et al. 2022). As with standard taxonomic nomenclature, the first word in the name (the genus) is capitalized, and the second word (the species epithet) is not. Since revised designations for members of the family *Iridoviridae* have only recently been approved, *Frog virus 3* will be re-named *Ranavirus ranal*, reflecting the genus of frog, *Rana* (now *Lithobates*), from which it was first isolated (Granoff et al. 1966).

J. K. Jancovich
Department of Biological Sciences, California State University, San Marcos, CA, USA

Q.-Y. Zhang
Institute of Hydrobiology, Chinese Academy of Sciences, Wuhan, Hubei, China

V. G. Chinchar (✉)
Department of Cell and Molecular Biology, The University of Mississippi Medical Center, Jackson, MS, USA
e-mail: vchinchar@umc.edu

© The Author(s) 2025
M. J. Gray, V. G. Chinchar (eds.), *Ranaviruses*,
https://doi.org/10.1007/978-3-031-64973-8_3

A table highlighting the proposed name changes for all ranavirus species is included in Chap. 2, "Ranavirus Taxonomy and Phylogeny," of this monograph (Waltzek et al. 2024). However, since at the time our current chapter was written this nomenclature had not yet been approved, we used here the previous non-binomial terms which matched the designations in the literature. Accordingly, a list of species names and abbreviations used in this chapter for the family members is found in Table 1.

The genus *Ranavirus* is one of seven genera within the family *Iridoviridae* (Table 1). The subfamily *Alphairidovirinae* is composed of three genera that infect ectothermic vertebrates (*Ranavirus, Megalocytivirus,* and *Lymphocystivirus*), whereas the subfamily *Betairidovirinae* contains four genera targeting invertebrates (*Chloriridovirus, Daphniairidovirus, Decapodiridovirus,* and *Iridovirus*). Whereas megalocytiviruses and lymphocystiviruses infect only fish, ranaviruses, despite their eponymous designation, target fish, amphibians, and reptiles. Moreover, indicative of their broad host range, some ranaviruses infect hosts from different vertebrate classes. For example, Bohle iridovirus (BIV) is capable of infecting both amphibians and fish (Moody and Owens 1994). More intriguing is the report that a strain of invertebrate iridescent virus 6 (IIV-6) designated Liz-CrIV has been found in a number of reptiles and amphibians. Studies with bearded dragons suggested that Liz-CrIV DNA replicated in vertebrate hosts but was not linked with disease (Papp and Marschang 2019). Moreover, the broad ranavirus host range is especially evident in vitro where ranaviruses infect cells from multiple vertebrate species, including mammals.

Iridovirids, a generic designation for all members of the family, possess an icosahedral capsid that encloses a linear, dsDNA genome (Chinchar et al. 2017a). As shown in Table 1, iridovirid genomes vary in size depending upon the specific virus and contain between approximately 100 and 250 putative open reading frames (ORFs). Moreover, iridovirid genomes, although physically linear, are circularly permutated and terminally redundant (Goorha and Murti 1982). Members of the subfamily *Alphairidovirinae*, aside from Singapore grouper iridovirus (SGIV), encode a DNA methyltransferase that targets each cytosine within CpG dinucleotides (Willis and Granoff 1980). GC content and genome size vary markedly among genera with lymphocystiviruses and all four genera within the *Betairidovirinae* displaying lower GC percentages and larger genome sizes (with the notable exception of LCDV-1) than megalocytiviruses and ranaviruses. As described in Chap. 2 "Ranavirus Taxonomy and Phylogeny" of the current monograph, phylogenetic analyses and establishment of higher order taxa by the ICTV indicate that members of the family *Iridoviridae* are linked to other Nuclear Cytoplasmic Large DNA Viruses (NCLDV) such as Pithoviruses, Marseilleviruses, and Ascoviruses within the order *Pimascovirales* and to poxviruses, African swine fever virus, phycodnaviruses, and mimiviruses within the phylum *Nucleocytoviricota* (Lefkowitz et al. 2018; Waltzek et al. 2024).

Although research prior to 1990 outlined key events in the ranavirus life cycle, the identity and function of most viral proteins was poorly understood. Beginning with the sequencing of the genomes of lymphocystis disease virus 1 (Tidona and

Table 1 Iridovirus taxonomy: viral genera and species

Subfamily	Genus	Species[a]	Size (bp)	No. ORFs	% G+C	GenBank accession number
Alphairidoviridae	*Lymphocystivirus*	*LCDV-1*	102,653	108	29	L63545
		LCDV-2	186,250	178	27	AY380826
		LCDV-3	208,501	183	33	KX643370
		LCDV-4	211,086	148	26	MN803438
	Megalocytivirus	*ISKNV*	111,362	117	55	AF371960
		RBIV	112,080	116	53	AY532606
		RSIV	112,414	93	53	BD143114
		OSGIV	112,636	116	54	AY894343
		TRBIV	110,104	115	55	GQ273492
		LYCIV	111,760	ND	ND	AY779031
		SDDV	131,129	135	37	MN562489
	Ranavirus	*ATV*	106,332	92	54	AY150217
		FV3	105,903	97	55	AY548484
		TFV	105,057	105	55	AF389451
		STIV	105,890	103	55	EU627010
		RGV	105,791	106	55	JQ654586
		EHNV	127,011	100	54	FJ433873
		ESV	127,732	136	54	JQ724856
		ENARV	116,726	97	55	MH665358
		CMTV	106,878	104	55	JQ231222
		ADRV	106,734	101	55	KC865735
		LMBV	99,290	99	52	MK681856
		SCRaV	99,405	105	52	OQ267588
		MSRaV	99,171	105	52	OQ267587
		SGIV	140,131	139	49	AY521625
		GIV	139,793	139	49	AY666015

(continued)

Table 1 (continued)

Subfamily	Genus	Species[a]	Size (bp)	No. ORFs	% G+C	GenBank accession number
Betairidoviridae	*Chloriridovirus*	*IIV-3*	191,132	126	48	DQ643392
		IIV-9	206,791	191	31	GQ918152
		IIV-22	197,693	167	29	HF920633
		IIV-25	204,815	177	30	HF920635
		AMIV	163,023	148	ND	KF938901
	Daphniairidovirus	*DIV-1*	288,858	367	39	PRJEB18974
	Decapodiridovirus	*SHIV*	165,809	170	35	MF599468
		CQIV	165,695	178	35	MF197913
	Iridovirus	*IIV-6*	212,482	211	29	AF303741
		IIV-31	220,222	203	35	HF920637

[a]Abbreviations used in the table are as follows: *LCDV-1* Lymphocystis disease virus 1, *LCDV-2* Lymphocystis disease virus 2, *LCDV-3* Lymphocystis disease virus-3, *LCDV-4* Lymphocystis disease virus-4, *ISKNV* Infectious spleen and kidney necrosis virus, *RBIV* rock bream iridovirus, *RSIV* red seabream iridovirus, *OSGIV* orange spotted grouper iridovirus, *TRBIV* turbot reddish body iridovirus, *LYCIV* large yellow croaker iridovirus, *SDDV* Scale drop disease virus, *ATV* Ambystoma tigrinum virus, *FV3* Frog virus 3, *LMBV* largemouth bass virus, *MSRaV* Micropterus salmoides ranavirus, *SCRaV* Siniperca chautsi ranavirus, *RGV* Rana grylio virus, *CMTV* Common midwife toad virus, *TFV* tiger frog virus, *STIV* soft-shelled turtle iridovirus, *ADRV* Andrias davidianus ranavirus, *EHNV* Epizootic haematopoietic necrosis virus, *ESV* European sheatfish virus, *ENARV* European North Atlantic ranavirus, *SGIV* Singapore grouper iridovirus, *GIV* grouper iridovirus, *IIV-3* Invertebrate iridescent virus type 3, *IIV-9* Invertebrate iridescent virus type 9, *IIV-22* Invertebrate iridescent virus type 22, *IIV-25* Invertebrate iridescent virus type 25, *AMIV* Anopheles minimus iridovirus, *DIV-1* Daphnia iridovirus-1, *SHIV* Shrimp hemocyte iridescent virus, *CQIV* Cherax quadricarnatus iridovirus, *IIV-6* Invertebrate iridescent virus type 6, *IIV-31* Invertebrate iridescent virus type 31. Abbreviations of viral names in *bold italic* type indicate species recognized by the International Committee on Taxonomy of Viruses; those in standard type are either tentative species or isolates of recognized species, *ND* not determined

Darai 1997), infectious skin and kidney disease virus (ISKNV) (He et al. 2001), tiger frog virus (TFV) (He et al. 2002), Ambystoma tigrinum virus (ATV) (Jancovich et al. 2003), and frog virus 3 (FV3) (Tan et al. 2004), a more complete understanding of iridovirid gene products and their functions began to emerge. The availability of complete genomic sequence information permitted analysis of overall genomic organization, protein sequence variation, and polymorphic regions and provided a facile way to determine viral gene function by directly targeting individual viral genes.

As a group, ranavirus genomes are among the smallest within the family *Iridoviridae* ranging in size from 99 to 140 kbp. They display a G+C content between 49% and 55%, are predicted to encode approximately 100–140 viral proteins, and are highly methylated (Table 1). Currently, five unique genomic organizations, divisible into three groups, have been identified by dot plot and phylogenetic analyses of completely sequenced ranavirus genomes (Chen et al. 2013; Yu et al. 2023). GIV-like ranaviruses (GIV and SGIV) comprise one group, largemouth bass virus (LMBV) (a *Santee-Cooper ranavirus* species) comprises another, and amphibian-like ranaviruses (ALRV) make up the third group. GIV- and LMBV-like viruses display only short segments of genomic co-linearity when compared to each other and other members of the genus, whereas members of the ALRV group, i.e., ATV-, CMTV-, EHNV-, ENARV-, and FV3-like ranaviruses, share longer regions of co-linearity (Yu et al. 2023). However, among the five ALRV species, inversions, deletions, and additions have occurred that distinguish one from the other (Fig. 1).

Ranavirus gene products have been identified by the in silico detection of ORFs greater than 120–150 nucleotides in length as well as by molecular and biochemical means such as SDS-polyacrylamide gel analysis of purified virions and virus-infected cells, microarray analysis of viral transcripts, and proteomic analysis of virions (Majji et al. 2009; Song et al. 2006; Willis et al. 1985). The function of about a third of the 92–139 ranavirus proteins has been inferred by identity/similarity to other known proteins/genes. Although the roles of the remaining genes are unknown, many are homologous to genes within the genus *Ranavirus* indicating that they play important roles in viral biogenesis. With few exceptions, all members of the family *Iridoviridae* contain 26 conserved structural and catalytic genes, e.g., the major capsid protein (MCP), a viral DNA polymerase (vDPOL), and the two largest subunits of the viral RNA polymerase (vPOL-IIα and vPOL-IIβ), whereas an additional 35–40 genes are conserved among members of the genus *Ranavirus* (Eaton et al. 2007; Jancovich et al. 2010; Price 2015). Since the latter are found only among ranaviruses, we hypothesize that they enhance viral replication in host tissues and cells and/or inhibit anti-viral immunity. Identification of ranavirus-specific genes may allow us to detect those that function in unique host environments and/or cause disease in a wide variety of hosts. Furthermore, their deletion may allow the generation of attenuated viruses that could prove useful as vaccines.

In addition to coding regions, ranavirus genomes contain palindromes, microsatellites, repeat regions, and areas of inter- and intragenic variation (Chen et al. 2013; Eaton et al. 2010; Jancovich et al. 2003; Lei et al. 2012b; Mavian et al. 2012; Morrison et al. 2014; Tan et al. 2004; Zhang and Gui 2015). Repeat and variable regions can be observed in genomic dot plots (Fig. 1) and may serve as sites that

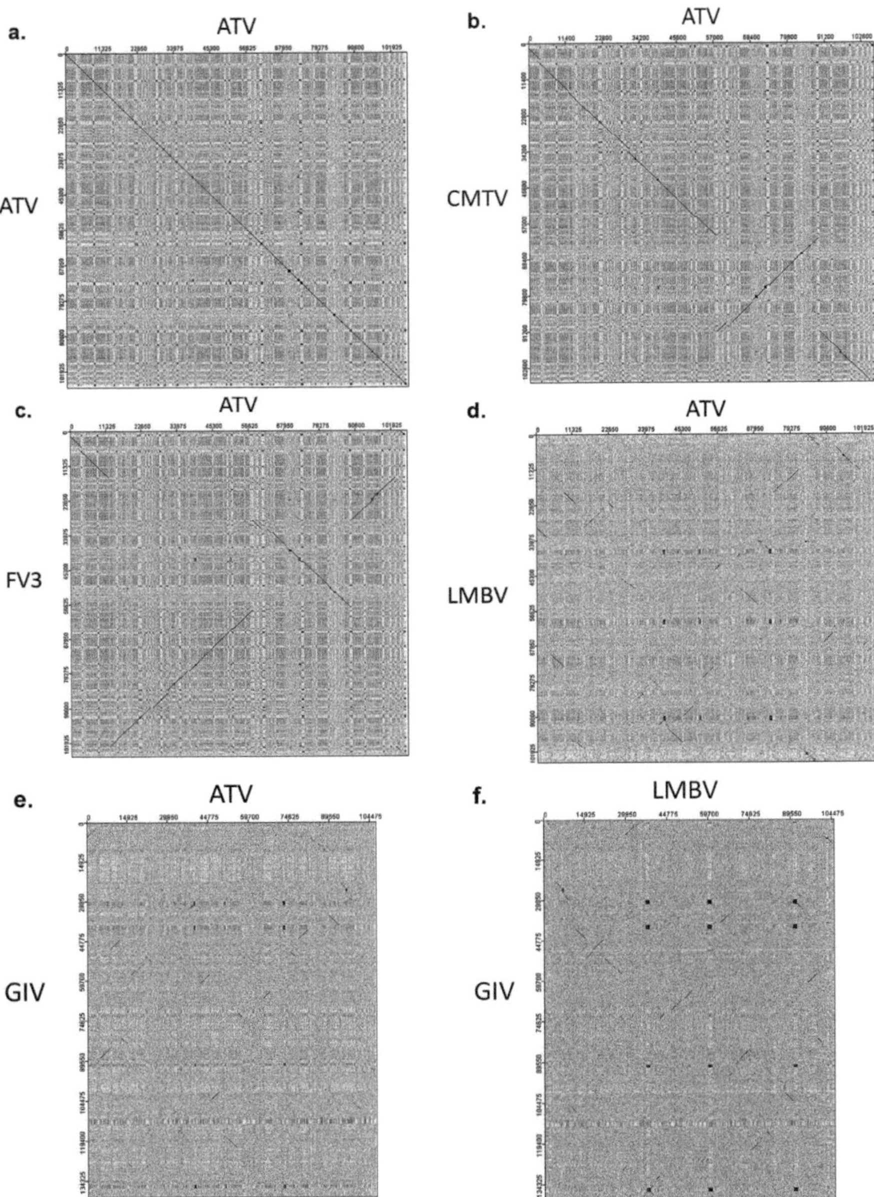

Fig. 1 Ranavirus genomic organization: Dotplot analysis of the genomic organization of representative members of the genus *Ranavirus*. The genomic sequence of ATV was compared to: (**a**) ATV; (**b**) CMTV; (**c**) FV3; (**d**) LMBV; (**e**) GIV. Panel (**f**) shows a plot between LMBV and GIV. Dot plots were generated using JDotter (Brodie et al. 2004). *Lines* on the plot indicate regions of sequence similarity/collinearity

facilitate recombination or regulate gene expression, whereas palindromic sequences at the 3′ end of viral messages may act as transcriptional termination signals (Kaur et al. 1995). In addition, comparisons of closely related FV3-like viruses that vary in virulence suggest that intragenic differences, as well as variation within repeated sequences, may influence viral pathogenesis (Morrison et al. 2014).

2 Ranavirus Replication Strategy

Here we discuss virus-encoded events that play direct roles in the production of infectious virions using, for the most part, FV3 as the model. In the following sections, we detail our current understanding of the impact of virus infection upon host cells, the interplay between virus and host at the cellular and immunological level, and then discuss ranavirus gene products and their function. Key events in ranavirus replication are indicated schematically in Fig. 2, and it is postulated that all iridoviruses replicate using essentially the same general strategy. Variations in the life cycles of different viral species may reflect how they interact with their hosts at the cellular and immunological levels. Additional information on ranavirus replication strategies can be found in several comprehensive reviews (Chinchar et al. 2009, 2011, 2017b; Goorha and Granoff 1979; Williams 1996; Williams et al. 2005; Willis et al. 1985). To facilitate understanding, a number of tables have been constructed. These include a list of select ranavirus homologs involved in viral replication and pathogenesis (Table 2), the putative elements of the ranavirus RNA transcription complex (Table 3), ranavirus replication- and transcription-related proteins detected by iPOND-MS analysis (Table 4), and ranavirus immune evasion proteins and the experimental approaches used to determine their function (Table 5).

2.1 Viral Entry

Ranaviruses are capable of entering and productively infecting a wide variety of cell types including those derived from both ectothermic (e.g., fish, reptiles, and amphibians) and endothermic (e.g., birds and mammals) vertebrates, provided that incubation temperatures do not exceed the viral maxima of 33 °C (Braunwald et al. 1979; Clark and Karzon 1967; Cordier et al. 1986; Gravell and Granoff 1970; Kelly 1975; Lopez et al. 1986; Morales et al. 2010; Pham et al. 2015). Above this temperature, infection is initiated but is limited to the expression of only early viral genes, suggesting that one or more key early proteins are unable to function at higher temperatures (Cordier et al. 1986; Gravell and Granoff 1970; Lopez et al. 1986). The ability to infect multiple hosts and cell types suggests that ranaviruses utilize a highly conserved cellular receptor and mechanism of entry or that multiple receptors and modes of entry are used. As with other members of the family, ranavirus particles are complex, multi-layered structures consisting of a core composed of the viral

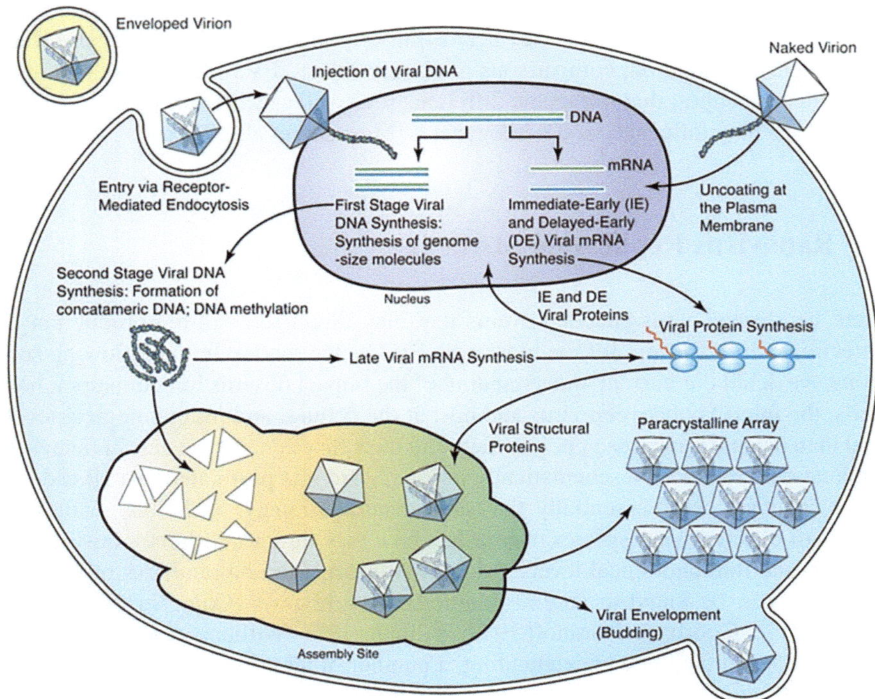

Fig. 2 Schematic diagram of ranavirus replication. Virions enter cells by one of several routes, two of which are shown here. Initial events in virus replication (early viral transcription and the synthesis of unit length genomes) take place within the nucleus. Viral genomes are subsequently transported into the cytoplasm where they are methylated and serve as templates for concatemer formation. Viral assembly sites contain viral DNA and a number of virus-encoded proteins and serve as the loci of virion formation. Newly synthesized virions are found free within the cytoplasm or as members of paracrystalline arrays, or seen budding from the plasma membrane and in the process acquiring an envelope

dsDNA genome associated with one or more virus-encoded proteins, an internal lipid membrane containing several intramembrane proteins, an icosahedral capsid composed primarily of a ~48 kDa major capsid protein, and (in those virions that are released by budding) a viral envelope derived from the plasma membrane containing one or more virus-encoded proteins (Darcy-Tripier et al. 1984; Devauchelle et al. 1985; Zhao et al. 2023). In appearance, ranavirus particles show marked similarity to the virions of African swine fever virus and members of the family *Phycodnaviridae* (Tulman et al. 2009; Wilson et al. 2009). In contrast to most other virus families, both non-enveloped and enveloped ranavirus particles are infectious. However, the specific infectivity of the latter is higher and may be a reflection of viral envelope proteins that enhance binding to cellular targets (Braunwald et al. 1979). Non-enveloped virions interact with the plasma membrane and, analogous to

Table 2 Select ranavirus catalytic, structural, virulence, and immune evasion genes

Category	Function[a]	FV3	RGV	ATV	ADRV	SGIV
Viral entry and virion structure	Entry/fusion; heparan binding	2L	2L	1L	2L	19L
	Major capsid protein (MCP)	90R	97R	14L	17L	72R
	Minor capsid protein (mCP) penton	31R	Δ33R	60R	80L	14L
	mCP-P1 (ZIP); virion stability; heparan binding	20R	22R	78R	87L (CMTV)	38L
	mCP-P2	NM[b]	NM	NM	NM	139R
	mCP-P3 (TmP)	74L	81L	30R	34R	137R
	mCP-P4	10R	NM	8R	11R	59L
	mCP-P5 (ANCHOR); myristylated membrane protein	53R	53R	51L	58L	88L
	mCP-P6	47L	47L	73L	65R	22L
	Putative transmembrane protein	40R	43R	67R	67L (CMTV)	NM
DNA synthesis	Viral DNA-dependent DNA polymerase (vDPOL)	60R	63R	44L	47L	128R
	Proliferating cell nuclear antigen (PCNA)	84R	91R	20L	23L	68L
	Viral single-stranded DNA binding protein (vSSBP) (p31K)	25R	27R	55R	85L	6R
	D5 - Helicase/primase	22R	24R	77L	88L	52L
	RAD2/recombination	95R	102R	10L	12L	97L
	DNA methyltransferase	83R	90R	21L	24L	NM
RNA synthesis	Large subunit of viral RNA transcriptase (vPOL-IIα) (Rpb1)	8R	9R	6R	9R	104L
	vPOL-IIβ (Rpb2)	62L	65L	43R	46R	73L
	Rbp5	30 (TFV)	NM	59R	82L	160L
	TFIIS	81R	88R	24L	27L	85R
	VETF	41R	44R	69R	68L	57L
Virulence and efficiency genes	Putative restriction-modification endonuclease; PD(D/E)xK nuclease-like	94L	NM	11R	13R	98R
	Ribonucleotide reductase (RR) alpha	38R	41R	65R	71L	64R
	RR beta	67L	73L	38R	42R	47L
	Dihydrofolate reductase (DHFR)[c]					
	dUTP phosphorylase (UTPase)	63R	67R	42L	44L	49L

(continued)

Table 2 (continued)

Category	Function[a]	FV3	RGV	ATV	ADRV	SGIV
	Ser/thr kinase	57R	60R	47L	51L	150L
	Thymidine kinase (TK)	85R	92R	19L	22L	67L
	Thymidylate synthase	88R (TFV)	NM	22L	25L	NM
	Fibroblast growth factor	NM	NM	NM	NM	144R/145R
	18K-like early protein	82R	89R	23L	26L	86R
	ICP-46	91R	98R	13L	16L	162L
Immune evasion	vCARD	64R	68R	40L	43L	48L
	Viral homolog of eukaryotic initiation factor 2 alpha (vIF-2α)	Δ26R	28R	57R	84L	NM
	Beta-hydroxysteroid dehydrogenase (β-HSD)	52L	52L	52R	60R	3R
	Bcl-1/Mcl-1	97R	105R	90R	101R	115R
	RNase III	80L	87L	25R	28R	84L
	Semaphorin	NM	NM	NM	NM	155R[d]
	Viral homolog of TNF receptor (vTNFR)	NM	NM	NM	NM	50L/51L/96R[e]
	Stingless	NM	NM	NM	NM	131
	LITAF	75L	82L	29R	33R	136R

The table displays homologs of the indicated proteins among the most studied ranaviruses. Viral genes are designated by their ORF number and whether they are full-size or truncated (Δ). Identity of homologous proteins was determined by BLASTP analysis, reference to compilations of homologous proteins (Eaton et al. 2007; Lei et al. 2012a, b), or annotated sequences of FV3 AY548484 (Tan et al. 2004), RGV JQ654586 (Lei et al. 2012b), ATV AY150217 (Jancovich et al. 2003), ADRV KC865735 (Chen et al. 2013), and SGIV AY521625 (Song et al. 2004). In cases where a match could not be found, a virus within the same species was used for comparison. Note that the distinction between virulence and catalytic genes is subtle, as a modification that increases the ability of, for example vPOL-II alpha, to synthesize viral transcripts will enhance replication as well as the ability to cause disease

[a]Genes that are considered among the 26 iridovirus core genes, i.e., found in nearly all sequenced members of the family, are underlined

[b]*NM* no match

[c]This gene was only seen in EHNV and short-finned eel ranavirus. However, homologs were detected in various fish species

[d]This gene and its homologs was seen in SGIV and various fish species

[e]Homologs were also found among megalocyti- and lymphocystiviruses

events with some other NCLDVs, viral cores are thought to be released into the cytoplasm following fusion of the plasma membrane and the internal viral membrane (de Souza et al. 2021; Van Etten and Dunigan 2016). In contrast, enveloped viruses are thought to enter cells by clathrin-mediated endocytosis, followed by release of non-enveloped virions into the cytoplasm. Virions, and presumably viral

Table 3 Putative structure of the iridovirid RNA transcription complex and associated transcription factors

Subunit	Source	Supporting evidence	References
Pol-IIα	Ranavirus-encoded	Sequence homology, asMO-mediated KD	Tan et al. (2004) and Sample et al. (2007)
Pol-IIβ	Ranavirus-encoded	Sequence homology, iPOND analysis	Tan et al. (2004) and Ke et al. (2022b)
Rpb 3	Host-encoded	iPond/MS analysis	Ke et al. (2022b)
Rpb5	Ranavirus-encoded	Affinity binding to FLAG-tagged vPOL-IIβ; sequence homology	Ke et al. (2022b) and Mirzakhanyan and Gershon (2017)
Rpb6	Host-encoded	iPond/MS analysis	Ke et al. (2022b)
Rpb7	Chloriridovirus-encoded	Sequence homology	Mirzakhanyan and Gershon (2017)
Rpb10	Chloriridovirus-encoded	Sequence homology	Mirzakhanyan and Gershon (2017)
Rpb11	Host-encoded	iPond/MS analysis	Ke et al. (2022b)
TFSII	Ranavirus-encoded	Sequence homology	Mirzakhanyan and Gershon (2020)
VETF	Ranavirus-encoded	Sequence homology	Mirzakhanyan and Gershon (2020)
VLTF2-like	Ranavirus-encoded	Sequence homology	Mirzakhanyan and Gershon (2020)
VLTF3-like	Ranavirus-encoded	Sequence homology	Mirzakhanyan and Gershon (2020)
TFIIB	Ranavirus-encoded	Deep mining	Mirzakhanyan and Gershon (2020)

Shown are the putative virus- and host-encoded subunits found within the iridovirid transcription complex. All noted subunits were found within members of the genus *Ranavirus* with the exception of Rpb 7 and 10 that were only detected within members of the genus *Chloriridovirus* and Rpb 3, 6, and 11 which were of host origin. See text for details

cores, are transported to the nuclear membrane where viral DNA is subsequently released into the nucleus (Braunwald et al. 1985; Gendrault et al. 1981). In addition, entry may also involve caveolae-mediated endocytosis or macropinocytosis (Guo et al. 2011, 2012; Huang et al. 2018; Jia et al. 2013). While the identity of specific viral and cellular proteins associated with viral entry has yet to be fully elucidated, recent work suggests that cellular class A scavenger receptor protein (SR-A) is utilized by FV3 during the cellular entry process (Vo et al. 2019). SR-As are found among vertebrates and have been associated with the entry of vaccinia virus and herpesviruses (MacLeod et al. 2013, 2015). In addition, ranavirus binding to heparan may be involved in the initial stages of virion entry (Ke et al. 2019). Future work to identify both cellular receptors and virus-specific proteins that facilitate iridovirid entry will help elucidate the promiscuous nature of ranavirus infections.

Table 4 Ranavirus replication- and transcription-related proteins detected by iPOND-MS analysis[a]

Event	Gene	Viral ORF (ADRV/RGV)
DNA replication	Viral DNA-dependent DNA polymerase (vDPOL)	**47L/63R**
	Proliferating cell nuclear antigen (vPCNA)	**23L/91R**
	Single-stranded DNA binding protein (vSSB)	85L/27R
	D5 family NTPase: putative viral helicase/primase	**88L/24R**
DNA modification and processing	Cytosine DNA methyltransferase	24L/90R
	FEN endonuclease (a RAD2 homolog) involved in replication, repair, and recombination	**12L/102R**
	DEAD-like helicase	**10L/10L**
	Ribonuclease III	**28R/87L**
RNA synthesis	vPOL-IIα	**9R/9R**
	vPOL-IIβ	**46R/65L**
	Transcription termination factor, Rho domain	79L/34R
Viral ORFs with homologs in other systems, but not clearly linked to viral replicative events	An RRV orf2-like protein	**68L/44R**
	Tyrosine kinase/lipopolysaccharide modifying enzyme	**83L/29R**
	2 cysteine adapter domain; protein kinase domain	**91L/21R**
	Putative AAA_ATPase	**96L/16R**
	ATPase-dependent protease	29L/86R
	SAP domain containing protein	62R/50L
	Erv/Air family protein	19L/95R
	US22 superfamily protein	**6R/6R**
ORFs with no known functions or conserved domains	Unknown	3L/3L
	Unknown	20R/94L
	Unknown	61L/51R
	Unknown	64R/48L
	Unknown	97L/15R

Table adapted from Ke et al. (2022b)

[a]Each virus was examined in triplicate and protein products identified in at least four of six independent assays are shown here. ORFs corresponding to iridovirid core genes are indicating in boldface type

Table 5 Ranaviral proteins involved in virus-mediated immune evasion

Viral protein	Strain	ORF	Characterization[a]	Function	References
Viral homolog of the eukaryotic translation initiation factor 2-alpha (vIF-2α)	ATV	57R	KO; HV	Inhibition of eIF-2α phosphorylation and degradation of PKR/PKZ	Jancovich and Jacobs (2011) Huynh et al. (2017)
	RCV-Z	ND[b]	EE	Inhibition of eIF-2α phosphorylation	Rothenburg et al. (2011)
	RGV	28R	SH	Inhibitor of IFN response	Lei et al. (2012a, b)
Caspase activation and recruitment domain (CARD)	FV3	63L	KO	Counteract IFN response and cellular apoptosis	De Jesus Andino et al. (2015)
	RGV	68R	SH	Putative apoptosis inhibitor	Lei et al. (2012a, b)
	GIV	27L	EE	Apoptosis inhibitor	Chen et al. (2015)
Beta-hydroxysteroid dehydrogenase (β-HSD)	FV3	52L	KO	Immune evasion	De Jesus Andino et al. (2015)
	RGV	52L	EE	Suppress virus-induced CPE	Sun et al. (2006)
Bcl-2	TFV	104R	EE	Inhibition of apoptosis	He et al. (2019)
	GIV	GIV66	EE	Inhibition of apoptosis	Lin et al. (2008) Banjara et al. (2018)
RNase III	ATV	25R	EE	Inhibitor of eIF-2α phosphorylation	Allen et al. (2017)
Semaphorin	SGIV	155R	EE	Attenuation of cellular immunity	Yan et al. (2014)
Viral homolog of the tumor necrosis factor receptor (vTNFR)	SGIV	VP51L VP96R	KO; EE EE	Enhance cell proliferation inhibit apoptosis	Yu et al. (2016), Yu et al. (2017) Huang et al. (2013)
Stingless	SGIV	VP131	EE	Degrade host Stimulator of Interferon Genes (STING)	Zhang et al. (2022a, b)

[a]*KO* knock-out deletion mutant virus, *EE* ectopic expression from plasmid, *HV* characterization in a heterologous virus system, *SH* sequence homology
[b]*ND* not determined

2.2 Nuclear Events

Within the nucleus, first stage viral DNA synthesis takes place as well as the synthesis of early viral transcripts. As with other DNA viruses, such as herpesviruses, ranaviruses utilize host RNA polymerase II to transcribe early viral messages (Goorha 1981). However, in contrast to herpesviruses, ranavirus transcription also requires the presence of one or more proteins designated virion-associated transcriptional transactivators (VATT). As a result, deproteinized viral genomic DNA cannot be transcribed and is not infectious (Willis et al. 1990; Willis and Granoff 1985; Willis and Thompson 1986). The precise identity of the VATT is not yet clear although recent work (described below) suggests four possible candidates. Viral gene expression takes place in an ordered temporal fashion. The first viral transcripts, termed "immediate-early" (IE), are synthesized using host RNA polymerase II (POL-II) and among their gene products are one or more proteins that are required for the synthesis of a second class of early transcripts, designated "delayed early" (DE) (Willis and Granoff 1978). As a group, IE and DE transcripts likely encode regulatory and virulence proteins as well as key catalytic proteins such as the two largest subunits of the viral RNA polymerase (vPOL-II) and the viral DNA polymerase (Majji et al. 2009). Following microarray analysis of FV3 gene expression, 33 IE and 22 DE transcripts, corresponding to approximately half of the FV3 coding potential, were identified (Majji et al. 2009). Similar levels of IE and DE gene products were seen with other ranaviruses (Chen et al. 2006; Teng et al. 2008). Whereas host POL-II is responsible for the transcription of IE (and perhaps DE) viral mRNAs, a novel viral transcriptase composed of vPOL-IIα and vPOL-IIβ along with additional host and viral subunits is responsible for transcription of late viral messages within cytoplasmic viral assembly sites (Tables 2 and 3; see below). The absence of methylated adenosine residues on late viral messages, an event thought to occur in the nucleus, suggests that late viral transcription takes place in the cytoplasm (Raghow and Granoff 1980). In addition to the requirement for vPOL-II, full expression of late viral transcription also requires de novo viral DNA synthesis. As a result, late transcription is markedly inhibited in the presence of DNA inhibitors such as phosphonoacetic acid (PAA) and cytosine arabinoside (araC) (Cheng 2014; Chinchar and Granoff 1984). Finally, as with host transcripts, viral transcripts are capped and methylated, but unlike their cellular counterparts, ranavirus mRNAs lack poly[A] tails and introns (Willis et al. 1985).

Because ranavirus virions do not contain a virion-associated DNA polymerase, viral genomic replication is catalyzed by the products of early viral gene expression. Several viral proteins, including the DNA-dependent DNA polymerase (vDPOL), a processivity factor with homology to Proliferating Cell Nuclear antigen (PCNA), a single-stranded DNA binding protein (vSSBP), a helicase/primase, and others, have been identified (see below) and contribute to the synthesis of genome-sized to twice genome-sized DNA molecules in the nucleus (Goorha 1982).

2.3 *Cytoplasmic Events*

Following its synthesis in the nucleus, newly synthesized viral DNA is transported into the cytoplasm where it is methylated by a virus-encoded cytosine-specific DNA methyltransferase (DMTase) (Willis et al. 1984; Willis and Granoff 1980). Each cytosine within the sequence CpG is targeted, leading to the methylation of 20–25% of cytosines and resulting in the highest level of DNA methylation seen among vertebrate viruses. However, despite this extraordinary level of methylation, the precise role that methylation plays in the viral life cycle is not known. Methylation has been suggested to protect viral genomic DNA from attack by a virus-encoded restriction-modification enzyme that targets unmethylated host DNA (Kaur et al. 1995). Consistent with this suggestion, FV3 infection in the presence of 5′-azacytidine (azaC), an inhibitor of DNA methylation, does not affect viral transcription or translation and only modestly decreases DNA synthesis, but results in a marked reduction in viral yield (Goorha et al. 1984). The reduction in viral yield was postulated to be due to a block in the packaging of genomic DNA caused by the generation of single-stranded breaks in unmethylated DNA by a virus-encoded restriction endonuclease. Alternatively, methylation may be an immune evasion mechanism that prevents recognition of viral genomic DNA by pattern recognition receptors such as TLR9 that sense unmethylated pathogen DNA and thus blocks activation of an immune response (Hoelzer et al. 2008; Krug et al. 2001, 2004). It would clearly be interesting to learn if unmethylated or undermethylated viral DNA, such as that synthesized in the presence of azaC or following gene knock-down (KD) or knock-out (KO) of the DMTase, triggers an increase in the expression of proinflammatory cytokines and/or interferon due to activation of the TLR9 pathway.

Within the cytoplasm, viral genomes serve as templates for second stage DNA synthesis in which large concatemers containing 10 or more copies of the viral genome are generated (Goorha 1982; Goorha and Dixit 1984) The precise structure of viral concatemers is not known. However, by analogy to phage T4, it is thought that concatemers are large, branched structures that are generated by a process of recombination (Goorha and Dixit 1984; Lo Piano et al. 2011; Mosig 1998). Using a collection of temperature-sensitive (ts) mutants, two complementation groups, defective in either 1st stage (nuclear) or 2nd stage (cytoplasmic) DNA synthesis, were identified (Chinchar and Granoff 1986; Goorha and Dixit 1984; Goorha et al. 1981). However, since viral genomes only encode a single viral DNA polymerase gene, it is likely that while one complementation group encodes the viral DNA polymerase, the second encodes a protein needed to translocate viral DNA from the nucleus to the cytoplasm or to mediate some other function, e.g., recombination, related to concatemer formation. If this hypothesis is correct, then since the ranavirus RAD2 homolog is suggested to play a role in repair and recombination, its KD or KO should block concatemer formation and limit DNA synthesis to the formation of only genome-sized molecules within the nucleus (Ke and Zhang 2022).

2.4 Virus Assembly

Virion formation takes place within morphologically distinct areas of the cytoplasm referred to as virus assembly sites (AS) or virus factories. Assembly sites are electron lucent areas of the cytoplasm that are devoid of cellular organelles (Murti et al. 1985, 1988; Zhang and Gui 2012). Unlike autophagosomes, ranavirus assembly sites are not enclosed within membranes but are surrounded by intermediate filaments, mitochondria, and ribosomes (Fig. 3). Both early and late viral proteins are found within assembly sites as well as viral DNA. Viral assembly sites were detected even when late viral transcription was blocked with an antisense morpholino oligonucleotide (asMO) targeting the largest subunit of vPOL-II (Fig. 3, upper middle panel; see below) or following infection with a ts mutant unable to synthesize late messages at non-permissive temperatures (Chinchar and Granoff 1984; Sample 2010; Sample et al. 2007). However, AS were not seen when cells were infected with a ts mutant unable to synthesize viral DNA at the non-permissive temperature

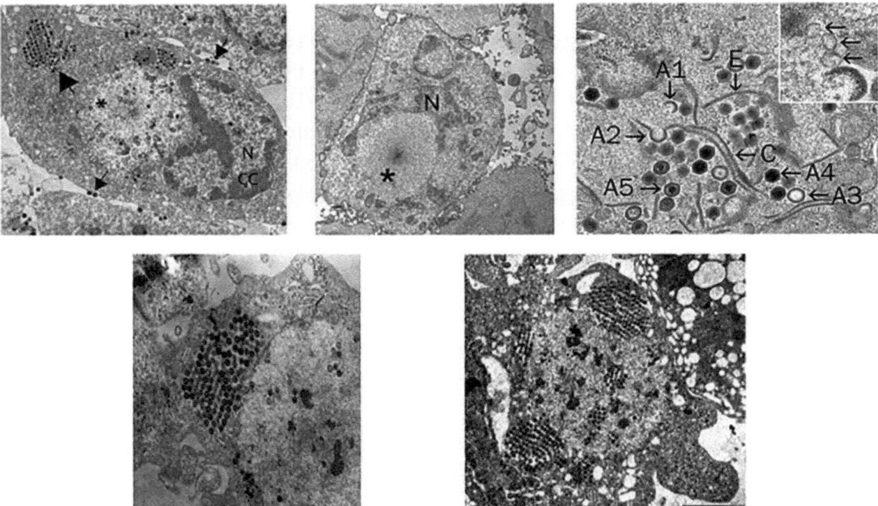

Fig. 3 Transmission electron micrographs of ranavirus infected cells. *Upper* and *lower panels* show typical virus-infected cells (FV3-infected FHM cells, *upper three panels;* ADRV-infected EPC and GSSC cells, *lower left* and *lower right*, respectively). In the upper left panel are shown the following: a nucleus (N) with evidence of chromatin condensation (CC), a well-defined viral assembly site (*), intracytoplasmic paracrystalline arrays (*thick arrow*), and virions budding from the plasma membrane (*thin arrow*). *Upper right panel* shows an enlargement of a viral assembly site displaying virions in various stages of assembly. Full (A4 and A5) and empty (A3) viral particles are shown as well as two possible intermediates (A1 and A2) and two aberrant forms (C and E). The *inset* indicates membranes (*arrows*), possible originating from the ER, that play a role in virion morphogenesis. The *middle top panel* shows FV3-infected FHM cells following exposure to an antisense morpholino oligonucleotide targeting the largest subunit of viral RNA polymerase II (vPOL-IIα). (TEMs courtesy of VG Chinchar, Q-Y Zhang, and co-workers)

or in cells infected with WT virus in the presence of araC (Cheng 2014). Collectively, these results suggest that early proteins and viral DNA are sufficient for AS formation and are consistent with similar findings in vaccinia virus (Chinchar et al. 1984; Hooda-Dhingra et al. 1989; Sample et al. 2007).

The specific steps required for the formation of infectious virions are poorly understood. Genetic analysis of ts mutants identified a minimum of 12 complementation groups that synthesized ostensibly all viral proteins (as indicated by SDS-polyacrylamide gel analysis of infected cell lysates) and viral DNA yet were unable to generate infectious virus particles (Chinchar and Granoff 1986). Transmission electron microscopic (TEM) analysis of these complementation groups identified several ts mutants in which virion structures did not form, whereas in others apparently complete, but non-infectious, particles were present (Purifoy et al. 1973; Sample 2010). Clearly multiple structural and catalytic proteins must be involved in the formation of infectious particles (see below). By analogy to African swine fever virus and other NCLDVs (Milrot et al. 2016; Mutsafi et al. 2014; Rouiller et al. 1998; Suarez et al. 2015; Tulman et al. 2009), virion assembly may involve binding of a myristylated viral protein (e.g., FV3 ORF 53R) to fragments of the endoplasmic reticulum, along with binding of the major capsid protein (MCP) to the opposite face (Whitley et al. 2010). Continued addition of 53R and MCP to membrane fragments is thought to result in the formation of crescent-shaped structures that eventually associate with viral DNA to form the virion. Consistent with this model, putative intermediates along the pathway to virion formation can be detected by transmission electron microscopy (Fig. 3, upper right panel). As discussed below, additional viral proteins are likely involved in this process but the details remain to be elucidated.

The process by which viral DNA is encapsidated has not yet been determined although, by analogy to phage T4, it is thought to take place through a process of headful packaging that generates circularly permuted, terminally redundant genomes (Goorha and Murti 1982). It is not known if ranavirus DNA enters through a unique virion portal as seen with some dsDNA viruses or by engulfment of viral DNA by the enlarging icosahedron (Cardone et al. 2007; Chang et al. 2007; Mutsafi et al. 2014). Moreover, although concatemeric DNA is thought to serve as the source of the DNA that is ultimately packaged into virions, how complete copies of the viral genome are resolved from concatemeric DNA is not known. Complete viral particles are present within viral assembly sites, free within the cytoplasm, as part of paracrystalline arrays, and seen budding from the plasma membrane (Fig. 3, upper left and lower panels). In cultured cells, most virions remain cell-associated and are released as naked, i.e., non-enveloped, particles following cell lysis. However, a variable number of particles bud from the plasma membrane and in the process acquire an envelope. The factors that determine whether a given particle remains cell-associated or enveloped are not known. It is possible that non-enveloped particles are more stable in the environment than enveloped virions, but that enveloped virions more efficiently transmit infections within an infected animal. Freeze-fracture electron microscopy indicates that mature FV3 particles display a Russian doll-like structure with a knobby, spherical core composed of viral DNA and associated proteins enclosed within a capsid composed primarily of a ~48 kDa major capsid protein (Darcy-Tripier et al. 1984; Devauchelle et al. 1985). Electron microscopy of *Chilo iridescent virus* (IIV-6; genus *Iridovirus*) identified three minor proteins (designated finger, zip, and anchor) associated with the virion, but did not link them with specific viral ORFs (Yan et al. 2009). However, as discussed below, recent cryo-EM analysis of SGIV identified four additional virus-encoded minor capsid proteins (mCPs) critical for viral structure formation (Zhao et al. 2023). Moreover, this work linked the seven mCPs to specific SGIV ORFs and supported the view that formation of the viral inner membrane involved association of the myristylated anchor protein (SGIV ORF 88L) with the endoplasmic reticulum (ER) (Table 2). Aside from the MCP and mCPs, other viral proteins, e.g., those responsible for translational shut-off and the transactivation of IE transcription, etc., are also associated with mature virions (Raghow and Granoff 1979; Willis and Granoff 1985). It is not known how these proteins come to be associated with the virion, nor is it known if one or more of the capsid proteins has additional catalytic functions.

3 Interactions Between Host and Virus Determine the Outcome of Infection

3.1 Cell Death: Necrosis, Apoptosis, Parapoptosis, and Autophagy

Ranavirus infections result in the rapid inhibition of host DNA, RNA, and protein synthesis and culminate in cell death (Goorha and Granoff 1979; Raghow and Granoff 1979; Willis et al. 1985). Interestingly, both infectious and non-infectious virions (i.e., heat- or UV-inactivated virus) trigger the turn off of cellular transcription and translation indicating that the "shut-off protein" is virion-associated and heat stable (Cordier et al. 1981; Raghow and Granoff 1979). Furthermore, virus-infected cells undergo apoptosis as shown by chromatin condensation and the fragmentation of cellular DNA (Chinchar et al. 2003). Apoptosis appears following either productive infection or infection by inactivated virions and may be a result of the prior inhibition of host cell macromolecular synthesis and/or the activation of protein kinase R (PKR) (Gil and Esteban 2000; Zuo et al. 2022). Ranavirus-induced apoptosis is dependent upon caspase activation and can be prevented by caspase inhibition (Chinchar et al. 2003). Apoptosis has also been observed in *Rana grylio* virus (RGV) infected fathead minnow (FHM) cells and has been linked to mitochondrial dysfunction (Huang et al. 2007).

In addition to apoptotic death, SGIV, the most phylogenetically distant member of the genus *Ranavirus*, triggers different forms of programmed cell death depending upon the infected cell type. In spleen cells from grouper (GS), the natural host for SGIV, infection triggers a nonapoptotic form of programmed cell death (PCD) designated parapoptosis (Huang et al. 2011). Parapoptosis is characterized by the appearance of cytoplasmic vacuoles, distended ER, and the absence of DNA fragmentation, apoptotic bodies, and caspase activation. In contrast, SGIV induces the typical form of apoptosis in non-host FHM cells characterized by caspase activation and DNA fragmentation and may be regulated by cellular microRNA miR-146a using a negative feedback mechanism allowing virus replication to continue (Ni et al. 2017). Furthermore, disruption of mitochondrial transmembrane potential and externalization of phosphatidylserine (PS) are not detected in GS cells but are seen in FHM cells after SGIV infection (Huang et al. 2011). Similar results were observed in GIV-infected grouper kidney (GK) cells. Likewise, infection with LMBV induced apoptosis in *Epithelioma papulosum* cyprini (EPC) cells and was regulated by the phosphatidylinositol 3-kinase (PI3K) and extracellular-signal-regulated kinase (ERK) signaling pathways (Huang et al. 2014b). In contrast, LMBV infection of a cell line derived from largemouth bass fin induced non-apoptotic cell death, indicating that cell death induced by LMBV, as that triggered by SGIV, may be cell type dependent (Yang et al. 2022). Utilizing a different ranavirus species, apoptosis with mitochondrial damage was observed in Chinese giant salamander iridovirus (CGSIV)-infected Chinese giant salamander muscle (GSM) cells (Li et al. 2019). Whether all ranaviruses trigger different forms of cell death in host versus non-host cells requires further study (Pham et al. 2012).

Autophagy is a highly conserved, lysosome-dependent process that removes excess or damaged organelles to maintain cellular homeostasis in connection with apoptosis. In view of that, autophagy has been implicated as a factor impacting the fate of intracellular pathogens, including viruses (Levine and Deretic 2007). Autophagy and apoptosis were induced in LMBV-infected EPC and MEF-1 cells and resulted in the inhibition of virus replication (Deng et al. 2022; Qi et al. 2016). While the exact mechanism of autophagy regulation in ranavirus infected cells is still unclear, this antiviral mechanism may play a significant role in inhibiting virus replication in infected cells.

3.2 Host Shut-off and the Selective Expression of Viral Gene Products

The identity of ranavirus virion-associated protein(s) responsible for the rapid inhibition of host transcription and translation and the subsequent inhibition of cellular DNA synthesis are not known. However, despite the rapid and profound inhibition of host macromolecular synthesis, ranavirus DNA replication, transcription, and translation remain unaffected and abundant levels of infectious viral particles are produced within 24 h or less (Willis et al. 1985). The maintenance of viral protein synthesis in the face of the marked inhibition of cellular translation is likely the result of several factors which have not yet been fully elucidated. In the first place, ranaviruses modulate the cellular eukaryotic translation initiation factor 2 alpha subunit (eIF-2α), a key component of translation initiation (Safer 1983; Sonenberg and Hinnebusch 2009), using viral homologs of eIF-2α (vIF-2α) and a virally encoded RNase III-like protein (see below). Cellular eIF-2α is targeted by protein kinase R (PKR), an IFN-inducible protein that binds to dsRNA. Binding of virus-derived dsRNA results in PKR activation by auto-phosphorylation and leads to the subsequent phosphorylation and inactivation of eIF-2α, thereby inhibiting translation in virally infected cells (Levin and London 1978). Most ranaviruses encode a homolog of cellular eIF-2α (vIF-2α) that is thought to act as a pseudosubstrate that binds and, in the case of ATV, degrades PKR, thus preventing the phosphorylation and subsequent inactivation of cellular eIF-2α and maintaining viral translation (Beattie et al. 1991; Huynh et al. 2017; Jancovich and Jacobs 2011). In addition, all ranaviruses encode an RNase III-like protein that is thought to degrade dsRNA, thereby preventing PKR activation and subsequent eIF-2α inactivation (Allen et al. 2017). In addition, ranaviruses synthesize abundant levels of highly efficient viral messages that outcompete host transcripts for access to the remaining cellular translational machinery and thus inhibit the translation of less efficient host messages (Chinchar and Yu 1990a, b). Collectively, although precise mechanisms have yet to be determined, these, and perhaps other, events lead to the shut-off of host translation and the onset and continuation of viral protein synthesis.

In contrast to translational shutoff, little is known about how ranaviruses selectively inhibit host transcription. Perhaps ranaviruses target host POL-II as infection progresses and rely upon vPOL-II to synthesize both early and late viral transcripts. If this model is correct, it would be instructive to determine whether early genes continue to be synthesized at later times using vPOL-II to catalyze their synthesis. A recent investigation of the ranavirus replication and transcription machinery in ADRV-infected cells showed that host POL-II subunits Rpb3, Rpb6, and Rpb11 and topoisomerase IIα/β (Top IIα/β) were recruited to viral DNA replicase and RNA transcriptase complexes (Ke et al. 2022b). This finding suggests that the recruitment of host proteins into viral DNA replicase and RNA transcriptase complexes (and thus away from their cellular counterparts) may be one of the strategies employed to inhibit cellular transcription and DNA replication. Furthermore, the presence of cellular Rpb 3, 6, and 11 and Top IIα/β within viral AS supports the suggestion that late viral transcription and DNA concatemer formation take place within viral assembly sites (Ke et al. 2022b). In addition to the recruitment of cellular proteins critical for both DNA replication and RNA synthesis into viral replication/transcription complexes, the inhibition of host DNA and RNA synthesis may also be due to the action of a virus-encoded endonuclease that is part of a restriction-modification system that degrades unmethylated host DNA. Degradation of host DNA will not only negatively impact host transcription and DNA replication, but the resulting nucleotides may also be re-utilized for the synthesis of viral DNA (Feighny et al. 1981; Kaur et al. 1995).

3.3 Interplay Between Host and Virus Determines the Outcome of Infection at the Organismal Level

Ranaviruses display a broad host range and are considered promiscuous pathogens of cold-blooded vertebrates (Chinchar and Waltzek 2014; Duffus et al. 2015). As the interplay between virus and host determines the outcome of infection, especially following heterologous infection, different ranavirus species induce varying degrees of pathogenicity in the same host. Infection studies using the Chinese giant salamander (CGS), a CGS cell line, and two ranavirus species illustrate this variability at the organismal level. Transcriptomic analysis of host and viral gene expression was performed following infection of CGS with ADRV, a virus that typically infects CGS, and RGV, a virus that normally infects anurans. Analysis of viral and host transcription in infected CGS revealed different replication dynamics for the two viruses and differential host responses (Ke et al. 2018). RGV replicated rapidly after entry compared to ADRV, although both displayed comparable levels of viral replication between days 3 and 6 post infection. Differential host responses were noted in RGV- and ADRV-infected animals and this may explain the different fates of the two viruses.

Virulence is a function of multiple factors but two that are likely most important involve those that enhance virus replication and those that block host immune defenses. Large DNA viruses such as herpesviruses and poxviruses encode numerous genes that fulfill these functions, and preliminary findings among ranaviruses suggest the existence of ranavirus homologs (Table 2) that perform similar functions (Johnston et al. 2005; Johnston and McFadden 2003; Lawler and Brady 2020; Lum and Cristea 2021; Yu et al. 2021). To put our later discussion of ranavirus gene function in context, we discuss events in vaccinia virus and herpesvirus here. In the case of viral genes that enhance replication, prime examples are viral homologs of the ribonucleotide reductase (RR) large (RR1) and small (RR2) subunits. Both vaccinia virus (VACV) and herpes simplex virus 1 (HSV1) encode RR genes. VACV RR2 subunits form functional complexes with host RR1 and ensure sufficient dNTPs for viral replication (Gammon et al. 2010). Likewise, HSV and pseudorabies virus-encoded RR convert ribonucleoside diphosphates into the corresponding deoxyribonucleotides and play key roles in viral DNA synthesis by maintaining dNTP pool sizes (Conner et al. 1994; Daikoku et al. 1991). Consistent with this key role in viral biogenesis, pseudorabies virus mutants defective in RR expression are avirulent in vivo (de Wind et al. 1993). However, in addition to their role in dNTP synthesis, HSV-1 and HSV-2 RR1 subunits also protect cells from apoptosis (Chabaud et al. 2007; Langelier et al. 2002). In contrast to the α- and β-herpesviruses, the RR1 subunit of murine cytomegalovirus is catalytically inactive and does not play a role in increasing dNTP pool sizes. Rather it has evolved a new function in which it inhibits RIP1, a cellular adaptor protein, and blocks signaling pathways involved in innate immunity and inflammation (Lembo and Brune 2009). The different roles of RR in various herpesvirus species indicates that caution must be used when inferring function from homology. Other viral genes that contribute to enhanced replication encode viral homologs of dUTPase and thymidine kinase.

RGV homologs of dUTPase and thymidine kinase have been identified as early viral genes (Zhao et al. 2007, 2009). However, although viewed as genes that could potentially enhance virus replication, infection of EPC cells by KO mutants targeting these two homologs showed one-step and multiple-step growth curves similar to that of WT virus suggesting that, at least in EPC cells, their function was not required for virus replication (Huang et al. 2016). An expanded discussion of ranavirus virulence proteins is found below (Sect. 4.3).

3.4 Anti-Viral Immunity

Both innate and adaptive immune systems are employed by various hosts to resist ranavirus infection. The host immune response to ranavirus infection has been most productively examined using a model developed by Robert and co-workers that pairs FV3, the best-characterized ranavirus at the molecular level, with *Xenopus laevis*, the amphibian with the most fully characterized immune system (Robert 2010; Ruiz and Robert 2023). Infection of immunocompetent adult frogs is manifested most prominently in the kidney with virus also seen in many other tissues including skin, intestine, liver, spleen, and brain at various points during infection (Grayfer et al. 2015b; Hauser et al. 2021; Wendel et al. 2017, 2018). In immunocompetent adults, these infections resolve within a few weeks with minimal mortality (Gantress et al. 2003). In contrast, infection of tadpoles, which possess limited MHC I expression, metamorphosing and post-metamorphic froglets, and immunocompromised adult animals display considerable morbidity and mortality (Gantress et al. 2003; Robert et al. 2005; Tweedell and Granoff 1968). In light of the fact that anti-ranaviral immunity is discussed at length in another chapter within this monograph (Grayfer et al. 2024), only a brief synopsis of the major facets of antiviral immunity is presented here.

Antiviral interferons (IFNs) play important roles in the host defense against ranaviruses. *Xenopus laevis* IFN types I and III were cloned and their expression analyzed following FV3 infection (Grayfer et al. 2014, 2015a). These investigators found that FV3-resistant adult frogs expressed high levels of type I IFN, whereas FV3-sensitive tadpoles expressed low levels of type I IFN, but high levels of type III IFN. Examination of IFN type I and III in the skin of *X. laevis* tadpoles and adult frogs following FV3 infection also showed that tadpoles mounted type III IFN responses but adult frogs adopted mainly the type I IFN response (Wendel et al. 2017, 2018). Interestingly, a robust type I and III IFN response was observed in the intestines of FV3-challenged tadpoles, but not in adults, and was thought to be mediated by the recruitment of myeloid cells into the intestines of infected tadpoles (Hauser et al. 2021). These observations are consistent with the finding that adult frogs exhibited greater intestinal levels of FV3 DNA than tadpoles and is also in line with the fact that tadpoles are more resistant to FV3 than froglets. In this case, resistance appears to be due to the elevated expression of endogenous retroviruses in myeloid cells within tadpole kidneys (Kalia et al. 2022).

In immunodeficient or immunocompromised *Xenopus*, infection begins in the kidney but becomes systemic and spreads to multiple organs including the liver, gastrointestinal tract, and skin (Gantress et al. 2003). Consistent with observations that fish could be protected from ranavirus- and megalocytivirus-induced disease by vaccination with inactivated virions or a DNA vaccine (Caipang et al. 2006; Ou-yang et al. 2012), antibody responses were shown to play a protective role in FV3 infections (Maniero et al. 2006). Consistent with that notion, vaccination of Chinese giant salamanders with a DNA vaccine expressing the ADRV 2L envelope protein triggered an antibody response and inhibited ADRV replication (Chen et al. 2018).

Likewise, cell-mediated immunity was shown to play an important role in protection from FV3-induced disease (Grayfer et al. 2015b; Morales and Robert 2007). Both conventional CD8+ cytotoxic T cells and MHC Class I-like/innate-like T (iT) cells play roles in host immunity to ranavirus infection (Grayfer et al. 2015b). In the latter case, nonclassical MHC class Ib XNC10-restricted invariant T (iT) cells (invariant Vα6 [iVα6]) were recruited from the spleen into the peritoneum during intraperitoneal FV3 infection. Transient depletion of iT cells reduced the effectiveness of the antiviral response indicating that *X. laevis* XNC10-restricted iVα6 T cells are important for the early anti-FV3 response (Edholm et al. 2015). Further study showed that upregulation of XNC10 restricted class I-like genes and the rapid recruitment of iVα6 T cells required virus replication and that a delay in the iVα6 T cell response, e.g., following XNC10 depletion, inhibited IFN and cytokine gene responses. Collectively these results indicated that iT cells serve as an immune surveillance system (Edholm et al. 2019). Thus, early activation of iT cells by nonpolymorphic MHC class I like molecules appears to play a crucial role in generating an efficient antiviral immune response.

Although macrophages likely play a critical role in immunity, they are susceptible to FV3 infection, and this may lead to two adverse consequences: the loss of their ability to process and present viral antigens and the generation of a cadre of persistently infected macrophages that maintain virus within an infected host (Morales et al. 2010). As shown by Robert and co-workers, quiescent FV3 residing in *X. levis* macrophages could be reactivated by stimulation with heat-killed *E. coli* or by the TLR5 ligand flagellin (Samanta et al. 2021). Further research in the Grayfer lab showed that colony-stimulating factor-1 (CSF-1) and interleukin-34 (IL-34)-stimulated macrophages displayed differential susceptibility to FV3. IL-34-stimulated macrophages expressed significantly greater levels of antiviral genes including IFN and IFN receptor genes and resisted FV3 infection, whereas CSF-1-stimulated macrophages were more susceptible to infection (Hossainey et al. 2021; Yaparla et al. 2018, 2019).

It is likely that multiple cellular genes play various roles in antiviral immunity and virus replication. Consistent with this suggestion, infection of FHM cells with either wild type FV3 or a vIF-2α knockout mutant induced multiple immune-response genes at the transcriptional level including IFN, IL-8, GILT, IRF-3 (Cheng et al. 2014). Likewise, vaccination of groupers with inactivated SGIV induced expression of numerous immune-related genes including Mx1, ISG15, IL-8, IL-1β, and MHC I/II indicating that the immune response is conserved among different fish species and similar to that seen in mammals (Ou-yang et al. 2012). Several grouper immune-related genes including tripartite motif-containing genes and IFN-induced transmembrane proteins possess anti-ranavirus activity (Zheng et al. 2022). Likewise, a series of MHC alleles were identified in Chinese giant salamanders (CGS), several of which were up-regulated following both in vivo and in vitro ADRV infection, indicating the important role of MHC in the CGS immune response (Zhu et al. 2014a). A cDNA library from the CGS thymus revealed 137 putative immune-related genes. Among them, IFN-inducible protein 6 (IFI6) was shown to have anti-ranavirus activity in cultured cells (Zhu et al. 2014b).

3.5 *Virus-Mediated Evasion of the Host Immune Response*

In addition to genes such as RR and dihydrofolate reductase (DHFR) that enhance viral replication, ranaviruses also encode genes that inhibit or counteract host immune responses (Tables 2 and 5). This finding is consistent with those seen in other viral systems (Garcia-Sastre 2017; Rojas et al. 2021; Seet et al. 2003). For example, poxviruses encode over two dozen immune evasion genes, and perhaps half of the genes encoded by human cytomegalovirus may be involved in immune evasion (Eberhardt et al. 2013; Lu and Zhang 2020; Yu et al. 2021). A variety of approaches have been used to identify and confirm the function of putative viral immune evasion proteins, and these and the results of those studies are discussed below. In addition to putative immune evasion proteins, there are other genes that are unique to specific virus species or genera and may act only within a specific host species, tissue, or cell type. For example, there are ~35 ranavirus core genes that might encode polypeptides that enhance viral replication or impair immunity within specific ectothermic hosts (Eaton et al. 2007; Price 2015). The challenge of identifying these proteins by similarity searches is considerable. Because the level of sequence similarity between viral and cellular homologs, even in mammalian systems, may be low, identification of a viral protein as a potential inhibitor of a specific function by homology alone is far from certain unless key sequence motifs are conserved.

4 Ranavirus Gene Products and Their Functions

Although much has been learned about iridovirid and ranavirus genes and their functions since the identification of the first ranavirus by Granoff and co-workers (1966) and the publication of the first iridovirid (Tidona and Darai 1997) and rana-virus (He et al. 2001; Jancovich et al. 2003; Tan et al. 2004) genomic sequences, the precise function of many ranavirus genes remains unknown. As noted above, viruses within the family *Iridoviridae* possess between approximately 100–250 ORFs capable of encoding proteins greater than 4–5 kDa (Chinchar et al. 2017a; Jancovich et al. 2015b). Within a given genus, there is a high level of sequence and gene conservation. For example, ranaviruses share up to 72 genes in common and display sequence identities within those genes ranging from 70% to nearly 100% (Eaton et al. 2007; Price 2015). However, between genera, sequence identity among homologous genes drops to 50% or less and the number of genes common to all members of the family falls to approximately 26 (Eaton et al. 2007; Jancovich et al. 2010; Lei et al. 2012b). Current thinking suggests that core genes, e.g., vDPOL, the two largest subunits of vPOL-II, the MCP, etc., encode catalytic functions needed for replication and/or structural proteins involved in virion assembly that are required in all host cells, whereas genes found within a given species or common to those in a given genera function to enhance replication in certain hosts or host cells or to evade host immunity. Below we describe research published from 2015 to 2023 that deepens our understanding of ranavirus biogenesis and viral gene function.

4.1 Approaches for Identifying Viral Gene Function

Various computational, genetic, biochemical, and molecular approaches have been used to identify ranavirus genes and their function. Herein we discuss these approaches and the insights they provide that enhance our understanding of viral gene function. Following genomic sequencing, in silico studies of ranavirus ORFs by BLAST analyses detected homologs among iridovirids, other virus families, and various host and non-host species. Based on these studies, putative functions have been assigned to more than one-third of ranavirus ORFs (Tan et al. 2004). However, because homology, including the presence of specific domains, does not provide definitive proof of function, additional studies are needed for confirmation.

Identification of ranavirus genes and their functions has advanced markedly since early attempts based on the isolation and characterization of FV3 temperature-sensitive (ts) and drug-resistance mutants (Chinchar and Granoff 1984, 1986; Purifoy et al. 1973). Although biochemical studies and analysis of ts mutants supported the two-stage model of viral DNA synthesis (Goorha 1982; Goorha and Dixit 1984), it was not possible to link a specific phenotypic defect with an individual viral gene due to the inability to use plasmids containing cloned FV3 restriction fragments to rescue ts mutants (VG Chinchar, unpublished). Furthermore, because the generation of ts mutants is a random process, it is not possible to target specific viral genes. However, as viral genomic sequence information became available, functions could be associated with specific genes by a variety of techniques. Knock down (KD) studies using both anti-sense morpholino oligonucleotides (asMOs) and small interfering RNAs (siRNAs) targeted specific ORFs and linked inhibition of viral gene expression with phenotypic changes to support functional associations (Sample et al. 2007; Whitley 2011;

Whitley et al. 2011; Xie et al. 2014). For example, KD of the largest subunit of the viral homolog of RNA polymerase II (vPOL-IIα) resulted in a marked reduction in late viral gene expression, indicating a role for vPOL-IIα in late viral transcription (Sample et al. 2007). Although KD studies have been useful in assessing gene function, KD must be verified by western blot or SDS-polyacrylamide gel electrophoresis of the targeted protein. However, in the absence of a verified reduction in the level of the targeted protein, interpretation of low or intermediate levels of viral replication following exposure to a given asMO or siRNA is equivocal. Alternatively, knockout (KO) studies, using homologous recombination to delete specific viral genes, avoid problems due to issues of incomplete KD and have identified several genes, likely playing key roles in viral immune evasion or serving an accessory function, that are non-essential for replication in vitro but required for growth in vivo (Allen et al. 2017; Aron et al. 2016; Chen et al. 2011; De Jesus Andino et al. 2015). In addition to KO mutants generated by homologous recombination, it may be possible to construct KO mutants using the CRISPR-Cas system as have been done with vaccinia and other DNA viruses (Yuan et al. 2015). Furthermore, although KO approaches are useful for identifying genes non-essential for replication in vitro, they cannot be used to target genes essential for virus replication because KO renders the resulting virus non-viable (Aron et al. 2016; Liu et al. 2023). To circumvent this limitation, the development of conditionally lethal mutants offers the possibility of targeting essential genes, i.e., those required for replication in all cells, by placing expression of the gene of interest under the control of an inducer such as IPTG (He et al. 2013). Alternatively, KO mutants may be constructed in cells expressing the gene of interest, and then virus replication compared in those cells with cells not expressing the gene to infer its role in replication.

In addition to the KD and KO approaches outlined above, gene function can also be explored through ectopic expression in which ranaviral proteins are expressed singly or in combination following transfection of plasmids capable of expressing these products in vitro. Once expressed, the subcellular location of those proteins could be determined, interactions with host and viral proteins identified, and the impact of their expression on viral replication and/or cellular functions studied. In one example, recombinant ADRV ORF 85L, an ORF with no known homologs, was shown to bind single-stranded DNA (ssDNA) in an electromobility shift assay (EMSA) supporting its role as a viral ssDNA binding protein (Ke et al. 2022b). Using iPOND (isolation of proteins on nascent DNA) methodology, proteins bound to viral DNA and thus putatively associated with viral DNA and RNA synthesis were identified following mass spectroscopic analysis (MS) (Ke et al. 2022b). Virion-associated proteins, both structural and catalytic, were also identified using gel analysis to separate proteins from purified virions by size and MS to determine to which specific viral gene they corresponded (Ince et al. 2010). Finally, sequence information can be used to clone and express viral proteins and to use these to generate antibodies that can then be used to track their intracellular location by immunofluorescence assay and their association with other cellular and viral products by co-immunoprecipitation. Below we discuss how these various approaches have been used to determine the roles of viral gene products in virus replication and virulence.

While all these approaches are useful, the elucidation of ranavirus gene products and their functions is complicated by both the number of different ranaviruses employed in these studies and the variety of host cells used to support virus replication. While ranaviruses share the majority of their genes in common, differences in gene content exist even among different isolates of the same species (Eaton et al. 2007; Jancovich et al. 2010). For example, as noted above, most ranaviruses possess a full-size copy of the viral homolog of eukaryotic translational initiation factor 2α (vIF-2α), whereas FV3 and a FV3-like virus isolated from soft-shell turtles contain only a truncated version of that ORF (Huang et al. 2009; Tan et al. 2004). Furthermore, because of insertions, deletions, and inversions within genomic DNA (Fig. 1), gene content and order vary among ranavirus species. Accordingly, ORF designations in one virus species do not match those in another, e.g., a myristylated membrane protein designated 53L in FV3 corresponds to SGIV 88L. To facilitate comparison, a table listing the ORF designations of the 26 core iridovirid genes found among 18 iridovirids representing 5 of the 7 recognized genera was constructed (Lei et al. 2012b). Based on this work and BLASTP analyses of the five ranavirus species/isolates most often subjected to molecular analysis, we constructed a table containing a selection of genes discussed in this review (Table 2). Furthermore, the use of different host cell lines creates another level of complexity due to the impact of accessory/efficiency genes and immune evasion genes on virus replication. The former are genes whose products enhance viral replication by alleviating defects in certain cells, whereas the latter encode products that antagonize various host immune responses. As a result of differences between cell types and organisms, expression of a given accessory or immune evasion gene may not be required in some cell lines, but absolutely required in others. Finally, ectopic expression studies, while useful in assessing gene function, may provide misleading information if levels of the expressed protein markedly exceed those synthesized in a normal infection.

In the following sections, we discuss ranavirus genes whose functions have been explored using these approaches. As some of these genes have been addressed in the first edition of this work (Jancovich et al. 2015a), we will focus for the most part on studies since that publication. In contrast to earlier studies which focused primarily on FV3, recent work has utilized additional species within the genus and several from other genera. To guide this discussion, several replicative events and categories will be considered including viral entry, DNA replication, viral transcription, and virion assembly, along with viral genes that augment host function and inhibit host immunity. It should be noted that even with these studies, half or more ranavirus genes remain ORFans whose functions are unknown or identified only as putative membrane or secreted proteins.

4.2 Viral Entry

Although several modes of entry have been detected among ranaviruses and other iridovirids (see above) and one specific host protein, the Class A Scavenger receptor (Vo et al. 2019), has been implicated in entry, no specific viral proteins have been definitively linked with entry. Further complicating the issue are the observations that both enveloped and non-enveloped (i.e., naked) virions are capable of entry, indicating that multiple proteins may play overlapping roles in entry (Braunwald et al. 1979, 1985). Despite this uncertainty, several viral proteins have been suggested to play critical roles in viral entry.

SGIV ORFs 88L and 19L were implicated in entry based on their designations as envelope proteins and ORF 19R's homology to the poxvirus entry-fusion complex G9/A16 (Yao et al. 2019; Yuan et al. 2016b). However, SGIV 88L is homologous to FV3 53R, a putative myristylated membrane protein, and, by analogy to p220 of African swine fever virus (ASFV), may play a role in virion assembly rather than entry (Andres et al. 1997; Salas and Andres 2013). Like other myristylated proteins, SGIV 88L is postulated to function in the recruitment and transformation of the ER membrane into the internal viral membrane (Zhao et al. 2023). SGIV 38L (the putative zip protein) and TFV 20R were linked to entry by possession of an RGD motif, which herpesvirus glycoprotein B uses to bind integrin and mediate viral entry (Garrigues et al. 2008; Wan et al. 2010; Wang et al. 2008; Zhao et al. 2023). In addition, RGV 43R was shown to be an envelope protein and thought to play a role in virion entry (Zeng et al. 2018). RGV 43R was found within the viral AS and a deletion mutant impacted viral entry but did not impact viral DNA levels at early times post infection. In addition, exposure of virions to anti-43R serum reduced viral replication, suggesting that 43R was on the exterior surface of the virion and played a role in entry. RGV 2L, a homolog of the aforementioned SGIV 19L, is a viral membrane protein that was found in viral factories co-localized with two other viral envelope proteins, RGV 22R and RGV 53R (He et al. 2014). Using a conditionally lethal mutant of 2L, He et al. (2014) found that virus titers were markedly reduced under non-permissive conditions attesting to the essential nature of this gene product. Lastly, cell surface heparan sulfate was shown to be involved in ADRV and RGV binding through interaction with ADRV-58L and RGV-53R, respectively (Ke et al. 2019). As with other viruses, initial interaction between virion surface proteins and heparan sulfate may lead to subsequent stronger binding to specific cellular receptors and facilitate viral entry (Ke et al. 2019). From the above discussion, it is not clear whether a single envelope or capsid protein is the principal viral receptor, or if multiple viral proteins (as is the case with vaccinia virus and herpesvirus) are required for viral entry (Howley et al. 2021).

4.3 Viral DNA Replication and Viral RNA Synthesis

Inspection of ranavirus genomes showed marked homology between the cellular and viral versions of vDPOL, vPOL-IIα, and vPOL-IIβ and provided strong putative evidence for their functions (Lei et al. 2012b; Tan et al. 2004). Work by Goorha and co-workers showed that early FV3 transcription required host RNA polymerase II, while late viral mRNA synthesis occurred even when host POL-II was inhibited by alpha-amanitin (Goorha 1981). Supporting the hypothesis that viral enzymes catalyzed late viral transcription, a role for vPOL-IIα in late gene transcription was experimentally confirmed by the observation that its KD using asMOs was accompanied by a marked drop in late viral gene expression (Sample et al. 2007). Interestingly, despite KD of vPOL-IIα and the marked decline in late gene expression, early viral transcription was unaffected and viral assembly sites, albeit devoid of virions, were readily detected (Fig. 3, upper middle panel) (Sample 2010; Sample et al. 2007). Moreover, since vDPOL is an early gene product (Majji et al. 2009), viral DNA synthesis was thought, but never formally shown, to be resistant to vPOL-IIα KD. Similar to vaccinia virus (Hooda-Dhingra et al. 1989), the presence of viral AS in the absence of late viral mRNA and protein synthesis supports the hypothesis that viral assembly site formation requires only viral DNA synthesis and one or more early proteins (Chinchar et al. 1984). Although the above confirms the role of vPOL-IIα in late viral transcription, the exact composition of the viral RNA polymerase remains to be determined. However, ongoing studies support the view that the viral transcriptase is a chimeric protein composed of both viral and host elements (see below). If correct, the presence of a chimeric enzyme composed of both host- and virus-encoded subunits is unprecedented (Mirzakhanyan and Gershon 2017).

Ranavirus DNA synthesis takes place via a two-stage process in which genomic-sized molecules are synthesized in the nucleus and large concatemers are generated within the cytoplasm (Goorha 1982). Despite synthesis of two distinct products, genomic sequence analysis identified only a single viral DNA polymerase gene suggesting that concatemer formation is catalyzed by the same enzyme responsible for the synthesis of genome-sized molecules within the nucleus (Tan et al. 2004). The earlier finding that two genetic complementation groups are associated with DNA synthesis (Chinchar and Granoff 1984; Goorha and Dixit 1984; Goorha et al. 1981), one targeting first stage and the other second stage DNA synthesis, suggests the existence of a second protein that either plays a role in the translocation of genome-size DNA molecules from the nucleus to the cytoplasm, and in whose absence DNA replication is confined to the production of genome-sized molecules within the nucleus, or one that plays some yet to be determined role in concatemer formation. As noted above, the RAD2 homolog may play a role in recombination and if so may be involved in concatemer formation (Goorha and Dixit 1984).

Aside from the three polymerase proteins discussed immediately above, a number of other viral ORFs were linked to viral DNA and RNA synthesis. Using iPOND analysis, recombinant virus-based affinity, EMSA (electrophoretic mobility shift assay), and NanoLuc complementation analyses, several viral ORFs were tentatively assigned functions (Ke et al. 2022b). In that study, two different ranavirus species were used: RGV (*Frog virus 3* species) and ADRV (*Common midwife toad virus* species). Forty-six ADRV and 38 RGV proteins bound nascent DNA and were identified by mass spectroscopy (iPOND-MS). Shown in Table 4 are 24 proteins that bound nascent DNA in 4 of 6 independent assays and 19 of which were assigned putative functions based on homology. It is not known if the remaining five proteins that displayed no homology to proteins in the database represent novel viral proteins not previously linked to DNA replication and/or transcription. While iPOND analysis provides a starting point for identifying ranavirus genes required for viral DNA and RNA synthesis, it may not provide a complete catalog of replicative and transcriptive proteins as some critical proteins were not detected by this assay. For example, a homolog of the transcriptional elongation factor TFIIS, FV3 ORF 82R, was not detected by iPOND analysis.

4.3.1 Proteins Linked to Viral DNA Replication

ADRV 85L and its RGV homolog 27R were shown to be viral single-stranded DNA binding proteins (vSSBP) that bound ssDNA from phage φχ174 in an EMSA (Ke et al. 2022b). Interestingly, ADRV 85L is a homolog of FV3 25R, a prominent early FV3 protein designated P31K that had previously been characterized as having no known function (Tan et al. 2004). In addition, NanoLuciferase complementation and co-immunoprecipitation assays were used to demonstrate interactions among viral proteins. Based on these analyses, the authors found (Fig. 4) that vDPOL interacted with multiple viral proteins including vPCNA, vSSBP, a D5-helicase/primase, and vFEN1(RAD2) along with host topoisomerase IIα, topoisomerase IIβ, and DNA ligase 3 (Ke et al. 2022b).

Among the viral proteins indicated above, vPCNA (viral proliferating cell nuclear antigen) is thought, by analogy to its mammalian counterparts, to function as a clamp protein and serve as a processivity factor during DNA replication and as a platform for other proteins interacting with the genome (Arbel et al. 2021; Boehm et al. 2016). An additional role for vPCNA has been suggested by Zeng and Zhang who showed that when expressed separately, vDPOL (RGV ORF 63R) and vPCNA (RGV ORF 91R) localized to the cytoplasm and nucleus, respectively, whereas when co-expressed vDPOL was found in both the cytoplasm and the nucleus (Zeng and Zhang 2019). Significantly, vDPOL was found within the nucleus co-localized with vPCNA. Inspection of the vPCNA sequence identified two nuclear localization signals (NLS) suggesting that translocation of vDPOL to the nucleus requires transport via interaction with vPCNA. In support of that hypothesis, fluorescent microscopy showed that a 91R construct lacking the two NLS (91RΔNLS-RFP) remained within the cytoplasm suggesting that the nuclear import of RGV 63R was dependent

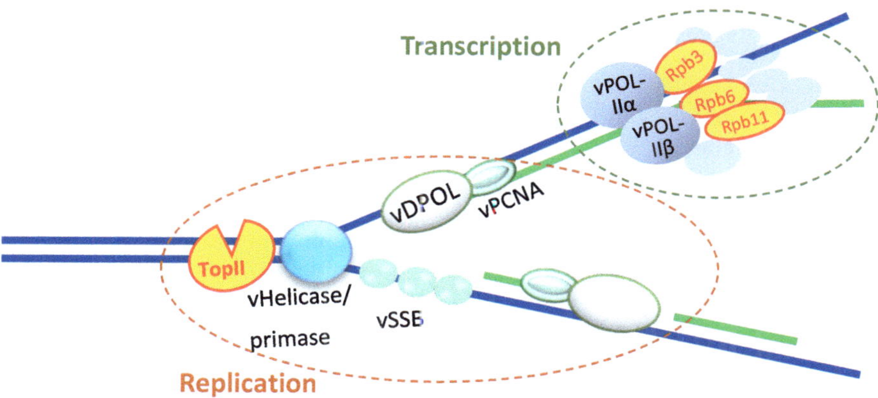

Fig. 4 Schematic diagram of ADRV genomic DNA replication and transcription complexes. The viral genome is indicated by blue and green (nascent DNA) lines. Host proteins are indicated by modules with yellow color and red font

on the NLSs of RGV 91R. Of the remaining three viral proteins identified by iPOND analyses, vSSBP is the aforementioned ssDNA binding protein, D5-Helicase/primase is hypothesized to be involved in unwinding the dsDNA genome, and vFEN1 (RAD2) is postulated to play a role in DNA repair, recombination, and concatemer formation (Ke and Zhang 2022). Interestingly, while ranaviruses appear to utilize host topoisomerase and ligases for viral DNA replication, invertebrate iridescent virus 6 (subfamily *Betairidoviridae*) encodes within its much larger genome a viral version of DNA topoisomerase II (Muller et al. 1999). Furthermore, while the above work suggests five viral products are responsible for DNA synthesis, at least 19 additional virus-encoded proteins are associated with viral DNA following iPOND analysis including a cytosine DNA methyltransferase, a SAP domain-containing protein, a putative resolvase, and a DEAD helicase (Ke and Zhang 2022) (Table 4). The specific roles, if any, of these proteins remain to be determined.

4.3.2 Proteins Associated with Viral mRNA Synthesis

Since, as with many other large DNA viruses, iridovirus transcription and viral DNA replication are coupled, iPOND-MS analysis and affinity binding of proteins to FLAG-tagged vPOL-IIβ identified proteins linked to transcription such as vPOL-IIα, a Rho domain termination factor, a putative ATPase-dependent protease, an IBR domain-containing protein, a viral homolog of Rpb5 (ADRV 82L) along with three host-encoded RNA polymerase II subunits, Rpb3, Rpb 6, and Rpb11 (Ke et al. 2022b). Additionally, virus-encoded polymerase subunits (Rpb 7 and 10) have been identified among non-ranavirus iridovirids along with four transcription factors (Mirzakhanyan and Gershon 2017). Because the ranavirus transcriptase is composed of both virus- and host-encoded proteins, its structure contrasts markedly with those seen in other large DNA viruses, e.g., baculovirus, vaccinia virus, and African swine fever virus, where all subunits of the viral enzyme are encoded by the virus (Cackett et al. 2020; Guarino et al. 1998). This result (shown schematically in Fig. 4 and in Tables 2 and 3) indicates that the viral transcriptase is likely a chimera composed of multiple viral and host-encoded proteins and supports previous gene designations inferred by homology. If correct, the chimeric nature of the ranavirus RNA polymerase is unique among DNA viruses.

Although early work by Willis and co-workers indicated that FV3 DNA was not infectious by itself and required one or more heat-labile virion-associated transcriptional transactivators (VATTs) to initiate immediate-early viral transcription (Willis and Granoff 1985; Willis and Thompson 1986), the identity of the VATT was not known. In confirmation of earlier studies suggesting that the VATT protein was heat-labile and virion associated, Xia et al. demonstrated that extracts from cells infected with viable and UV-inactivated virus, but not heat-inactivated virus, were able to drive luciferase expression controlled by a promoter upstream from the immediate-early ICP46 gene (Xia et al. 2019). Furthermore, they showed by EMSA that four SGIV virion proteins bound the ICP46 promoter and subsequently identified them by mass spectroscopy as VP12, VP39, VP57, and VP72R. All but VP12 are iridovirus core proteins. VP39 is a putative serine/threonine protein kinase, VP72R the major capsid protein (MCP), and VP57 of unknown function. While their work tentatively identified putative VATTs, the inclusion of MCP may be adventitious due to its abundance. Moreover, attempts to activate transcription following expression of these gene products either singly or in combination were not successful suggesting that additional virion and/or host proteins are need for IE gene transcription.

In addition to the above experimental work and homology searches based on comparison of primary amino acid sequences, a "deep mining" approach based on identifying three-dimensional structural homology has recently been used to annotate ORFs from multiple NCLDVs (Mirzakhanyan and Gershon 2020). Deep mining is built on the premise that determining the function of a given protein by homology based on the primary amino acid sequence alone may not be possible but that elucidation of higher order structure may uncover structural similarities that point to possible function. Thus, a protein whose primary structure differed markedly from functionally similar proteins would be incorrectly annotated. Using deep mining along with more traditional approaches, additional ranavirus ORFs linked to viral DNA replication and transcription were identified (Mirzakhanyan and Gershon 2017, 2020).

4.3.3 Additional Viral Proteins Linked to Nucleic Acid Metabolism

Deep mining detected the presence in FV3 of a type IV restriction endonuclease (RE) that is thought to distinguish between methylated and unmethylated DNA and to cleave the latter (Mirzakhanyan and Gershon 2020). It is instructive to note that 10 newly predicted type IV RE identified by deep mining had previously been uncharacterized or mis-characterized. The conservation of secondary structure within the PD(D/E)xK catalytic domain of these REs facilitated identification by deep mining. Mirzakhanyan and Gershon suggest that FV3 ORF 94L is a candidate restriction modification enzyme and, if so, is postulated to degrade DNA lacking methylated cytosines. Since every cytosine within the sequence CpG is methylated by the FV3 cytosine DNA methyltransferase, viral DNA is protected from cleavage, whereas unmethylated host DNA is susceptible to cleavage. The ability of the viral RE to cleave unmethylated DNA may explain the aforementioned inability to rescue ts mutants using plasmids bearing FV3 restriction fragments as unmethylated plasmid DNA might be cleaved before rescue is achieved (Willis et al. 1979). Given the putative identification of FV3 94L as a Type IV RE, it would be instructive to determine how its KD or KO impacts viral replication. Previously FV3 ORF 94L had been identified as one of the 26 core genes but, unlike most of the other core genes, was not assigned a function. Furthermore, it is not clear if it functions as a RE as genomes within the subfamily *Betairidovirinae* are not methylated. ATV ORF 11R, a homolog of FV3 94L, was unable to be deleted from the virus by homologous recombination suggesting this viral gene is important for viral replication (Aron et al. 2016).

ADRV 12L, a putative Rad2 family member, was shown to be essential for virus replication (Ke and Zhang 2022). Rad2 family proteins are structure-specific nucleases and are required during DNA metabolism to repair DNA damage. Accordingly, Ke and Zhang (2022) postulated that ADRV 12L (and its FV3 homolog 95R) function in DNA recombination and repair. Consistent with this hypothesis, they showed that ADRV 12L co-localized with newly synthesized viral DNA and the single-stranded DNA binding protein 85L within viral assembly sites. Furthermore, they constructed a 12L KO mutant and showed that it resulted in a 100-fold reduction in viral yield and generated a small plaque phenotype. In addition, they found that the 12L KO displayed a markedly reduced ability to promote homologous recombination in vitro. Collectively, these results suggest that ADRV 12L plays an import role in viral replication by promoting homologous recombination and concatemer formation. Moreover, the presence of Rad2 and newly synthesized DNA within viral assembly sites supports the suggestion that concatemer formation takes place within assembly sites. Preliminary work indicates that additional ranavirus proteins are required for replication in vitro. Experiments using asMOs targeting FV3 ORFs 9L (DEAD/H helicase), 41R (vaccinia early transcription factor), 46R (neurofilament protein), 32R (neurofilament protein), and 38R (RR large subunit) resulted in viral yields reduced by more than 70% and support their roles as essential proteins (Cheng 2014; Whitley 2011).

In contrast, a viral homolog of dihydrofolate reductase (DHFR) was determined to be non-essential for viral replication (Martin et al. 2015). Mammalian DHFR plays a key role in nucleic acid synthesis, and the presence of a viral homolog (vDHFR) may be a way for virus-infected cells to maintain adequate pools of nucleic acid precursors (Trimble et al. 1988). While this may be the case in some host-virus systems, vDHFR is not found in all ranaviruses and even where it is present, as it is in European sheatfish virus, DHFR KO mutants replicated well in cell culture and generated the same level of mortality following zebrafish challenge as wild type virus (Martin et al. 2015). Two additional genes, thymidine kinase (TK) and RGV 67R (dUTPase), are also non-essential for replication in vitro despite the roles that both play in nucleic acid synthesis in mammalian cells (Huang et al. 2014a, 2016).

Aside from the roles that various viral proteins play in viral DNA replication and transcription, multiple host proteins likely also play critical roles. In one example, cellular MutS homolog 2 (MSH2), a mismatch repair protein, was shown to play a role in ADRV replication and transcription (Ke et al. 2022a). MSH2 expression was maintained in ADRV-infected cells and MSH2 protein was found to localize to viral factories and viral nascent DNA. Co-immunoprecipitation found that MSH2 interacted with ADRV 35L, 47L (vDPOL), and 98R. Moreover, KD of MSH2 expression via siRNA reduced late ADRV gene expression, and a MSH2 KO generated using CRISPR/Cas9 methodology reduced viral titers, genome replication, and late ranavirus transcription. Clearly MSH2 is only one of many host gene products that plays a critical role in ranavirus replication. The presence or absence of any one of these cellular products will likely determine the permissiveness of a given cell line to virus replication. In general, new ranavirus isolates are constantly being identified, and the accompanying functional genomics and phylogenomics will deepen our understanding of ranavirus replication and transcription (Zhang et al. 2022a, b; Yu et al. 2023).

4.4   Virion Assembly

The formation of ranavirus virions, like that of poxviruses, African swine fever virus, and other NCLDVs, takes place within morphologically distinct cytoplasmic regions designated viral assembly sites (AS) or viral factories (VF) (Murti et al. 1988; Netherton et al. 2007). In the electron microscope, ranavirus AS appear as large, electron lucent regions, devoid of cellular organelles and containing virions in various stages of assembly (Netherton et al. 2007). Furthermore, IFA using antibodies targeted to a variety of viral gene products and stains such as DAPI, Hoechst, and Feulgen that detect DNA show AS to be regions of viral protein and DNA accumulation (Chinchar et al. 1984). As with ASFV, ranavirus AS resemble aggresomes and are typically surrounded by mitochondria and intermediate filaments consisting of vimentin (Chen et al. 1986; Netherton et al. 2007). It is unclear if mitochondria migrate to AS and provide energy for virion assembly or if they and intermediate filaments are simply pushed aside by the expanding AS and cluster at the periphery. Furthermore, the vimentin cage that surrounds AS may prevent host components from entering the AS and interfering virion morphogenesis.

In contrast to other large icosahedral DNA viruses such as African swine fever, phycodnavirus, and mimivirus (Milrot et al. 2016; Mutsafi et al. 2014; Suarez et al. 2015), little is known about the stages of ranavirus virion morphogenesis. By analogy to ASFV, it is thought that binding of the MCP (FV3 90R) to one face of

ER-derived membrane fragments and binding of a myristylated membrane protein (FV3 53R) on the opposite face of the same membrane initiates formation of the virion shell (Salas and Andres 2013). Similar to the myristylated ASFV protein p220, FV3 53R is thought to form an inner capsid layer with the internal lipid membrane sandwiched between it and the MCP in the outer shell (Andres et al. 1997, 2002). KD of FV3 53R led to a marked decrease in viral titer and is supported by the observation that a conditionally lethal mutant of the corresponding RGV homolog incubated under non-permissive conditions also resulted in a marked reduction in virion formation as monitored by TEM (He et al. 2013; Whitley et al. 2010). Likewise, an approximately 90% reduction in MCP expression following asMO-targeting of its cognate mRNA blocked virion formation and resulted in the formation of linear structures termed atypical elements. The generation of atypical elements may reflect the deposition of one or more viral structural proteins on ER membrane fragments that, in the absence of the MCP, fail to form icosahedral viral particles (Sample et al. 2007). In addition to the MCP, three minor capsid proteins (mCP), zip, finger, and anchor, were visualized within virions by TEM (Pintilie et al. 2019; Yan et al. 2009). Recently, based on the information from cryo-EM mapping and proteomic analysis of purified SGIV virions, four additional mCPs were identified and linked to specific SGIV ORFs (Zhao et al. 2023) (Table 2). Among the identifications made, VP14 was shown to be the penton protein found at the apex of the 12 vertices, VP38 the zip protein that stabilizes the capsid shell, VP88 the anchor protein whose myristylated moiety binds ER membrane fragments and contributes to the formation of the internal virion membrane, and VP22 a protein that may be involved in genome condensation and formation of the inner shell. Furthermore, the authors speculated that mCPs may associate with the MCP and form precursors that contribute to virion assembly. BLASTP analysis identified homologs of 6 of the 7 SGIV mCPs among FV3, RGV, ADRV, and ATV, indicating that these proteins are important for virion formation (Table 2).

In addition, Wang et al. identified a putative scaffold protein, SGIV 75L, similar to the herpesvirus protein VP22 that is located between the viral capsid shell and the DNA core. Furthermore, they showed that KD of 75L expression by asMOs blocked encapsidation, reduced thickness of the core shell, and resulted in a reduction in viral titer (Wang et al. 2015a). Additional viral envelope proteins (RGV 2L, 22R, and 50L) were shown to co-localize with AS (He et al. 2014; Lei et al. 2012a; Zhao et al. 2008). Whether these proteins play catalytic or structural roles in virion assembly or whether they are involved in viral entry by binding to host cell receptors remains to be elucidated. MS analysis of purified SGIV virions detected 73–89 virion-associated proteins (Yao et al. 2019; Zhao et al. 2023). Many of these, e.g., the major capsid protein and a protein with conserved domains similar to the poxvirus entry fusion complex protein, are likely structural components of the virion. However, others, e.g., 3-beta hydroxysteroid dehydrogenase, dUTPase, ribonucleotide reductase, and DNA polymerase, might represent virion-associated catalytic proteins or simply proteins adventitiously associated with virions. Lastly, although Gooha and Murti hypothesized that encapsulation of the ranavirus genome generates circularly permuted and terminally redundant genomes by a process of headful packaging (Goorha and Murti 1982), the precise mechanism of DNA encapsidation is not understood even among better studied NCLDVs (Mutsafi et al. 2014; Suarez et al. 2015). Moreover, it is not known if there is a specific viral portal involved in ranavirus genome entry (or egress) or the identities of virion proteins involved in genome packaging.

4.5 Viral Immune Evasion Proteins

Aside from the formation of infectious virions, successful viral replication also requires escape from innate and acquired immunity at both the cellular and organismal level. Thus, to achieve full replicative potential, ranaviruses encode multiple proteins that allow them to circumvent host immunity. For example, SGIV counteracts the poly [I:C]-induced expression of multiple interferon signaling and inflammatory genes such as STING, MAVS, NFKB1, etc. (Wang et al. 2023). Here we highlight several key ranavirus immune evasion proteins including vIF-2α, semaphorin, vTNFR, βHSD, vCARD, Bcl-2-family member proteins, viral proteins that degrade STING-TBK1, RNase III, virus-encoded microRNAs (v-miR) and viral mimics of IFN regulatory factors (IRFs), IFN receptors (IFNR), IL-8, and IL-10R. A listing of putative ranavirus immune evasion proteins can be found in Table 2 and evidence supporting specific functions can be found in Table 5.

vIF-2α vIF-2α is a homolog of eukaryotic translational initiation factor 2α and the poxvirus protein K3L (Beattie et al. 1991). In vaccinia virus-infected cells, K3L, acting as a pseudosubstrate, is postulated to bind dsRNA-activated protein kinase R (PKR), prevent the phosphorylation and subsequent inactivation of eIF-2α, and maintain viral protein synthesis in infected cells (Beattie et al. 1991; Gal-Ben-Ari et al. 2019). In addition, to blocking phosphorylation of eIF-2α, vIF-2α may inhibit apoptosis and the synthesis of inflammatory cytokines (Gal-Ben-Ari et al. 2019). Surprisingly, although vIF-2α is present in most ranaviruses, it is N-terminally truncated in FV3 and STIV suggesting that alternative mechanisms are present to maintain viral translation (Huang et al. 2009; Tan et al. 2004). Using a yeast system, Rothenburg et al. showed that ectopic expression of the vIF-2α homolog of Rana catesbeiana virus-Z (RCV-Z, a *Frog virus 3* species) blocked the growth inhibitory effects of both human and zebrafish PKR and prevented phosphorylation and inactivation of eIF-2α (Rothenburg et al. 2011). Interestingly, they found that the C-terminal domain of RCV-Z vIF-2α was required to prevent eIF-2α phosphorylation. Furthermore, Jancovich and Jacobs showed that vIF-2α from ATV blocked phosphorylation of eIF-2α and degraded fish PKZ, an interferon-inducible enzyme closely related to mammalian PKR (Jancovich and Jacobs 2011). In addition, degradation of PKR was observed after expression of the ATV vIF-2α homolog within a vaccinia virus E3L KO mutant (Huynh et al. 2017). Therefore, expression of ATV vIF-2α leads to the degradation of both piscine (PKZ) and mammalian (PKR) eIF-2α kinases ensuring that protein synthesis continues in virus infected cells. As noted above, the FV3 vIF-2α homolog is truncated at its N-terminus and lacks the VIRVDxxKGYIDL sequence, found in K3L, cellular vIF-2α, and other ranavirus vIF-2α homologs, which is thought to bind PKR and prevent the phosphorylation of eIF-2α (Beattie et al. 1991; Kawagishi-Kobayashi et al. 1997). In cells infected with WT FV3, phosphorylation of eIF-2α was observed yet virus-specific protein synthesis continued suggesting that a full-length vIF-2α protein is not required to maintain viral translation (Chinchar and Dholakia 1989). Moreover, despite the absence of its N-terminus which contains this critical binding motif, an FV3 vIF-2α KO mutant is attenuated for growth in vivo resulting in reduced mortality, suggesting that the C-terminal portion of the FV3 eIF-2α homolog contains a function required for full virulence (Chen et al. 2011). Mutational analysis of viral vIF-2α in a ranavirus system will be required to shed light into the functional domains required to inhibit eIF-2α phosphorylation.

Semaphorins Semaphorins comprise a large protein family involved in axonal steering, organogenesis, neoplastic transformation, and immunity. SGIV ORF 155R encodes a 575 amino acid protein containing a ~370 aa N-terminal Sema domain (vSema). vSema, when ectopically expressed, enhanced viral replication in vitro and altered cytoskeletal structure by disrupting microfilament organization. In addition, vSema attenuated cellular immunity by decreasing expression of IL-8, IL-15, TNF-α, and MITA, a mediator of IRF3 activation (Yan et al. 2014).

TNF Receptor Homologs Viral homologs of the tumor necrosis factor receptor (vTNFR) have been identified in poxviruses, herpesviruses, and iridoviruses and are thought to play a role in evading host immune and inflammatory responses (Lu et al. 2021). Three TNFR homologs have been identified in SGIV corresponding to ORFs VP50, VP51, and VP96 (Huang et al. 2013; Yu et al. 2016). VP96 and VP51 were characterized and shown to differ in the number of cysteine-rich domains they possess, their time of expression, and their subcellular location suggesting they might play different roles at various stages of virus infection (Yu et al. 2016). Despite these differences, expression of both VP96 and VP51 inhibited apoptosis and increased viral replication. Infection with a KO mutant (SGIVΔVP51) reduced replication in vitro and decreased cumulative mortalities in fish challenged with SGIVΔVP51 compared to WT-SGIV (Yu et al. 2017). Furthermore, VP51 deletion led to increased apoptosis and reduced the expression of pro-inflammatory cytokines in vitro.

β-Hydroxysteroid Dehydrogenase The role of ranaviral homologs of β-hydroxysteroid dehydrogenase (βHSD) remains to be fully elucidated. Although poxvirus βHSD homologs have been postulated to play a role in suppressing host immunity (Reading et al. 2003), their function among ranaviruses is still unclear. In RGV, a virus closely related to FV3, a 355 amino acid βHSD homolog was identified as an immediate-early protein and shown to colocalize with mitochondria. When over-expressed, vβHSD was found to suppress RGV-induced cytopathic effect (Sun et al. 2006). In another viral system, a FV3 KO mutant (FV3Δ52L) targeting vβHSD replicated as well as wtFV3 in non-amphibian fathead minnow (FHM) and baby hamster kidney (BHK) cells but displayed markedly lower levels of replication in amphibian A6 cells (De Jesus Andino et al. 2015). Consistent with its reduced replication in amphibian A6 cells, FV3Δ52L displayed markedly lower levels of mortality and viral replication following infection of *Xenopus laevis* tadpoles compared to WT virus. Lastly, the level of apoptosis seen following infection with FV3Δ52R, although higher than in uninfected cells, was the same as that seen following WT infection (De Jesus Andino et al. 2015) suggesting that vβHSD does not block apoptosis. Interestingly, genomic sequencing of ATV strains identified a mutation in the 5′ end of the vβHSD ORF that truncates the protein (Epstein and Storfer 2015) suggesting that the vβHSD may not be required for ATV infections.

CARD-Containing Viral Proteins Proteins containing a Caspase Activation and Recruitment Domain (CARD) belong to the Death Domain Superfamily (DDS) and play critical roles in apoptosis, inflammation, necrosis, and immune signaling (Kim et al. 2019). Assembly of caspase-activated complexes such as the inflammasome takes place via DDS interactions and formation of these complexes can be inhibited by proteins containing only the CARD domain, termed "CARD-only" proteins. Several ranaviruses encode proteins that possess the CARD motif (termed vCARD) and are postulated to inhibit host immunity and the induction of apoptosis (He et al. 2002; Jancovich et al. 2010; Lei et al. 2012b; Tan et al. 2004). Examination of a FV3 vCARD KO mutant (FV3Δ63L) showed, as was the case with the aforementioned βHSD KO (FV3Δ52L), that replication in non-amphibian cell cultures was not impaired, whereas replication was reduced in amphibian A6 cells. Furthermore, following infection of *Xenopus laevis* tadpoles, markedly lower levels of virus replication and mortality were observed (De Jesus Andino et al. 2015). Infection with FV3Δ63L resulted in increased apoptosis compared to WT FV3, suggesting that one of the roles of vCARD is to impair the induction of apoptosis and thus enhance virus replication. Consistent with the above results, a vCARD protein from grouper iridovirus blocked apoptosis by inhibiting caspases 8 and 9 (Chen et al. 2015).

Bcl-2 Family Proteins Another approach by which viruses prevent premature apoptosis involves the possession of functional and structural homologs of Bcl-2 family members (Banjara et al. 2018). Viruses harboring Bcl-2 family homologs include members of the *Adenoviridae, Herpesviridae, Asfarviridae,* and *Poxviridae* families (Banjara et al. 2018). Interestingly, whereas Bcl-2 homologs were detected among adenoviruses, herpesviruses, and African swine fever by their primary amino acid sequence, Bcl-2 homologs were only detected among poxviruses after their secondary structure was determined. Recently, Bcl-2 family members were also identified in GIV and SGIV (Banjara et al. 2018; Yao et al. 2019), as well as in FV3 (ORF 97R), tiger frog virus (TFV ORF 104R), and other closely related ranavirus species (He et al. 2019). GIV ORF 66R was shown previously to localize to the outer mitochondrial membrane and inhibit Bcl-2 mediated apoptosis in UV-irradiated cells (Lin et al. 2008). Subsequently, determination of the crystal structure of GIV 66R found that it is a dimeric Bcl-2-like protein that suppressed apoptosis by sequestering Bim, a potent trigger of apoptosis (Banjara et al. 2018). Interestingly, BLASTP analysis failed, with the exception of SGIV, to find GIV 66R homologs among other ranaviruses but found similarity to Bcl-2-like proteins in a variety of fish species. Despite the apparent absence of Bcl-2 homologs among other irido-virids when GIV 66R was used as the query, homology to another member of the Bcl-2 family, myeloid cell leukemia protein 1 (Mcl-1), was detected by BLASTP analysis in TFV, FV3, and in a number of other ranaviruses (Tan et al. 2004). One such Mcl-1 protein, TFV ORF 104R, was shown to be a viral structural protein that co-localized with mitochondria. When overexpressed, ORF 104R suppressed cyto-chrome C release and the activities of caspases-3 and -9 (He et al. 2019). Moreover, TFV ORF 104R interacted with cellular VDAC2, one of a family of proteins involved in the release of apoptotic proteins from the outer mitochondrial mem-brane, and through this interaction was postulated to regulate apoptosis.

In addition to the effect of Bcl-2-like proteins on apoptosis, Bcl-2 like proteins may also play critical roles in the regulation of autophagy, cellular senescence, inflammation, and other events that control cell survival and death (Chong et al. 2020). Autophagy plays an antiviral role during GIV and LMBV infections (Qi et al. 2016). Accordingly, antiviral autophagic effects were triggered by GIV and LMBV. In view of that, three viral proteins (VP48, VP122, and VP132) from the closely related SGIV were found to bind autophagy related gene 5 (Atg5) and inhibit autophagy in GS (grouper spleen) cells by affecting LC3 conversion (Li et al. 2020). Whether the three aforementioned proteins are also members of the Bcl-2 family or block the onset of autophagy in a different manner is not known.

Degradation of STING and TBK1 Viruses, including ranaviruses, encode a myr-iad of proteins that inhibit interferon production (Garcia-Sastre 2017; Johnston and McFadden 2003, 2004; Rojas et al. 2021; Seet et al. 2003). SGIV, the most diverse member of the genus *Ranavirus*, encodes VP131 ("stingless") a protein with an Ig-like domain that degrades STING and TBK1 and inhibits interferon synthesis (Zhang et al. 2022a, b). The cGAS-STING (cytosolic **c**yclic **GMP-AMP S**ynthase - **ST**imulator of **IN**terferon **Ge**nes) pathway is initiated by the binding of cGAS to dsDNA. Subsequent steps involve the recruitment and autophosphorylation of TBK1, phosphorylation of IRF3, the induction of IFN, and the synthesis of an array of IFN stimulated genes, pro-apoptotic genes, and cytokines (Decout et al. 2021). Zhang et al. showed that overexpression of GFP-VP131 enhanced viral replication, whereas its knockdown reduced viral replication in vitro (Zhang et al. 2022a, b). They subsequently demonstrated that overexpression of GFP-VP131 inhibited IFN promoter activity and showed that VP131 interacted with and degraded STING and TBK1. Their work demonstrated that evasion of a host antiviral response could be achieved by targeting two key components (STING and TBK1) of the signaling IFN pathway. Interestingly BLASTP and PSI-BLAST failed to detect VP131 homologs among any other virus, including GIV (VG Chinchar, unpublished). Recently a sec-ond SGIV protein (VP122) was found to target STING by blocking STING-STING interaction. Furthermore, in contrast to VP131, VP122 homologs were found among several ranaviruses (Xu et al. 2023).

Virus-Encoded MicroRNAs (v-miR) v-miRs are small non-coding RNAs that are found within multiple virus families, particularly DNA viruses (Mishra et al. 2019). V-miRs act to silence host immune-related genes and to post-transcriptionally regulate gene expression by targeting 3′-untranslated regions (UTR) within mRNA. Given their small size and ability to target multiple genes, v-miRs could be an efficient way for viruses with limited genomic capacity to modulate host and viral gene expression. There is growing evidence that ranaviruses also encode v-miRs that regulate host and viral gene expression and play a role in the evasion of host antiviral immunity. For example, 11 of the 16 novel SGIV-encoded v-miRs identified by Illumina/Solexa deep-sequencing were present and functional in SGIV-infected grouper cells when examined by stem-loop quantitative RT-PCR and luciferase reporter assays (Yan et al. 2011). One v-miR, SGIV miR-homoHSV, attenuated SGIV-induced apoptosis thereby enhancing virus replication (Guo et al. 2013). In tiger frog virus, Yuan et al. combined computational prediction and deep sequencing to detect 24 novel TFV-encoded v-miRs and to verify the expression of the vast majority of them in TFV-infected ZF4 and HepG2 cells by qRT-PCR (Yuan et al. 2016a). Among this group, TFV-miR-11, -13, and -14 were shown to be essential for TFV replication (Yuan et al. 2016a). In a recent study, Tian and co-workers conducted whole transcriptomic analysis (RNA-Seq) of both viral and cellular transcripts in FV3-infected X. *laevis* (Tian et al. 2021b). FV3 expression was monitored in samples from infected intestine, liver, spleen, lung, and kidney tissue and found to arise from both coding sequences and non-coding intergenic regions. They also identified FV3 ORFs containing domains that mimicked those found in immune-related proteins which could potentially interfere with host IFN signaling (see below). Furthermore, they detected an array of putative v-miRs residing in five intergenic regions located upstream of several highly transcribed genes that encoded components of the *Xenopus* IFN signaling pathway. Sites targeted by these v-miRs were detected within the 3′-UTR of *Xenopus* IFN receptor subunits (particularly ifnar2.S and il10rb.S) and interferon regulatory factors (particularly irf1, irf5, irf2.L, irf7.L, and irf10.L). The potential effect of these v-miR was demonstrated by the differential expression of IFN-related target genes in infected tissue samples and validated using siRNAs containing sequences identical to several of the key v-miRs identified in this study (Tian et al. 2021b). Another study discovered 43 v-miR by profiling small RNA sequencing libraries from FV3-infected Xela DS2 cells, a *Xenopus laevis* dorsal skin epithelial-like cell line, at 24- and 72-hours post-infection (hpi). Differential comparison demonstrated that 15 of the 43 v-miRs were upregulated at 24 hpi, while 18 were upregulated at 72 hpi. FV3 v-miRs targeted host genes involved in key antiviral signaling pathways in addition to broadly impacting other host cell functions (Todd and Katzenback 2021). Collectively, these studies identified multiple ranavirus-encoded v-miRs and provided a comprehensive genome-wide analysis of non-coding regulatory mechanisms that control both viral and host gene expression.

Viral Mimics of Host IFN Regulatory Factors (IRFs), IFN Receptors (IFNRs), and IL-10 Receptors (IL-10R) In the course of transcriptomic analyses designed to validate the expression and determine the temporal class of FV3 transcripts following infection of adult *X. laevis*, Tian and co-workers identified multiple FV3 ORFs that contained conserved domains found in host IRFs, IFNR, and IL-10R (Tian et al. 2021a). The identified ORFs and their cellular homologs include FV3 13L (vIRF3), 19R (vIRF3), 27R (vIRF8), 42L (vIRF6), 41R (vIRF4A), 66L (vIFNLR1), 59L (vIL-10R), and 82R (vIRF8). The authors suggest that ranaviruses such as FV3 have acquired these mimics to interfere with host IFN signaling and that their presence explains the differential sensitivity of adult and larval frogs to FV3 challenge (Gantress et al. 2003; Tweedell and Granoff 1968). Following asMO-mediated knockdown of the expression of FV3 ORF 82, a putative homolog of IRF8, which encodes an abundant 18 kDa protein, no difference was noted in replication in cultured fathead minnow cells (Sample et al. 2007). Furthermore, infection of cultured cells with a KO mutant (FV3Δ18K) also had little to no effect on replication in vitro but resulted in markedly less mortality when *X. laevis* tadpoles were challenged indicating that, although the 18K protein was not essential for replication in vitro, it may play a role in subverting host immunity in an infected amphibian (Chen et al. 2011). Finally, homology between FV3 18K and IRF8 may not explain all of the former's role in viral replication since other work suggests that 18K also enhances cell growth and virus replication (see below).

LITAF (LPS-Induced TNFα Factor) Host cell LITAF (cLITAF) is a transcription factor that activates the expression of TNFα and regulates apoptosis and inflammation (Chen et al. 2021). In addition, cLITAF plays a role in the ubiquitin-mediated lysosomal degradation pathway (Chen et al. 2016). Several ranaviruses encode LITAF homologs (e.g., SGIV ORF 136, FV3 ORF 75L, and TFV ORF 80L). Interestingly, vLITAFs are about half the size of their cellular counterparts and lack the N-terminus which contains two PPXY motifs. TFV ORF 80L encodes an 84 amino acid peptide that is 96% identical to that of FV3 ORF 75L and 60% identical to SGIV ORF 136 (Chen et al. 2016). Data concerning the function of vLITAF is, however, conflicting as overexpression of SGIV ORF 136 leads to increased apoptosis (Huang et al. 2008), which should decrease viral replication, whereas siRNA-mediated KD of TFV ORF 80L results in a reduction in viral replication (Chen et al. 2016). FV3 ORF 75L has been shown to co-localize with cLITAF within late endosomes and lysosomes (Eaton et al. 2013). If cLITAF plays a role in intracellular transport, then vLITAF/cLITAF interaction might block cLITAF's function and impair endosomal sorting and trafficking (Chen et al. 2016). If, on the other hand, the primary role of cellular LITAF is to facilitate TNFα expression, then cLITAF/vLITAF interaction might sequester the cellular component and inhibit host cell immunity.

RNase III RNase III is one of the 26 iridovirus core genes (Eaton et al. 2007) and, as shown using the rock bream iridovirus (genus *Megalocytivirus*) homolog, has the ability to degrade dsRNA in a Mg+2 dependent manner (Zenke and Kim 2008). Moreover, RNase III from sweet potato chlorotic stunt crinivirus (SPCSV) and Heliothis virescens ascovirus (HvAV), the latter of which is in the same taxonomic order (*Pimascovirales*) as the iridovirids, suppress RNA interference (RNAi) pathways (Weinheimer et al. 2015). Likewise, ectopic expression of pike-perch iridovirus (genus *Ranavirus*) RNase III in plants and nematodes led to the cleavage of small dsRNA interfering molecules (siRNA) and suppressed RNAi (Weinheimer et al. 2015). To further explore the role of ranavirus RNAse III, Allen et al. treated cells ectopically expressing ATV RNase III (ATV ORF 25R) with the IFN inducer poly I:C and monitored interferon synthesis and the phosphorylation states of eIF-2α and PKR (Allen et al. 2017). They found that IFN synthesis and phosphorylation of both eIF-2α and PKR were reduced suggesting that RNase III degraded poly I:C and blocked the induction of an IFN response. Next, using a RNase III KO mutant (ATVΔ25R) to challenge tiger salamanders, they observed markedly less mortality in salamanders challenged with the KO mutant than with WT ATV (Allen et al. 2017). In contrast to these findings, IIV6 (genus *Iridovirus*) encodes two proteins containing motifs that might play a role in blocking RNAi. IIV6 ORF 142R encodes a putative RNase III, whereas ORF 340R possesses a dsRNA binding domain (dsRBD) (Bronkhorst et al. 2014). Bronkhorst and co-workers showed that ORF 142R did not suppress RNAi and suggested that it may be involved in processing viral or cellular RNAs, whereas ORF 340R bound long dsRNA and prevented Dicer-2 from processing these molecules into siRNAs. Furthermore, 340R bound siRNA and blocked its assembly into the RNA-induced signaling complex (RISC). In a subsequent study, the authors generated a KO mutant (IIV6Δ340R) and confirmed that the product of ORF 340R did not affect siRNA generation but blocked siRNA from associating with RISC (Bronkhorst et al. 2019). Based on these findings, iridovirid RNase III proteins may have multiple functions that support viral replication and enhance immune evasion and pathogenicity in vivo.

4.6 Viral Gene Products That Enhance Viral Replication: Viral Efficiency Proteins

There are several ranavirus proteins that are thought to enhance viral replication by increasing nucleotide pool levels or enhancing the cellular environment for replication. Among these are ORFs encoding homologs of the two subunits of ribonucleotide reductase (RRα and RRβ), thymidylate synthase (TS), thymidine kinase (TK), deoxyuridine triphosphatase (dUTPase), dihydrofolate reductase (DHFR), viral insulin-like proteins (VILP), ERV, and others. RRα, RRβ, TS, TK, DUTPase, and DHFR play important roles in maintaining dNTP pools by encoding viral enzymes used in the salvage or de novo pathways of dNTP synthesis (Stillman 2013). Moreover, aside from its role in dTTP synthesis, dUTPase also blocks incorporation of dUTP into viral DNA (Vertessy and Toth 2009). Interestingly these putative viral genes are not found in all ranaviruses. For example, DHFR is found in EHNV, but not in FV3 and other amphibian-like ranaviruses (Jancovich et al. 2010; Tan et al. 2004). Furthermore, vDHFR's marked sequence similarity to piscine DHFR suggests it may have been acquired by horizonal gene transfer (VG Chinchar, unpublished; Jancovich et al. 2010).

ERV1   Among the 26 iridovirid core genes is ERV1 (Essential for Respiration and Viability), a homolog of a multifunctional cellular protein involved in DNA replication, protein synthesis, and protein folding (Holmgren 1985). ERV1 homologs are found among other NCLDVs including poxviruses, chlorella virus, and African swine fever virus where they are thought to play a role in virion assembly by maintaining the cytosol in a reduced state (Cobbold et al. 2007). ERV1 homologs, among them RGV ORF 88R, share a highly conserved Cys-X-X-Cys motif, a motif associated with redox activity and found within thioredoxin and glutaredoxin. RGV ORF 88R is a late viral protein that although present in both the nucleus and cytoplasm was not seen in either the nucleolus or the viral assembly site. RNAi-mediated KD of ORF 88R failed to reduce virus replication in vitro, causing Ke et al. to speculate that residual levels of ORF 88R or existing cellular redox proteins might be sufficient to maintain virus replication (Ke et al. 2009). However, consistent with an essential role for ERV1, KO of the ASFV ERV1 homolog markedly reduced viral replication in vitro and virulence in vivo (Lewis et al. 2000). Interestingly, ASFV virions generated in cells infected with the ASFVΔERV1 KO mutant showed an acentric nucleoid structure, suggesting that alterations in the redox state within the cytoplasm adversely impacted virion assembly (Lewis et al. 2000). Interestingly, an FV3 ts mutant (ts12814) that failed to generate infectious virus at the non-permissive temperature also displayed acentric nucleoids, suggesting its defect may lie within the FV3 ERV1 gene (Sample 2010).

ICP 18K Homologs   Ranaviruses, but not other genera within the family *Iridoviridae*, encode an 18 kDa immediate early protein that is synthesized in large amounts during infection. Amino acid identity was high (>95%) among amphibian ranavirus 18K homologs, but markedly lower (<50%) when compared to SGIV. KD and KO of FV3 18K (ORF 82R) did not negatively impact replication in FHM, *Xenopus* A6, and BHK cells but markedly reduced viral replication and mortality in vivo (Chen et al. 2011; De Jesus Andino et al. 2015; Sample et al. 2007). The SGIV 18K homolog (ORF 86R) displayed a punctate cytoplasmic distribution but was not detected within viral assembly sites (Xia et al. 2009). Overexpression of SGIV ORF 86R promoted the growth of grouper embryonic cells and contributed to enhanced SGIV replication (Xia et al. 2009). asMO-mediated KD of SGIV ORF

86R did not impact virus replication but changed the expression levels of some host proteins (Wang et al. 2015b). Using a BioID approach, which detects proteins that interact with SGIV 18K fused to biotin ligase, Wu et al. identified 112 interacting cellular proteins (Wu et al. 2021). Among this number, 13 cell cycle regulatory proteins including CDK1 were present along with 11 cell adhesion-related proteins. Collectively, the data suggest that 18K modulates a number of cellular processes including those affecting the cell cycle and cell adhesion.

ICP-46 Homologs In contrast to 18K, which is found only among members of the genus *Ranavirus*, ICP-46 is present among all iridovirids as well as among ascoviruses and Marseilles virus. The SGIV homolog of ICP-46 (ORF 162L) is an immediate early protein that is found within the cytoplasm. When ectopically expressed, SGIV ICP-46, like 18K, drives cellular proliferation and enhances viral yields (Xia et al. 2010). However, elucidation of the exact mechanism by which ICP-46 promotes cell growth and virus replication remains to be determined.

Viral Insulin-Like Peptides (VILPs) VILPs have been detected among isolates of both LCDV (genus *Lymphocystivirus*) and SGIV (Altindis et al. 2018). VILPs are 30–56% identical to human insulin A- and B-chains and all 6 cysteine residues involved in the formation of insulin tertiary structure are conserved. The SGIV VILP homolog (ORF 62R) is an early gene product that is present mainly within the cytoplasm (Yan et al. 2013). Overexpression of SGIV VILP enhanced grouper cell growth by promoting the G1/S phase transition and infection of cells overexpressing VILP resulted in 10-fold higher levels of virus replication (Yan et al. 2013). Chemically synthesized VILPs bind human and murine insulin growth factor-1 (IGF-1)/insulin receptors, stimulate receptor autophosphorylation and downstream signaling, and identify an antagonist of potential medical importance (Zhang et al. 2021). Interestingly, analyses of human fecal viromes identified DNA sequences belonging to LCDV-1 and SGIV suggesting that humans may carry these viruses and that VILPs produced by them could be involved in triggering, or protecting from, disease (Altindis et al. 2018). IGF-like proteins were also found in Siniperca chuatsi ranavirus and Micropterus salmoides ranavirus (Yu et al. 2023).

Viral SAP Domain-Containing Proteins Virus-encoded proteins containing SAP domains have been identified among all ranaviruses. Lei et al. partially characterized RGV ORF 50L encoding a 499 amino acid SAP domain protein that appeared to undergo post-translational modification (Lei et al. 2012a). RGV 50L is an NLS-containing, immediate-early protein which immunofluorescence assays showed was present in the cytoplasm, nucleus, and assembly site. In cells ectopically expressing 50L and subsequently infected with RGV, levels of RGV 53R mRNA were elevated suggesting that 50L upregulated viral transcription. Based on these studies, RGV 50L is considered a viral structural protein that may play roles in virion assembly and transcription.

Viral AAA-ATPase Domain Containing Proteins ADRV 96L, a virus-encoded AAA-ATPase domain containing protein, has been identified as one of the irido-virus core gene products. ADRV 96L displays homology with the poxvirus A32 protein, which has been reported to be involved in DNA packaging. Purified recombinant ADRV 96L displayed ATPase activity and overexpression of ADRV 96L in Chinese giant salamander thymus cells (GSTC) promoted cell growth (Zhang and Zhang 2018). However, generating a 96L-KO ADRV was not possible, indicating that 96L may be essential for virus replication.

5 Concluding Thoughts

Although the general outline of ranavirus replicative events has remained essentially the same since the first edition of this monograph in 2015, knowledge of the roles of specific viral catalytic and structural proteins, the host anti-viral immune response, and virus-encoded anti-immune countermeasures have grown markedly. Furthermore, aside from earlier studies that focused primarily on FV3, recent studies utilized additional ranavirus strains and species as well as representatives from other genera. These studies confirmed the pathways utilized by ranaviruses to generate infectious virus particles and facilitated identification of viral protein function. Given the growing commercial importance of various fish species (e.g., grouper, red seabream, mandarin fish, etc.), many of which are targeted by ranaviruses and other iridovirids, and the ecological importance of multiple amphibian, piscine, and reptilian species, understanding how ranavirus proteins control virus replication and modulate the host immune response will be critical in protecting vulnerable species. Clearly, additional studies are required as investigators work to precisely elucidate viral gene function and incorporate that information into the narrative of virus replication and virion formation. The putative identification of viral gene products involved in DNA synthesis and metabolism, RNA transcription, virion assembly, and immune evasion will suggest critical targets for further study. Possible targets include the following: DNA synthesis and metabolism, RAD2, PCNA, DMTase, and the viral restriction and modification endonuclease; RNA synthesis, virus- and host-encoded Rpb subunits along with candidate VATT polypeptides; virion assembly, the seven currently identified mCPs; viral immune evasion, vIF-2α, Bcl-2, vSema, and viral mimics of key immune proteins; and viral efficiency proteins such as ICP46, ERV, and 18K. To cite but one example, KD or KO of viral mCPs will determine if their presence is needed for the assembly of infectious virions and whether their absence results in the accumulation of virion intermediates or virions bearing observable defects. A systematic use of the KD and KO strategies, as well as the other approaches described above, will allow elucidation of ranavirus gene function and provide a more complete understanding of critical events in viral replication and pathogenesis.

Amphibians and fish are especially vulnerable to environmental and anthropogenic stressors, e.g., changes in land use, intensive aquacultural practices, agricultural chemicals, etc., which increase their susceptibility to ranavirus-induced disease (Brunner et al. 2015). Although studies of a "frog virus" may be viewed by some of little importance when compared to diseases triggered by known human pathogens such as HIV-1, SARS-CoV-2, pandemic influenza A viruses, and others, ranavirus studies have benefits beyond the commercial (i.e., aquacultural and maricultural industries) and ecological. "Unusual organisms" such as ranaviruses deserve to be studied not simply because they are unusual, but because they may better illuminate aspects of replication that are common to many other organisms or may uncover pathways absent or hidden among better studied viruses. Furthermore, by providing a window into the amphibian (and by extension lower vertebrate) immune system, ranavirus studies will expand our understanding of the anti-viral immune response of lower vertebrates and provide insight into the origins of the vertebrate immune system. The extensive studies currently performed in Asia in megalocytivirus- and ranavirus-infected fish mirror those conducted in the FV3-*Xenopus laevis* system and will yield critical information about piscine anti-viral immunity. Collectively, this information will be critical in developing therapies to confront ongoing diseases and may speed the development of vaccines to prevent further outbreaks among commercially and ecologically important fish and amphibian species.

Acknowledgements We would like to thank Yongming Sang and Leon Grayfer, respectively, for helpful suggestions regarding viral miRNAs and the ranavirus anti-viral immune response along with Richard Condit and Craig Brunetti for thoughtful critiques of the manuscript. We thank the following organizations for providing funds to support Open Access publishing of the second edition: University of Tennessee's (UT) Open Publishing Support Fund, UT School of Natural Resources, UT Center for Wildlife Health, Washington State University Libraries, University of Mississippi Medical Center, Gordon State College, Association of Reptile and Amphibian Veterinarians, and the Global Ranavirus Consortium.

References

Allen AG, Morgans S, Smith E, Aron MM, Jancovich JK (2017) The Ambystoma tigrinum virus (ATV) RNase III gene can modulate host PKR activation and interferon production. Virology 511:300–308

Altindis E, Cai W, Sakaguchi M, Zhang F, GuoXiao W, Liu F, De Meyts P, Gelfanov V, Pan H, DiMarchi R, Kahn CR (2018) Viral insulin-like peptides activate human insulin and IGF-1 receptor signaling: a paradigm shift for host-microbe interactions. Proc Natl Acad Sci USA 115:2461–2466

Andres G, Simon-Mateo C, Vinuela E (1997) Assembly of African swine fever virus: role of polyprotein pp220. J Virol 71:2331–2341

Andres G, Alejo A, Salas J, Salas ML (2002) African swine fever virus polyproteins pp220 and pp62 assemble into the core shell. J Virol 76:12473–12482

Arbel M, Choudhary K, Tfilin O, Kupiec M (2021) PCNA loaders and unloaders-one ring that rules them all. Genes (Basel) 12:1812

Aron MM, Allen AG, Kromer M, Galvez H, Vigil B, Jancovich JK (2016) Identification of essential and non-essential genes in Ambystoma tigrinum virus. Virus Res 217:107–114

Banjara S, Mao J, Ryan TM, Caria S, Kvansakul M (2018) Grouper iridovirus GIV66 is a Bcl-2 protein that inhibits apoptosis by exclusively sequestering Bim. J Biol Chem 293:5464–5477

Beattie E, Tartaglia J, Paoletti E (1991) Vaccinia virus-encoded eIF-2 alpha homolog abrogates the antiviral effect of interferon. Virology 183:419–422

Boehm EM, Gildenberg MS, Washington MT (2016) The many roles of PCNA in Eukaryotic DNA replication. Enzyme 39:231–254

Braunwald J, Tripier F, Kirn A (1979) Comparison of the properties of enveloped and naked frog virus 3 (FV 3) particles. J Gen Virol 45:673–682

Braunwald J, Nonnenmacher H, Tripier-Darcy F (1985) Ultrastructural and biochemical study of frog virus 3 uptake by BHK-21 cells. J Gen Virol 66(Pt 2):283–293

Brodie R, Roper RL, Upton C. (2004) JDotter a JAVA based interface to multiple display plots generated by dotter. Bioinformatics 20: 279–281

Bronkhorst AW, van Cleef KW, Venselaar H, van Rij RP (2014) A dsRNA-binding protein of a complex invertebrate DNA virus suppresses the Drosophila RNAi response. Nucleic Acids Res 42:12237–12248

Bronkhorst AW, Vogels R, Overheul GJ, Pennings B, Gausson-Dorey V, Miesen P, van Rij RP (2019) A DNA virus-encoded immune antagonist fully masks the potent antiviral activity of RNAi in Drosophila. Proc Natl Acad Sci USA 116:24296–24302

Brunner JL, Storfer A, Gray MJ, Hoverman JT (2015) Ranavirus ecology and evolution: from epidemiology to extinction. In: Gray MJ, Chinchar VG (eds) Ranaviruses. Springer International Publishing

Cackett G, Sykora M, Werner F (2020) Transcriptome view of a killer: African swine fever virus. Biochem Soc Trans 48:1569–1581

Caipang CM, Takano T, Hirono I, Aoki T (2006) Genetic vaccines protect red seabream, Pagrus major, upon challenge with red seabream iridovirus (RSIV). Fish Shellfish Immunol 21:130–138

Cardone G, Winkler DC, Trus BL, Cheng N, Heuser JE, Newcomb WW, Brown JC, Steven AC (2007) Visualization of the herpes simplex virus portal in situ by cryo-electron tomography. Virology 361:426–434

Chabaud S, Sasseville AM, Elahi SM, Caron A, Dufour F, Massie B, Langelier Y (2007) The ribonucleotide reductase domain of the R1 subunit of herpes simplex virus type 2 ribonucleotide reductase is essential for R1 antiapoptotic function. J Gen Virol 88:384–394

Chang JT, Schmid MF, Rixon FJ, Chiu W (2007) Electron cryotomography reveals the portal in the herpesvirus capsid. J Virol 81:2065–2068

Chen M, Goorha R, Murti KG (1986) Interaction of frog virus 3 with the cytomatrix. IV. Phosphorylation of vimentin precedes the reorganization of intermediate filaments around the virus assembly sites. J Gen Virol 67(Pt 5):915–922

Chen LM, Wang F, Song WJ, Hew CL (2006) Temporal and differential gene expression of Singapore grouper iridovirus. J Gen Virol 87:2907–2915

Chen GC, Ward BM, Yu KH, Chinchar VG, Robert J (2011) Improved knockout methodology reveals that frog virus 3 mutants lacking either the 18K immediate-early gene or the truncated vIF-2 alpha gene are defective for replication and growth in vivo. J Virol 85:11131–11138

Chen ZY, Gui JF, Gao XC, Pei C, Hong YJ, Zhang QY (2013) Genome architecture changes and major gene variations of Andrias davidianus ranavirus (ADRV). Vet Res 44:101

Chen CW, Wu MS, Huang YJ, Lin PW, Shih CJ, Lin FP, Chang CY (2015) Iridovirus CARD protein inhibits apoptosis through intrinsic and extrinsic pathways. PLoS One 10:e0129071

Chen YS, Chen NN, Qin XW, Mi S, He J, Lin YF, Gao MS, Weng SP, Guo CJ, He JG (2016) Tiger frog virus ORF080L protein interacts with LITAF and impairs EGF-induced EGFR degradation. Virus Res 217:133–142

Chen ZY, Li T, Gao XC, Wang CF, Zhang QY (2018) Protective immunity induced by DNA vaccination against Ranavirus infection in Chinese Giant Salamander Andrias davidianus. Viruses 10:52

Chen K, Tian J, Wang J, Jia Z, Zhang Q, Huang W, Zhao X, Gao Z, Gao Q, Zou J (2021) Lipopolysaccharide-induced TNFalpha factor (LITAF) promotes inflammatory responses and activates apoptosis in zebrafish Danio rerio. Gene 780:145487

Cheng KH (2014) Transcriptional, taxonomic, and biochemical studies of two ranavirus (frog virus 3 and Bohle iridovirus-like agent) infecting amphibians. University of Mississippi Medical Center, Jackson

Cheng K, Escalon BL, Robert J, Chinchar VG, Garcia-Reyero N (2014) Differential transcription of fathead minnow immune-related genes following infection with frog virus 3, an emerging pathogen of ectothermic vertebrates. Virology 456:77–86

Chinchar VG, Dholakia JN (1989) Frog virus 3-induced translational shut-off: activation of an eIF-2 kinase in virus-infected cells. Virus Res 14:207–223

Chinchar VG, Granoff A (1984) Isolation and characterization of a frog virus 3 variant resistant to phosphonoacetate: genetic evidence for a virus-specific DNA polymerase. Virology 138:357–361

Chinchar VG, Granoff A (1986) Temperature-sensitive mutants of frog virus 3: biochemical and genetic characterization. J Virol 58:192–202

Chinchar VG, Waltzek TB (2014) Ranaviruses: Not Just for Frogs. PLoS Pathog 10:e1003850

Chinchar VG, Yu W (1990a) Frog virus 3-mediated translational shut-off: frog virus 3 messages are translationally more efficient than host and heterologous viral messages under conditions of increased translational stress. Virus Res 16:163–174

Chinchar VG, Yu W (1990b) Translational efficiency: iridovirus early mRNAs outcompete tobacco mosaic virus message in vitro. Biochem Biophys Res Commun 172:1357–1363

Chinchar VG, Goorha R, Granoff A (1984) Early proteins are required for the formation of frog virus 3 assembly sites. Virology 135:148–156

Chinchar VG, Bryan L, Wang J, Long S, Chinchar GD (2003) Induction of apoptosis in frog virus 3-infected cells. Virology 306:303–312

Chinchar VG, Hyatt A, Miyazaki T, Williams T (2009) Family Iridoviridae: poor viral relations no longer. Curr Top Microbiol 328:123–170

Chinchar VG, Yu KH, Jancovich JK (2011) The molecular biology of Frog Virus 3 and other Iridoviruses infecting cold-blooded vertebrates. Viruses-Basel 3:1959–1985

Chinchar VG, Hick P, Ince IA, Jancovich JK, Marschang R, Qin QW, Subramaniam K, Waltzek TB, Whittington R, Williams T, Zhang QY, Consortium IR (2017a) ICTV virus taxonomy profile: Iridoviridae. J Gen Virol 98:890–891

Chinchar VG, Waltzek TB, Subramaniam K (2017b) Ranaviruses and other members of the family Iridoviridae: their place in the virosphere. Virology 511:259–271

Chong SJF, Marchi S, Petroni G, Kroemer G, Galluzzi L, Pervaiz S (2020) Noncanonical cell fate regulation by Bcl-2 proteins. Trends Cell Biol 30:537–555

Clark HF, Karzon DT (1967) Terrapene heart (TH-1), a continuous cell line from the heart of the box turtle Terrapene carolina. Exp Cell Res 48:263–268

Cobbold C, Windsor M, Parsley J, Baldwin B, Wileman T (2007) Reduced redox potential of the cytosol is important for African swine fever virus capsid assembly and maturation. J Gen Virol 88:77–85

Conner J, Marsden H, Clements BH (1994) Ribonucleotide reductase of herpesviruses. Rev Med Virol 4:25–34

Cordier O, Aubertin AM, Lopez C, Tondre L (1981) Inhibition de la traduction par le FV3: action des proteines virales de structure solubilisees sur la synthese proteique in vivo et in vitro. Ann Virol (Inst Pasteur) 132 E:25–39

Cordier O, Tondre L, Aubertin AM, Kirn A (1986) Restriction of frog virus 3 polypeptide synthesis to immediate early and delayed early species by supraoptimal temperatures. Virology 152:355–364

Daikoku T, Yamamoto N, Maeno K, Nishiyama Y (1991) Role of viral ribonucleotide reductase in the increase of dTTP pool size in herpes simplex virus-infected Vero cells. J Virol 72:1441–1444

Darcy-Tripier F, Nermut MV, Braunwald J, Williams LD (1984) The organization of frog virus 3 as revealed by freeze-etching. Virology 138:287–299

De Jesus Andino F, Grayfer L, Chen GC, Chinchar VG, Edholm ES, Robert J (2015) Characterization of Frog Virus 3 knockout mutants lacking putative virulence genes. Virology 485:162–170

de Souza GAP, Queiroz VF, Coelho LFL, Abrahao JS (2021) Alohomora! What the entry mechanisms tell us about the evolution and diversification of giant viruses and their hosts. Curr Opin Virol 47:79–85

de Wind N, Berns A, Gielkens A, Kimman T (1993) Ribonucleotide reductase-deficient mutants of pseudorabies virus are avirulent for pigs and induce partial protective immunity. J Gen Virol 74(Pt 3):351–359

Decout A, Katz JD, Venkatraman S, Ablasser A (2021) The cGAS-STING pathway as a therapeutic target in inflammatory diseases. Nat Rev Immunol 21:548–569

Deng L, Feng Y, OuYang P, Chen D, Huang X, Guo H, Deng H, Fang J, Lai W, Geng Y (2022) Autophagy induced by largemouth bass virus inhibits virus replication and apoptosis in epithelioma papulosum cyprini cells. Fish Shellfish Immunol 123:489–495

Devauchelle G, Stoltz DB, Darcy-Tripier F (1985) Comparative ultrastructure of iridoviridae. Curr Top Microbiol Immunol 116:1–21

Duffus AJ, Waltzek T, Stöhr A, Allender M, Gotesman M, Whittington R, Hick P, Hines M, Marschang R (2015) Distribution and host range of Ranaviruses. In: Gray MJ, Chinchar VG (eds) Ranaviruses. Springer International Publishing, pp 9–57

Eaton HE, Metcalf J, Penny E, Tcherepanov V, Upton C, Brunetti CR (2007) Comparative genomic analysis of the family Iridoviridae: re-annotating and defining the core set of iridovirus genes. Virol J 4:11

Eaton HE, Ring BA, Brunetti CR (2010) The genomic diversity and phylogenetic relationship in the family Iridoviridae. Viruses-Basel 2:1458–1475

Eaton HE, Lacerda AF, Desrochers G, Metcalf J, Angers A, Brunetti CR (2013) Cellular LITAF interacts with Frog Virus 3 75L protein and alters its subcellular localization. J Virol 87:716–723

Eberhardt MK, Deshpande A, Chang WL, Barthold SW, Walter MR, Barry PA (2013) Vaccination against a virus-encoded cytokine significantly restricts viral challenge. J Virol 87:11323–11331

Edholm ES, Grayfer L, De Jesus Andino F, Robert J (2015) Nonclassical MHC-restricted invariant Valpha6 T cells are critical for efficient early innate antiviral immunity in the Amphibian Xenopus laevis. J Immunol 195:576–586

Edholm EI, De Jesus Andino F, Yim J, Woo K, Robert J (2019) Critical role of an MHC Class I-Like/Innate-Like T cell immune surveillance system in host defense against Ranavirus (Frog Virus 3) infection. Viruses 11:330

Epstein B, Storfer A (2015) Comparative genomics of an emerging amphibian virus. G3 (Bethesda) 6:15–27

Feighny RJ, Henry BE 2nd, Pagano JS (1981) Epstein-Barr virus-induced deoxynuclease and the reutilization of host-cell DNA degradation products in viral DNA replication. Virology 115:395–400

Gal-Ben-Ari S, Barrera I, Ehrlich M, Rosenblum K (2019) PKR: a kinase to remember. Front Mol Neurosci 11:480

Gammon DB, Gowrsihankar, Duraffour S, Andrei G, Upton C, Evans DH (2010) Vaccinia virus-encoded ribonucleotide reductase subunits are differentially required for replication and pathogenesis. PLoS Pathog 6:e1000984

Gantress J, Maniero GD, Cohen N, Robert J (2003) Development and characterization of a model system to study amphibian immune responses to iridoviruses. Virology 311:254–262

Garcia-Sastre A (2017) Ten strategies of interferon evasion by viruses. Cell Host Microbe 22:176–184

Garrigues HJ, Rubinchikova YE, Dipersio CM, Rose TM (2008) Integrin alphaVbeta3 Binds to the RGD motif of glycoprotein B of Kaposi's sarcoma-associated herpesvirus and functions as an RGD-dependent entry receptor. J Virol 82:1570–1580

Gendrault JL, Steffan AM, Bingen A, Kirn A (1981) Penetration and uncoating of frog virus 3 (FV3) in cultured rat Kupffer cells. Virology 112:375–384

Gil J, Esteban M (2000) Induction of apoptosis by the dsRNA-dependent protein kinase (PKR): mechanism of action. Apoptosis 5:107–114

Goorha R (1981) Frog virus 3 requires RNA polymerase II for its replication. J Virol 37:496–499

Goorha R (1982) Frog virus 3 DNA replication occurs in two stages. J Virol 43:519–528

Goorha R, Dixit P (1984) A temperature-sensitive (TS) mutant of frog virus 3 (FV3) is defective in second-stage DNA replication. Virology 136:186–195

Goorha R, Granoff A (1979) Icosahedral cytoplasmic deoxyriboviruses. In: Fraenkel-Conrat H, Wagner RR (eds) Comprehensive virology. Plenum Press, New York, pp 347–399

Goorha R, Murti KG (1982) The genome of frog virus 3, an animal DNA virus, is circularly permuted and terminally redundant. Proc Natl Acad Sci USA 79:248–252

Goorha R, Willis DB, Granoff A, Naegele RF (1981) Characterization of a temperature-sensitive mutant of frog virus 3 defective in DNA replication. Virology 112:40–48

Goorha R, Granoff A, Willis DB, Murti KG (1984) The role of DNA methylation in virus replication: inhibition of frog virus 3 replication by 5-azacytidine. Virology 138:94–102

Granoff A, Came PE, Breeze DC (1966) Viruses and renal carcinoma of Rana pipiens. I. The isolation and properties of virus from normal and tumor tissue. Virology 29:133–148

Gravell M, Granoff A (1970) Viruses and renal carcinoma of Rana pipiens: IX. The influence of temperature and host cell on replication of frog polyhedral cytoplasmic deoxyribovirus (PCDV). Virology 41:596–602

Grayfer L, De Jesus Andino F, Robert J (2014) The amphibian (Xenopus laevis) Type I interferon response to Frog Virus 3: new insight into Ranavirus pathogenicity. J Virol 88:5766–5777

Grayfer L, Andino FD, Robert J (2015a) Prominent amphibian (Xenopus laevis) Tadpole Type III interferon response to the Frog Virus 3 Ranavirus. J Virol 89:5072–5082

Grayfer L, Edholm E-S, De Jesús Andino F, Chinchar VG, Robert J (2015b) Ranavirus host immunity and immune evasion. In: Gray MJ, Chinchar VG (eds) Ranaviruses. Springer International Publishing, pp 141–170

Grayfer L, Edholm E-S, Chinchar VG, Sang Y, Robert J (2024) Immune defenses against ranavirus infections. In: Chinchar VG, Gray MJ (eds) Ranaviruses: lethal pathogens of ectothermic vertebrates. Springer

Guarino LA, Xu B, Jin J, Dong W (1998) A virus-encoded RNA polymerase purified from baculovirus-infected cells. J Virol 72:7985–7991

Guo CJ, Liu D, Wu YY, Yang XB, Yang LS, Mi S, Huang YX, Luo YW, Jia KT, Liu ZY, Chen WJ, Weng SP, Yu XQ, He JG (2011) Entry of Tiger Frog Virus (an Iridovirus) into HepG2 cells via a pH-dependent, atypical, Caveola-mediated endocytosis pathway. J Virol 85:6416–6426

Guo CJ, Wu YY, Yang LS, Yang XB, He J, Mi S, Jia KT, Weng SP, Yu XQ, He JG (2012) Infectious spleen and kidney necrosis virus (a fish iridovirus) enters Mandarin fish fry cells via caveola-dependent endocytosis. J Virol 86:2621–2631

Guo C, Yan Y, Cui H, Huang X, Qin Q (2013) miR-homoHSV of Singapore grouper iridovirus (SGIV) inhibits expression of the SGIV pro-apoptotic factor LITAF and attenuates cell death. PLoS One 8:e83027

Hauser KA, Singer JC, Hossainey MRH, Moore TE, Wendel ES, Yaparla A, Kalia N, Grayfer L (2021) Amphibian (Xenopus laevis) Tadpoles and adult frogs differ in their antiviral responses to intestinal Frog Virus 3 infections. Front Immunol 12:737403

He JG, Deng M, Weng SP, Li Z, Zhou SY, Long QX, Wang XZ, Chan SM (2001) Complete genome analysis of the mandarin fish infectious spleen and kidney necrosis iridovirus. Virology 291:126–139

He JG, Lu L, Deng M, He HH, Weng SP, Wang XH, Zhou SY, Long QX, Wang XZ, Chan SM (2002) Sequence analysis of the complete genome of an iridovirus isolated from the tiger frog. Virology 292:185–197

He LB, Gao XC, Ke F, Zhang QY (2013) A conditional lethal mutation in Rana grylio virus ORF 53R resulted in a marked reduction in virion formation. Virus Res 177:194–200

He LB, Ke F, Wang J, Gao XC, Zhang QY (2014) Rana grylio virus (RGV) envelope protein 2L: subcellular localization and essential roles in virus infectivity revealed by conditional lethal mutant. J Gen Virol 95:679–690

He J, Mi S, Qin XW, Weng SP, Guo CJ, He JG (2019) Tiger frog virus ORF104R interacts with cellular VDAC2 to inhibit cell apoptosis. Fish Shellfish Immunol 92:889–896

Hoelzer K, Shackelton LA, Parrish CR (2008) Presence and role of cytosine methylation in DNA viruses of animals. Nucleic Acids Res 36:2825–2837

Holmgren A (1985) Thioredoxin. Annu Rev Biochem 54:237–271

Hooda-Dhingra U, Thompson CL, Condit RC (1989) Detailed phenotypic characterization of five temperature-sensitive mutants in the 22- and 147-kilodalton subunits of vaccinia virus DNA-dependent RNA polymerase. J Virol 63:714–729

Hossainey MRH, Yaparla A, Hauser KA, Moore TE, Grayfer L (2021) The roles of amphibian (Xenopus laevis) macrophages during chronic Frog Virus 3 infections. Viruses 13:2299

Howley PM, Knipe DM, Cohen JL, Damania BA (2021) Fields virology: DNA viruses, 7th edn. Lippincott Williams & Wilkins, Philadelphia

Huang YH, Huang XH, Gui JF, Zhang QY (2007) Mitochondrion-mediated apoptosis induced by Rana grylio virus infection in fish cells. Apoptosis 12:1569–1577

Huang X, Huang Y, Gong J, Yan Y, Qin Q (2008) Identification and characterization of a putative lipopolysaccharide-induced TNF-alpha factor (LITAF) homolog from Singapore grouper iridovirus. Biochem Biophys Res Commun 373:140–145

Huang YH, Huang XH, Liu H, Gong J, Ouyang ZL, Cui HC, Cao JH, Zhao YT, Wang XJ, Jiang YL, Qin QW (2009) Complete sequence determination of a novel reptile iridovirus isolated from soft-shelled turtle and evolutionary analysis of Iridoviridae. BMC Genomics 10:224

Huang XH, Huang YH, Ouyang ZL, Xu LX, Yan Y, Cui HC, Han X, Qin QW (2011) Singapore grouper iridovirus, a large DNA virus, induces nonapoptotic cell death by a cell type dependent fashion and evokes ERK signaling. Apoptosis 16:831–845

Huang XH, Huang YH, Cai J, Wei SN, Gao R, Qin QW (2013) Identification and characterization of a tumor necrosis factor receptor like protein encoded by Singapore grouper iridovirus. Virus Res 178:340–348

Huang X, Pei C, He LB, Zhang QY (2014a) The construction of a novel recombinant virus Delta67R-RGV and preliminary analyses the function of the 67R gene. Bing Du Xue Bao 30:495–501

Huang XH, Wang W, Huang YH, Xu LW, Qin QW (2014b) Involvement of the PI3K and ERK signaling pathways in largemouth bass virus-induced apoptosis and viral replication. Fish Shellfish Immun 41:371–379

Huang XH, Fang J, Chen ZY, Zhang QY (2016) Rana grylio virus TK and DUT gene locus could be simultaneously used for foreign gene expression. Virus Res 214:33–38

Huang Y, Huang X, Wang S, Yu Y, Ni S, Qin Q (2018) Soft-shelled turtle iridovirus enters cells via cholesterol-dependent, clathrin-mediated endocytosis as well as macropinocytosis. Arch Virol 163:3023–3033

Huynh TP, Jancovich JK, Tripuraneni L, Heck MC, Langland JO, Jacobs BL (2017) Characterization of a PKR inhibitor from the pathogenic ranavirus, Ambystoma tigrinum virus, using a heterologous vaccinia virus system. Virology 511:290–299

Ince IA, Boeren SA, van Oers MM, Vervoort JJM, Vlak JM (2010) Proteomic analysis of Chilo iridescent virus. Virology 405:253–258

Jancovich JK, Jacobs BL (2011) Innate immune evasion mediated by the Ambystoma tigrinum virus Eukaryotic translation initiation factor 2 alpha homologue. J Virol 85:5061–5069

Jancovich JK, Mao J, Chinchar VG, Wyatt C, Case ST, Kumar S, Valente G, Subramanian S, Davidson EW, Collins JP, Jacobs BL (2003) Genomic sequence of a ranavirus (family Iridoviridae) associated with salamander mortalities in North America. Virology 316:90–103

Jancovich JK, Bremont M, Touchman JW, Jacobs BL (2010) Evidence for multiple recent host species shifts among the Ranaviruses (Family Iridoviridae). J Virol 84:2636–2647

Jancovich J, Qin Q, Zhang QY, Chinchar VG (2015a) Ranavirus replication: molecular, cellular, and immunological events. In: Gray MJ, Chinchar VG (eds) Ranaviruses. Springer International Publishing, pp 105–139

Jancovich J, Steckler N, Waltzek T (2015b) Ranavirus taxonomy and phylogeny. In: Gray MJ, Chinchar VG (eds) Ranaviruses. Springer International Publishing, pp 59–70

Jia KT, Wu YY, Liu ZY, Mi S, Zheng YW, He J, Weng SP, Li SC, He JG, Guo CJ (2013) Mandarin fish caveolin 1 interaction with major capsid protein of infectious spleen and kidney necrosis virus and its role in early stages of infection. J Virol 87:3027–3038

Johnston JB, McFadden G (2003) Poxvirus immunomodulatory strategies: current perspectives. J Virol 77:6093–6100

Johnston JB, McFadden G (2004) Technical knockout: understanding poxvirus pathogenesis by selectively deleting viral immunomodulatory genes. Cell Microbiol 6:695–705

Johnston JB, Barrett JW, Nazarian SH, Goodwin M, Ricciuto D, Wang G, McFadden G (2005) A poxvirus-encoded pyrin domain protein interacts with ASC-1 to inhibit host inflammatory and apoptotic responses to infection. Immunity 23:587–598

Kalia N, Hauser KA, Burton S, Hossainey MRH, Zelle M, Horb ME, Grayfer L (2022) Endogenous retroviruses augment amphibian (Xenopus laevis) tadpole antiviral protection. J Virol 96:e0063422

Kaur K, Rohozinski J, Goorha R (1995) Identification and characterization of the frog virus 3 DNA methyltransferase gene. J Gen Virol 76(Pt 8):1937–1943

Kawagishi-Kobayashi M, Silverman JB, Ung TL, Dever TE (1997) Regulation of the protein kinase PKR by the vaccinia virus pseudosubstrate inhibitor K3L is dependent on residues conserved between the K3L protein and the PKR substrate eIF2alpha. Mol Cell Biol 17:4146–4158

Ke F, Zhang QY (2022) ADRV 12L: a Ranaviral Putative Rad2 family protein involved in DNA recombination and repair. Viruses 14:908

Ke F, Zhao L, Zhang QY (2009) Cloning, expression and subcellular distribution of a *Rana grylio* virus late gene encoding ERV1 homologue. Molec Biol Rep 36:1651–1659

Ke F, Gui JF, Chen ZY, Li T, Lei CK, Wang ZH, Zhang QY (2018) Divergent transcriptomic responses underlying the ranaviruses-amphibian interaction processes on interspecies infection of Chinese giant salamander. BMC Genomics 19:211

Ke F, Wang ZH, Ming CY, Zhang QY (2019) Ranaviruses bind cells from different species through interaction with Heparan sulfate. Viruses 11:593

Ke F, Wang RB, Wang ZH, Zhang QY (2022a) Andrias davidianus Ranavirus (ADRV) genome replicate efficiently by engaging cellular mismatch repair protein MSH2. Viruses 14:952

Ke F, Yu XD, Wang ZH, Gui JF, Zhang QY (2022b) Replication and transcription machinery for ranaviruses: components, correlation, and functional architecture. Cell Biosci 12:6

Kelly DC (1975) Frog virus 3 replication: electron microscope observations on the sequence of infection in chick embryo fibroblasts. J Gen Virol 26:71–86

Kim CM, Ha HJ, Kwon S, Jeong JH, Lee SH, Kim YG, Lee CS, Lee JH, Park HH (2019) Structural transformation-mediated dimerization of caspase recruitment domain revealed by the crystal structure of CARD-only protein in frog virus 3. J Struct Biol 205:189–195

Krug A, Towarowski A, Britsch S, Rothenfusser S, Hornung V, Bals R, Giese T, Engelmann H, Endres S, Krieg AM, Hartmann G (2001) Toll-like receptor expression reveals CpG DNA as a unique microbial stimulus for plasmacytoid dendritic cells which synergizes with CD40 ligand to induce high amounts of IL-12. Eur J Immunol 31:3026–3037

Krug A, Luker GD, Barchet W, Leib DA, Akira S, Colonna M (2004) Herpes simplex virus type 1 activates murine natural interferon-producing cells through toll-like receptor 9. Blood 103:1433–1437

Langelier Y, Bergeron S, Chabaud S, Lippens J, Guilbault C, Sasseville AM, Denis S, Mosser DD, Massie B (2002) The R1 subunit of herpes simplex virus ribonucleotide reductase protects cells against apoptosis at, or upstream of, caspase-8 activation. J Gen Virol 83:2779–2789

Lawler C, Brady G (2020) Poxviral targeting of interferon regulatory factor activation. Viruses 12:1191

Lefkowitz EJ, Dempsey DM, Hendrickson RC, Orton RJ, Siddell SG, Smith DB (2018) Virus taxonomy: the database of the International Committee on Taxonomy of Viruses (ICTV). Nucleic Acids Res 46:D708–D717

Lei XY, Ou T, Zhang QY (2012a) Rana grylio Virus (RGV) 50L is associated with viral matrix and exhibited two distribution patterns. PLoS One 7:e43033

Lei XY, Ou T, Zhu RL, Zhang QY (2012b) Sequencing and analysis of the complete genome of Rana grylio virus (RGV). Arch Virol 157:1559–1564

Lembo D, Brune W (2009) Tinkering with a viral ribonucleotide reductase. Trends Biochem Sci 34:25–32

Levin D, London IM (1978) Regulation of protein synthesis: activation by double-stranded RNA of a protein kinase that phosphorylates eukaryotic initiation factor 2. Proc Natl Acad Sci USA 75:1121–1125

Levine B, Deretic V (2007) Unveiling the roles of autophagy in innate and adaptive immunity. Nat Rev Immunol 7:767–777

Lewis T, Zsak L, Burrage TG, Lu Z, Kutish GF, Neilan JG, Rock DL (2000) An African swine fever virus ERV1-ALR homologue, 9GL, affects virion maturation and viral growth in macrophages and viral virulence in swine. J Virol 74:1275–1285

Li Y, Jiang N, Fan Y, Zhou Y, Liu W, Xue M, Meng Y, Zeng L (2019) Chinese Giant Salamander (Andrias davidianus) Iridovirus infection leads to apoptotic cell death through mitochondrial damage, caspases activation, and expression of apoptotic-related genes. Int J Mol Sci 20:6149

Li C, Wang L, Liu J, Yu Y, Huang Y, Huang X, Wei J, Qin Q (2020) Singapore Grouper Iridovirus (SGIV) inhibited autophagy for efficient viral replication. Front Microbiol 11:1446

Lin PW, Huang YJ, John JA, Chang YN, Yuan CH, Chen WY, Yeh CH, Shen ST, Lin FP, Tsui WH, Chang CY (2008) Iridovirus Bcl-2 protein inhibits apoptosis in the early stage of viral infection. Apoptosis 13:165–176

Liu Z, Xie D, Nong S, Wu Y, Huang S, He X, Zhou T, Li W (2023) Chinese Giant Salamander Iridovirus 025L is a viral essential gene. Viruses 15:617

Lo Piano A, Martinez-Jimenez MI, Zecchi L, Ayora S (2011) Recombination-dependent concatemeric viral DNA replication. Virus Res 160:1–14

Lopez C, Aubertin AM, Tondre L, Kirn A (1986) Thermosensitivity of frog virus 3 genome expression: defect in early transcription. Virology 152:365–374

Lu Y, Zhang L (2020) DNA-sensing antiviral innate immunity in poxvirus infection. Front Immunol 11:1637

Lu JF, Jin TC, Zhou T, Lu XJ, Chen JP, Chen J (2021) Identification and characterization of a tumor necrosis factor receptor like protein encoded by Cyprinid Herpesvirus 2. Dev Comp Immunol 116:103930

Lum KK, Cristea IM (2021) Host innate immune response and viral immune evasion during Alphaherpesvirus infection. Curr Issues Mol Biol 42:635–686

MacLeod DT, Nakatsuji T, Yamasaki K, Kobzik L, Gallo RL (2013) HSV-1 exploits the innate immune scavenger receptor MARCO to enhance epithelial adsorption and infection. Nat Commun 4:1963

MacLeod DT, Nakatsuji T, Wang Z, di Nardo A, Gallo RL (2015) Vaccinia virus binds to the scavenger receptor MARCO on the surface of keratinocytes. J Invest Dermatol 135:142–150

Majji S, Thodima V, Sample R, Whitley D, Deng Y, Mao J, Chinchar VG (2009) Transcriptome analysis of Frog virus 3, the type species of the genus Ranavirus, family Iridoviridae. Virology 391:293–303

Maniero GD, Morales H, Gantress J, Robert J (2006) Generation of a long-lasting, protective, and neutralizing antibody response to the ranavirus FV3 by the frog Xenopus. Dev Comp Immunol 30:649–657

Martin V, Mavian C, Bueno AL, de Molina A, Diaz E, Andres G, Alcami A, Alejo A (2015) Establishment of a Zebrafish infection model for the study of wild-type and recombinant European Sheatfish virus. J Virol 89:10702–10706

Mavian C, Lopez-Bueno A, Balseiro A, Casais R, Alcami A, Alejo A (2012) The genome sequence of the emerging common midwife toad virus identifies an evolutionary intermediate within Ranaviruses. J Virol 86:3617–3625

Milrot E, Mutsafi Y, Fridmann-Sirkis Y, Shimoni E, Rechav K, Gurnon JR, Van Etten JL, Minsky A (2016) Virus-host interactions: insights from the replication cycle of the large Paramecium bursaria chlorella virus. Cell Microbiol 18:3–16

Mirzakhanyan Y, Gershon PD (2017) Multisubunit DNA-dependent RNA polymerases from vaccinia virus and other nucleocytoplasmic large-DNA viruses: impressions from the age of structure. Microbiol Mol Biol Rev 81:e00010

Mirzakhanyan Y, Gershon PD (2020) Structure-based deep mining reveals first-time annotations for 46 percent of the dark annotation space of the 9,671-member Superproteome of the nucleocytoplasmic large DNA viruses. J Virol 94:e00854

Mishra R, Kumar A, Ingle H, Kumar H (2019) The interplay between viral-derived miRNAs and host immunity during infection. Front Immunol 10:3079

Moody NJG, Owens L (1994) Experimental demonstration of pathogenicity of a frog virus, Bohle iridovirus, for a fish species, barramundi *Lates Calcarifer*. Disea Aquat Organ 18:95–102

Morales HD, Robert J (2007) Characterization of primary and memory CD8 T-cell responses against ranavirus (FV3) in Xenopus laevis. J Virol 81:2240–2248

Morales HD, Abramowitz L, Gertz J, Sowa J, Vogel A, Robert J (2010) Innate immune responses and permissiveness to Ranavirus infection of peritoneal leukocytes in the frog Xenopus laevis. J Virol 84:4912–4922

Morrison EA, Garner S, Echaubard P, Lesbarreres D, Kyle CJ, Brunetti CR (2014) Complete genome analysis of a frog virus 3 (FV3) isolate and sequence comparison with isolates of differing levels of virulence. Virol J 11:46

Mosig G (1998) Recombination and recombination-dependent DNA replication in bacteriophage T4. Annu Rev Genet 32:379–413

Muller K, Tidona CA, Darai G (1999) Identification of a gene cluster within the genome of Chilo iridescent virus encoding enzymes involved in viral DNA replication and processing. Virus Genes 18:243–264

Murti KG, Goorha R, Chen M (1985) Interaction of frog virus 3 with the cytoskeleton. Curr Top Microbiol Immunol 116:107–131

Murti KG, Goorha R, Klymkowsky MW (1988) A functional role for intermediate filaments in the formation of frog virus 3 assembly sites. Virology 162:264–269

Mutsafi Y, Fridmann-Sirkis Y, Milrot E, Hevroni L, Minsky A (2014) Infection cycles of large DNA viruses: emerging themes and underlying questions. Virology 466-467:3–14

Netherton C, Moffat K, Brooks E, Wileman T (2007) A guide to viral inclusions, membrane rearrangements, factories, and viroplasm produced during virus replication. Adv Virus Res 70:101–182

Ni SW, Yan Y, Cui HC, Yu YP, Huang YH, Qin QW (2017) Fish miR-146a promotes Singapore grouper iridovirus infection by regulating cell apoptosis and NF-kappa B activation. J Gen Virol 98:1489–1499

Ou-yang Z, Wang P, Huang X, Cai J, Huang Y, Wei S, Ji H, Wei J, Zhou Y, Qin Q (2012) Immunogenicity and protective effects of inactivated Singapore grouper iridovirus (SGIV) vaccines in orange-spotted grouper, Epinephelus coioides. Dev Comp Immunol 38:254–261

Papp T, Marschang RE (2019) Detection and characterization of invertebrate Iridoviruses found in reptiles and prey insects in Europe over the past two decades. Viruses 11:600

Pham PH, Lai YS, Lee FFY, Bols NC, Chiou PP (2012) Differential viral propagation and induction of apoptosis by grouper iridovirus (GIV) in cell lines from three non-host species. Virus Res 167:16–25

Pham PH, Huang YJ, Mosser DD, Bols NC (2015) Use of cell lines and primary cultures to explore the capacity of rainbow trout to be a host for frog virus 3 (FV3). In Vitro Cell Dev-An 51:894–904

Pintilie G, Chen DH, Tran BN, Jakana J, Wu J, Hew CL, Chiu W (2019) Segmentation and comparative modeling in an 8.6-A Cryo-EM map of the Singapore Grouper Iridovirus. Structure 27:1561–1569 e1564

Price SJ (2015) Comparative genomics of amphibian-like Ranaviruses, nucleocytoplasmic large DNA viruses of Poikilotherms. Evol Bioinformatics Online 11:71–82

Purifoy D, Naegele RF, Granoff A (1973) Viruses and renal carcinoma of Rana pipiens. XIV. Temperature-sensitive mutants of frog virus 3 with defective encapsidation. Virology 54:525–535

Qi H, Yi Y, Weng S, Zou W, He J, Dong C (2016) Differential autophagic effects triggered by five different vertebrate iridoviruses in a common, highly permissive mandarinfish fry (MFF-1) cell model. Fish Shellfish Immunol 49:407–419

Raghow R, Granoff A (1979) Macromolecular synthesis in cells infected by frog virus 3. X. Inhibition of cellular protein synthesis by heat-inactivated virus. Virology 98:319–327

Raghow R, Granoff A (1980) Macromolecular synthesis in cells infected by frog virus 3. XIV. Characterization of the methylated nucleotide sequences in viral messenger RNAs. Virology 107:283–294

Reading PC, Moore JB, Smith GL (2003) Steroid hormone synthesis by vaccinia virus suppresses the inflammatory response to infection. J Exp Med 197:1269–1278

Robert J (2010) Emerging Ranaviral infectious diseases and amphibian decline. Diversity 2:314

Robert J, Morales H, Buck W, Cohen N, Marr S, Gantress J (2005) Adaptive immunity and histopathology in frog virus 3-infected Xenopus. Virology 332:667–675

Rojas JM, Alejo A, Martin V, Sevilla N (2021) Viral pathogen-induced mechanisms to antagonize mammalian interferon (IFN) signaling pathway. Cell Mol Life Sci 78:1423–1444

Rothenburg S, Chinchar VG, Dever TE (2011) Characterization of a ranavirus inhibitor of the antiviral protein kinase PKR. BMC Microbiol 11:56

Rouiller I, Brookes SM, Hyatt AD, Windsor M, Wileman T (1998) African swine fever virus is wrapped by the endoplasmic reticulum. J Virol 72:2373–2387

Ruiz VL, Robert J (2023) The amphibian immune system. Philos Trans R Soc Lond Ser B Biol Sci 378:20220123

Safer B (1983) 2B or not 2B: regulation of the catalytic utilization of eIF-2. Cell 33:7–8

Salas ML, Andres G (2013) African swine fever virus morphogenesis. Virus Res 173:29–41

Samanta M, Yim J, De Jesus Andino F, Paiola M, Robert J (2021) TLR5-mediated reactivation of quiescent Ranavirus FV3 in Xenopus peritoneal macrophages. J Virol 95:e00215

Sample R (2010) Elucidation of Frog Virus 3 gene function and pathways of Virion formation. University of Mississippi Medical Center, Jackson

Sample R, Bryan L, Long S, Majji S, Hoskins G, Sinning A, Olivier J, Chinchar VG (2007) Inhibition of iridovirus protein synthesis and virus replication by antisense morpholino oligonucleotides targeted to the major capsid protein, the 18 kDa immediate-early protein, and a viral homolog of RNA polymerase II. Virology 358:311–320

Seet BT, Johnston JB, Brunetti CR, Barrett JW, Everett H, Cameron C, Sypula J, Nazarian SH, Lucas A, McFadden G (2003) Poxviruses and immune evasion. Annu Rev Immunol 21:377–423

Sonenberg N, Hinnebusch AG (2009) Regulation of translation initiation in eukaryotes: mechanisms and biological targets. Cell 136:731–745

Song W, Lin Q, Joshi SB, Lim TK, Hew CL (2006) Proteomic studies of the Singapore grouper iridovirus. Molec Cell Proteom 5:256–264

Song WJ, Qin Q, Qiu J, Huang CH, Wang F, and Hew CL. (2004) Functional genomic analysis of Singapore grouper iridovirus: Complete sequence determination and proteomic analysis. J. Virol. 78: 12576–12590

Stillman B (2013) Deoxynucleoside triphosphate (dNTP) synthesis and destruction regulate the replication of both cell and virus genomes. Proc Natl Acad Sci USA 110:14120–14121

Suarez C, Andres G, Kolovou A, Hoppe S, Salas ML, Walther P, Krijnse Locker J (2015) African swine fever virus assembles a single membrane derived from rupture of the endoplasmic reticulum. Cell Microbiol 17:1683–1698

Sun W, Huang YH, Zhao Z, Gui JF, Zhang QY (2006) Characterization of the Rana grylio virus 3beta-hydroxysteroid dehydrogenase and its novel role in suppressing virus-induced cytopathic effect. Biochem Biophys Res Commun 351:44–50

Tan WGH, Barkman TJ, Chinchar VG, Essani K (2004) Comparative genomic analyses of frog virus 3, type species of the genus Ranavirus (family Iridoviridae). Virology 323:70–84

Teng Y, Hou Z, Gong J, Liu H, Xie X, Zhang L, Chen X, Qin QW (2008) Whole-genome transcriptional profiles of a novel marine fish iridovirus, Singapore grouper iridovirus (SGIV) in virus-infected grouper spleen cell cultures and in orange-spotted grouper, Epinephulus coioides. Virology 377:39–48

Tian Y, De Jesus Andino F, Khwatenge CN, Li J, Robert J, Sang Y (2021a) Virus-targeted transcriptomic analyses implicate ranaviral interaction with host interferon response in Frog virus 3-infected frog tissues. Viruses 13:1325

Tian Y, Khwatenge CN, Li J, De Jesus Andino F, Robert J, Sang Y (2021b) Targeted transcriptomics of Frog virus 3 in infected frog tissues reveal non-coding regulatory elements and microRNAs in the ranaviral genome and their potential interaction with host immune response. Front Immunol 12:705253

Tidona CA, Darai G (1997) The complete DNA sequence of lymphocystis disease virus. Virology 230:207–216

Todd LA, Katzenback BA (2021) Discovery of frog virus 3 microRNAs and their roles in evasion of host antiviral responses. bioRxiv

Toenshoff ER, Fields PD, Bourgeois YX, Ebert D (2018) The end of a 60-year riddle: identification and genomic characterization of an Iridovirus, the causative agent of white fat cell disease in Zooplankton. G3 (Bethesda) 8:1259–1272

Trimble JJ, Murthy SC, Bakker A, Grassmann R, Desrosiers RC (1988) A gene for dihydrofolate reductase in a herpesvirus. Science 239:1145–1147

Tulman ER, Delhon GA, Ku BK, Rock DL (2009) African swine fever virus. Curr Top Microbiol Immunol 328:43–87

Tweedell K, Granoff A (1968) Viruses and renal carcinoma of Rana pipiens. V. Effect of frog virus 3 on developing frog embryos and larvae. J Natl Cancer Inst 40:407–410

Van Etten JL, Dunigan DD (2016) Giant Chloroviruses: five easy questions. PLoS Pathog 12:e1005751

Vertessy BG, Toth J (2009) Keeping uracil out of DNA: physiological role, structure and catalytic mechanism of dUTPases. Acc Chem Res 42:97–106

Vo NTK, Guerreiro M, Yaparla A, Grayfer L, DeWitte-Orr SJ (2019) Class A scavenger receptors are used by Frog Virus 3 during its cellular entry. Viruses 11:93

Waltzek TB, Subramaniam K, Jancovich JK (2024) Ranavirus taxonomy and phylogeny. In: Chichar VG, Gray MJ (eds) IN: Ranaviruses: lethal pathogens of ectotermic vertebrates. Springer

Wan QJ, Gong J, Huang XH, Huang YH, Zhou S, Ou-Yang ZL, Cao JH, Ye LL, Qin QW (2010) Identification and characterization of a novel capsid protein encoded by Singapore grouper iridovirus ORF038L. Arch Virol 155:351–359

Wang Q, Luo YW, Xie JF, Dong CF, Weng SP, Ai HS, Lu L, Yang XQ, Yu XQ, He JG (2008) Identification of two novel membrane proteins from the Tiger frog virus (TFV). Virus Res 136:35–42

Wang F, Liu Y, Zhu Y, Ngoc Tran B, Wu J, Leong Hew C (2015a) Singapore Grouper Iridovirus ORF75R is a Scaffold protein essential for viral assembly. Sci Rep 5:13151

Wang F, Zhu Y, Hew CL (2015b) Quantitative study of proteomic alterations in a Zebrafish (danio rerio) cell line infected with the Singapore Grouper Iridovirus (SGIV). Virus Res 199:62–67

Wang W, Zhang Y, Guo X, Xu W, Qin Q, Huang Y (2023) Singapore grouper iridovirus infection counteracts poly I:C induced antiviral immune response in vitro. Fish Shellfish Immunol 135:108685

Weinheimer I, Jiu YM, Rajamaki ML, Matilainen O, Kallijarvi J, Cuellar WJ, Lu R, Saarma M, Holmberg CI, Jantti J, Valkonen JPT (2015) Suppression of RNAi by dsRNA-degrading RNaseIII enzymes of viruses in animals and plants. PLoS Pathog 11:e1004711

Wendel ES, Yaparla A, Koubourli DV, Grayfer L (2017) Amphibian (Xenopus laevis) tadpoles and adult frogs mount distinct interferon responses to the Frog Virus 3 ranavirus. Virology 503:12–20

Wendel ES, Yaparla A, Melnyk MLS, Koubourli DV, Grayfer L (2018) Amphibian (Xenopus laevis) Tadpoles and adult frogs differ in their use of expanded repertoires of Type I and Type III interferon Cytokines. Viruses 10:372

Whitley DJS (2011) Determinations of ranavirus gene function using an antisense morpholino-mediated approach. Ph.D. Dissertation

Whitley DS, Yu K, Sample RC, Sinning A, Henegar J, Norcross E, Chinchar VG (2010) Frog virus 3 ORF 53R, a putative myristoylated membrane protein, is essential for virus replication in vitro. Virology 405:448–456

Whitley DS, Sample RC, Sinning AR, Henegar J, Chinchar VG (2011) Antisense approaches for elucidating ranavirus gene function in an infected fish cell line. Dev Comp Immunol 35:937–948

Williams T (1996) The iridoviruses. Adv Virus Res 46:345–412

Williams T, Barbosa-Solomieu V, Chinchar VG (2005) A decade of advances in iridovirus research. Adv Virus Res 65:173–248

Willis DB, Granoff A (1978) Macromolecular synthesis in cells infected by frog virus 3. IX. Two temporal classes of early viral RNA. Virology 86:443–453

Willis DB, Granoff A (1980) Frog virus 3 DNA is heavily methylated at CpG sequences. Virology 107:250–257

Willis DB, Granoff A (1985) Transactivation of an immediate-early frog virus 3 promoter by a virion protein. J Virol 56:495–501

Willis DB, Thompson JP (1986) The Iridovirus frog virus 3: a model for trans-acting proteins. Microbiol Sci 3:59–63

Willis DB, Goorha R, Granoff A (1979) Nongenetic reactivation of frog virus 3 DNA. Virology 98:476–479

Willis DB, Goorha R, Granoff A (1984) DNA methyltransferase induced by frog virus 3. J Virol 49:86–91

Willis DB, Goorha R, Chinchar VG (1985) Macromolecular synthesis in cells infected by frog virus 3. Curr Top Microbiol Immunol 116:77–106

Willis DB, Essani K, Goorha R, Thompson JP, Granoff A (1990) Transcription of a methylated DNA virus, nucleic acid methylation. Alan R. Liss, Inc, pp 139–151

Wilson WH, Van Etten JL, Allen MJ (2009) The Phycodnaviridae: the story of how tiny giants rule the world. Curr Top Microbiol Immunol 328:1–42

Wu G, Lin Q, Lim TK, Zhang Y, Aweya JJ, Zhu J, Yao D (2021) The interactome of Singapore grouper iridovirus protein ICP18 as revealed by proximity-dependent BioID approach. Virus Res 291:198218

Xia L, Cao J, Huang X, Qin Q (2009) Characterization of Singapore grouper iridovirus (SGIV) ORF086R, a putative homolog of ICP18 involved in cell growth control and virus replication. Arch Virol 154:1409–1416

Xia L, Liang H, Huang Y, Ou-Yang Z, Qin Q (2010) Identification and characterization of Singapore grouper iridovirus (SGIV) ORF162L, an immediate-early gene involved in cell growth control and viral replication. Virus Res 147:30–39

Xia LQ, Chen JL, Zhang HL, Cai J, Zhou S, Lu YS (2019) Identification of virion-associated transcriptional transactivator (VATT) of SGIV ICP46 promoter and their binding site on promoter. Virol J 16:110

Xie JF, Lai YX, Huang LJ, Huang RQ, Yang SW, Shi Y, Weng SP, Zhang Y, He JG (2014) Genome-wide analyses of proliferation-important genes of Iridovirus-tiger frog virus by RNAi. Virus Res 189:214–225

Xu L, Liu M, Chen H, Zhang L, Xu Q, Zhan Z, Xu Z, Liu S, Wu S, Zhang X, Qin Q, Wei J (2023) Singapore grouper iridovirus VP122 targets grouper STING to evade the interferon immune response. Fish Shellfish Immunol 140:108990

Yan X, Yu Z, Zhang P, Battisti AJ, Holdaway HA, Chipman PR, Bajaj C, Bergoin M, Rossmann MG, Baker TS (2009) The capsid proteins of a large, icosahedral dsDNA virus. J Mol Biol 385:1287–1299

Yan Y, Cui HC, Jiang SS, Huang YH, Huang XH, Wei SN, Xu WY, Qin QW (2011) Identification of a novel marine fish virus, Singapore grouper iridovirus-encoded microRNAs expressed in grouper cells by Solexa sequencing. Plos One 6:e19148

Yan Y, Cui H, Guo C, Li J, Huang X, Wei J, Qin Q (2013) An insulin-like growth factor homologue of Singapore grouper iridovirus modulates cell proliferation, apoptosis and enhances viral replication. J Gen Virol 94:2759–2770

Yan Y, Cui H, Guo C, Wei J, Huang Y, Li L, Qin Q (2014) Singapore grouper iridovirus-encoded semaphorin homolog (SGIV-sema) contributes to viral replication, cytoskeleton reorganization and inhibition of cellular immune responses. J Gen Virol 95:1144

Yang J, Xu W, Wang W, Pan Z, Qin Q, Huang X, Huang Y (2022) Largemouth bass virus infection induced non-apoptotic cell death in MsF cells. Viruses 14:1568

Yao D, Liu Y, Chen X, Lim TK, Wang L, Aweya JJ, Zhang Y, Lin Q (2019) In-depth proteomic profiling of the Singapore grouper iridovirus virion. Arch Virol 164:1889–1895

Yaparla A, Popovic M, Grayfer L (2018) Differentiation-dependent antiviral capacities of amphibian (Xenopus laevis) macrophages. J Biol Chem 293:1736–1744

Yaparla A, Docter-Loeb H, Melnyk MLS, Batheja A, Grayfer L (2019) The amphibian (Xenopus laevis) colony-stimulating factor-1 and interleukin-34-derived macrophages possess disparate pathogen recognition capacities. Dev Comp Immunol 98:89–97

Yu Y, Huang Y, Wei S, Li P, Zhou L, Ni S, Huang X, Qin Q (2016) A tumour necrosis factor receptor-like protein encoded by Singapore grouper iridovirus modulates cell proliferation, apoptosis and viral replication. J Gen Virol 97:756–766

Yu Y, Huang Y, Ni S, Zhou L, Liu J, Zhang J, Zhang X, Hu Y, Huang X, Qin Q (2017) Singapore grouper iridovirus (SGIV) TNFR homolog VP51 functions as a virulence factor via modulating host inflammation response. Virology 511:280–289

Yu H, Bruneau RC, Brennan G, Rothenburg S (2021) Battle Royale: innate recognition of Poxviruses and viral immune evasion. Biomedicines 9:765

Yu XD, Ke F, Zhang QY, Gui JF (2023) Genome characteristics of two Ranavirus isolates from Mandarin fish and largemouth bass. Pathogens 12:730

Yuan M, Zhang W, Wang J, Al Yaghchi C, Ahmed J, Chard L, Lemoine NR, Wang Y (2015) Efficiently editing the vaccinia virus genome by using the CRISPR-Cas9 system. J Virol 89:5176–5179

Yuan JM, Chen YS, He J, Weng SP, Guo CJ, He JG (2016a) Identification and differential expression analysis of MicroRNAs encoded by Tiger Frog Virus in cross-species infection in vitro. Virol J 13:73

Yuan YM, Wang YZ, Liu QZ, Zhu F, Hong YH (2016b) Singapore grouper iridovirus protein VP088 is essential for viral infectivity. Sci Rep 6:31170

Zeng X-T, Zhang QY (2019) Interaction between two Iridovirus Core proteins and their effects on Ranavirus (RGV) replication in cells from different species. Viruses 11:416

Zeng XT, Gao XC, Zhang QY (2018) Rana grylio virus 43R encodes an envelope protein involved in virus entry. Virus Genes 54:779–791

Zenke K, Kim KH (2008) Functional characterization of the RNase III gene of rock bream iridovirus. Arch Virol 153:1651–1656

Zerbini FM, Siddell SG, Mushegian AR, Walker PJ, Lefkowitz EJ, Adriaenssens EM, Alfenas-Zerbini P, Dutilh BE, Garcia ML, Junglen S, Krupovic M, Kuhn JH, Lambert AJ, Lobocka M, Oksanen HM, Robertson DL, Rubino L, Sabanadzovic S, Simmonds P, Suzuki N, Van Doorslaer K, Vandamme AM, Varsani A (2022) Differentiating between viruses and virus species by writing their names correctly. Arch Virol 167:1231–1234

Zhang QY, Gui JF (2012) Atlas of aquatic viruses and viral diseases. Science Press, Beijing

Zhang QY, Gui JF (2015) Virus genomes and virus-host interactions in aquaculture animals. Sci China Life Sci 58:156–169

Zhang R, Zhang QY (2018) Adenosine triphosphatase activity and cell growth promotion of Andrias davidianus ranavirus 96L-encoded protein (ADRV-96L). Microbiol China 45:1090–1099

Zhang F, Altindis E, Kahn CR, DiMarchi RD, Gelfanov V (2021) A viral insulin-like peptide is a natural competitive antagonist of the human IGF-1 receptor. Mol Metab 53:101316

Zhang Y, Gao X, Yang X, Wang Y, Wang W, Huang X, Qin Q, Huang Y (2022a) Singapore Grouper Iridovirus VP131 drives degradation of STING-TBK1 pathway proteins and negatively regulates antiviral innate immunity. J Virol 96:e0068222

Zhang QY, Ke F, Gui L, Zhao Z (2022b) Recent insights into aquatic viruses: emerging and reemerging pathogens, molecular features, biological effects, and novel investigative approaches. Water Biol Secur 1:100062

Zhao Z, Ke F, Gui JF, Zhang QY (2007) Characterization of an early gene encoding for dUTPase in Rana grylio virus. Virus Res 123:128–137

Zhao Z, Ke F, Huang YH, Zhao JG, Gui JF, Zhang QY (2008) Identification and characterization of a novel envelope protein in Rana grylio virus. J Gen Virol 89:1866–1872

Zhao Z, Ke F, Shi Y, Zhou GZ, Gui JF, Zhang QY (2009) Rana grylio virus thymidine kinase gene: an early gene of iridovirus encoding for a cytoplasmic protein. Virus Genes 38:345–352

Zhao Z, Huang Y, Liu C, Zhu D, Gao S, Liu S, Peng R, Zhang Y, Huang X, Qi J, Wong CCL, Zhang X, Wang P, Qin Q, Gao GF (2023) Near-atomic architecture of Singapore grouper iridovirus and implications for giant virus assembly. Nat Commun 14:2050

Zheng J, Zhi L, Wang W, Ni N, Huang Y, Qin Q, Huang X (2022) Fish TRIM21 exhibits antiviral activity against grouper iridovirus and nodavirus infection. Fish Shellfish Immunol 127:956–964

Zhu R, Chen ZY, Wang J, Yuan JD, Liao XY, Gui JF, Zhang QY (2014a) Extensive diversification of MHC in Chinese giant salamanders Andrias davidianus (Anda-MHC) reveals novel splice variants. Dev Comp Immunol 42:311–322

Zhu R, Chen ZY, Wang J, Yuan JD, Liao XY, Gui JF, Zhang QY (2014b) Thymus cDNA library survey uncovers novel features of immune molecules in Chinese giant salamander Andrias davidianus. Dev Comp Immunol 46:413–422

Zuo W, Wakimoto M, Kozaiwa N, Shirasaka Y, Oh SW, Fujiwara S, Miyachi H, Kogure A, Kato H, Fujita T (2022) PKR and TLR3 trigger distinct signals that coordinate the induction of antiviral apoptosis. Cell Death Dis 13:707

Immune Defenses Against Ranavirus Infections

Leon Grayfer, Eva-Stina Edholm, V. Gregory Chinchar, Yongming Sang, and Jacques Robert

1 Introduction

Infections of ectothermic vertebrates by members of the genus *Ranavirus* (RV; family *Iridoviridae*) and the resulting disease outbreaks and die-offs among wild and farmed populations have escalated at alarming rates over a decade and raised considerable concerns. While it is apparent that individual teleost, amphibian, and reptile species vary in their susceptibility to these pathogens, the immune and viral determinants of ranaviral diseases are at present unclear. In fact, with the rapid rise in both the prevalence of ranavirus infections and the remarkable capacity of these viruses to infect new hosts, ranaviruses such as frog virus 3 (FV3) are now considered to be a potential global threat to ectothermic populations (Gray and Miller 2013). There is a pressing need to determine whether the susceptibility of a given ectothermic species reflects its inability to mount a protective antiviral immune response or the capacity of the ranavirus to overcome otherwise intact immune barriers. Indeed, ranaviruses appear to possess an array of immune evasion and host

L. Grayfer (✉)
Department of Biological Sciences, George Washington University, Washington, DC, USA
e-mail: leon_grayfer@gwu.edu

E.-S. Edholm
The Arctic University of Norway, Norwegian College of Fishery Science, Tomsø, Norway

V. G Chinchar
University of Mississippi Medical Center, Department of Cell and Molecular Biology, Jackson, MS, USA

Y. Sang
Department of Agricultural and Environmental Sciences, College of Agriculture, Tenessee State University, Nashville, NT, USA

J. Robert
University of Rochester Medical Center, Department of Microbiology and Immunology, and Environmental Medicine, Rochester, Rochester, NY, USA

M. J. Gray, V. G. Chinchar (eds.), *Ranaviruses*,
https://doi.org/10.1007/978-3-031-64973-8_4

modulation mechanisms (Grayfer et al. 2012). Thus, a more thorough examination of the ranavirus-host immune interface at the molecular and cellular levels is necessary to devise potential preventative measures against these viral agents.

Ectothermic vertebrates possess complex immune systems that are reminiscent of those seen in mammals, but which have been shaped by very distinct evolutionary pressures, and thus require careful examination in the context of species-specific host-pathogen interactions (Robert and Ohta 2009). A great example of this are the innate antiviral immune defenses of ectothermic vertebrates, which are distinct from those described in mammals. The exact contribution and efficacies of these immune responses await further characterization in the context of infections with pathogens such as ranaviruses.

This chapter summarizes recent advances in our understanding of the contributions of innate and adaptive immune responses to the elimination and/or progression of ranaviral infections in poikilotherms, the impacts of environmental pollutants on amphibian anti-ranaviral defenses as well as recently discovered putative ranavirus immune evasion strategies.

2 Infections of Bony Fish and Urodele Amphibians

There is substantial literature documenting innate and inflammatory responses to ranavirus infections in bony fish. Infection of the *Epithilioma papulosum cyprini* (EPC) teleost cell line with four distinct ranaviruses, FV3, European catfish virus (ECV), doctor fish virus (DFV), and epizootic haematopoietic necrosis virus (EHNV), resulted in distinct inflammatory gene expression profiles (Holopainen et al. 2012). Specifically, EHNV and FV3 elicited expression of the hallmark pro-inflammatory genes, tumor necrosis factor-alpha (TNFα) and interleukin-1-beta (IL-1β), whereas ECV and DFV induced the transient expression of a generally immunosuppressive gene, encoding transforming growth factor-beta (TGFβ) (Holopainen et al. 2012). Interestingly, all four viruses elicited expression of apoptotic components and β2-microglobulin. The latter is critical for surface major histocompatibility complex (MHC) class I expression and cytotoxic T cell function, suggesting that FV3 infections of teleosts may elicit adaptive immune responses.

Consistent with the notion that ectothermic hosts mount broad inflammatory responses to ranaviruses, a comprehensive microarray analysis of axolotls (*Ambystoma mexicanum*) infected with Ambystoma tigrinum virus (ATV) revealed the upregulation of numerous hallmark pro-inflammatory and innate immune gene components in the spleens and lungs of these animals (Cotter et al. 2008). These genes included (but were not limited to) phagocytic receptors and intracellular components, cytokine signaling molecules, complement components, NADPH oxidase subunits (myeloid enzyme catalyzing the reactive oxygen antimicrobial response), and myloperoxidase (granulocyte/monocyte enzyme catalyzing the production of hypochlorous acid and other reactive radical species) (Cotter et al. 2008). In contrast to what has been observed in *X. laevis* infected with FV3 (Morales et al. 2010;

Morales and Robert 2007), no lymphocyte proliferation genes were upregulated in response to ATV infections (Cotter et al. 2008). This lack of an efficient adaptive response in this species may explain why ATV is so lethal to urodeles. Alternatively, these observations may reflect effective ATV immune evasion.

A recent study examined the capacities of hellbender (*Cryptobranchus allegani-ensis alleganiensis*) skin mucus secretions to inactivate a frog virus 3-like ranavirus and the *Batrachochytrium dendrobatidis* (*Bd*) chytrid fungus (Cusaac et al. 2021). While these hellbender mucosomes had no effect on *Bd*, they resulted in a 40% inactivation of the ranavirus, suggesting the presence of antiviral components in hellbender skin secretions. Notably, skin secretions collected from hellbenders and enriched for small cationic peptides inhibited growth of *Bd* and *B. salamandriv-orans* (Hardman et al. 2023). Future studies using disparate approaches and examining distinct components of amphibian skin secretions should grant greater insights into both the antifungal and antiviral properties of distinct amphibian skin products.

The Chinese giant salamander iridovirus (GSIV) is another ranavirus that has emerged as an important pathogen infecting the Chinese giant salamander (*Andrias davidianus*; the world's largest living amphibian), causing high mortality and severe economic losses in aquaculture (Dong et al. 2011; Geng et al. 2011; Lu et al. 2020). The acute decline and economic importance of *A. davidianus* has stimulated research on the immune system and antiviral immune responses in this species (reviewed in (Jiang et al. 2021)). Notably and in contrast to the general perception that salamander susceptibility to ranaviruses is due to sub-optimal immune responses, studies in *A. davidianus* have revealed that these animals possess a robust arsenal of innate immune antiviral factors (Meng et al. 2022; Wang et al. 2020; Xu et al. 2020). Moreover, they also appear to possess a well-developed adaptive immune arm, capable of mounting effective adaptive T and B cell immune responses (Jiang et al. 2018). This is best demonstrated by successful vaccination with self-assembled recombinant GSIV capsid produced in a yeast expression system (Chen et al. 2018; Zhou et al. 2015). Immunization with these virus-like particles gener-ates long-lasting neutralizing antibodies, presumably in a T cell-dependent manner; providing over 50% protection from GSIV infections (Chen et al. 2018). These studies suggest that *A. davidianus* susceptibility may be due in large part to GSIV virulence and immune evasion factors rather than generally ineffective host responses.

3 Infections of Anuran Amphibians and Innate Immune Responses to Ranaviruses

3.1 Antimicrobial Peptide Responses to Ranaviral Infection

Antimicrobial peptides (AMPs) are an important component of anuran (frogs and toads) and urodele (salamanders and newts) innate immune defenses, providing protection to skin mucosa against a variety of pathogens. These small molecules are synthesized and stored in cutaneous granular glands and secreted into mucus in response to stress or injury (Rollins-Smith 2009; Rollins-Smith et al. 2005). AMPs are thought to contribute to amphibian defenses against ranaviruses. For example, esculentin-2P (E2P) and ranatuerin-2P (R2P), two AMPs isolated from *Rana pipiens,* are capable of inactivating both FV3 and channel catfish virus (CCV) within minutes and at temperatures as low as 0 °C (Chinchar et al. 2001). This suggests that direct interaction of these peptides with this virus, rather than inhibition of viral replication is responsible for the antiviral protection. The capacities of AMPs to mediate their effects across a broad range of temperatures presumably reflects the ectothermic nature of the amphibian hosts generating them. Notably, 50 μM of E2P or R2P was sufficient for 99% inactivation of CCV, whereas a ten times greater concentration of either peptide was necessary to achieve 90% inactivation of FV3 (Chinchar et al. 2001). It was postulated that the greater resistance of FV3 to AMP-mediated inactivation reflects the difficulty of antimicrobial peptides to target the inner lipid membrane beneath the FV3 capsid. Presumably, this inner membrane requires disruption for viral inactivation to occur.

Other antimicrobial peptides, including Ranatuerin-2YJ, Dybowskin-YJb, Dybowskin-YJa, Temperin-YJa, and Temperin-YJb have been identified and cloned from the skin of *Rana dybowskii* infected with Rana grylio virus (RGV) (Yang et al. 2012). Interestingly, all these peptides conferred concentration-dependent inhibition of RGV plaque formation, while viral clearance coincided with increased expression of these genes (Yang et al. 2012).

3.2   *Proinflammatory Responses to Ranavirus Infections*

Ranaviral infections are widely associated with prominent host inflammatory responses. Indeed, akin to mammalian viral infections, ranavirus-elicited inflammatory responses represent a double-edged sword as they are both critical for viral clearance but may also exacerbate ranavirus-mediated disease and adversely affect host survival. As it stands, there is substantial documentation of innate immune responses and associated inflammation to ranavirus infections across a range of poikilothermic host species (Carey et al. 1999; Chen and Robert 2011; Grayfer et al. 2014; Jancovich and Jacobs 2011; Morales et al. 2010).

The African clawed frog, *Xenopus laevis,* is the foremost model of amphibian immunology and has served as the leading research platform for investigating amphibian host-ranavirus interactions, using FV3 as the prototypic ranavirus. The majority of the in vivo *X. laevis* FV3 infection studies have relied on intraperitoneal (IP) infections, which are thought to be comparable to the more physiologically relevant water bath exposures (Robert et al. 2011). While the former tends to be more convenient and consistent for immunological studies (Morales et al. 2010), it should be noted that amphibian skin and gastro-intestinal tract both represent important immunological barriers and FV3 target sites (see below). Using an intraperitoneal inoculation infection model, Morales et al. (2010) delineated the sequential progression of the innate and adaptive immune responses of adult *X. laevis* throughout the course of FV3 infection. In *X. laevis* adults, histochemical and flow cytometric analyses revealed that activated mononuclear and polymorphonuclear phagocytes are recruited to, and heavily represented within peritoneal exudates as early as one day following infection (Morales et al. 2010). This work also noted peritoneal recruitment and accumulation of natural killer (NK) cells by 3 days after infection, whereas lymphocyte recruitment, including the increased presence of T cells, occurred later to peak at 6 days post-FV3 challenge (Morales et al. 2010). Notably, the rapid accumulation of peritoneal leukocytes coincided with substantially elevated inflammatory gene expression. In particular, significant increases in the expression of the proinflammatory TNF-α cytokine gene as early as 1 day post infection and persisting to 3 days after FV3 exposure were observed (Morales et al. 2010). Expression of the IL-1β gene, encoding another proinflammatory cytokine, was elevated at days 1 through 6 of FV3 challenge, while the anti-inflammatory arginase-1 (Arg-1), a marker associated with alternatively polarized (M2) macrophages (Joerink et al. 2006b, c), was elevated at day 1 post-viral challenge and subsequently decreased (Morales et al. 2010). Together these findings suggest an effective and well-coordinated frog antiviral immune response, with sequential recruitment of innate and adaptive immune cell effectors and corresponding immune gene activation. The elevated level of Arg-1 gene expression at 1-day post-infection may be reflective of resident, rather than recruited inflammatory myeloid populations. Indeed, after FV3 peritoneal inoculation, we consistently observed elevated mRNA transcripts for macrophage and granulocyte colony-stimulating factor receptors (CSF1R and CSF3R, respectively), indicative of accumulating myeloid infiltrates (unpublished observation). Notably, the elevated expression of CSF1R (and CSF3R) within peritoneal leukocytes (PLs) is typically accompanied by significantly increased expression of the M1 macrophage marker, inducible nitric oxide synthase (iNOS), which catalyzes the production of the antimicrobial nitric oxide by inflammatory macrophages (L. Grayfer, F.D.J. Andino and J. Robert, personal observations). This supports the observation of decreased Arg-1 expression with the onset of an inflammatory state within the peritoneum and indicates that Arg-1 and iNOS functions are antithetic across multiple groups of vertebrates (Joerink et al. 2006a, b, c; Wiegertjes and Forlenza 2010).

Earlier research suggested that compared to adult frogs, *X. laevis* tadpoles possess less robust and delayed proinflammatory responses to FV3 (De Jesus Andino et al. 2012). This was marked by FV3-infected tadpole peritoneal leukocytes, splenocytes, and kidney (major site of viral replication) exhibiting much less pronounced and considerably delayed expression of hallmark inflammatory genes (TNFα, IL-1β, IFNγ). It should be noted that Xenopus tadpole baseline mRNA expression of TNFα is greater than that seen in adult frogs, although it remains unclear if this corresponds to functional differences. Nevertheless, successful immunological control of viral infections typically depends on antiviral interferon (IFN) responses (see Sects. 4.1 and 4.2) rather than inflammatory mediators, which in some instances may exacerbate viral infections (see Sect. 3.3). A greater understanding of amphibian immune defenses will provide clearer perspectives into the relative importance of these immune components throughout infections and at different developmental stages.

Amphibian susceptibility to ranaviruses varies considerably among species, their respective stages of development, and even between different populations of the same species (Miller et al. 2011). These differences likely result from multiple complex determinants including host and ranavirus genetic variability, epigenetic control, or resistance (see below) as well as the respective host immune status. Extensive immunological studies of *X. laevis* suggest that tadpoles rely on distinct immune responses compared to adult frogs. The tadpole adaptive immune system appears to be less developed than that of adult frogs (e.g., more limited T cell and antibody responses than adult frogs) and for some time it was assumed that this explained why pre-metamorphic and metamorphic animals are more likely to succumb to ranavirus infections (Bayley et al. 2013; Grayfer et al. 2014; Hoverman et al. 2010; Reeve et al. 2013). This notion, however, was not universally accepted, as several reports indicated that at least some metamorphic and post-metamorphic amphibian species are more susceptible to ranaviruses than their respective larval forms. Concurrently, it has become more apparent that tadpoles do mount effective innate and adaptive antiviral responses, albeit distinct from those engaged by their adult counterparts (Grayfer et al. 2014; Hauser et al. 2021; Wendel et al. 2017, 2018).

The molecular mechanisms controlling these development-associated susceptibility differences are only now starting to emerge (see Sects. 4.3 and 4.4). We will revisit this concept later in this chapter but would like to underline here that for some time now we have observed that irrespective of infection route, FV3-infected *X. laevis* tadpoles bear lower viral loads than adult frogs (Grayfer et al. 2014; Kalia et al. 2022; Wendel et al. 2017), contradicting the previous belief that larval amphibians are less effective at controlling these viral infections.

3.3 Aberrant Inflammation and Ranavirus-Mediated Pathology

An appropriate, timely resolution of inflammation is just as important as the induction and progression of this response because a prolonged inflammatory response increases the risk of tissue damage and ultimately host death (Fullerton et al. 2013). Although sparse, there is evidence suggesting that ranavirus infections may exacerbate inflammatory responses, accounting for some of the observed ranavirus pathology. For example, in 1997 a novel iridovirus was isolated in Saskatchewan, Canada from larval tiger salamanders (Bollinger et al. 1999). These animals suffered from exacerbated inflammation, necrosis, and characteristic ranavirus-induced cytoplasmic inclusions within splenic, renal, lymphoid, and hematopoietic tissues (Bollinger et al. 1999). Similarly, whole populations of ranavirus-infected, green-striped tree dragons (*Japalura splendidum*) exhibited systemic hemorrhaging, necrosis, granulomatous, and necrotic inflammation, as well as severe renal pathology, hyperanemia, and extensive hepatic damage, culminating in mass mortality. Ranavirus infection of pythons suggests that inflammation may be a determinant of ranaviral pathology (Hyatt et al. 2002). Mortality of largemouth bass infected intraperitoneally with largemouth bass virus (LMBV) is believed to result from virally-induced inflammation and associated necrosis (Zilberg et al. 2000). Consistent with these inflammatory symptoms, juvenile bass inoculated with LMBV exhibited corkscrew swimming and distended abdomens (Zilberg et al. 2000). Notably, the deeper tissues of infected fish were unaffected, bringing into question whether virus-induced damage was due to target cell accessibility or the limitations of LMBV cell tropism. The latter suggests that inflammation and necrotic damage resulting in mortality may be due to primary injuries at the initial sites of infection. Indeed, the above observations are reminiscent of earlier studies of FV3 infections in rodents (Gut et al. 1981; Kirn et al. 1980, 1982) in which, despite the inability of FV3 to replicate at 37 °C (Aubertin et al. 1973), the initial viral inoculum was responsible for extensive inflammation, necrosis, and liver damage.

Mindful of the idea that *X. laevis* adults presumably mount effective anti-ranaviral responses leading to viral clearance, Grayfer et al. (2014) were intrigued to find that at least during acute infections, adult frogs possessed significantly greater FV3 loads (1–2 orders of magnitude higher) than tadpoles, which are thought to be more susceptible to FV3 infection. Notably, a temperature dependency on viral loads was observed in wood frog tadpoles. Specifically infected animals housed at 25 °C succumbed to infection more rapidly and displayed greater viral loads than animals maintained at 15 °C (J. Chaney and M. Gray, personal observations). These results suggest that FV3 virulence is not strictly dependent on the magnitude of viral replication or on viral loads. Additionally, immunocompetent tadpoles may be more vulnerable to ranaviral virulence factors and other environmental parameters than adults. In support of this hypothesis, although tadpoles pre-stimulated with recombinant *X. laevis* type I interferon (rXlIFN I, see Sect. 4.2) possessed viral loads several logs lower than adults, they nonetheless succumbed to FV3 infection (Grayfer et al. 2014). Furthermore, despite lower FV3 loads, these IFN-treated tadpoles experienced damage to multiple organs, including extensive loss of tissue architecture and cellular organization through necrosis and apoptosis, albeit without extensive leukocyte infiltration (Grayfer et al. 2014). Therefore, even at markedly reduced viral loads, ranaviruses may confer irreversible tissue damage in tadpoles relatively early in infection, resulting primarily from virus-mediated cytopathology rather than from viral replication. Indeed, as described by the rodent models of FV3 (see Sect. 5.1), ranaviruses may trigger toxic and potentially lethal effects, irrespective of their capacity to replicate within their host cells (Gendrault et al. 1981). This notion has recently been substantiated by grouper iridovirus (GIV) studies. Replication-deficient, UV-inactivated GIV induced apoptosis in two of the three infected cell lines (Pham et al. 2012). Similarly, heat- and UV-inactivated FV3 elicits FHM cell apoptosis and inhibits host RNA and protein synthesis (Chinchar et al. 2003; Raghow and Granoff 1979). Based on these reports, we hypothesize that inoculation of animals with sufficient inactivated virus will induce toxicity in the absence of virus replication. If this hypothesis is correct for other members of the genus and family, we may need to consider that these viruses are more pathogenic than previously thought.

4 Antiviral Immune Responses to Ranavirus Infections

4.1 Antiviral Interferons of Ectothermic Vertebrates

The IFN response provides a significant contribution to antiviral immunity. IFN responses generally arise as the result of recognition of viral products through an array of host pathogen recognition receptors (PRRs), including toll-like receptors (TLRs), retinoic acid-inducible gene 1-(RIG-I)-like receptors (RLRs), and cytosolic DNA sensors (Baum and Garcia-Sastre 2010; Sadler and Williams 2008). This branch of antiviral immunity consists of four classes of IFN cytokines, type I, II, and recently designated III and IV IFNs (Sadler and Williams 2008). IFNγ, the only type II IFN of mammals (bony fish possess multiple type II IFNs (Grayfer et al. 2010)), plays multiple immune and antiviral roles, whereas IFN-I and IFN-III (and probably the understudied IFN-IV, in animals that encode this cytokine) function predominantly as antiviral molecules. Mammalian IFN-I possesses broad cellular specificities, whereas IFN-III targets specific cell subsets, such as at mucosal barriers (Levraud et al. 2007; Zou et al. 2007). While the distinct receptor systems utilized by different IFN types dictate cell specificity, IFN-I and –III types activate overlapping downstream Janus kinase (JAK) and Signal Transducer and Activator of Transcription (STAT) signaling pathways, culminating in similar antiviral outcomes (Sadler and Williams 2008) including the induction of a gamut of antiviral genes, known as IFN-stimulated genes (ISGs).

Mammalian IFN responses are well studied whereas ectothermic vertebrate species such as those subject to ranavirus infection possess unique and much less understood IFN systems. Higher vertebrates such as reptiles, birds, and mammals encode intronless type I IFNs and type III IFNs with five exon/four intron organization (Robertsen 2006; Zou and Secombes 2011). Bony fish exclusively encode type I IFNs bearing five exon/four intron organization (Chang et al. 2009; Qi et al. 2010; Robertsen 2006; Zou and Secombes 2011; Zou et al. 2007) whereas cartilaginous fish (Redmond et al. 2019) and amphibians (Tian et al. 2019) encode both intronless and intron-containing type I and type III IFN genes.

Bony fish IFNs are subdivided into two groups (group I: 2C; group II: 4C) based on cysteine patterns (Sun et al. 2009; Zou et al. 2007), and further classified into four groups (IFNa-d) according to phylogeny (Chang et al. 2009; Sun et al. 2009). Importantly, while multiple distinct mammalian IFNs confer their biological roles through the same receptor complex (Li et al. 2008; Samuel 2001), fish group I and II IFNs signal through unique receptor complexes (Aggad et al. 2009). Functional studies have been performed predominantly on those group I fish IFNs (Aggad et al. 2009; Altmann et al. 2003; Long et al. 2004; Lopez-Munoz et al. 2009; Robertsen et al. 2003; Zou et al. 2007), and it has been demonstrated that these IFNs differ in their capacities to establish cellular antiviral states (Aggad et al. 2009; Levraud et al. 2007; Li et al. 2010; Lopez-Munoz et al. 2009). For example, salmonid IFNs a-d possess different transcriptional regulation patterns and distinct antiviral functions, as some of these cytokines elicit potent antiviral responses while others are believed not to have antiviral functions at all (Svingerud et al. 2012). The type II IFN systems of amphibians and reptiles remain largely uncharacterized, whereas those of bony fish appear to be much more complex than that of mammals (Zou and Secombes 2011), and will not be addressed further here.

The mammalian IFN-III complex is comprised of interferon lambda (IFNλ) -1, -2, and -3 (also designated as IL-28A, IL-28B, and IL-29). These molecules are encoded by five exon/four intron gene transcripts and signal through a receptor system composed of the interferon lambda receptor-1 (IFNλR1) and interleukin-10 receptor-2 (IL-10R2; reviewed in reference (Kotenko 2011)). While bona fide type III IFNs either do not exist, or have not yet been identified in bony fish, amphibians possess both type I IFNs with the same five exon/four intron gene organization as their fish counterparts, as well as true type III IFNs (Qi et al. 2010). This is especially relevant when considering that amphibians are key evolutionary intermediates between fish and mammals and inhabit both aquatic and terrestrial habitats. In fact, a hallmark characteristic of fish and amphibian type I IFNs is the five exon/four intron genomic organization, which is distinct from the reptile, avian, and mammalian intronless type I IFNs (Robertsen 2006; Robertsen et al. 2003; Sun et al. 2009). Critical annotation of published Xenopus genome assemblies permitted a greater perspective into the underestimated complexity of the amphibian IFN system. Studies demonstrated that *Xenopus* genomes contain an IFN complex with intronless and intron-containing type I and III IFN types, with these respective genes appearing to undergo active expansion and positive or purifying selection (Krause 2016; Sang et al. 2016). In two Xenopus species, *X. tropicalis* and *X. laevis*, 37 and 26 intronless IFN-like genes (Xt/XlIFNX of either IFN-I or -III) were characterized, respectively (Sang et al. 2016; Shields et al. 2019; Tian et al. 2019). Several of the intronless IFN genes include a short intron of about 50 bp, which supports the hypothesis that those short amphibian intron-containing IFN genes represent intermediates of retro(trans)position events leading to intronless IFN genes. Hence, the updated repertoire of IFN gene loci in *X. tropicalis* is comprised of 14 intron-containing IFNs (7 IFN-I, 1 IFN-II, and 6 IFN-III) and the expansion of 37 intronless IFN-like genes (36 IFN-I and 1 IFN-III). *X. laevis* encode 18 intron-containing (7 IFN-I, 1 IFN-II, 9 IFN-III, and 1 IFN-IV) and 24 intronless IFN genes (22 IFN-I and 2 IFN-III) (Adeyemi et al. 2023; Chen et al. 2022a). To date, the *Xenopus* IFN complex represents the highest molecular diversity of the IFN gene composition across all examined vertebrate species, which include all six molecular types (i.e., intron-containing and intronless IFN-I/-III and intron-containing IFN-II/-IV; see below about IFN-IV).

Cross-species and genome-wide comparisons of IFN gene compositions in other amphibian species illustrate the molecular diversity of the *Xenopus* IFN complex, indicating a near chaotic diversification in either redundancy or expansion of antiviral IFN-I and IFN-III types. This diversity includes molecular redundancy of IFN-I genes, having only 1–2 intron-containing IFN-III genes in the two examined caecilian species. There appears to be molecular redundancy of IFN-I genes with expansion of intron-containing or intronless IFN-III in two frog species, respectively.

There is moderate expansion (>3 fold than the IFN-coding gene numbers in the ancestral fish) of both intron-containing and intronless IFN-I and IFN-III gene subtypes in ten anuran species while five anuran species exhibit rather extensive IFN expansions (>10 fold). Notably, recent findings indicate the presence of a IFN-IV gene in *Xenopus,* the product of which confers IFN-I/III-like antiviral responses (Adeyemi et al. 2023; Chen et al. 2022b). This diversity suggests that the evolution of amphibian IFN genes, especially the intronless IFNs, seem to happen in a species/clade-independent manner, perhaps resulting from environment-specific pressures (Robertsen 2006; Robertsen et al. 2003; Sun et al. 2009). Previously, there has been substantial debate regarding the precise phylogenetic relationship of fish IFN-I to higher vertebrate IFN-I and -III. Fish cytokines exhibit exon/intron gene organization similar to that of mammalian type III IFNs, yet possess hallmarks of higher vertebrate type I IFNs such as conserved cysteine positioning and (with the exception of the catfish IFN-I) a C-terminal CAWE motif, a conserved sequence motif found within nearly all IFNs (Lutfalla et al. 2003; Qi et al. 2010; Robertsen 2006; Zou et al. 2007). It will be interesting to learn what are the respective roles of these molecules in fish and amphibian antiviral immunity to ranaviruses, particularly considering that fish appear to only have type I IFNs, whereas frogs generally possess intron-containing and intronless forms of both IFN types I and III (Qi et al. 2010).

Comprehensive in silico comparisons of intronless and intron-containing type I IFN genes encoded by the Tibetan frog *Nanorana parkeri* to those expressed by *X. laevis* led to the proposition that the intronless IFN genes of these two amphibian species arose from independent retroposition events (Gan et al. 2018). In turn, this suggests that intronless type I IFNs encoded by amniotes likely also arose from a distinct retroposition event to those that gave rise to these amphibian intronless IFNs. Moreover, recombinant forms of these *N. parkeri* IFNs elicited robust ISG expression and anti-FV3 protection in the *X. laevis* kidney-derived A6 cell line, but only when the cell line was expressing the *N. parkeri* IFNAR1. This further underlines the distinct evolutionary origins and species-specific IFN ligand-receptor interactions of the *X. laevis* and *N. parkeri* intronless IFNs. This study calls for caution about interpreting comparative research that relies on chimeric cross-species functional studies without taking into consideration possible divergent evolution of species-specific protein-protein interactions.

It was recently discovered that cartilaginous and bony fishes, amphibians, reptiles, birds, and monotremes, but not therian mammals, encode a type IV IFN, IFNυ (Chen et al. 2022b). This cytokine is encoded by a 6-exon (5 coding exons) transcript and ligates a heteromeric receptor complex comprising of an IFNυ-specific IFNUR1 and a shared IL10R2 (Chen et al. 2022b). While the zebrafish version of IFNυ confers antiviral protection, the type IV IFNs of other vertebrates, including amphibians, await characterization (Chen et al. 2022b). Genome-wide annotation across amphibian species indicates that IFN-IV gene homologs exist in three *Xenopus* species, although only *X. laevis* encodes what appears to be a functional IFN-IV gene, with the other two examined amphibian species possessing what presently appear to be pseudogenes, which may imply functional redundancies between IFN-IV with IFN-I and/or IFN-III, thus permitting the loss of the former (72).

4.2 *Interferon Responses to Ranavirus Infections*

As described above, an important antiviral gene product synthesized during the interferon response is the Myxovirus resistance (Mx) protein (Samuel 2001). Mx proteins are believed to be pivotal in the establishment of the antiviral state conferred by IFN (Samuel 2001). Mx proteins are high molecular weight GTPases belonging to the dynamin superfamily and are known to facilitate intracellular membrane remodeling as well as intracellular trafficking (Kochs et al. 2005). As in mammals, teleost Mx proteins function as antiviral mediators, with distinct Mx isoforms from different species conferring somewhat unique antiviral effects. To date the Mx of most, but not all fish species have proven ineffective in preventing infection by various members of the family *Iridoviridae*. For example, Japanese flounder Mx is capable of inhibiting the replication of two species of rhabdovirus, but is incapable of inhibiting replication of red seabream iridovirus (RSIV, genus *Megalocytivirus*; family *Iridoviridae*) (Caipang et al. 2003). Similarly, Barramundi Mx inhibits replication of the nodavirus, viral nervous necrosis virus (VNNV) and of infectious pancreatic necrosis virus (IPNV) but fails to show antiviral effects against Taiwan grouper iridovirus (TGIV) (Wu et al. 2012; Wu and Chi 2007). Likewise, Senegalese sole Mx confers antiviral effects against IPNV and viral hemorrhagic septicemia virus (VHSV, family *Rhabdoviridae)*, but not against European sheatfish virus (ESV) ranavirus (Alvarez-Torres et al. 2013). Finally, rainbow trout Mx1 is antiviral towards IPNV, salmonid alpha virus (SAV, *Togaviridae)*, and infectious hematopoetic necrosis virus (IHNV, *Rhabdoviridae)*, but is not effective at blocking replication of the ranavirus EHNV (Lester et al. 2012; Trobridge et al. 1997). Possibly, the host antiviral responses coevolved with local ranaviral isolates. Thus, the inadequacy of antiviral components such as Mx1 in dealing with foreign ranaviral isolates may culminate in a global threat represented by geographically distant ranavirus strains introduced by sub-clinically infected hosts naturally migrating or being imported by international trade.

Perhaps the most commercially and aquaculturally important fish species in southern Europe is the gilthead seabream, at least in part because of its natural resistance to most viral pathogens (Cano et al. 2006; Cano et al. 2009). In fact, the only viral disease affecting commercial seabream populations is lymphocystis disease virus (LCDV, genus *Lymphocystivirus*, family *Iridoviridae*) (Leiva-Rebollo et al. 2021). Interestingly, seabream possess at least three Mx proteins. One Mx isoform effectively inhibits replication of VHSV and LCDV, a second Mx molecule effectively inhibits replication of ESV and LCDV, and the third is protective against

VHSV (Alvarez-Torres et al. 2013; Fernandez-Trujillo et al. 2013). This represents the first example of a teleost Mx molecule effectively inhibiting DNA virus infection. This is interesting considering that LCDV nonetheless plagues this species. It is noteworthy that in contrast to the mortality caused by many members of the family *Iridoviridae*, seabream effectively clear LCDV infections, although it is believed that they may asymptomatically harbor the virus. Thus, the efficacy of the teleost IFN/Mx response may well dictate the susceptibility of individual fish species to highly virulent pathogens such as iridoviruses. Notably, many fish species are infected by, and clear LCDV. Since these infections involve fish skin (Leibovitz 1980), systemic antiviral responses such as Mx may be less important to the resolution of LCDV.

Japanese flounder IFN-inducible transmembrane (IFITM) protein is upregulated in response to Rana grylio virus (RGV) infection (Zhu et al. 2013). Furthermore, thanks to overexpression and siRNA knockdown studies, flounder IFITM1 has been shown to play an important role in the cellular antiviral response to RGV (Zhu et al. 2013). IFITM1 functions by suppressing viral-host cell entry and targeting the Golgi apparatus (Zhu et al. 2013).

As further research on these individual antiviral components is conducted, it will likely become more apparent that there are key, previously unknown factors that participate in antiviral response of all vertebrate species, not just ectotherms. We propose that these individual IFN-elicited antiviral components are most likely interdependent on entire networks of other IFN-regulated molecules. Thus, the relative potency of the antiviral IFN response, at both cellular and whole organism levels, presumably relies on the balance of numerous cellular and molecular components. Since different fish and amphibian species are now known to possess very distinct repertoires of antiviral effector molecules, it is not surprising that these disparate organisms display very different susceptibilities to similar pathogens.

Microarray analysis of axolotls infected with ATV revealed that in addition to a multifaceted inflammatory gene response, these animals also upregulate expression of multiple antiviral interferon responsive genes (Cotter et al. 2008). Among the numerous genes elicited by ranavirus infection were Mx1 genes, antiviral helicases, interferon regulatory factors, an IFITM, and a ribonuclease (Cotter et al. 2008). However, the genes encoding axolotl type I and type III IFN remain to be identified. It will be important to delineate the precise repertoire(s) of antiviral IFNs present within the axolotl genome and examine the transcriptional regulation, as well as functional roles of these moieties during immune responses against ranaviruses such as ATV.

Microarray analyses of fathead minnow (FHM) cells challenged with either wild type (WT) FV3 or an FV3 knock-out (KO) mutant lacking the truncated vIF-2α gene have been reported (Cheng et al. 2014). Infection with WT FV3 resulted in the upregulation of numerous immune related genes by 8 hours post infection including interleukin (IL)-8, type I interferon (IFN), IFN regulatory factor (IRF)—1, -2, and -3, and IL-1β, amongst others. For the most part, similar genes were upregulated in cells infected with the FV3 KO mutant, but the magnitude of the induction was generally lower (Cheng et al. 2014).

4.3 Tissue- and Developmental Stage-Specific IFN Responses to Ranavirus Infections

Before the discovery that amphibians encode highly expanded repertoires of intron and intronless type I and type III IFNs (see Sect. 4.1), the functional roles of a type I (now known as IFN7) and a type III (now known as IFNL3) IFN was examined in the context of *X. laevis* FV3 infections. As mentioned above, since tadpoles had previously been assumed to be very permissive to ranavirus infections, it was surprising to discover that FV3-challenged tadpoles mounted timely antiviral IFN responses to kidney (a central site of FV3 replication) infections, which took the form of elevated type III (IFNL3) but not type I IFN (IFN7) gene expression (Grayfer et al. 2015). Conversely, adult frogs responded to FV3 infections by favoring type I over type III IFN responses. While both recombinant (r)IFN7 and rIFNL3 showed potent antiviral activity, rIFN7 conferred greater anti-FV3 protection to infected tadpoles than rIFNL3 (Grayfer et al. 2015). In mammals, type I IFN receptors are more ubiquitously expressed than type III IFN receptors, so if this aspect of vertebrate IFN biology is evolutionarily conserved, the above observation could be explained by frog type I IFNs being able to elicit more systemic antiviral responses than type III IFNs.

The amphibian skin is one of the first points of contact with aquatic pathogens such as ranaviruses. Findings from the Grayfer lab (Wendel et al. 2017, 2018) indicate that akin to results from kidney infection studies (Grayfer et al. 2015), *X. laevis* tadpoles likewise preferentially mount type III IFN responses to skin FV3 infections whereas adult frogs rely on type I IFN responses to deal with skin FV3 infections. Interestingly, subcutaneous injections of rIFN7 or rIFNL3 prior to tadpole FV3 challenge resulted in enhanced, albeit short-lived protection against FV3. Moreover, rIFN7 and rIFNL3 elicit the expression of distinct IFN-stimulated genes (Grayfer et al. 2015), suggesting at least some non-overlapping antiviral roles for these respective immune mediators.

In addition to skin, amphibian gastrointestinal tract represents a major target of FV3, often leading to hemorrhaging and necrosis within this tissue (Robert et al. 2011). As discusses above, amphibian (*X. laevis*) tadpoles and adults exhibit distinct immune responses and capacities to control tissue specific FV3 infections. In turn, antiviral interferon (IFN) cytokine (type I and III) responses represent a cornerstone of vertebrate antiviral immunity, and recent findings indicate that FV3-infected tadpoles generate more robust intestinal type I and III IFN responses than adult frogs (Hauser et al. 2021). These IFN responses appear to be mediated by tadpole myeloid cells that are recruited into FV3-infected intestines, presumably in response to these viral infections. These or similar myeloid cell type(s) populate *X. laevis* intestines during metamorphosis and reside within the intestines of healthy adult frogs, presumably in anticipation of infections. It will be interesting to learn how metamorphosis, changing microbiomes that accompany changing diets and habitats influence the observed differences between tadpole and adult frog immune responses in general, and to ranaviruses in particular.

Tian et al. (2021b) recently examined immune gene expression in developing and adult *X. laevis* at baseline and following challenge with wild type FV3 and mutant FV3, defective for a Caspase Activation and Recruitment Domain (CARD)-containing gene (Δ64R) (Tian et al. 2021a). The examined tissues included the intestine, kidney, liver, muscle, spleen, and thymus from froglets and adult frogs. Almost all IFN ligand and IFN-receptor genes displayed tissue-dependent differential expression (see Sect. 4.1). IFN receptor genes for all three IFN types were more broadly expressed than IFN ligand genes across the examined tissues and at different developmental stages. Only some type I and II IFN receptor genes but none of the IFN genes, were highly expressed in fertilized eggs. Three intronless IFN-I genes (XaIFNX8-10) were highly expressed during early tadpole development, as early as Nieuwkoop and Faber; (NF) stages 25 or 35 (Zahn et al. 2022). Most IFN-I and IFN-III genes were differentially regulated in various organs up to froglets at NF stage 55/56 (X55/56) to adult frog (about 1 year-old). In addition, multiple IFN genes, particularly XaIFN1-7, XaIFNX12/15/16/21/22, and XaIFNL4/L9/LX1/LX2, were upregulated following WT FV3 infections and even more robustly induced by the Δ64R FV3. In addition to IFN genes, differential expression of most critical immune gene families, including those encoding inflammatory cytokines/chemokines, MHC genes, toll-like receptors, transcription factors critically involved in immunomodulation, and lectin genes across different developmental stages and tissue types, were observed. Moreover, WT FV3 but not Δ64R FV3 actively suppressed the expression of multiple IFN gene types and in particularly intronless IFNs.

4.4   Epigenetic Control of Tadpole IFN Responses

A large proportion of higher vertebrate genomes is comprised of endogenous retroviruses (ERVs), which appear to play intimate roles in eliciting the expression of antiviral IFNs, at least in part through the production of dsRNAs (Chiappinelli et al. 2017). Mammalian ERVs are expressed at higher levels during early development but are much more stringently controlled or are entirely silenced in adult tissues by means of epigenetic modifications (Eiden 2008). These silenced ERVs may be reactivated in both mammals and fish by epigenetic modifiers such as DNA methylation inhibitors (DNMTis), resulting in robust IFN responses (Chernyavskaya et al. 2017; Chiappinelli et al. 2017; Roulois et al. 2015). While a handful of putative *X. laevis* ERVs are present in GenBank, Xen-1 is the only ERV characterized to date within *X. laevis* and *X. tropicalis* genomes (Kambol et al. 2003).

As described above, anuran tadpoles were thought to be significantly more susceptible to ranaviruses than adult frogs of the same species (Hoverman et al.

2010; Landsberg et al. 2013; Reeve et al. 2013). However, this notion is challenged by converging observations that FV3-infected *X. laevis* tadpoles bear significantly lower FV3 loads across several tissues including the kidney, an important site of FV3 replication (Grayfer et al. 2015; Hauser et al. 2021; Wendel et al. 2017). It is notable that it takes up to several weeks for experimentally infected *X. laevis* tadpoles to die from FV3 infections, supporting the evidence that tadpoles possess antiviral defenses (Gantress et al. 2003). As discussed above, these defenses take the form (at least in part, see adaptive immunity section) of type III antiviral interferon responses to FV3 and differ from the predominantly type I IFN responses mounted by adult *X. laevis* (Grayfer et al. 2015; Wendel et al. 2017, 2018). This discrepancy was recently resolved by demonstrating that *X. laevis* tadpoles are intrinsically more resistant to FV3 kidney infections than cohort-matched metamorphic or post-metamorphic froglets, with this antiviral resistance being epigenetically conferred by *X. laevis* ERVs (Kalia et al. 2022). These findings indicate that akin to what has been reported in mammals, enhancing *X. laevis* ERV gene expression activates cellular double-stranded RNA-sensing pathways, culminating in increased IFN gene expression and greater anti-FV3 protection of the A6 kidney cell line. Moreover, the Grayfer lab found that large esterase-positive myeloid-lineage cells home to tadpole kidneys via the granulocyte growth/chemotactic factor, colony-stimulating factor-3 (CSF-3, aka granulocyte-colony stimulating factor, G-CSF). In the tadpole kidneys, these cells express high levels of ERVs, which trigger their expression of antiviral IFNs, accounting for the anti-FV3 protection observed within tadpole kidneys. Interestingly, ERV expression and the parallel IFN gene expression are lost with metamorphosis, rendering animals more susceptible to FV3. This loss of ERV/IFN expression explains, at least in part, why FV3-infected tadpoles succumb to infection as they initiate metamorphosis.

It is notable that adult *X. laevis* IL-34-macrophages, which are integral to *X. laevis* antiviral IFN responses (see Sect. 5.3), also possess significantly greater ERV expression than other bone marrow-derived cell types. We hypothesize that ERVs have been similarly coopted across evolution to mitigate antiviral immunity across vertebrate species. Likewise, we believe that viral pathogens have co-evolved with their hosts to overcome such antiviral mechanisms and FV3 is a great example of this. Corticosteroid hormones facilitate tadpole transition into metamorphosis, while FV3 encodes a β-hydroxysteroid dehydrogenase homolog (52L), which is thought to serve as an immune evasion mechanism (De Jesus Andino et al. 2015). Indeed, tadpoles infected with a mutant FV3 defective for this 52L/β-hydroxysteroid dehydrogenase homolog are significantly slower to begin metamorphosis than cohort-matched tadpoles infected with wild type FV3 (Kalia et al. 2022). This suggests that FV3 has coevolved to circumvent anti-viral resistance within the tadpole kidney by promoting metamorphosis and thus diminishing ERV expression and the IFN response.

5 The Complex Roles of Macrophage-Lineage Cells During Ranavirus Infections

5.1 Inferences from Murine Models of FV3 Infection

The involvement of macrophage-lineage cells in ranavirus infections may be inferred from studies conducted over 40 years ago using rodent models of hepatitis (Gut et al. 1981; Kirn et al. 1980, 1982). These early studies revealed that Kupffer cells (liver-resident macrophages) were the principal targets of FV3 infection wherein deaths of FV3-infected animals were linked to the loss of hepatic clearance, culminating in severe hepatitis (Gut et al. 1981). These studies also implicated inflammation as a contributor to FV3-mediated pathology, including extensive leukotriene release by Kupffer cells (Hagmann et al. 1987). Inhibition of leukotriene synthesis in FV3-infected animals dramatically reduced virus elicited hepatic damage (Hagmann et al. 1987), suggesting that pathology was largely due to these animals' inflammatory responses.

FV3 is not a mammalian pathogen and, with the exception of expression of select early genes (Lopez et al. 1986), does not replicate at 37 °C (Aubertin et al. 1973). However, this murine work nonetheless supports the current hypothesis that because of their high phagocytic and endocytic activity, macrophage-lineage cells are integral targets for ranavirus infection. In fact, the inability of these pathogens to replicate at 37 °C may be advantageous for investigating the mechanisms of FV3 cell entry. In cultured rat Kupffer cells, viral particles appeared in phagocytic vacuoles and endocytic compartments promptly following FV3 infection (Gendrault et al. 1981). Moreover, a substantial proportion of FV3 virions that attached to cells displayed viral capsid-host membrane fusion and release of viral core contents into cell cytoplasm (Gendrault et al. 1981). FV3 also readily infects multiple mammalian cell lines including HeLa, RAW, TM4, and BHK, as long as the incubation temperature does not exceed the viral temperature maximum of ~30 °C. These observations suggest that the underlying mechanisms governing ranavirus entry are promiscuous enough to facilitate infection of cells from organisms as evolutionarily diverged as mammals, fish, and amphibians. In line with this reasoning, it is likely that cells of the myeloid lineage serve as ranaviral targets precisely because of their robust uptake of extracellular materials, facilitated by an array of endocytic/phagocytic surface receptors, several of which likely recognize and bind ranaviruses. This notion has been supported by recent studies showing that FV3 exploits class A scavenger receptors for cell entry (Vo et al. 2019). In turn, this feature of vertebrate professional phagocytes may have been targeted as a ranaviral infection strategy and may explain why ranavirus can successfully cross host species boundaries. Furthermore, because ranaviruses cannot replicate at mammalian body temperatures, the above-described literature implies that the pathological events seen in FV3-infected rodents are not the result of full virus replication. Instead, cell death is presumably triggered by preformed lytic factors enclosed within FV3 virions or the expression of early viral gene products (Lopez et al. 1986). Indeed, FV3 infection of mammalian cells induces rapid cellular RNA, DNA, and protein synthesis arrest (Elharrar et al. 1973). Furthermore, factors solubilized from FV3 virions result in cellular toxicity and inhibit host macromolecular synthesis (Aubertin et al. 1973; Kirn et al. 1972). Thus, the extensive viral burdens to which these animals were exposed together with the tissue damage resulting from FV3 toxicity may have together compounded the robust inflammatory responses, which undoubtedly contributed to the observed liver pathology reported in these early studies.

5.2 Amphibian Macrophages and Ranaviral Persistence

Increasing evidence from natural ranavirus infections of amphibians supports the idea that macrophages are important not only for antiviral defense, but also to ranavirus infection strategies. Notably, *X. laevis*-FV3 infection studies indicate that FV3 persists within frogs for several months following the resolution of apparent disease (Gantress et al. 2003; Hossainey et al. 2021b; Robert et al. 2014; Samanta et al. 2021). This suggests that FV3 possesses a strategy for chronically persisting within its amphibian hosts and evading complete immunological clearance. Macrophage-lineage cells are important candidates as reservoirs for persisting FV3, based on numerous observations that FV3 effectively infects frog peritoneal macrophages both in vitro and in vivo (Robert et al. 2007). FV3 persists within these cells for up to 12 days and undergoes active viral gene expression, with nominal changes to the overall pathogen loads within these macrophages (Robert et al. 2007). Transmission electron microscopic analyses of these infected cells confirmed the presence of FV3 virus within these cells, in a presumably infectious form (Morales et al. 2010). It is intuitive that akin to other viral pathogens, FV3 would target terminally differentiated, long-lived immune cell types like macrophages as a means of persistence and immune evasion. In this respect, FV3-macrophage interactions are reminiscent of the HIV-1-macrophage relationship in which viral particles accumulate within these myeloid cells as a mechanism of dissemination to T cells (Coiras et al. 2009; Goodenow et al. 2003; Gousset et al. 2008; Groot et al. 2008). Interestingly, when peritoneal MHC class II$^+$/CSF1R$^+$ macrophages are isolated from previously FV3 infected but asymptomatic frogs and stimulated in culture with a TLR5 ligand, flagellin, active FV3 replication and production of infectious particles is readily initiated (Samanta et al. 2021). This contrasts with minimal viral replication and virion production seen in unstimulated peritoneal macrophages or neutrophils. Other TLR ligands, including bacterial RNA, known to stimulate TLR22, were unable to increase FV3 replication.

5.3 Dichotomous Roles of Frog Macrophages During Frog Virus 3 Infections

The commitment, differentiation, survival, and functionality of macrophage-lineage cells are controlled by the colony-stimulating factor-1 (CSF-1; M-CSF) receptor (Hanington et al. 2007; Pixley and Stanley 2004; Wang et al. 2008), which is ligated by CSF-1 and interleukin-34 (IL-34) (Hashimoto et al. 2011; Lin et al. 2008). Using the *Xenopus laevis* model, it was established that amphibian macrophages differentiated by IL-34 and CSF-1 are morphologically, transcriptionally, and functionally distinct from each other (Grayfer and Robert 2014, 2015; Yaparla et al. 2018, 2021). More importantly, frog IL-34-derived macrophages confer extensive anti-FV3 protection, whereas macrophages differentiated by CSF-1 are highly susceptible to FV3 in vitro and exacerbate FV3 infections in vivo (Grayfer and Robert 2014, 2015; Yaparla et al. 2018, 2021). These findings indicate that the antiviral nature of frog IL-34-macrophages stem from their robust production of type I and type III IFN cytokines (see Sect. 4.1) and heightened expression of several antiviral restriction factors (Yaparla et al. 2018). This is reminiscent of mammalian plasmacytoid dendritic cells (DCs), which are important producers of antiviral IFN cytokines. Converging findings indicate that frog IL-34-macrophages share many features of not only mammalian dendritic cells, but also with frog-derived dendritic cells (Hossainey et al. 2022). These frog CSF-1- and IL-34-derived macrophage populations are equally susceptible to FV3 entry while IL-34-macrophages are highly effective at eliminating a large proportion of infiltrating virus (Grayfer and Robert 2014, 2015; Yaparla et al. 2018). By stark contrast, FV3-infected CSF-1-macrophages appear to be much more susceptible to FV3 replication and establish immuno-suppressive states within the kidneys of infected animals towards viral persistence (Hossainey et al. 2021a). Akin to the functional dichotomy of these frog macrophages, concurrent studies indicate that human peripheral blood monocyte-derived IL-34-macrophages are likewise more resistant to HIV-1 than human CSF-1-macrophages (Paquin-Proulx et al. 2018). It is intriguing to consider what aspects of macrophage biology are evolutionarily converged, and which mechanisms diverged to meet species-specific physiological and immune pressures.

5.4 Chronic and Asymptomatic Ranavirus Infections

While adult anurans (chiefly studied in *X. laevis*) clear primary FV3 infections, residual virus persists within kidney and myeloid cells (Robert et al. 2007, 2014; Samanta et al. 2021), presumably turning these animals into chronic reservoirs for prolonged environmental dissemination. Indeed, persisting asymptomatic infections have been noted for several ranavirus-resistant amphibian species (Bielby et al. 2021; Love et al. 2016; Stohr et al. 2013). The transcriptional status of the persisting FV3 has not been well defined, with lingering questions regarding whether the virus becomes quiescent or maintains some transcriptional activity during such prolonged/chronic infections. Whereas FV3 gene expression could not be detected in peritoneal macrophages of older/larger chronically infected animals using conventional PCR (Robert et al. 2014), viral gene expression was detected in chronically-infected froglets using quantitative PCR (Hossainey et al. 2021a; Yaparla et al. 2021). The latter approach allows for much greater sensitivity and thus permits detection of much fewer viral transcripts. Indeed, FV3 gene expression in chronically infected froglets was several logs lower than that detected during acute infections, and we anticipate these low expression levels would not be detectable by conventional PCR. Of course, it is possible that FV3 becomes quiescent in older, more resistant *X. laevis,* while remaining transcriptionally active in relatively more susceptible juvenile froglets. It is notable that in context of HIV-1 latency, macrophages are relatively more restrictive to viral replication and virus-induced apoptosis, thus becoming notable reservoirs for HIV-1 (Kumar et al. 2014). Perhaps something akin to this is taking place with amphibian FV3-infected macrophages.

Infectious virus has been detected in both chronically infected (i.e., for 3 weeks or more) older animals as well as juvenile *X. laevis* (Hossainey et al. 2021a; Robert et al. 2014). In addition, viral transcription and viral loads vary across experimental individuals and likely vary among cell types (e.g., macrophages versus kidney epithelial cells). It is important to consider that viral latency is a complex, incompletely understood area of active research and that distinct types of viruses undergo disparate types of latency. For example, HIV-1 infects and integrates into the genomes of several leukocyte subsets including $CD4^+$ T cells and $CD4^+$ macrophages, undergoing low level replication in the relatively more susceptible T cells while becoming transcriptionally quiescent in the HIV-1-restrictive macrophages (reviewed in (Darcis et al. 2018; Pagani et al. 2022)). Conversely, during latency of herpesviruses (most thoroughly studied in context of human herpes simplex viruses-1 and -2), circular viral genomes are maintained, unintegrated in host cell nuclei in absence of all viral gene expression except for the non-coding latency-associated transcript (LAT), which is not required for latency initiation but is important for reactivation of active herpesvirus infections (reviewed in (Kennedy et al. 2015)). These latent herpesvirus infections may be reactivated by immune perturbations, resulting in robust virus productions. Akin to these reactivated herpesvirus infections, several studies indicate that chronic, potentially quiescent FV3 infections of asymptomatic frogs are exacerbated, or potentially reactivated by inflammatory stimuli such as heat-killed *E. coli* (Robert et al. 2014) or TLR5 agonists (Samanta et al. 2021). In turn, frogs appear to be less effective at controlling such reactivated FV3 infections, leading to higher death rates of adult frogs than seen with primary infections (Robert et al. 2014). Interestingly, peritoneal CSF1R+ macrophages from chronically FV3-infected frogs do not express the FV3 major capsid protein (a late gene transcript) unless these cells are stimulated with the TLR5 ligand, flagellin (Samanta et al. 2021). Furthermore, enriching frogs for CSF-1-macrophages exacerbates chronic FV3 infections, culminating in greater viral loads at later infection times, more pronounced kidney pathology, and robust immunosuppressive states within infected animal kidneys (Hossainey et al. 2021a). In line with recent observations, we anticipate that FV3 polarizes myeloid subsets such as CSF-1-macrophages towards production of immunosuppressive mediators like interleukin-10 and transforming growth factor-beta (Hossainey et al. 2021a), thereby facilitating immune evasion and chronic persistence in frog tissues like the kidney. How inflammatory pathways are involved in this process and whether specific viral genes control FV3 persistence remains to be elucidated.

FV3 results in robust inflammatory responses, while many other more extensively studied viruses have coopted host inflammatory responses towards their replication strategies (Zhao et al. 2015). Thus, we anticipate that FV3 has likewise evolved to usurp the amphibian host inflammatory components towards their own infection strategies. As such, we propose that inflammatory stimuli may result in the mobilization and/or shifts in FV3-laden frog macrophages away from antiviral states (akin to IL-34-macrophages), thereby resulting in dissemination of the virus and greater replication within the reservoir macrophages (akin to CSF-1-macrophages) and the newly infected cells. Latency is defined as a persisting viral infection with highly restricted viral gene expression and complete absence of infectious viral particle production (Speck and Ganem 2010). As such, we propose that FV3 and presumably other ranaviruses do not undergo true latency but rather persist at low, transcriptionally reduced levels in naturally and virally manipulated, permissive myeloid cells, and possibly as of yet unidentified other immune and/or non-immune cell types. Persisting ranavirus infections have been noted in several distinct asymptomatic, often ranavirus-resistant amphibian populations in the wild (Bielby et al. 2021; Love et al. 2016; Stohr et al. 2013), exhibiting similar quiescent infections involving macrophages, with a risk of increased replication leading to sudden outbreaks (Bielby et al. 2021; Love et al. 2016; Stohr et al. 2013).

5.5 Ranavirus Infections of Macrophages of Other Ectothermic Vertebrates

Like many other pathogens, ranaviruses presumably overcome macrophage antimicrobial barriers, at which point these cells become vehicles for both viral dissemination and persistence. However, exploitation of macrophage-lineage cells as vectors of viral dissemination and persistence does not appear to be confined to FV3, as other members of the genus *Ranavirus* and family *Iridoviridae* have also adopted this mechanism of host infiltration and immune evasion. For example, an iridovirus-like pathogen infects sheatfish kidney macrophages and is capable of down-regulating phorbol myristate acetate-elicited reactive oxygen production by these cells in vitro (Siwicki et al. 1999). Likewise, following infection with Taiwan grouper Iridovirus (TGIV), elevated numbers of phosphatase-positive, highly phagocytic basophilic, and eosinophilic mononuclear leukocytes were detected (Chao et al. 2004). TGIV genomic DNA was found only within the nuclei of mononuclear phagocytes at early times after infection, whereas at later times it was seen in both nuclear and cytosolic compartments concurrent with these cells losing their phago-cytic capacities (Chao et al. 2004). Clearly, TGIV has evolved intricate and tempo-rally regulated strategies for overcoming and utilizing the very immune cells that would presumably be coordinating the antiviral immune response. It is probable that the strategy of invading mononuclear phagocytes as a means of immune eva-sion and dissemination is a strategy shared by all vertebrate iridoviruses. Further development of in vitro primary macrophage cultures derived from relevant host species and infection models will provide additional insight into these infection strategies.

6 Adaptive Immune Responses to Ranavirus Infections

Anti-ranavirus immune responses of ectothermic vertebrates are multifaceted, com-plex, and poorly understood. However, it is becoming evident that clearance of rana-viruses is heavily contingent on successful adaptive immune responses, which have been investigated to date almost exclusively in *X. laevis*.

6.1 *Antibody Responses to Ranavirus Infection*

The amphibian organization and usage of the immunoglobulin (Ig) heavy and light chain loci are reminiscent of their mammalian counterparts, including V-(D)-J rear-rangements, class-switch recombination, somatic hypermutation, and affinity matu-ration (Du Pasquier et al. 1989, 2000; Hsu 1998). As in mammals, the *Xenopus* Ig class-switch from IgM to IgY (the IgG functional analog) is thymus-dependent and requires T cell-B cell collaboration (Blomberg et al. 1980; Turner and Manning 1974). Although affinity maturation of amphibian IgY results in only a ten-fold increase, as compared to the ten thousand-fold increase seen with mammalian IgGs, it has been clearly demonstrated that *Xenopus* humoral immunity is a significant contributing factor to anti-ranaviral immune responses, particularly in adult frogs (Maniero et al. 2006).

Following secondary FV3 infection of *X. laevis* adults, animals produce substan-tial amounts of virus-specific IgY, first detectable 1 week after infection and peak-ing around 3 weeks after viral challenge (Gantress et al. 2003). Indeed, frogs re-infected (in the absence of adjuvant) with FV3 up to 15 months post primary infection, develop anti-FV3 specific IgY antibodies in a thymus-dependent manner, which are detectable from 10 days up to 8 weeks post re-immunization (Maniero et al. 2006). Notably, FV3 is effectively neutralized by exposure to these immune sera in vitro (Maniero et al. 2006). In addition, administration of immune sera to naturally susceptible *X. laevis* tadpoles immediately preceding FV3 infection con-fers partial, but significant passive protection against the virus (Maniero et al. 2006).

Clearly, the amphibian antibody response is integral to the clearance of ranavirus infections, while the extent to which this particular immune facet contributes to the ultimate anti-ranaviral immunity seen in *Xenopus* adults remains to be determined. These results are consistent with findings with red seabream iridiovirus (RSIV, genus *Megalocytivirus*), wherein vaccination with inactivated virions protected fish from subsequent viral challenge (Caipang et al. 2006; Nakajima and Kunita 2005; Nakajima et al. 1999).

6.2 T Cell Responses and Immunological Memory Against Ranavirus Infections

X. laevis tadpoles express suboptimal levels of classical polymorphic MHC class I (class Ia) molecules (Du Pasquier et al. 1989) and yet their splenocytes include bona fide $CD8^+$ T cells that express the pan-T *Xenopus* cell-surface marker CD5 (Jurgens et al. 1995) and exhibit fully rearranged $TCR\alpha/\beta$ transcripts (Horton et al. 1998). It is possible that suboptimal class Ia protein expression in tadpoles is sufficient to facilitate T cell differentiation and selection pathways, distinct from those of post-metamorphic animals. Conversely, tadpole T cell ontogeny and selection may rely more heavily on non-polymorphic nonclassical MHC class I (nonclassical class Ib) molecules (see Sect. 6.3), which are expressed at greater levels in *X. laevis* tadpoles (Edholm et al. 2013). Indeed, in the absence of optimal class Ia-mediated T cell selection, larval $CD8^+$ T cells may possess a more restricted antigen-binding repertoire, possibly reflected in the relative susceptibility of tadpoles to ranaviruses. As discussed later in this section, there are distinct nonclassical class I-mediated T cell selection mechanisms and T cell subsets that may complement conventional class Ia-restricted CD8 T cells in tadpoles.

In contrast to tadpoles, adult *X. laevis* display conventional class Ia-restricted $CD8^+$ cytotoxic T cell populations. Despite the absence of available antibodies, $CD4^+$ T helper cells are also likely present owing to the presence of all genes involved in differentiation and function of $CD4^+$ T cells, the expression of the CD4 gene in $CD8^-/CD5^+$ cells, and the MHC class II-dependent proliferation response obtained by mixed lymphocyte reaction (Du Pasquier et al. 1989). The requirement of T cells for FV3 clearance in *X. laevis* adults has been demonstrated by using sublethal γ-irradiation, which depletes thymus-derived T cells. Irradiated, T cell-depleted adult frogs cannot control FV3 and succumb to infection (Robert et al. 2005). Furthermore, depletion of *X. laevis* $CD8^+$ T cells by administration of an anti-*X. laevis* CD8 monoclonal antibody (Ab) also substantially increases adult frog susceptibility to FV3 infections (Robert et al. 2005). These $CD8^+$ T cell-depleted, FV3-infected animals experienced severe edema and hemorrhaging, extensive elevation of viral loads and succumbed to infections, whereas control cohorts effectively cleared the virus (Robert et al. 2005). Intuitively, cytotoxic $CD8^+$ T cell responses are critical for effective FV3 clearance. Intriguingly, administration of the anti-CD8 Ab to tadpoles did not result in either CD8⁺ T cell depletion or in increased susceptibility to FV3 (Robert et al. 2005), again emphasizing the unconventional nature of the tadpole T cell populations.

Frogs re-infected with FV3 exhibit expedited viral clearance concomitant with earlier proliferation of CD5⁺CD8⁺ splenocytes and faster infiltration (3 vs 6 dpi) of the kidney, a central site of ranavirus replication (Morales and Robert 2007). This not only underlines the importance of CD8⁺ T cells in ranaviral clearance, but also indicates the presence of a T cell memory responses to ranavirus re-infections in adult *X. laevis*. Interestingly, while kidney infiltration by CD8⁺ cells was substantially accelerated upon secondary FV3 challenge compared to primary infections, the numbers of recruited CD8⁺ cells were substantially lower during this second immune event (Morales and Robert 2007). This could be attributed to a higher frequency of T cell precursor infiltration during the primary response and/or the presence of fewer, but more effective, CD8⁺ memory T cells upon re-infection. It cannot be excluded that this modest secondary response is an inherent property of the evolutionarily primordial amphibian adaptive immune system, considering the relatively meager degree of *X. laevis* T cell expansion seen following immunological challenges and the absence of draining lymph nodes (Du Pasquier et al. 1989). Alternatively, this modest secondary CD8⁺ T cell response could be accounted for by the recruitment and immune involvement of additional effector populations during subsequent anti-ranavirus responses. In support of this notion, during the secondary anti-FV3 response there is the rapid and robust recruitment and kidney infiltration of CD8⁻ MHCII⁺ immune populations, which may be B cells, CD4⁺ T cells, or CD8⁻ nonclassical MHC class I-restricted invariant T cell populations (see Sect. 6.3).

The importance of T cell-mediated adaptive immunity in amphibian resistance against ranaviruses is reinforced by immunogenetic studies. For example, variation of MHC-II polymorphism associated with the intensity of ranavirus infection has been characterized in wood frog (*Rana sylvatica*) tadpoles (Savage et al. 2019). In addition, selection of particular MHC-I haplotypes in wild common frog (*Rana temporaria*) populations exposed to ranavirus have been shown, wherein MHC heterozygosity and some groups of alleles were significantly associated with lower ranavirus infection intensities (Teacher et al. 2009). While this clearly implicates a role for T cell immunity in anti-ranaviral protection, it remains to be determined whether and how specific MHC alleles confer susceptibility or resistance to ranaviruses.

6.3 Roles of Non-classical MHC-Restricted T Cells in Ranavirus Immunity

Nonclassical MHC class I molecules exhibit structural similarities to class Ia molecules, but typically possess limited tissue distribution and no, to very limited, polymorphisms (Flajnik and Kasahara 2001). In mammals, some of these surface glycoproteins are involved in the differentiation and functional regulation of distinct subsets of preset or innate-like T (iT) cells, including CD1d-restricted iNKT cells and MR1 restricted mucosal associated iT (MAIT) cells (Bendelac et al. 1997; Harly et al. 2022; Matsuda and Gapin 2005; Mayassi et al. 2021). Both of these lymphocyte populations undergo unconventional differentiation pathways in the thymus, exhibit very limited T cell receptor (TCR) rearrangements, and participate in early antimicrobial and antiviral immune responses (Behar and Porcelli 2007; Choi et al. 2008; Cohen et al. 2009; Le Bourhis et al. 2010).

It is intriguing that although *Xenopus* tadpoles are naturally deficient for the polymorphic MHC-Ia (Du Pasquier et al. 1989), they express a variety of non-classical MHC-I genes (XNCs; now renamed mhc1-uba; (Dimitrakopoulou et al. 2023)). In addition, deep sequencing of T cell receptors (TCR) expressed by splenic T cells in 2 week-old *X. laevis* tadpoles has revealed a very limited repertoire diversity suggesting a preponderant reliance on iT cells during early larval stages (145). Notably, the nonclassical MHC-I XNC10 (mhc1-uba10) that is typically expressed in thymi and spleens (Goyos et al. 2009, 2011) is required for both the development and function of a prominent iT cell subset, critical for host resistance against FV3 (Edholm et al. 2013). Using a recombinant XNC10 tetramer as well as reverse genetics combining transgenesis and RNA interference, it was determined that the iT cell population restricted by XNC10 is CD8$^-$/CD4$^-$, expresses a semi-invariant TCR consisting of an invariant TCRα rearrangement (iVα6-Jα1.43) combined with a limited TCRβ repertoire, and fails to develop in the absence of, or with diminished, XNC10 expression (Banach et al. 2016; Edholm et al. 2013). Notably, transgenic animals with effectively RNAi-silenced thymic and splenic XNC10 expression failed to develop this iT cell subset. Moreover, they were also significantly more susceptible to, and more readily succumbed to FV3 infections (Edholm et al. 2013), indicating that these iVα6T cells are important in tadpole anti-ranaviral defenses. The critical role of iVα6T cells in controlling FV3 infection was further demonstrated by treating tadpoles with XNC10 tetramers at the time of infection, which specifically and transiently depletes these cells in vivo. FV3 infection of XNC10 tetramer-treated tadpoles resulted in marked increases in viral loads and death rates (Edholm et al. 2019). Interestingly, iVα6T cell deficiency induced by both reverse genetic and XNC10-tetramer treatment was accompanied with delayed innate immune responses, suggesting that these iT cells act at least in part by regulating innate immune defenses.

Although the TCR repertoire of conventional T cells is markedly more diversified in adult *X. laevis*, with an estimated 10^{10} different rearrangements (Robert and Edholm 2014), iT cells are also present as minor subsets. Indeed, (Edholm et al. 2015) identified a XNC10-dependent iVα6T cell population in adult frogs and showed that an iVα6T cell defect similarly results in delayed anti-FV3 innate interferon and cytokine gene responses, increased viral load and more kidney tissue damage (Edholm et al. 2015).

It is noteworthy that deep sequencing analysis of tadpole TCRα revealed that *X. laevis* tadpoles possess several additional predominant iT cell populations (Edholm et al. 2013), which are possibly XNC (mhc-uba) restricted and likely participate in immune responses against pathogens. Thus, it stands to reason that during primary and secondary anti-FV3 responses, these lymphocyte subsets may be among the CD8$^-$ kidney infiltrating immune populations (Morales and Robert 2007) discussed above.

7 Impacts of Environmental Pollutants on Amphibian Anti-ranaviral Immunity

Presumably because of the complexities in its development and regulation, the immune system is particularly susceptible to perturbation by water contaminants such as endocrine disrupting chemicals. While water pollution by chemicals such as pesticides has been reported to impact amphibian fitness, physiology (Carrasco et al. 2021; Davidson et al. 2007; Sparling et al. 2015) and development (Boone and Bridges-Britton 2006; Hayes et al. 2006; Relyea and Jones 2009), the potential negative impacts of water contaminants on host immunity and resistance to ranaviruses has been generally overlooked. This is particularly relevant since exposure of mammals to various endocrine disruptor chemicals during early development has induced long-lasting immune defects that often persist into adulthood. Considering that the aquatic development of amphibians directly exposes them to waterborne contaminants, the long-term influence of these pollutants on their immune systems is likely. In turn, these immune perturbations may be significant contributors to amphibian susceptibility to ranavirus infection.

Using *X. laevis* as an experimental model, it has been demonstrated that certain herbicides (atrazine) and insecticides (carbaryl) contaminating water at low, but ecologically relevant concentrations induce dramatic acute and persistent defects in antiviral immune responses (De Jesus Andino et al. 2017; Sifkarovski et al. 2014). Furthermore, exposure of tadpoles to a mixture of endocrine disruptors associated with unconventional oil and gas extraction induce acute immune perturbations which persist into the adult lives of these animals (Robert et al. 2018).

More recently, the effects of chemical pollutants with thyroid disrupting activity have been investigated. Thyroid hormone strictly controls metamorphosis during which the immune system undergoes major remodeling and differentiation (Rollins-Smith et al. 1997; Sachs and Buchholz 2019; Shibata et al. 2020). Thus, this transition is particularly sensitive to environmental perturbation. Exposure of tadpoles to a mixture of 4 chemicals with thyroid-disrupting activity at doses found in the environment, delayed metamorphosis and induced hypertrophy-like pathology in animal thyroid glands (McGuire et al. 2021). Moreover, these exposures resulted in short-term perturbations to animal thyroid hormone axes and to thymocyte differentiation (McGuire et al. 2021). Furthermore, tadpole exposure to these chemicals was detrimental to formation of T cell immunity in post-metamorphic life of exposed animals, manifesting in reduced frequencies of surface MHC-II$^+$ splenic lymphocytes and weakened anti-viral immune responses (McGuire and Robert 2022).

8 The FV3 Transcriptome and Putative Ranavirus Immune Evasion Strategies

Transcriptomes of several tissues from FV3-infected *X. laevis* have been characterized at 1-, 3-, and 6-days post infection (Tian et al. 2019, 2021b). In the intestine, liver, spleen, lung, and especially kidney from infected frogs, FV3-targeted transcriptomic analyses mapped reads covering the full FV3-genome at ~10× depth on both positive and negative strands. By contrast, reads were only mapped to partial genomic regions in the thymus, skin, and muscle samples from the infected animals. Extensive analyses demonstrated the expression of almost all the 98 annotated protein-coding ORFs in the reference FV3 genome (GenBank accession no. NC_005946.1) and their differential expression in a tissue-, virus-, and temporal class-dependent manner. Further analyses identified several putative FV3 ORFs that encode hypothetical proteins containing viral mimics of conserved domains found in IFN regulatory factors (IRFs, including FV3 ORF19R and ORF82R) and IFN receptors (including FV3 ORF59L), indicating that these viral proteins may serve as potential ranaviral mechanisms to interfere with IFN-mediated antiviral signaling (Tian et al. 2019).

Additionally, FV3 transcriptomic analyses detected significant transcription of non-coding intergenic regions next to highly expressed coding genes in FV3 genome. The enrichment of various *cis*-regulatory elements (*CREs*) in these intergenic regions corresponded to transcriptomic profiles of highly expressed coding genes. In silico analyses suggest that these intergenic regions immediately upstream of some highly expressed FV3 genes have potentially evolved to enhance targeting and even silencing of host transcription factors, including IRFs, NF-κB, and STATs that are critical in the host antiviral regulation concentrated on IFN-signaling. Moreover, computational screening of the FV3 genome and matching to transcriptomic analysis allowed the detection of the enrichment of putative viral microRNA (v-miRNA) sequences in more than five intergenic regions along the FV3 genome (Tian et al. 2021b; Todd and Katzenback 2021). Remarkably, an array of these virus-derived miRNAs are predicted to target the 3'-UTR regions of *Xenopus* genes involved in IFN-dependent immune responses, notably those encoding IFN receptor subunits and IFN-regulatory factors. Clearly, these pathogens have extensively co-evolved with, and have adapted to, evade the immune responses of their amphibian hosts. With the refinement of state-of-the-art sequencing and bioinformatic approaches, we stand to gain a much more in-depth understanding of the coevolutionary pathways of amphibian immune systems and ranaviruses.

In addition to the viral mimics and v-miRNAs discussed above, ranaviruses, as well as other members of the family, encode multiple proteins whose function is to inhibit host anti-viral immunity. These viral gene products include, but are likely not limited to, homologs of eIF-2α, semaphorin, tumor necrosis factor receptor (TNFR), beta-hydroxysteroid dehydrogenase (βHSD), caspase activation and recruitment domain (CARD) containing proteins, and Bcl-2-like proteins, as well as viral proteins that degrade STING and TBK1 and inhibit IFN synthesis. Viral immune evasion proteins are discussed more extensively in the chapter of this book, authored by Jancovich et al. (2024, this volume). Furthermore, study of viral immune evasion proteins is important not only because it illustrates the importance of specific host elements in response to ranavirus infection, but because, as shown in studies utilizing KO mutants of putative immune evasion genes, generation of KO mutants might lead to the development of protective vaccines.

9 Concluding Remarks and Future Directions

It is evident from studies described here that anti-ranaviral immunity is multifaceted, complex, as well as being species- and developmental stage-specific. Also evident are the many gaps in our understanding of the immune response to these pathogens as well as possible defects in the host's capacity to mount effective responses that contain and eliminate these infections. As with other DNA viruses such as poxviruses, ranaviruses, and members of other genera within the family *Iridoviridae* have evolved numerous, highly efficient strategies for evading, and even utilizing host immune components, to achieve persistence, facilitate dissemination, and expand host range. Clearly, ranaviruses encode many putative gene products, which represent both potential virulence factors and promising targets for future therapeutic interventions.

While it is easy to dismiss ectothermic vertebrate immune systems as either primitive or simply functionally analogous to those of mammals, there is a growing literature suggesting otherwise. It is through the fundamental understanding of the physiological and ecological pressures governing these unique immune systems that we may begin to comprehend ranavirus infection strategies and the immune systems that may or may not have adequately co-evolved to stop them.

The investigation of ranavirus infection and immune subversion strategies should be approached not only by considering well-defined mammalian pathogens, but also by acknowledging the possibility that ranaviruses may represent unique viral pathogens. In contrast to the majority of endothermic vertebrate pathogens, ranaviruses are extraordinary in their ability to overcome cell and host tropism barriers, while their mechanisms of pathogenicity appear to be at least partially much less dependent on viral burdens. In turn, the immune systems of ectothermic hosts have evolved as the result of and are subject to different physiological and pathogenic pressures than those that have shaped mammalian immune systems. It stands to reason that ranavirus pathogens have co-evolved with these unique immune systems. Thus, we must garner greater insights into both to fully understand either.

Acknowledgments We would like to thank Matthew J. Gray and V. Gregory Chinchar for organizing and coordinating the compilation of this book.

LG was supported by grants from the National Science Foundation, IOS: 1749427, 2131061, and 2147466.

JR was supported by grant of the National Institute of Allergy and Infectious Diseases at the National Institutes of Health R24-AI059830; and the National Science Foundation, IOS: 1456213 and 1754274.

ESE was supported by the Tromsø Research Foundation Starting Grant.

We thank Dr. Louise Rollins-Smith and Dr. Nicholas Cohen, for their critical review and helpful suggestions that have improved this book chapter. We thank the following organizations for providing funds to support Open Access publishing of the second edition: University of Tennessee's (UT) Open Publishing Support Fund, UT School of Natural Resources, UT Center for Wildlife Health, Washington State University Libraries, University of Mississippi Medical Center, Gordon State College, Association of Reptile and Amphibian Veterinarians, and the Global Ranavirus Consortium.

References

Adeyemi OD, Tian Y, Khwatenge CN, Grayfer L, Sang Y (2023) Molecular diversity and functional implication of amphibian interferon complex: remarking immune adaptation in vertebrate evolution. Dev Comp Immunol 140:104624

Aggad D, Mazel M, Boudinot P, Mogensen KE, Hamming OJ, Hartmann R, Kotenko S, Herbomel P, Lutfalla G, Levraud JP (2009) The two groups of zebrafish virus-induced interferons signal via distinct receptors with specific and shared chains. J Immunol 183:3924–3931

Altmann SM, Mellon MT, Distel DL, Kim CH (2003) Molecular and functional analysis of an interferon gene from the zebrafish, Danio rerio. J Virol 77:1992–2002

Alvarez-Torres D, Garcia-Rosado E, Fernandez-Trujillo MA, Bejar J, Alvarez MC, Borrego JJ, Alonso MC (2013) Antiviral specificity of the Solea senegalensis Mx protein constitutively expressed in CHSE-214 cells. Marine Biotechnol (New York, NY) 15:125–132

Aubertin AM, Hirth C, Travo C, Nonnenmacher H, Kirn A (1973) Preparation and properties of an inhibitory extract from frog virus 3 particles. J Virol 11:694–701

Banach M, Edholm ES, Robert J (2016) Exploring the functions of nonclassical MHC class Ib genes in Xenopus laevis by the CRISPR/Cas9 system. Dev Biol 426:261

Baum A, Garcia-Sastre A (2010) Induction of type I interferon by RNA viruses: cellular receptors and their substrates. Amino Acids 38:1283–1299

Bayley AE, Hill BJ, Feist SW (2013) Susceptibility of the European common frog Rana temporaria to a panel of ranavirus isolates from fish and amphibian hosts. Dis Aquat Org 103:171–183

Behar SM, Porcelli SA (2007) CD1-restricted T cells in host defense to infectious diseases. Curr Top Microbiol Immunol 314:215–250

Bendelac A, Rivera MN, Park SH, Roark JH (1997) Mouse CD1-specific NK1 T cells: development, specificity, and function. Annu Rev Immunol 15:535–562

Bielby J, Price SJ, Monsalve-CarcaNo C, Bosch J (2021) Host contribution to parasite persistence is consistent between parasites and over time, but varies spatially. Ecol Appl 31:e02256

Blomberg B, Bernard CC, Du Pasquier L (1980) In vitro evidence for T-B lymphocyte collaboration in the clawed toad, Xenopus. Eur J Immunol 10:869–876

Bollinger TK, Mao J, Schock D, Brigham RM, Chinchar VG (1999) Pathology, isolation, and preliminary molecular characterization of a novel iridovirus from tiger salamanders in Saskatchewan. J Wildl Dis 35:413–429

Boone MD, Bridges-Britton CM (2006) Examining multiple sublethal contaminants on the gray treefrog (Hyla versicolor): effects of an insecticide, herbicide, and fertilizer. Environ Toxicol Chem 25:3261–3265

Caipang CM, Hirono I, Aoki T (2003) In vitro inhibition of fish rhabdoviruses by Japanese flounder, Paralichthys olivaceus Mx. Virology 317:373–382

Caipang CM, Takano T, Hirono I, Aoki T (2006) Genetic vaccines protect red seabream, Pagrus major, upon challenge with red seabream iridovirus (RSIV). Fish Shellfish Immunol 21:130–138

Cano I, Alonso MC, Garcia-Rosado E, Saint-Jean SR, Castro D, Borrego JJ (2006) Detection of lymphocystis disease virus (LCDV) in asymptomatic cultured gilt-head seabream (Sparus aurata, L.) using an immunoblot technique. Vet Microbiol 113:137–141

Cano I, Ferro P, Alonso MC, Sarasquete C, Garcia-Rosado E, Borrego JJ, Castro D (2009) Application of in situ detection techniques to determine the systemic condition of lymphocystis disease virus infection in cultured gilt-head seabream, Sparus aurata L. J Fish Dis 32:143–150

Carey C, Cohen N, Rollins-Smith L (1999) Amphibian declines: an immunological perspective. Dev Comp Immunol 23:459–472

Carrasco GH, de Souza MB, de Souza Santos LR (2021) Effect of multiple stressors and population decline of frogs. Environ Sci Pollut Res Int 28:59519–59527

Chang M, Nie P, Collet B, Secombes CJ, Zou J (2009) Identification of an additional two-cysteine containing type I interferon in rainbow trout Oncorhynchus mykiss provides evidence of a major gene duplication event within this gene family in teleosts. Immunogenetics 61:315–325

Chao CB, Chen CY, Lai YY, Lin CS, Huang HT (2004) Histological, ultrastructural, and in situ hybridization study on enlarged cells in grouper Epinephelus hybrids infected by grouper iridovirus in Taiwan (TGIV). Dis Aquat Org 58:127–142

Chen G, Robert J (2011) Antiviral immunity in amphibians. Viruses 3:2065–2086

Chen ZY, Li T, Gao XC, Wang CF, Zhang QY (2018) Protective immunity induced by DNA vaccination against Ranavirus infection in Chinese Giant Salamander Andrias davidianus. Viruses 10:52

Chen F, El-Naccache DW, Ponessa JJ, Lemenze A, Espinosa V, Wu W, Lothstein K, Jin L, Antao O, Weinstein JS, Damani-Yokota P, Khanna K, Murray PJ, Rivera A, Siracusa MC, Gause WC (2022a) Helminth resistance is mediated by differential activation of recruited monocyte-derived alveolar macrophages and arginine depletion. Cell Rep 38:110215

Chen SN, Gan Z, Hou J, Yang YC, Huang L, Huang B, Wang S, Nie P (2022b) Identification and establishment of type IV interferon and the characterization of interferon-upsilon including its class II cytokine receptors IFN-upsilonR1 and IL-10R2. Nat Commun 13:999

Cheng K, Escalon BL, Robert J, Chinchar VG, Garcia-Reyero N (2014) Differential transcription of fathead minnow immune-related genes following infection with frog virus 3, an emerging pathogen of ectothermic vertebrates. Virology 456-457:77–86

Chernyavskaya Y, Mudbhary R, Zhang C, Tokarz D, Jacob V, Gopinath S, Sun X, Wang S, Magnani E, Madakashira BP, Yoder JA, Hoshida Y, Sadler KC (2017) Loss of DNA methylation in zebrafish embryos activates retrotransposons to trigger antiviral signaling. Development 144:2925–2939

Chiappinelli KB, Strissel PL, Desrichard A, Li H, Henke C, Akman B, Hein A, Rote NS, Cope LM, Snyder A, Makarov V, Budhu S, Slamon DJ, Wolchok JD, Pardoll DM, Beckmann MW, Zahnow CA, Merghoub T, Chan TA, Baylin SB, Strick R (2017) Inhibiting DNA methylation causes an interferon response in cancer via dsRNA including endogenous retroviruses. Cell 169:361

Chinchar VG, Wang J, Murti G, Carey C, Rollins-Smith L (2001) Inactivation of frog virus 3 and channel catfish virus by esculentin-2P and ranatuerin-2P, two antimicrobial peptides isolated from frog skin. Virology 288:351–357

Chinchar VG, Bryan L, Wang J, Long S, Chinchar GD (2003) Induction of apoptosis in frog virus 3-infected cells. Virology 306:303–312

Choi HJ, Xu H, Geng Y, Colmone A, Cho H, Wang CR (2008) Bacterial infection alters the kinetics and function of iNKT cell responses. J Leukoc Biol 84:1462–1471

Cohen NR, Garg S, Brenner MB (2009) Antigen presentation by CD1 lipids, T cells, and NKT cells in microbial immunity. Adv Immunol 102:1–94

Coiras M, Lopez-Huertas MR, Perez-Olmeda M, Alcami J (2009) Understanding HIV-1 latency provides clues for the eradication of long-term reservoirs. Nat Rev Microbiol 7:798–812

Cotter JD, Storfer A, Page RB, Beachy CK, Voss SR (2008) Transcriptional response of Mexican axolotls to Ambystoma tigrinum virus (ATV) infection. BMC Genomics 9:493

Cusaac JPW, Carter ED, Woodhams DC, Robert J, Spatz JA, Howard JL, Lillard C, Graham AW, Hill RD, Reinsch S, McGinnity D, Reeves B, Bemis D, Wilkes RP, Sutton WB, Waltzek TB, Hardman RH, Miller DL, Gray MJ (2021) Emerging pathogens and a current-use pesticide: potential impacts on eastern hellbenders. J Aquat Anim Health 33:24–32

Darcis G, Van Driessche B, Bouchat S, Kirchhoff F, Van Lint C (2018) Molecular control of HIV and SIV latency. Curr Top Microbiol Immunol 417:1–22

Davidson C, Benard MF, Shaffer HB, Parker JM, O'Leary C, Conlon JM, Rollins-Smith LA (2007) Effects of chytrid and carbaryl exposure on survival, growth and skin peptide defenses in foothill yellow-legged frogs. Environ Sci Technol 41:1771–1776

De Jesus Andino F, Chen G, Li Z, Grayfer L, Robert J (2012) Susceptibility of Xenopus laevis tadpoles to infection by the ranavirus Frog-Virus 3 correlates with a reduced and delayed innate immune response in comparison with adult frogs. Virology 432:435–443

De Jesus Andino J, Grayfer L, Chen G, Chinchar VG, Edholm ES, Robert J (2015) Characterization of Frog Virus 3 knockout mutants lacking putative virulence genes. Virology 485:162–170

De Jesus Andino F, Lawrence BP, Robert J (2017) Long term effects of carbaryl exposure on antiviral immune responses in Xenopus laevis. Chemosphere 170:169–175

Dimitrakopoulou D, Khwatenge CN, James-Zorn C, Paiola M, Bellin EW, Tian Y, Sundararaj N, Polak EJ, Grayfer L, Barnard D, Ohta Y, Horb M, Sang Y, Robert J (2023) Advances in the Xenopus immunome: diversification, expansion, and contraction. Dev Comp Immunol 145:104734

Dong W, Zhang X, Yang C, An J, Qin J, Song F, Zeng W (2011) Iridovirus infection in Chinese giant salamanders, China, 2010. Emerg Infect Dis 17:2388–2389

Du Pasquier L, Schwager J, Flajnik MF (1989) The immune system of Xenopus. Annu Rev Immunol 7:251–275

Du Pasquier L, Robert J, Courtet M, Mussmann R (2000) B-cell development in the amphibian Xenopus. Immunol Rev 175:201–213

Edholm ES, Albertorio Saez LM, Gill AL, Gill SR, Grayfer L, Haynes N, Myers JR, Robert J (2013) Nonclassical MHC class I-dependent invariant T cells are evolutionarily conserved and prominent from early development in amphibians. Proc Natl Acad Sci USA 110:14342–14347

Edholm ES, Grayfer L, De Jesus Andino F, Robert J (2015) Nonclassical MHC-restricted invariant Valpha6 T cells are critical for efficient early innate antiviral immunity in the amphibian Xenopus laevis. J Immunol 195:576–586

Edholm EI, De Jesus Andino F, Yim J, Woo K, Robert J (2019) Critical role of an MHC class I-Like/Innate-Like T cell immune surveillance system in host defense against Ranavirus (Frog Virus 3) infection. Viruses 11:330

Eiden MV (2008) Endogenous retroviruses--aiding and abetting genomic plasticity. Cell Mol Life Sci 65:3325–3328

Elharrar M, Hirth C, Blanc J, Kirn A (1973) Pathogenesis of the toxic hepatitis of mice provoked by FV 3 (frog virus 3): inhibition of liver macromolecular synthesis (author's transl). Biochim Biophys Acta 319:91–102

Fernandez-Trujillo MA, Garcia-Rosado E, Alonso MC, Castro D, Alvarez MC, Bejar J (2013) Mx1, Mx2 and Mx3 proteins from the gilthead seabream (Sparus aurata) show in vitro antiviral activity against RNA and DNA viruses. Mol Immunol 56:630–636

Flajnik MF, Kasahara M (2001) Comparative genomics of the MHC: glimpses into the evolution of the adaptive immune system. Immunity 15:351–362

Fullerton JN, O'Brien AJ, Gilroy DW (2013) Pathways mediating resolution of inflammation: when enough is too much. J Pathol 231:8–20

Gan Z, Yang YC, Chen SN, Hou J, Laghari ZA, Huang B, Li N, Nie P (2018) Unique composition of Intronless and Intron-containing Type I IFNs in the Tibetan Frog Nanorana parkeri provides new evidence to support independent Retroposition hypothesis for Type I IFN genes in amphibians. J Immunol 201:3329–3342

Gantress J, Maniero GD, Cohen N, Robert J (2003) Development and characterization of a model system to study amphibian immune responses to iridoviruses. Virology 311:254–262

Gendrault JL, Steffan AM, Bingen A, Kirn A (1981) Penetration and uncoating of frog virus 3 (FV3) in cultured rat Kupffer cells. Virology 112:375–384

Geng Y, Wang KY, Zhou ZY, Li CW, Wang J, He M, Yin ZQ, Lai WM (2011) First report of a ranavirus associated with morbidity and mortality in farmed Chinese giant salamanders (Andrias davidianus). J Comp Pathol 145:95–102

Goodenow MM, Rose SL, Tuttle DL, Sleasman JW (2003) HIV-1 fitness and macrophages. J Leukoc Biol 74:657–666

Gousset K, Ablan SD, Coren LV, Ono A, Soheilian F, Nagashima K, Ott DE, Freed EO (2008) Real-time visualization of HIV-1 GAG trafficking in infected macrophages. PLoS Pathog 4:e1000015

Goyos A, Ohta Y, Guselnikov S, Robert J (2009) Novel nonclassical MHC class Ib genes associated with CD8 T cell development and thymic tumors. Mol Immunol 46:1775–1786

Goyos A, Sowa J, Ohta Y, Robert J (2011) Remarkable conservation of distinct nonclassical MHC class I lineages in divergent amphibian species. J Immunol 186:372–381

Gray MJ, Miller DL (2013) Rise of ranavirus: an emerging pathogen threatens ectothermic vertebrates. Wildlife Profess 7:51–55

Grayfer L, Robert J (2014) Divergent antiviral roles of amphibian (Xenopus laevis) macrophages elicited by colony-stimulating factor-1 and interleukin-34. J Leukoc Biol 96:1143–1153

Grayfer L, Robert J (2015) Distinct functional roles of amphibian (Xenopus laevis) colony-stimulating factor-1- and interleukin-34-derived macrophages. J Leukoc Biol 98:641–649

Grayfer L, Garcia EG, Belosevic M (2010) Comparison of macrophage antimicrobial responses induced by type II interferons of the goldfish (Carassius auratus L.). J Biol Chem 285:23537–23547

Grayfer L, Andino Fde J, Chen G, Chinchar GV, Robert J (2012) Immune evasion strategies of ranaviruses and innate immune responses to these emerging pathogens. Viruses 4:1075–1092

Grayfer L, De Jesus Andino F, Robert J (2014) The amphibian (Xenopus laevis) type I interferon response to frog virus 3: new insight into ranavirus pathogenicity. J Virol 88:5766–5777

Grayfer L, De Jesus Andino F, Robert J (2015) Prominent amphibian (Xenopus laevis) tadpole type III interferon response to the frog virus 3 ranavirus. J Virol 89:5072–5082

Groot F, Welsch S, Sattentau QJ (2008) Efficient HIV-1 transmission from macrophages to T cells across transient virological synapses. Blood 111:4660–4663

Gut JP, Anton M, Bingen A, Vetter JM, Kirn A (1981) Frog virus 3 induces a fatal hepatitis in rats. Lab Investig J Tech Methods Pathol 45:218–228

Hagmann W, Steffan AM, Kirn A, Keppler D (1987) Leukotrienes as mediators in frog virus 3-induced hepatitis in rats. Hepatology (Baltimore, Md) 7:732–736

Hanington PC, Wang T, Secombes CJ, Belosevic M (2007) Growth factors of lower vertebrates: characterization of goldfish (Carassius auratus L.) macrophage colony-stimulating factor-1. J Biol Chem 282:31865–31872

Hardman RH, Reinert LK, Irwin KJ, Oziminski K, Rollins-Smith L, Miller DL (2023) Disease state associated with chronic toe lesions in hellbenders may alter anti-chytrid skin defenses. Sci Rep 13:1982

Harly C, Robert J, Legoux F, Lantz O (2022) γδ T, NKT, and MAIT cells during evolution: redundancy or specialized functions? J Immunol 209:217–225

Hashimoto D, Miller J, Merad M (2011) Dendritic cell and macrophage heterogeneity in vivo. Immunity 35:323–335

Hauser K, Singer J, Hossainey MRH, Moore T, Wendel ES, Yaparla A, Kalia N, Grayfer L (2021) Amphibian (Xenopus laevis) tadpoles and adult frogs differ in their antiviral responses to intestinal Frog Virus 3 infections. Front Immunol. https://doi.org/10.3389/fimmu.2021.737403

Hayes TB, Stuart AA, Mendoza M, Collins A, Noriega N, Vonk A, Johnston G, Liu R, Kpodzo D (2006) Characterization of atrazine-induced gonadal malformations in African clawed frogs (Xenopus laevis) and comparisons with effects of an androgen antagonist (cyproterone acetate) and exogenous estrogen (17beta-estradiol): support for the demasculinization/feminization hypothesis. Environ Health Perspect 114(Suppl 1):134–141

Holopainen R, Tapiovaara H, Honkanen J (2012) Expression analysis of immune response genes in fish epithelial cells following ranavirus infection. Fish Shellfish Immunol 32:1095–1105

Horton JD, Horton TL, Dzialo R, Gravenor I, Minter R, Ritchie P, Gartland L, Watson MD, Cooper MD (1998) T-cell and natural killer cell development in thymectomized Xenopus. Immunol Rev 166:245–258

Hossainey MRH, Yaparla A, Hauser K, Moore T, Grayfer L (2021a) The roles of amphibian (Xenopus laevis) macrophages during chronic Frog Virus 3 infections. Viruses 13:2299–2313

Hossainey MRH, Yaparla A, Hauser KA, Moore TE, Grayfer L (2021b) The roles of amphibian (Xenopus laevis) macrophages during chronic Frog Virus 3 infections. Viruses 13:2299

Hossainey MRH, Hauser K, Garvey CN, Kalia N, Garvey JN, Grayfer L (2022) A perspective into the relationships between amphibian (Xenopus laevis) myeloid cell subsets. Philos Trans R Soc B. in press

Hoverman JT, Gray MJ, Miller DL (2010) Anuran susceptibilities to ranaviruses: role of species identity, exposure route, and a novel virus isolate. Dis Aquat Org 89:97–107

Hsu E (1998) Mutation, selection, and memory in B lymphocytes of exothermic vertebrates. Immunol Rev 162:25–36

Hyatt AD, Williamson M, Coupar BE, Middleton D, Hengstberger SG, Gould AR, Selleck P, Wise TG, Kattenbelt J, Cunningham AA, Lee J (2002) First identification of a ranavirus from green pythons (Chondropython viridis). J Wildl Dis 38:239–252

Jancovich JK, Jacobs BL (2011) Innate immune evasion mediated by the Ambystoma tigrinum virus eukaryotic translation initiation factor 2alpha homologue. J Virol 85:5061–5069

Jiang N, Fan Y, Zhou Y, Liu W, Robert J, Zeng L (2018) Rag1 and rag2 gene expressions identify lymphopoietic tissues in juvenile and adult Chinese giant salamander (Andrias davidianus). Dev Comp Immunol 87:24–35

Jiang N, Fan Y, Zhou Y, Meng Y, Liu W, Li Y, Xue M, Robert J, Zeng L (2021) The immune system and the antiviral responses in Chinese Giant Salamander, Andrias davidianus. Front Immunol 12:718627

Joerink M, Forlenza M, Ribeiro CM, de Vries BJ, Savelkoul HF, Wiegertjes GF (2006a) Differential macrophage polarisation during parasitic infections in common carp (Cyprinus carpio L.). Fish Shellfish Immunol 21:561–571

Joerink M, Ribeiro CM, Stet RJ, Hermsen T, Savelkoul HF, Wiegertjes GF (2006b) Head kidney-derived macrophages of common carp (Cyprinus carpio L.) show plasticity and functional polarization upon differential stimulation. J Immunol 177:61–69

Joerink M, Savelkoul HF, Wiegertjes GF (2006c) Evolutionary conservation of alternative activation of macrophages: structural and functional characterization of arginase 1 and 2 in carp (Cyprinus carpio L.). Mol Immunol 43:1116–1128

Jurgens JB, Gartland LA, Du Pasquier L, Horton JD, Gobel TW, Cooper MD (1995) Identification of a candidate CD5 homologue in the amphibian Xenopus laevis. J Immunol 155:4218–4223

Kalia N, Hauser KA, Burton S, Hossainey MRH, Zelle M, Horb ME, Grayfer L (2022) Endogenous retroviruses augment amphibian (Xenopus laevis) tadpole antiviral protection. J Virol 96:e0063422

Kambol R, Kabat P, Tristem M (2003) Complete nucleotide sequence of an endogenous retrovirus from the amphibian, Xenopus laevis. Virology 311:1–6

Kennedy PG, Rovnak J, Badani H, Cohrs RJ (2015) A comparison of herpes simplex virus type 1 and varicella-zoster virus latency and reactivation. J Gen Virol 96:1581–1602

Kirn A, Gut JP, Elharrar M (1972) FV3 (Frog Virus 3) toxicity for the mouse. La Nouvelle presse medicale 1:1943

Kirn A, Steffan AM, Bingen A (1980) Inhibition of erythrophagocytosis by cultured rat Kupffer cells infected with frog virus 3. J Reticuloendothel Soc 28:381–388

Kirn A, Bingen A, Steffan AM, Wild MT, Keller F, Cinqualbre J (1982) Endocytic capacities of Kupffer cells isolated from the human adult liver. Hepatology (Baltimore, Md) 2:216–222

Kochs G, Reichelt M, Danino D, Hinshaw JE, Haller O (2005) Assay and functional analysis of dynamin-like Mx proteins. Methods Enzymol 404:632–643

Kotenko SV (2011) IFN-lambdas. Curr Opin Immunol 23:583–590

Krause CD (2016) Intron loss in interferon genes follows a distinct set of stages, and may confer an evolutionary advantage. Cytokine 83:193–205

Kumar A, Abbas W, Herbein G (2014) HIV-1 latency in monocytes/macrophages. Viruses 6:1837–1860

Landsberg JH, Kiryu Y, Tabuchi M, Waltzek TB, Enge KM, Reintjes-Tolen S, Preston A, Pessier AP (2013) Co-infection by alveolate parasites and frog virus 3-like ranavirus during an amphibian larval mortality event in Florida, USA. Dis Aquat Org 105:89–99

Le Bourhis L, Martin E, Peguillet I, Guihot A, Froux N, Core M, Levy E, Dusseaux M, Meyssonnier V, Premel V, Ngo C, Riteau B, Duban L, Robert D, Huang S, Rottman M, Soudais C, Lantz O (2010) Antimicrobial activity of mucosal-associated invariant T cells. Nat Immunol 11:701–708

Leibovitz L (1980) Lymphocystis disease. J Am Vet Med Assoc 176:202

Leiva-Rebollo R, Castro D, Moreno P, Borrego JJ, Labella AM (2021) Evaluation of Gilthead Seabream (Sparus aurata) immune response after LCDV-Sa DNA vaccination. Animals (Basel) 11:1613

Lester K, Hall M, Urquhart K, Gahlawat S, Collet B (2012) Development of an in vitro system to measure the sensitivity to the antiviral Mx protein of fish viruses. J Virol Methods 182:1–8

Levraud JP, Boudinot P, Colin I, Benmansour A, Peyrieras N, Herbomel P, Lutfalla G (2007) Identification of the zebrafish IFN receptor: implications for the origin of the vertebrate IFN system. J Immunol 178:4385–4394

Li Z, Strunk JJ, Lamken P, Piehler J, Walz T (2008) The EM structure of a type I interferon-receptor complex reveals a novel mechanism for cytokine signaling. J Mol Biol 377:715–724

Li Z, Xu X, Huang L, Wu J, Lu Q, Xiang Z, Liao J, Weng S, Yu X, He J (2010) Administration of recombinant IFN1 protects zebrafish (Danio rerio) from ISKNV infection. Fish Shellfish Immunol 29:399–406

Lin H, Lee E, Hestir K, Leo C, Huang M, Bosch E, Halenbeck R, Wu G, Zhou A, Behrens D, Hollenbaugh D, Linnemann T, Qin M, Wong J, Chu K, Doberstein SK, Williams LT (2008) Discovery of a cytokine and its receptor by functional screening of the extracellular proteome. Science 320:807–811

Long S, Wilson M, Bengten E, Bryan L, Clem LW, Miller NW, Chinchar VG (2004) Identification of a cDNA encoding channel catfish interferon. Dev Comp Immunol 28:97–111

Lopez C, Aubertin AM, Tondre L, Kirn A (1986) Thermosensitivity of frog virus 3 genome expression: defect in early transcripti. Virol 152:365–374

Lopez-Munoz A, Roca FJ, Meseguer J, Mulero V (2009) New insights into the evolution of IFNs: zebrafish group II IFNs induce a rapid and transient expression of IFN-dependent genes and display powerful antiviral activities. J Immunol 182:3440–3449

Love CN, Winzeler ME, Beasley R, Scott DE, Nunziata SO, Lance SL (2016) Patterns of amphibian infection prevalence across wetlands on the Savannah River Site, South Carolina, USA. Dis Aquat Org 121:1–14

Lu C, Chai J, Murphy RW, Che J (2020) Giant salamanders: farmed yet endangered. Science (New York, N.Y.) 367:989

Lutfalla G, Roest Crollius H, Stange-Thomann N, Jaillon O, Mogensen K, Monneron D (2003) Comparative genomic analysis reveals independent expansion of a lineage-specific gene family in vertebrates: the class II cytokine receptors and their ligands in mammals and fish. BMC Genomics 4:29

Maniero GD, Morales H, Gantress J, Robert J (2006) Generation of a long-lasting, protective, and neutralizing antibody response to the ranavirus FV3 by the frog Xenopus. Dev Comp Immunol 30:649–657

Matsuda JL, Gapin L (2005) Developmental program of mouse Valpha14i NKT cells. Curr Opin Immunol 17:122–130

Mayassi T, Barreiro LB, Rossjohn J, Jabri B (2021) A multilayered immune system through the lens of unconventional T cells. Nature 595:501–510

McGuire CC, Robert JR (2022) Developmental exposure to thyroid disrupting chemical mixtures alters metamorphosis and post-metamorphic thymocyte differentiation. Curr Res Toxicol 3:100094

McGuire CC, Lawrence BP, Robert J (2021) Thyroid disrupting chemicals in mixture Perturb Thymocyte differentiation in Xenopus laevis Tadpoles. Toxicol Sci 181:262–272

Meng Y, Fan Y, Jiang N, Xue M, Li Y, Liu W, Zeng L, Zhou Y (2022) Four Mx genes identified in Andrias davidianus and characterization of their response to Chinese Giant Salamander Iridovirus infection. Animals (Basel) 12:2147

Miller D, Gray M, Storfer A (2011) Ecopathology of ranaviruses infecting amphibians. Viruses 3:2351–2373

Morales HD, Robert J (2007) Characterization of primary and memory CD8 T-cell responses against ranavirus (FV3) in Xenopus laevis. J Virol 81:2240–2248

Morales HD, Abramowitz L, Gertz J, Sowa J, Vogel A, Robert J (2010) Innate immune responses and permissiveness to ranavirus infection of peritoneal leukocytes in the frog Xenopus laevis. J Virol 84:4912–4922

Nakajima K, Kunita J (2005) Red sea bream iridoviral disease. Uirusu 55:115–125

Nakajima K, Maeno Y, Honda A, Yokoyama K, Tooriyama T, Manabe S (1999) Effectiveness of a vaccine against red sea bream iridoviral disease in a field trial test. Dis Aquat Org 36:73–75

Pagani I, Demela P, Ghezzi S, Vicenzi E, Pizzato M, Poli G (2022) Host restriction factors modulating HIV latency and replication in macrophages. Int J Mol Sci 23:3021

Paquin-Proulx D, Greenspun BC, Kitchen SM, Saraiva Raposo RA, Nixon DF, Grayfer L (2018) Human interleukin-34-derived macrophages have increased resistance to HIV-1 infection. Cytokine 111:272–277

Pham PH, Lai YS, Lee FF, Bols NC, Chiou PP (2012) Differential viral propagation and induction of apoptosis by grouper iridovirus (GIV) in cell lines from three non-host species. Virus Res 167:16–25

Pixley FJ, Stanley ER (2004) CSF-1 regulation of the wandering macrophage: complexity in action. Trends Cell Biol 14:628–638

Qi Z, Nie P, Secombes CJ, Zou J (2010) Intron-containing type I and type III IFN coexist in amphibians: refuting the concept that a retroposition event gave rise to type I IFNs. J Immunol 184:5038–5046

Raghow R, Granoff A (1979) Macromolecular synthesis in cells infected by frog virus 3. X. Inhibition of cellular protein synthesis by heat-inactivated virus. Virology 98:319–327

Redmond AK, Zou J, Secombes CJ, Macqueen DJ, Dooley H (2019) Discovery of all three types in Cartilaginous fishes enables phylogenetic resolution of the origins and evolution of interferons. Front Immunol 10:1558

Reeve BC, Crespi EJ, Whipps CM, Brunner JL (2013) Natural stressors and ranavirus susceptibility in larval wood frogs (Rana sylvatica). EcoHealth 10:190–200

Relyea RA, Jones DK (2009) The toxicity of Roundup Original Max to 13 species of larval amphibians. Environ Toxicol Chem 28:2004–2008

Robert J, Edholm ES (2014) A prominent role for invariant T cells in the amphibian Xenopus laevis tadpoles. Immunogenetics 66:513

Robert J, Ohta Y (2009) Comparative and developmental study of the immune system in Xenopus. Dev Dyn 238:1249–1270

Robert J, Morales H, Buck W, Cohen N, Marr S, Gantress J (2005) Adaptive immunity and histopathology in frog virus 3-infected Xenopus. Virology 332:667–675

Robert J, Abramowitz L, Gantress J, Morales HD (2007) Xenopus laevis: a possible vector of Ranavirus infection? J Wildl Dis 43:645–652

Robert J, George E, De Jesus Andino F, Chen G (2011) Waterborne infectivity of the Ranavirus frog virus 3 in Xenopus laevis. Virology 417:410–417

Robert J, Grayfer L, Edholm ES, Ward B, De Jesus Andino F (2014) Inflammation-induced reactivation of the ranavirus Frog Virus 3 in asymptomatic Xenopus laevis. PLoS One 9:e112904

Robert J, McGuire CC, Kim F, Nagel SC, Price SJ, Lawrence BP, De Jesus Andino F (2018) Water contaminants associated with unconventional oil and gas extraction cause immunotoxicity to amphibian tadpoles. Toxicol Sci 166:39

Robertsen B (2006) The interferon system of teleost fish. Fish Shellfish Immunol 20:172–191

Robertsen B, Bergan V, Rokenes T, Larsen R, Albuquerque A (2003) Atlantic salmon interferon genes: cloning, sequence analysis, expression, and biological activity. J Interf Cytokine Res 23:601–612

Rollins-Smith LA (2009) The role of amphibian antimicrobial peptides in protection of amphibians from pathogens linked to global amphibian declines. Biochim Biophys Acta 1788:1593–1599

Rollins-Smith LA, Flajnik MF, Blair PJ, Davis AT, Green WF (1997) Involvement of thyroid hormones in the expression of MHC class I antigens during ontogeny in Xenopus. Dev Immunol 5:133–144

Rollins-Smith LA, Reinert LK, O'Leary CJ, Houston LE, Woodhams DC (2005) Antimicrobial peptide defenses in amphibian skin. Integr Comp Biol 45:137–142

Roulois D, Loo Yau H, Singhania R, Wang Y, Danesh A, Shen SY, Han H, Liang G, Jones PA, Pugh TJ, O'Brien C, De Carvalho DD (2015) DNA-Demethylating agents target colorectal cancer cells by inducing viral mimicry by endogenous transcripts. Cell 162:961–973

Sachs LM, Buchholz DR (2019) Insufficiency of thyroid hormone in frog metamorphosis and the role of glucocorticoids. Front Endocrinol (Lausanne) 10:287

Sadler AJ, Williams BR (2008) Interferon-inducible antiviral effectors. Nat Rev 8:559–568

Samanta M, Yim J, De Jesus Andino F, Paiola M, Robert J (2021) TLR5-mediated reactivation of quiescent Ranavirus FV3 in Xenopus peritoneal macrophages. J Virol 95:e00215

Samuel CE (2001) Antiviral actions of interferons. Clin Microbiol Rev 14:778–809, table of contents

Sang Y, Liu Q, Lee J, Ma W, McVey DS, Blecha F (2016) Expansion of amphibian intronless interferons revises the paradigm for interferon evolution and functional diversity. Sci Rep 6:29072

Savage AE, Muletz-Wolz CR, Campbell Grant EH, Fleischer RC, Mulder KP (2019) Functional variation at an expressed MHC class IIβ locus associates with Ranavirus infection intensity in larval anuran populations. Immunogenetics 71:335–346

Shibata Y, Wen L, Okada M, Shi YB (2020) Organ-specific requirements for thyroid hormone receptor ensure temporal coordination of tissue-specific transformations and completion of Xenopus metamorphosis. Thyroid 30:300–313

Shields LE, Jennings J, Liu Q, Lee J, Ma W, Blecha F, Miller LC, Sang Y (2019) Cross-species genome-wide analysis reveals molecular and functional diversity of the unconventional interferon-omega subtype. Front Immunol 10:1431

Sifkarovski J, Grayfer L, De Jesus Andino F, Lawrence BP, Robert J (2014) Negative effects of low dose atrazine exposure on the development of effective immunity to FV3 in Xenopus laevis. Dev Comp Immunol 47:52–58

Siwicki AK, Pozet F, Morand M, Volatier C, Terech-Majewska E (1999) Effects of iridovirus-like agent on the cell-mediated immunity in sheatfish (Silurus glanis)--an in vitro study. Virus Res 63:115–119

Sparling DW, Bickham J, Cowman D, Fellers GM, Lacher T, Matson CW, McConnell L (2015) In situ effects of pesticides on amphibians in the Sierra Nevada. Ecotoxicology 24:262–278

Speck SH, Ganem D (2010) Viral latency and its regulation: lessons from the gamma-herpesviruses. Cell Host Microbe 8:100–115

Stohr AC, Hoffmann A, Papp T, Robert N, Pruvost NB, Reyer HU, Marschang RE (2013) Long-term study of an infection with ranaviruses in a group of edible frogs (Pelophylax kl. esculentus) and partial characterization of two viruses based on four genomic regions. Vet J 197:238–244

Sun B, Robertsen B, Wang Z, Liu B (2009) Identification of an Atlantic salmon IFN multigene cluster encoding three IFN subtypes with very different expression properties. Dev Comp Immunol 33:547–558

Svingerud T, Solstad T, Sun B, Nyrud ML, Kileng O, Greiner-Tollersrud L, Robertsen B (2012) Atlantic salmon type I IFN subtypes show differences in antiviral activity and cell-dependent expression: evidence for high IFNb/IFNc-producing cells in fish lymphoid tissues. J Immunol 189:5912–5923

Teacher AG, Garner TW, Nichols RA (2009) Evidence for directional selection at a novel major histocompatibility class I marker in wild common frogs (Rana temporaria) exposed to a viral pathogen (Ranavirus). PLoS One 4:e4616

Tian Y, Jennings J, Gong Y, Sang Y (2019) Xenopus interferon complex: inscribing the amphibiotic adaption and species-specific pathogenic pressure in vertebrate evolution? Cells 9:67

Tian Y, De Jesus Andino F, Khwatenge CN, Li J, Robert J, Sang Y (2021a) Virus-targeted transcriptomic analyses implicate Ranaviral interaction with host interferon response in Frog Virus 3-infected frog tissues. Viruses 13:1325

Tian Y, Khwatenge CN, Li J, De Jesus Andino F, Robert J, Sang Y (2021b) Targeted transcriptomics of Frog Virus 3 in infected frog tissues reveal non-coding regulatory elements and microRNAs in the Ranaviral genome and their potential interaction with host immune response. Front Immunol 12:705253

Todd LA, Katzenback BA (2021) Discovery of frog virus 3 microRNAs and their roles in evasion of host antiviral responses. 2021 2021.09.17.460379

Trobridge GD, Chiou PP, Leong JA (1997) Cloning of the rainbow trout (Oncorhynchus mykiss) Mx2 and Mx3 cDNAs and characterization of trout Mx protein expression in salmon cells. J Virol 71:5304–5311

Turner RJ, Manning MJ (1974) Thymic dependence of amphibian antibody responses. Eur J Immunol 4:343–346

Vo NTK, Guerreiro M, Yaparla A, Grayfer L, DeWitte-Orr SJ (2019) Class A Scavenger Receptors are used by Frog Virus 3 during its cellular entry. Viruses 11:93

Wang T, Hanington PC, Belosevic M, Secombes CJ (2008) Two macrophage colony-stimulating factor genes exist in fish that differ in gene organization and are differentially expressed. J Immunol 181:3310–3322

Wang LZ, Xu YP, Zhou YL, Liu ZP, Li B, Gu WB, Zhao XF, Dong WR, Shu MA (2020) The first evidence of four transcripts from two Interleukin 18 genes in animal and their involvement in immune responses in the largest amphibian Andrias davidianus. Dev Comp Immunol 106:103598

Wendel ES, Yaparla A, Koubourli DV, Grayfer L (2017) Amphibian (Xenopus laevis) tadpoles and adult frogs mount distinct interferon responses to the Frog Virus 3 ranavirus. Virology 503:12–20

Wendel ES, Yaparla A, Melnyk MLS, Koubourli DV, Grayfer L (2018) Amphibian (Xenopus laevis) Tadpoles and adult frogs differ in their use of expanded repertoires of Type I and Type III interferon cytokines. Viruses 10:372

Wiegertjes GF, Forlenza M (2010) Nitrosative stress during infection-induced inflammation in fish: lessons from a host-parasite infection model. Curr Pharm Des 16:4194–4202

Wu YC, Chi SC (2007) Cloning and analysis of antiviral activity of a barramundi (Lates calcarifer) Mx gene. Fish Shellfish Immunol 23:97–108

Wu MS, Chen CW, Liu YC, Huang HH, Lin CH, Tzeng CS, Chang CY (2012) Transcriptional analysis of orange-spotted grouper reacting to experimental grouper iridovirus infection. Dev Comp Immunol 37:233–242

Xu YP, Wang LZ, Zhou YL, Xiao Y, Gu WB, Li B, Zhao XF, Dong WR, Shu MA (2020) Identification and functional analysis of two interferon regulatory factor 3 genes and their involvement in antiviral immune responses in the Chinese giant salamander Andrias davidianus. Dev Comp Immunol 110:103710

Yang SJ, Xiao XH, Xu YG, Li DD, Chai LH, Zhang JY (2012) Induction of antimicrobial peptides from Rana dybowskii under Rana grylio virus stress, and bioactivity analysis. Can J Microbiol 58:848–855

Yaparla A, Popovic M, Grayfer L (2018) Differentiation-dependent antiviral capacities of amphibian (Xenopus laevis) macrophages. J Biol Chem 293:1736–1744

Yaparla A, Hossainey MRH, Uzzaman Z, Lugo A, Hauser K, Rollins-Smith LA, Grayfer L (2021) The roles of colony stimulating factor-1 and interleukin-34 macrophages during chytrid infections. PLoS One under review

Zahn N, James-Zorn C, Ponferrada VG, Adams DS, Grzymkowski J, Buchholz DR, Nascone-Yoder NM, Horb M, Moody SA, Vize PD, Zorn AM (2022) Normal table of Xenopus development: a new graphical resource. Development (Cambridge, England) 149:dev200356

Zhao J, He S, Minassian A, Li J, Feng P (2015) Recent advances on viral manipulation of NF-kappaB signaling pathway. Curr Opin Virol 15:103–111

Zhou Y, Fan Y, LaPatra SE, Ma J, Xu J, Meng Y, Jiang N, Zeng L (2015) Protective immunity of a Pichia pastoris expressed recombinant iridovirus major capsid protein in the Chinese giant salamander, Andrias davidianus. Vaccine 33:5662–5669

Zhu R, Wang J, Lei XY, Gui JF, Zhang QY (2013) Evidence for Paralichthys olivaceus IFITM1 antiviral effect by impeding viral entry into target cells. Fish Shellfish Immunol 35:918–926

Zilberg D, Grizzle JM, Plumb JA (2000) Preliminary description of lesions in juvenile largemouth bass injected with largemouth bass virus. Dis Aquat Org 39:143–146

Zou J, Secombes CJ (2011) Teleost fish interferons and their role in immunity. Dev Comp Immunol 35:1376–1387

Zou J, Tafalla C, Truckle J, Secombes CJ (2007) Identification of a second group of type I IFNs in fish sheds light on IFN evolution in vertebrates. J Immunol 179:3859–3871

Characterization, Pathogenesis, and Immuno-Biological Control of Singapore Grouper Iridovirus (SGIV)

Youhua Huang, Shaowen Wang, Xiaohong Huang, Jingguang Wei, and Qiwei Qin

Singapore grouper iridovirus (SGIV) was firstly isolated from diseased cultured groupers in Singapore and has been identified as a novel species within the genus *Ranavirus* (family *Iridoviridae*; subfamily *Alphairidovirinae*). SGIV infection causes considerable morbidity and mortality in many economically important fish species, such as grouper and seabass. In this chapter, we describe virus isolation in cell culture, virion purification, ultrastructural analysis, virion morphogenesis, and molecular identification of SGIV. SGIV has been molecularly characterized based on the SGIV genome, transcriptome, proteome, and viral miRNAs. Various aspects of pathogenesis resulting from SGIV infection were investigated, including cytopathology, virus entry and transport, paraptosis, autophagy, and signaling pathways. Functions of host immune and metabolism-related genes during SGIV infection are evaluated and discussed. Immuno-biological control strategies, including antibody-based flow cytometry and microfluidic chip detection technology, loop-mediated isothermal amplification (LAMP), and nucleic acid aptamer detection methods, were developed. Efficient SGIV vaccines have also been developed. These research approaches provide the basis for a better understanding of the pathogenesis of SGIV and other ranaviruses and offer technical support to control fish ranaviruses.

Y. Huang · S. Wang · X. Huang · J. Wei · Q. Qin (✉)
College of Marine Sciences, South China Agricultural University, Guangzhou, China
e-mail: qinqw@scau.edu.cn

© The Author(s) 2025
M. J. Gray, V. G. Chinchar (eds.), *Ranaviruses*,
https://doi.org/10.1007/978-3-031-64973-8_5

1 Isolation and Identification of SGIV

1.1 Establishment of Susceptible and Permissive Cell Lines from Marine Fish for the Propagation of SGIV

The generation of cultured cell lines has not only contributed to theoretical studies of viruses infecting fish and fish anti-viral responses, but also to the development of vaccines against virus diseases (Lakra et al. 2011). To date, many cell lines derived from marine fish have been established (Huang et al. 2011d; Guo et al. 2015). Seven cell lines from *E. awoara* have been established, including from the eye, fin, heart, kidney, liver, and brain (Lai et al. 2000, 2003). In addition, four cell lines from *E. coioides* including brain, fin, eye, and spleen (Parameswaran et al. 2007; Wen et al. 2009; Chi et al. 2010; Qin et al. 2006); six cell lines from *E. akaara* including spleen (EAGS), liver (EAGL), kidney (EAGK), brain (EAGB), swim bladder (EAGSB), and snout (Huang et al. 2009; Zhou et al. 2007; Ou-Yang et al. 2010; Gong et al. 2011; Huang et al. 2011d); and three cell lines from the head kidney (ELHK), snout (ELGSN), and heart (ELGH) of the giant grouper (*E. lanceolatus*) were developed (Liu et al. 2021b; Huang et al. 2014; Guo et al. 2015). The majority of grouper cell lines grow well in L-15 medium containing 10% fetal bovine serum (FBS) at 28 °C. Four cell lines from mid-kidney (TOK), caudal fin (TOCF), head kidney (TOHK), and brain (TOGB) of the golden pompano, *Trachinotus ovatus*, were also developed (Zhou et al. 2017; Wei et al. 2018; Li et al. 2016a, 2017). The susceptibility and permissiveness of marine fish cell lines to SGIV were demonstrated by the occurrence of cytopathic effect (CPE), the increased transcription of viral genes, and the production of virus particles. It has been reported that all cell lines from *E. akaara, E. lanceolatus,* and *Trachinotus ovatus* were susceptible to SGIV. Some of the above-mentioned grouper cell lines have been widely used for the isolation and purification of SGIV, studies on virus-host interactions, functional analysis of viral genes, and the production of inactivated vaccines. For example, SGIV Hainan strain (SGIV-HN) was isolated and identified from diseased groupers in Hainan province by coculturing affected tissue with EAGS cells (Wei et al. 2019). SGIV infection in EAGS cells altered the morphology and structure of microtubules, including their rearrangement into a ring-like structure around the nucleus, and aggregations around the virus assembly sites (Huang et al. 2009). In addition, in vitro SGIV infection could be used to investigate host defense mechanisms in response to virus infection (Guo et al. 2022a, b; Wang et al. 2023).

1.2 Isolation and Identification of Singapore Grouper Iridovirus (SGIV)

In 1994, an infectious disease, known as Sleepy Grouper Disease (SGD), resulted in mass mortality events among net-caged farmed brown-spotted groupers (*Epinephelus tauvina*) in Singapore, Indonesia, and Malaysia (Chua et al. 1994). However, the virus was not isolated or genetically characterized at that time. In 1998, the same disease occurred again in fish farms in Singapore following importation of brown-spotted grouper fry from other SE Asian countries and the virus was successfully isolated (Qin et al. 2003). The virus infected and induced CPE in a marine fish cell line (GP) derived from the hatching embryo of grouper, *E. tauvina* (Qin et al. 2003). The virus was characterized as a member of the genus *Ranavirus* based on its morphological, biochemical and genetic properties, and named Singapore grouper iridovirus (SGIV) (Qin et al. 2003). Based on phylogenetic analysis, SGIV is currently viewed as the most divergent species within the genus *Ranavirus* (Chinchar et al. 2017).

1.3   Biochemical, Ultrastructure, and Morphogenesis of SGIV

SGIV displays an icosahedral virion with a diameter of 200 nm. The size of the intracellular nucleocapsid is 154 nm between the opposite sides or 176 nm between opposite vertices with an inner electron-dense core of 93 nm. SGIV proliferated well in several grouper cell lines, and high levels of infectious virus were generated. SGIV was sensitive to acid, ether, and heat treatments. Moreover, detergent treatment triggered the loss of viral lipids, capsid proteins, and infectivity (Wu et al. 2010). SGIV replication was inhibited by 5-iodo-2-deoxyuridine (IUDR), indicating that the SGIV genome is composed of DNA. The SGIV major capsid protein (MCP) has a mol wt of 49 kDa and shares 73.33% sequence identity with largemouth bass virus (LMBV) and 72.79% identity with frog virus 3 (FV3) (Qin et al. 2003). The morphogenesis and the ultrastructure of SGIV were studied by electron microscopy (EM) (Qin et al. 2001). EM observation confirmed that the morphogenesis and the ultrastructure of SGIV were similar to that of another iridoviruses (Berthiaume et al. 1984; Heppell and Berthiaume 1992). The purified SGIV particles showed a three-layered structure composed of an external lipoprotein envelope, an inner protein capsid, and a lipid-containing internal membrane. A regular array of surface capsid subunits was observed after removal of the outer envelope with detergent. In SGIV-infected cells, various stages of virus amplification, maturation, and assembly were detected in the cytoplasm. Mature virions were postulated to form by insertion of electron-dense core material, i.e., viral DNA, into partly formed capsids. The nucleocapsids were found within the assembly sites as pseudocrystalline arrays or scattered individually. In later stages of infection, nucleocapsids were enveloped and released by budding from the plasma membrane. These enveloped virus particles could enter neighboring cells by endocytosis to start the next round infection (Qin et al. 2001).

2 Molecular Characterization

2.1 Multiple Omics of SGIV

Annotation of the viral genomic sequence is crucial to understanding the virus's origin and structure. This information contributes greatly to the development of vaccines, antiviral drugs, and efficient preventive strategies. The full length SGIV genome (Genbank accession No. AY521625) consists of 140,131 nucleotides with a G+C content of 48.64%, which was slightly less than that of TFV (55.01%), ISKNV (54.78%), and ATV (54.02%), but more than that of LCDV-1 (29.07%) and CIV (28.63%). A total of 162 open reading frames (ORFs) coding for proteins varying in size from 41 to 1268 amino acids were identified. These potential ORFs were associated with different biological processes, such as DNA replication and transcription, nucleotide metabolism, protein processing, manipulation of cellular responses, and virus-host interaction. Genetic analysis demonstrated that 77 of these ORFs exhibited homologies to known virus genes, and 23 of these matched functional iridovirus proteins. In addition, 42 of them contained putative conserved domains or signatures, which were detected in the NCBI CD-Search database and PROSITE database (Song et al. 2004).

SGIV proteins from purified virions were analyzed and identified by SDS-PAGE followed by MALDI-TOF mass spectrometry. Proteomic results showed that 26 proteins matched theoretical SGIV ORFs, and 20 of these were considered novel or previously unidentified genes (Song et al. 2006). SGIV structural proteins were further identified using SDS-assisted tube-gel digestion and DOC-assisted in-solution digestion coupled with LC-ESI-MS/MS. A total of 90 SGIV structural proteins including 24 newly reported proteins were identified. Additionally, several post-translationally modified proteins (PTMs) were identified, including 26 N-terminally acetylated proteins, 3 phosphorylated proteins, and 1 myristoylated protein. Of note, 47 of the proteins that were identified were predicted to contain conserved domains (Yao et al. 2019). Furthermore, the envelope proteins of SGIV were also identified by 1-DE-MALDI-TOF/TOF-MS/MS and LC-MALDI-TOF/TOF-MS/MS analysis. A total of 19 virus-encoded envelope proteins were identified from the envelope fractions of purified SGIV virions. Among them, 3 and 10 viral envelope proteins were uniquely identified by 1-DE-MALDI and LC-MALDI, respectively, whereas 6 proteins were identified by both workflows, including ORF015L, ORF016L, ORF019R, ORF088L, ORF090L, and ORF102L (Zhou et al. 2011b). Subsequently, ORF016L, ORF019R, ORF088L, and ORF039L were confirmed as viral envelope proteins by immunoelectron microscopy (Zhou et al. 2011b; Huang et al. 2013; Zhang et al. 2015).

According to the genomic and proteomic annotation, the transcriptional profiles of viral genes in SGIV-infected grouper spleen (GS) cells were evaluated using DNA microarray analysis. The results showed that 3 temporal kinetic classes (15 immediate-early, 89 early, and 53 late genes) were classified during in vitro infection by their dependence on de novo protein synthesis and viral DNA synthesis (Chen et al. 2006; Teng et al. 2008).

2.2 The Functions of SGIV Encoded Proteins

Using a variety of approaches, great progress has been made in determining the virion or subcellular localization and function of SGIV proteins. To date, 19 viral genes were cloned and characterized as summarized below (Table 1).

VP088 VP088 was verified as a late gene and encoded an envelope protein. GFP-tagged VP088 is a cytoplasmic protein as VP88GFP is distributed evenly in the cytoplasm of the host cell. However, in SGIV-infected cells, viral assembly site (VAS) formation is initiated before the aggregation of VP088 (Yuan and Hong 2016). Deletion of VP088 significantly suppresses SGIV infection without altering viral gene expression and host responses. Quantitative analysis of the genome copy numbers of host cells and virions revealed that VP088 deletion dramatically reduced SGIV infectivity by inhibiting virus entry without altering viral pathogenicity, genome stability and replication, and progeny virus release (Yuan et al. 2016). Moreover, antibody against VP088 displayed neutralized activity, suggesting that VP088 represented an ideal target for SGIV intervention (Zhou et al. 2011b).

VP019 As a late expressed viral protein, VP019 was initially found in the cytoplasm in a punctate pattern, and then aggregated within the virus assembly site at the late stage of SGIV infection, suggesting that VP019 was involved in virus assembly. In addition, western blot assay and electron microscopic observation revealed that SGIV VP019 was associated with the viral envelope (Huang et al. 2013). Furthermore, VP019 was viewed as a possible vaccine candidate as 75.0% and 28.7% relative percent survival (RPS) was achieved when grouper were immunized with 90 µg pcDNA3.1-VP019 or oral suspensions of *Bacillus subtilis* spores displaying VP019 protein (Yu et al. 2019; Liang et al. 2023).

VP075 VP075 encoded a cytoplasmic protein which accumulated within viral assembly sites during SGIV infection. Immunogold-labeling indicated that it was localized between the viral capsid shell and DNA core. Knockdown of VP075 by antisense morpholino oligonucleotides resulted in the reduction of core shell thickness, the failure of DNA encapsidation, and a low yield of infectious particles. VP075 was phosphorylated at multiple sites in SGIV-infected cell lysates and virions, but the vast majority of VP075 in virions was the dephosphorylated isoform. Moreover, VP075 could be phosphorylated in vitro by the SGIV structural protein VP039, which contained a serine/threonine protein kinase catalytic domain. Addition of ATP and Mg(2+) into purified virions prompted extensive phosphorylation of structural proteins and release of VP075 from virions (Wang et al. 2015).

VP018 VP018, which is conserved among vertebrate iridoviruses, is an abundant virion protein. Knockdown of VP018 expression resulted in a reduction in the expression of viral late genes and inhibition of viral particle assembly in grouper embryonic cells. Western blotting with phosphoserine-specific antibody indicated that serine phosphorylation was significantly enhanced for proteins of molecular mass 17–32 kDa when VP018 expression was eliminated. Two-dimensional gel

Table 1 Assessment of SGIV gene function

Gene (ORF)	Phenotype and function	References
ORF008L	Exerted a role in protection or stabilization of viral genome, DNA binding ability, protected DNA from degradation by DNase I in vitro	Wang et al. (2013)
ORF018R	An abundant virion protein, conserved among vertebrate iridoviruses, a putative Ser/Thr kinase; played critical roles in virion assembly, expression of viral late genes, and serine/threonine phosphorylation	Wang et al. (2008)
ORF019R	A late expressed viral gene, a viral envelope protein, involved in virus assembly, a possible vaccine candidate	Huang et al. (2013) and Liang et al. (2023)
ORF038L	A late viral gene, a viral protein with an RGD motif; might play a role in entry	Wan et al. (2010)
ORF039L	A core gene of the family *Iridoviridae*, a viral envelope protein	Zhang et al. (2015)
ORF048L	A putative homolog containing a caspase recruitment domain (CARD), enhanced SGIV replication and inhibited apoptosis and autophagy	Zhang et al. (2023c)
ORF049L	A dUTP pyrophosphatase (dUTPase) homolog, a viral early gene, contains a nuclear export signal	Gong et al. (2010)
ORF051L	A putative homolog of TNFR, a viral IE gene, functioned as a critical virulence factor affecting cell proliferation, regulation of host cell apoptosis and the inflammatory response	Yu et al. (2016b, 2017b)
ORF062R	A novel insulin like growth factor, stimulated cell growth and virus replication by promoting G1/S transition; overexpression led to increased apoptosis in non-host cells	Yan et al. (2013)
ORF075R	A cytoplasmic protein, involved in viral assembly, could be phosphorylated in vitro by the structural protein VP39	Wang et al. 2015)
ORF088L	A late viral gene, a viral envelope protein, played a role in viral entry	Zhou et al. (2011b), Yuan and Hong (2016), Yuan et al. (2016)
ORF086R	A homolog of ICP18, a viral IE gene, promoted the growth of grouper embryonic cells and contributed to SGIV replication; interacted with CDK1	Xia et al. (2009) and Wu et al. (2021)
ORF096R	A putative homolog of TNFR, contributed to viral replication by modulating the host apoptotic response and cell proliferation	Huang et al. (2013)
ORF122L	Promoted SGIV replication by negatively regulating the interferon immune response	Xu et al. (2023)
ORF131R	A homolog with immunoglobulin (Ig)-like domains, a proviral factor, negatively regulates the IFN response by inhibiting EcSTING-EcTBK1 signaling for viral evasion	Zhang et al. (2022)
ORF136R	A putative LITAF homolog, played crucial roles in cell death by inducing apoptosis	Huang et al. (2008)

(continued)

Table 1 (continued)

Gene (ORF)	Phenotype and function	References
ORF155R	A novel semaphorin homologue gene, a viral early gene, promoted viral replication and attenuated the cellular immune response	Yan et al. (2014)
ORF158L	A histone H3/H4 chaperon, interacts with the histone H3/H4 complex and H3, involved in both the regulation and the expression of histone H3 and H3 methylation; facilitates viral replication	Tran et al. (2011)
ORF162L	A putative homolog of ICP46, a viral IE gene, a structural protein of the nucleocapsid, promoted cell growth and contributed to SGIV replication, four putative VATTs (VP12, VP39, VP57, and MCP) interacted with the SGIV ICP46 promoter	Xia et al. (2010, 2019)

electrophoresis showed that numerous protein spots were shifted to a lower pI and higher molecular mass as a result of the loss of VP018 function. Among them, three proteins with enhanced phosphorylation were identified, including VP049 (dUT-Pase), VP075, and VP086 (Wang et al. 2008).

VP136 (LITAF) VP136 encoded a putative LITAF homolog and distributed predominantly in the cytoplasm in association with mitochondria. Reverse transcription-PCR (RT-PCR) and western blot analyses of SGIV-infected cells revealed that LITAF was an early viral gene. Overexpression of LITAF induced apoptosis, as shown by an increased number of apoptotic bodies, depolarization of mitochondrial membrane potential (DeltaPsi(m)), and activation of caspase-3. Furthermore, NF-κB and NFAT activities were increased in cells expressing SGIV LITAF (Huang et al. 2008).

VP096 VP096 encoded a putative homolog of TNFR that contained two extracellular cysteine-rich domains (CRDs) with four or six conserved cysteine residues but lacked the transmembrane domain at the C-terminus. VP096 was identified as an early (E) gene and localized in the cytoplasm in transfected or infected cells. Overexpression of VP096 in vitro enhanced cell proliferation and improved cell survival against SGIV infection. Furthermore, virus infection induced apoptosis and caspase-3 activity was inhibited in VP096 expressing fathead minnow (FHM) cells compared to control cells (Huang et al. 2013).

VP155 VP155 encoded a novel semaphorin homologue gene which contained a ~370 aa N-terminal Sema domain, a conserved plexin-semaphorin-integrin (PSI) domain, and an immunoglobulin (Ig)-like domain near the C terminus. VP155 was an early gene product and was predominantly distributed in the cytoplasm with a speckled and clubbed pattern of appearance. VP155 promoted viral replication in vitro and attenuated the cellular immune response, with no effect on the proliferation of host cells. Intriguingly, ectopically expressed VP155 altered the cytoskeletal structure of fish cells, characterized by a circumferential ring of microtubules near the nucleus and a disrupted microfilament organization (Yan et al. 2014).

VP062 VP062 encoded a novel insulin like growth factor with a variable number of tandem repeats (VNTR) at the 3′-end. VP062 was an early transcribed gene that was distributed predominantly in the cytoplasm with a diffused granular appearance. Overexpression of VP062 stimulated the growth of GP cells by promoting G1/S phase transition, which was at least partially dependent on its C-terminal VNTR locus. Furthermore, VP062 promoted SGIV replication in grouper cells. In addition, overexpression of VP062 slightly facilitated apoptosis in SGIV-infected FHM cells (Yan et al. 2013).

VP051 VP051 encoded a cytoplasmic protein. Overexpression of VP051 in vitro enhanced cell proliferation and affected cell cycle progression via altering the G1/S transition. Furthermore, VP051 overexpression improved cell viability during SGIV infection by inhibiting virus-induced apoptosis, as evidenced by the reduction of apoptotic bodies and the decrease of caspase-3 activation. Overexpression of VP051 increased viral replication and the expression of genes encoding the viral major capsid protein (MCP) and cell proliferation-promoting ICP-18 (Yu et al. 2016b). Deletion of VP51 via knockout (SGIVΔ051) resulted in decreased virulence as evidenced by reduced replication in vitro and decreased cumulative mortalities in SGIV-Δ051 challenged grouper compared to WT-SGIV. Moreover, VP051 deletion significantly increased virus induced apoptosis and reduced the expression of pro-inflammatory cytokines in vitro. In addition, the expression of several pro-inflammatory genes was decreased in SGIV-Δ051 infected grouper compared to WT-SGIV. Thus, SGIV VP051 was postulated to function as a critical virulence factor via regulation of host cell apoptosis and the inflammatory response (Yu et al. 2017a).

VP131 VP131 localized predominantly within the endoplasmic reticulum (ER). Ectopic expression of GFP-VP131 significantly enhanced SGIV replication, while VP131 knockdown decreased viral infection in vitro, suggesting that VP131 functioned as a proviral factor during SGIV infection. Overexpression of GFP-VP131 inhibited interferon (IFN)-1 promoter activity and the mRNA level of IFN-related genes induced by poly(I:C) such as *Epinephelus coioides* cyclic GMP/AMP synthase (EccGAS)/stimulator of IFN genes (EcSTING), TANK-binding kinase 1 (EcTBK1), or melanoma differentiation-associated gene 5 (EcMDA5), whereas activation induced by mitochondrial antiviral signaling protein (EcMAVS) was not affected. Moreover, VP131 interacted with EcSTING and degraded EcSTING through both the autophagy-lysosomal and ubiquitin-proteasome pathways. Further analysis indicated that VP131 induces K63-linked Ub-proteasomal degradation of EcSTING. In addition, GFP-VP131 inhibited EcSTING- or EcTBK1-induced antiviral activity upon red-spotted grouper nervous necrosis virus (RGNNV) infection. Thus, SGIV VP131 was hypothesized to negatively regulate the IFN response by inhibiting EcSTING-EcTBK1 signaling for viral evasion (Zhang et al. 2022).

VP049 (dUTP pyrophosphatase, dUTPase) dUTPase is a ubiquitous enzyme responsible for regulating dUTP concentrations (whose incorporation into DNA would be deleterious for virus replication) and raising levels of dTTP via the salvage pathway (Kato et al. 2014). SGIV VP049 encodes a dUTPase homolog, a peptide of 155 amino acids containing five conserved motifs. VP049 was characterized as an early gene product. Subcellular location analysis showed that VP049 is a cytoplasmic protein, and its nuclear export signal (NES) was crucial for its translocation from the nucleus to the cytoplasm (Gong et al. 2010).

VP086 (SGIV ICP18) ICP18 homologs can be found in all ranaviruses but not in other sequenced large DNA viruses. SGIV ICP18 is an immediate-early gene and begins to be transcribed as early as 2 hr post-infection (p.i.). Subcellular localization analysis revealed that the SGIV ICP18 displayed a finely punctate cytoplasmic pattern. Furthermore, overexpression of SGIV ICP18 promoted the growth of grouper embryonic cells and contributed to SGIV replication (Xia et al. 2009). Furthermore, the potential functions of ICP18 were further explored by identifying its interactors using a proximity-dependent BioID method. A total of 112 cellular proteins that potentially bind to ICP18 were identified by mass spectrometric analysis. Further analysis revealed that the identified ICP18-interacting proteins modulate cellular processes, such as the cell cycle and adhesion. In addition, interaction between ICP18 and one of its candidate interactors, i.e., cyclin-dependent kinase1 (CDK1), was confirmed using co-immunoprecipitation (co-IP) and pull-down assays (Wu et al. 2021).

VP162 (ICP46) Based on sequence analysis, ICP46 was identified as a core gene of the family *Iridoviridae*. SGIV ORF162L encodes a putative homolog of ICP46, which is classified as an immediate-early (IE) gene. ICP46, as a structural protein of the nucleocapsid, was distributed predominantly in the cytoplasm. The overexpression of SGIV ICP46 promoted the growth of GP cells and contributed to SGIV replication (Xia et al. 2010). Furthermore, the transcription start site (TSS) and 5'-untranslated region (5'-UTR) of SGIV ICP46 were determined using 5' RACE. Identification of putative virion-associated transcriptional transactivator(s) (VATT) that interacted with the SGIV ICP46 promoter and their binding site on the promoter was determined by electrophoretic mobility shift assays (EMSA), DNA pull-down assays, and mass spectrometry (MS). Four putative VATTs (VP12, VP39, VP57, and MCP) that bound the SGIV ICP46 promoter and might be involved in promoter activation were identified. In addition, the binding site of the VATTs on the promoter was determined to be a 10-base DNA sequence between the transcriptional start site (TSS) and a presumptive downstream promoter element (DPE) (Xia et al. 2019).

VP039 A homolog of SGIV VP039 was found in all sequenced iridoviruses and is considered a core gene of the family *Iridoviridae*. SGIV VP039 was classified as a late viral gene. Subcellular analysis showed that VP039 was mainly located in the cytoplasm, especially within viral assembly sites during the late stages of SGIV infection. Previous mass spectrometric analyses suggested that VP039 was an envelope protein, and immunogold electron microscopy further confirmed that assessment. In addition, a mouse anti-VP039 polyclonal antibody exhibited SGIV-neutralizing activity in vitro, suggesting that VP039 is involved in SGIV infection (Zhang et al. 2015).

VP038 VP038 encoded a 170-aa peptide containing an RGD motif. The VP038 gene was identified as a late viral gene. Subcellular localization analysis showed that VP038 was predominately distributed in the cytoplasm and gradually recruited to virus assembly sites. The results of immunogold electron microscopy indicated that SGIV VP038 was a viral capsid protein, consistent with previous mass spectrometric analyses. Like VP039, antibodies specific for SGIV VP038 exhibited substantial SGIV-neutralizing activity in vitro (Wan et al. 2010).

VP158 VP158 was observed in nuclei and virus assembly sites in SGIV infected cells. Crystal structure determination demonstrated that VP158 partially exhibits a structural resemblance to the histone binding region of anti-silencing factor 1 (Asf1), a histone H3/H4 chaperon. Isothermal titration calorimetry (ITC) experimentation showed interaction of VP158 with the histone H3/H4 complex and H3. Moreover, mutant analysis indicated that Arg67 and Ala93 of VP158 were key residues for its interaction with histone H3. Subsequently, ORF158L knockdown and iTRAQ (isobaric tags/mass spectrometry for relative and absolute quantification) analysis demonstrated that VP158 might be involved in both the regulation and the expression of histone H3 and H3 methylation. Thus, VP158 was thought to function as a histone H3 chaperon, enabling it to control host cellular gene expression and facilitate viral replication (Tran et al. 2011).

2.3 SGIV Encoded MicroRNAs (miRNAs)

In addition to functional viral proteins, SGIV encoded microRNAs (miRNAs) might also play important roles during virus infection. Using Illumina/Solexa deep-sequencing technology, 16 novel SGIV-encoded miRNAs were identified by a computational pipeline, including a miRNA that shared a similar sequence to herpesvirus miRNA HSV2-miR-H4-5p, which suggests miRNAs are present within iridovirids. These 16 miRNAs are encoded by sequences found throughout the SGIV genome, whereas 3 of them are located within the ORF057L region. Some SGIV-encoded miRNAs showed marked sequence and length heterogeneity at their 3′ and/or 5′ ends that could modulate their functions. Expression levels and potential biological activities of these viral miRNAs were examined by stem-loop quantitative RT-PCR and luciferase reporter assay, respectively, and 11 of these viral miRNAs might be functional during SGIV infection (Yan et al. 2011).

Among these miRNAs, miR-homoHSV, which targets SGIV VP136, a viral gene that encodes the pro-apoptotic lipopolysaccharide-induced TNF-α (LITAF)-like factor, can attenuate SGIV-induced cell death. miR-homoHSV suppressed exogenous and endogenous SGIV LITAF expression and inhibited SGIV LITAF-induced apoptosis. Therefore, miR-homoHSV was postulated to facilitate viral replication by down-regulating the expression of the pro-apoptotic viral gene LITAF and decreasing virus induced cell death (Guo et al. 2013). In addition, SGIV-miR-13 was demonstrated to function in a negative regulatory manner to restrict early viral replication via targeting SGIV MCP (Yan et al. 2015). Therefore, further detailed investigation of SGIV encoded miRNAs may provide new insights into the mechanism underlying iridovirus pathogenesis.

3 Pathogenesis of SGIV

3.1 Pathology Induced by SGIV Infection

Histopathological observations revealed that the characteristic symptoms caused by SGIV infection were splenomegaly and splenic hemorrhage accompanied by a significantly enlarged kidney, suggesting that the spleen and kidney were important target organs for SGIV infection (Qin et al. 2001). In situ hybridization using the SGIV MCP gene as a probe was performed to detect SGIV replication in the kidney, spleen, liver, intestine, stomach, and gills from naturally infected fish. The results showed strong hybridization signals in the kidney, followed by the spleen, liver, intestine, stomach, and gills. Furthermore, the viral signals were located specifically within epithelial, endothelial, and sub-endothelial hypertrophic cells in all detected tissues (Huang et al. 2004).

3.2 Virus Entry and Transport Tracking of SGIV

It has been reported that the entry of an iridovirus into host cells primarily utilizes endocytotic pathways mediated by clathrin or caveolin. For example, both tiger frog virus (TFV) and infectious spleen and kidney necrosis virus (ISKNV) entered cells via caveola-dependent endocytosis but not clathrin-mediated endocytosis (Guo et al. 2011, 2012a). Crucial events during SGIV entry were investigated using a combination of single-virus tracking and biochemical assays. SGIV infection was strongly inhibited when cells were pretreated with drugs blocking clathrin-mediated endocytosis and macropinocytosis. However, disruption of cellular cholesterol by methylcyclodextrin had no effect on virus infection suggesting that caveola-dependent endocytosis was not involved in cell entry (Wang et al. 2014). Further studies indicated that grouper clathrin light chains (EcCLCa) participated in the internalization of viral particles and delivery of particles to early endosomes, not in SGIV attachment (Wang et al. 2019). In addition, the intracellular transport of SGIV requires microtubules and actin filaments (Wang et al. 2014). Moreover, SGIV entered cells in a pH-dependent manner, and the Rab protein played a key role during SGIV entry. For example, the small GTPase Rab5c played roles in both SGIV infection and host immunity. Notably, EcRab5c disruption impaired crucial events at the early stage of SGIV infection, including virus binding, entry, transport, and host immune response (Wang et al. 2020). Ultimately, overexpression of EcRab20 significantly inhibited SGIV infection, whereas overexpression of dominant-negative (DN) EcRab20 (DN EcRab20) promoted SGIV infection. However, EcRab20 or DN EcRab20 had no effect on the entry of SGIV. Further studies indicated that EcRab20 affected SGIV infection by positively regulating the interferon immune and inflammatory response (Wang et al. 2022). Additionally, research has shown that SGIV VP69 and VP101 may interact with endosomal Rab7 and participate in virus entry (Fu et al. 2014).

The velocity and force of a single virion during SGIV invagination were monitored by developing and using "force tracking," a novel ultrafast (at the microsecond level) tracking technique. These results unambiguously revealed that the maximum velocity during cell entry of a single SGIV virion by membrane invagination is approximately 200 nm s (-1), the endocytic force is approximately 60.8 ± 18.5 pN, and the binding energy density increases with the engulfment depth (Pan et al. 2015). Taken together, those studies provided new insight into the dynamic process of SGIV entry via endocytosis and the mechanism of membrane invagination at the single-particle level.

3.3 Viral Assembly of SGIV

Determining viral structures at high resolution and exploring the capsid assembly process could provide pivotal information for identifying structure-function relationships and may be helpful for better understanding the mechanism of virus pathogenesis. The cryo-EM structure of the SGIV capsid at 3.5 Å resolution was explored using a block-based reconstruction method. Combining the cryo-EM map with viral proteomics, crosslinking mass spectrometry (CXMS), and fluorescence colocalization assays, the major capsid protein (MCP) and seven minor capsid proteins (mCPs) were identified. The atomic models for an asymmetric unit (ASU) of the viral capsid constituted by different copies of these capsid proteins were further determined. VP072, the MCP, is the most abundant virion component. Similar to the MCPs of nucleocytoviricota viruses (NCVs), the trunk region of VP072 displays a typical double jelly roll (JR) fold (JR-1 and JR-2), and each JR fold comprises two β-sheets formed by eight antiparallel β-strands. Three VP72 protomers trimerize to form a pseudo-hexameric capsomer. In addition, fluorescence colocalization assays showed that the anchor protein embedded within the viral inner membrane also colocalized with the endoplasmic reticulum (ER), supporting the hypothesis that the biogenesis of the inner membrane is associated with the ER. The observation that mCPs colocalized with the MCP within the cytoplasm of SGIV-infected cells before assembly into viral particles suggested that the virus could use diverse building blocks formed by mCP and MCP complexes for capsid assembly. These findings contributed to understanding the capsid assembly of the icosahedral NCVs and provided new clues for developing drugs and vaccines to prevent iridovirus infections (Zhao et al. 2023; Pintilie et al. 2019).

3.4   Programmed Cell Death (PCD) Induced by SGIV Infection

Virus induced programmed cell death (PCD), including apoptosis, autophagy, necroptosis, and other forms of PCD, play a critical role in the pathogenesis of virus diseases. SGIV triggers different forms of programmed cell death depending upon the cell type. In cultured grouper cells, a natural host for SGIV, infection triggers a nonapoptotic form of programmed cell death designated paraptosis. Paraptosis is characterized by the appearance of cytoplasmic vacuoles, distended endoplasmic reticulum, and the absence of DNA fragmentation, apoptotic bodies, and caspase activation. In contrast, in cells cultured from a non-host finfish, e.g., FHM, SGIV induces the typical form of apoptosis characterized by caspase activation and DNA fragmentation. Furthermore, disruption of mitochondrial transmembrane potential and externalization of phosphatidylserine (PS) are not detected in grouper cells but are seen in FHM cells after SGIV infection (Huang et al. 2011a). Similar results were obtained in grouper iridovirus (GIV)-infected grouper kidney cells (Pham et al. 2012).

Aside from apoptosis, virus infection could impact autophagy, which functions as either a pro-viral or antiviral pathway, depending on the virus and host cells. Based on the changes of LC3-II, Beclin1, and p-mTOR levels in infected cells, we found that SGIV inhibited autophagy in grouper spleen (GS) cells, which enhanced virus replication. Further studies showed that SGIV developed at least two strategies to inhibit autophagy: (1) increasing the cytoplasmic p53 level and (2) encoding viral proteins (VP48, VP122, VP132) that competitively bind autophagy related gene 5 and affect LC3 conversion (Li et al. 2020). In addition, a series of autophagy-related genes (Atgs) exerted negative regulation on the antiviral immune response against SGIV infection, including Atg5, Atg12, and Atg16L (Li et al. 2019a, b, c). A recent study has provided evidence that the exogenous addition of palmitic acid promoted SGIV replication by decreasing autophagy flux in GS cells via suppression of autophagic degradation (Yu et al. 2020). These results provide an antiviral strategy for SGIV from the perspective of autophagy.

3.5   Signaling Pathways Involved in SGIV Infection

To better clarify molecular events during SGIV infection, differentially expressed genes (DEGs) in SGIV-infected spleen were analyzed by high-throughput sequencing. The results showed that some genes enriched in intracellular signaling pathways, such as mitogen-activated protein kinase (MAPK) and the ubiquitin-proteasome system (UPS), were significantly upregulated after SGIV infection, suggesting that those pathways might be involved in the process of virus infection (Huang et al. 2011a, b, c). In host cell lines, the increased phosphorylation of extracellular signal-regulated kinase (ERK), mitogen-activated protein kinase (p38 MAPK), and c-Jun N-terminal kinase (JNK) occurred during active replication of SGIV. Moreover, downstream key molecules (MEK1/2, MSK-1, ELK-1, c-Jun, MAPKAPK-2, and ATF-2) were also activated during SGV infection, indicating that SGIV replication activated the ERK, JNK, and p38 MAPK signaling pathways (Huang et al. 2011b). Notably, using specific inhibitors, further studies showed that MAPK cascades were involved in SGIV replication and paraptosis induced by SGIV infection. Moreover, transcription of grouper immune genes, such as interferon regulatory factor 1 (IRF1), interleukin-8 (IL-8), and tumor necrosis factor alpha (TNF-α) was regulated by JNK, while only TNF-α was regulated by p38 MAPK. The JNK pathway is proposed to be important for SGIV replication and modulates inflammatory responses during virus infection (Huang et al. 2011b). Subsequently, MAPK signaling pathway related genes, including MAP kinase kinase (MKK) 7, p38 MAPK, JNK1, JNK2, JNK3, c-Jun, MAP kinase phosphatase (MKP) 1, were cloned, and their roles in SGIV infection were simultaneously elucidated (Guo et al. 2023; Guo et al. 2016d; Cai et al. 2011). Functional research analysis showed that JNK molecules not only regulated apoptosis but also participated in regulating the immune response (Guo et al. 2016a, b, c). Interestingly, Ecc-Jun accumulated within virus assembly sites during the late stage of SGIV infection, suggesting that Ecc-Jun might interact with viral proteins to regulate SGIV replication (Wei et al. 2015).

The ubiquitin-proteasome system (UPS) serves as the major intracellular pathway for protein degradation and plays crucial roles in several cellular processes. Numerous studies showed that UPS is required for successful viral replication by affecting viral entry, viral gene transcription, virion assembly, virion release, and immune evasion (Banks et al. 2003; Wang et al. 2016a; Casorla-Pérez et al. 2017). Data from transcriptomic analysis of DEGs illustrated that the expression of 65 genes within the UPS pathway, impacting ubiquitin encoding, ubiquitination, deubiquitination, and proteasome formation, were up- or down-regulated during SGIV infection. Inhibition of proteasome function decreased SGIV replication in vitro, confirming that UPS was required for SGIV infection. Sixty-two differentially expressed proteins (15 viral proteins and 47 host proteins) were identified as potential target molecules regulated by UPS during SGIV infection. Among them, host proteins were involved in ubiquitin mediated protein degradation, metabolism, cytoskeleton, macromolecular biosynthesis, and signal transduction (Huang et al.

2018). The tripartite motif (TRIM) gene family is a highly conserved group of E3 ubiquitin ligase proteins involved in multiple biological processes. Several TRIMs were cloned from grouper, and their roles in fish virus replication were characterized. For example, grouper TRIM13, TRIM62, and TRIM35 were found to negatively regulate the antiviral immune response against nodavirus (Huang et al. 2016a, b, 2017; Yang et al. 2016b), whereas grouper TRIM32, TRIM23, TRIM21, and TRIM82 acted as antiviral factors against iridovirus and nodavirus infection, respectively (Yu et al. 2017a; Lv et al. 2019; Zhi et al. 2022; Zheng et al. 2022a).

4 Host Immune Defense and the Metabolic Response Against SGIV Infection

4.1 Immune Defense Against SGIV Infection

To systematically track the host responses against SGIV infection, comprehensive transcriptomic analysis in vitro and in vivo was carried out using RNA-seq. In the spleen, differentially expressed genes (DEGs) associated with mitogen-activated protein kinase (MAPK), chemokines, toll-like receptors (TLR), and the RIG-I signaling pathway were all significantly impacted in response to SGIV infection (Huang et al. 2011a). Consistently, in grouper spleen cells, KEGG pathways related to the immune response, cell growth, and death pathways were significantly affected. In addition, RIG-I-like receptor signaling, TLR signaling, NOD-like receptor signaling, the cytosolic DNA sensing pathway, cytokine-cytokine receptor interaction, mTOR signaling pathway, Wnt signaling pathway, and MAPK signaling pathway were also significantly altered during the SGIV infection. Moreover, several adapters and signaling molecules, such as MAVS, TRIM25, MITA, TBK1, MDA5, and TRAF2 were significantly down-regulated upon SGIV infection (Guo et al. 2023). These results suggest that SGIV infection blocks the host interferon response in vitro. Next, the regulatory roles of SGIV infection on the poly[I:C] induced immune response were investigated in detail. SGIV infection significantly decreased poly [I:C] induced interferon promoter activation. Moreover, transcriptomic analysis showed that SGIV infection reduced the expression of 1479 up-regulated genes and increased the expression of 297 down-regulated genes in poly [I:C] transfected cells. The differentially expressed genes (DEGs) down-regulated by SGIV were directly related to immunity, inflammation, and viral infection. For example, consistent with the qRT-PCR findings, expression levels of interferon signaling and inflammation-related genes, including cGAS, STING, TBK1, MAVS, TNF, IRAK4, and NOD2, were up-regulated by poly [I:C] stimulation, but all were significantly down-regulated upon SGIV infection. Thus, it is postulated that SGIV infection counteracts the poly [I:C] induced antiviral immune response and enabled SGIV to escape host immune surveillance (Wang et al. 2023).

Elucidation of the roles of host genes during virus infection will contribute significantly to understanding the potential mechanism underlying the manipulation of the host response by SGIV. To date, certain host genes (discussed immediately below) have been demonstrated to exert crucial roles in response to SGIV infection.

RIG-I-like receptors Melanoma differentiation-associated gene 5 (MDA5), a critical member of the retinoic acid-inducible gene I (RIG-I)-like receptor (RLR) family, recognizes viral dsRNA and initiates an antiviral response in host cells. Grouper MDA5 (EcMDA5) is mostly localized within the cytoplasm of grouper spleen (GS) cells. Interestingly, during SGIV infection, the distribution pattern of EcMDA5 was significantly altered, suggesting that EcMDA5 might interact with viral proteins during virus infection. Ectopic expression of EcMDA5 in vitro delayed viral cytopathic effect (CPE) and significantly inhibited SGIV gene transcription. Moreover, overexpression of EcMDA5 significantly increased interferon (IFN) and IFN-stimulated response element (ISRE) promoter activities in a dose-dependent manner and enhanced the expression of IRF3, IRF7, and TRAF6. In addition, the transcription level of proinflammatory factors, including TNF-α, IL-6, and IL-8, was differently altered in EcMDA5 overexpressing cells. The regulatory roles of proinflammatory cytokines might contribute greatly to the antiviral activity of EcMDA5 against SGIV replication (Huang et al. 2016c).

As another important member of the RLR family, the laboratory of genetics and physiology 2 protein (LGP2) played crucial roles in modulating the cellular antiviral response during viral infection (Yoneyama et al. 2005). Grouper LGP2 (EcLGP2) contained three conserved domains: a DEAD/DEAH box helicase domain, a helicase superfamily C-terminal domain, and a C-terminal domain of RIG-I. Subcellular localization analysis showed that EcLGP2 distributed throughout the cytoplasm in grouper cells. Notably, the intracellular distribution of EcLGP2 was altered during late stages of SGIV infection. Moreover, overexpression of EcLGP2 in vitro significantly enhanced SGIV replication. Further studies indicated that EcLGP2 overexpression decreased the expression level of interferon related molecules or effectors, including IRF3, IRF7, ISG15, IFP35, MXI, MXII, and MDA5, suggesting that the negative feedback exerted on the interferon immune response by EcLGP2 might contribute to the enhancement of virus replication. Moreover, the expression levels of pro-inflammatory cytokines, including IL-8 and TNFα, were significantly decreased, but that of IL-6 was increased by the ectopic expression of EcLGP2. Thus, grouper LGP2 might negatively regulate the antiviral innate immune response during iridovirus infection (Yu et al. 2016c).

DNA sensors and effectors Cyclic GMP-AMP synthase (cGAS) is one of the classical pattern recognition receptors that recognize mainly intracytoplasmic DNA. Grouper cGAS (EccGAS) contained a Mab-21 typical structural domain, was uniformly distributed in the cytoplasm, and colocalized with the endoplasmic reticulum and mitochondria. Silencing of EccGAS inhibited the replication of Singapore grouper iridovirus (SGIV) in grouper spleen (GS) cells and enhanced the expression of interferon-related factors. Furthermore, EccGAS interacted with EcSTING,

EcTAK1, EcTBK1, and EcIRF3, and its overexpression negatively regulated EcSTING-induced interferon response (Zhang et al. 2023a).

DEAD (Asp-Glu-Ala-Asp)-box polypeptide 41 (DDX41) is a member of the DEXDc family of helicases that has recently been identified to be a crucial intracellular DNA sensor that triggers multiple signaling molecules to activate the type I interferon response. Grouper DDX41 (EcDDX41) was localized within the nucleus, and its overexpression significantly inhibited SGIV replication in GS cells. EcDDX41 overexpression significantly increased the expression of antiviral and inflammatory cytokine genes, including interferon regulatory factor genes (e.g., IRF1, IRF2, IRF3, and IRF7), interferon induced genes (e.g., ISG15, ISG56, IFP35, Viperin, and MXI), and pro-inflammatory cytokine genes (e.g., TNFα, IL-1β, and IL-8). Moreover, EcDDX41 positively regulated the mitochondrial antiviral-signaling protein (MAVS) and TANK-binding kinase 1 (TBK1)-induced interferon immune response but not the STING induced interferon immune response in grouper cells (Liu et al. 2019).

Stimulator of interferon genes (STING, also known as MITA, ERIS, MPYS, or TMEM173) has been identified as a central component in the innate immune response to cytosolic DNA and RNA derived from different pathogens. Grouper STING (EcSTING) encoded an ER-localized protein. Reporter gene assay showed that EcSTING activated ISRE and zebrafish type I IFN and type III IFN promoters in vitro. Mutant analysis showed that the phosphorylation of serine residues S379 and S387 mediated IFN promoter activity. The viral cytopathic effect and viral protein synthesis were significantly inhibited by overexpression of EcSTING, and the inhibitory effect was abolished in serine residue S379 and S387 mutant transfected cells. Thus, EcSTING might be an important regulator of the grouper's innate immune response against iridovirus infection (Huang et al. 2015).

TANK-binding kinase-1 (TBK1) was initially demonstrated to act as a critical integrator involved in coordinating the activation of IRF3 and NF-κB in innate antiviral immunity. Besides its antiviral activity, TBK1 was also found to be involved in other signaling pathways, such as oncogenesis and autophagy. Grouper TBK1 (EcTBK1) localized primarily in the cytoplasm of GS cells and ectopic expression of EcTBK1 significantly inhibited SGIV replication. Furthermore, EcTBK1 overexpression significantly increased the expression levels of interferon related cytokines and pro-inflammatory factors. In addition, the overexpression of EcTBK1 increased the IRF3- and IRF7-regulated interferon promoter ISRE and IFN activity and the regulatory effect on interferon immune response was dependent on its kinase domain (Hu et al. 2018).

Interferon Stimulated Genes

Interferon-stimulated genes (ISGs) encode direct-acting antiviral effectors or molecules with the potential to positively or negatively regulate IFN signaling and other host responses. Among them, interferon-induced proteins with tetratricopeptide repeats (IFIT) are strongly induced by type I interferon (IFN), double-stranded RNAs, and virus infection. IFIT1 has antiviral activity against both DNA and RNA viruses, such as Japanese encephalitis virus (JEV), West Nile virus (WNV), human papillomaviruses (HPV), hepatitis C virus (HCV), human cytomegalovirus (HCMV), and lymphocytic choriomeningitis virus (LCMV) via different mechanisms (Fensterl and Sen 2015). Grouper IFIT1 (EcIFIT1) is localized throughout the cytoplasm in transfected cells. EcIFIT1 overexpression significantly suppressed the replication of SGIV as demonstrated by decreases in cytopathic effect (CPE), viral gene transcription, and virus titers. Further studies showed that the ectopic expression of EcIFIT1 increased the transcription level of IFN-related molecules, including IFN regulatory factor (IRF) 3, IRF7, IFN stimulated gene (ISG) 15, and myxovirus resistance gene (MX) I. In addition, flow cytometry analysis suggested that EcIFIT1 overexpression affected cell cycle progression by mediating S/G2 transition (Zhang et al. 2018). Recently, TRIM proteins, which contained three conserved domains, RING (Really Interesting New Gene), B-Box, and coiled-coil (RBCC), have attracted attention as ISGs because of their crucial roles in antiviral immunity (Levraud et al. 2019). Multiple TRIM genes were cloned, and their roles in SGIV replication were elucidated, such as grouper TRIM8, TRIM16L, TRIM21, TRIM23, TRIM25, and TRIM32 (Yu et al. 2016a, 2017a; Yang et al. 2016a, b). For example, TRIM21 exerted an antiviral role via various mechanisms following infection with different viruses. During coxsackievirus B3 infection, TRIM21 interacted with MAVS and positively regulated the IRF3-induced type I interferon immune response. In contrast, TRIM21 was demonstrated to negatively regulate the dsDNA-induced interferon immune response by targeting and degrading DDX41 (Zhang et al. 2013). Grouper TRIM21 (EcTRIM21) was distributed in the cytoplasm in a punctate manner. Overexpression of EcTRIM21 in vitro significantly inhibited SGIV replication, as evidenced by the decreased severity of cytopathic effect (CPE) and the reduced expression levels of viral core genes. Consistently, the knockdown of EcTRIM21 by small interfering RNA (siRNA) promoted the replication of SGIV in vitro. Furthermore, EcTRIM21 overexpression increased both interferon (IFN) and interferon stimulated response element (ISRE) promoter activities. In addition, the transcription levels of IFN signaling related molecules were positively regulated by EcTRIM21 overexpression (Zheng et al. 2022a).

TRIM23, also named ADP ribosylation factor domain protein 1 (ARD1), has been reported to interact with TBK1, which is known to regulate the cellular IFN response. In addition, TRIM23 was confirmed to be involved in antiviral innate and inflammatory responses mediated by toll-like receptor 3 (TLR3) and RIG-I/MDA5. Grouper TRIM23 (EcTRIM23) exhibited either diffuse or aggregated localization in grouper cells. EcTRIM23 overexpression in vitro significantly inhibited SGIV replication as evidenced by delayed CPE and the decreased expression of viral core genes. EcTRIM23 significantly increased the expression levels of interferon (IFN)-related signaling molecules and pro-inflammatory cytokines, as well as the promoter activities of IFN and NF-κB. After co-transfection, TANK binding kinase 1 (TBK1), TNF receptor associated factor (TRAF) 3, TRAF4, TRAF5, and TRAF6 interacted with EcTRIM23 in grouper cells. Moreover, these proteins could be recruited and co-localized with EcTRIM23 in vitro, suggesting that EcTRIM23 might exert an antiviral function by positively regulating the TBK1- and TRAF-induced host IFN response (Zhi et al. 2022).

TRIM8 has been demonstrated to interact with a protein inhibitor of activated STAT3 (PIAS3), and the overexpression of TRIM8 abolished the negative effect of PIAS3 on the signal transducer and activator of transcription 3 (STAT3), either by degradation of PIAS3 or exclusion of PIAS3 from the nucleus. Grouper TRIM8 (EcTRIM8) showed different distribution patterns in grouper spleen (GS) cells, including punctate fluorescence evenly situated throughout the cytoplasm and bright aggregates. The ectopic expression of EcTRIM8 in vitro significantly inhibited the replication of SGIV as evidenced by the reduction in the severity of CPE and the significant decrease in viral gene transcription and protein synthesis. Moreover, the transcription of proinflammatory factors and interferon-related immune factors were differently regulated by EcTRIM8 during SGIV infection. In addition, overexpression of EcTRIM8 significantly increased the transcription of IRF3 and IRF7 and enhanced IRF3 or IRF7 induced interferon-stimulated response element (ISRE) promoter activity. Thus, the antiviral actions of EcTRIM8 might be due to its regulatory effects on the proinflammatory cytokines and interferon related transcription factors in response to fish viruses (Huang et al. 2016a, b).

Interestingly, a new subset of TRIM genes termed finTRIM (fish novel TRIM) has recently been predicted to share a low identity with mammal TRIM homologs but exert multiple roles in response to virus infection. Grouper finTRIM82 (EcfinTRIM82) was localized in the cytoplasm of grouper spleen (GS) cells. The ectopic expression of EcfinTRIM82 significantly enhanced the replication of SGIV in GS cells, evidenced by increased expression of viral genes, including the SGIV MCP and VP19. Furthermore, the ectopic expression of EcfinTRIM82 significantly decreased the expression of interferon (IFN)-related signaling molecules, including interferon regulatory factor 3 (IRF3), IRF7, interferon stimulated gene 15 (ISG15), ISG56, IFP35, and MXI, suggesting that EcfinTRIM82 regulated viral replication via the negative regulation of the host IFN response. Furthermore, the ectopic expression of EcfinTRIM82 significantly weakened the MDA5, MITA, and MAVS-induced IFN response by inhibiting the transcription of interferon related cytokines and the promoter activity of IFN (Lv et al. 2019).

Antimicrobial peptides (AMPs) Antimicrobial peptides (AMPs) are considered as one of the key protective components of all animals, providing the first line of defense against a wide range of microbial pathogens (Smith et al. 2000). Defensins are a group of small antimicrobial peptides playing an important role in innate host defense (White et al. 1995). Grouper β-defensin (EcDefensin) showed inhibitory effects on the infection and replication of SGIV. After infection with SGIV, EcDefensin was partly translocated from the cytoplasm to the nucleus. Furthermore, the remaining cytoplasmic EcDefensin gathered around the viral assembly site during the late stages of viral replication. Both synthetic peptide EcDefensin and cells over-expressing EcDefensin reduced the infectivity of SGIV by directly interacting with viral particles and affecting host cells (Guo et al. 2012b). Moreover, EcDefensin enhanced the host antiviral immune responses by activating the expression of Mx and IL-1β (Guo et al. 2012b). In addition, hepcidin is a multifunctional hormone oligopeptide that not only plays a regulatory role in iron metabolism, but also participates in regulating immunity in teleost animals. Grouper hepcidins (EC-hepcidin1 and EC-hepcidin2) have been demonstrated to show apparent antiviral activity against SGIV infection (Zhou et al. 2011a).

TNF receptor associated factors (TRAFs) TNF receptor associated factors (TRAFs) are a family of proteins primarily involved in the regulation of inflammation, antiviral responses, and apoptosis (Ji et al. 2000). In grouper, TRAF5 overexpression promotes SGIV replication during viral infection in vitro and promotes the activity of interferon-β, interferon-sensitive response element, and NF-κB (Sun et al. 2021). However, TRAF6 significantly repressed the transcription of SGIV genes in GS cells. When TRAF6 was co-transfected with IRF3 or IRF7, TRAF6 inhibited IRF-induced IFN-β activation (Wei et al. 2017). Thus, TRAFs could regulate the cellular antiviral response differently and play important roles in the host immune response against SGIV infection.

Autophagy-related genes Autophagy is a process of cellular degradation and recycling that is conserved in eukaryotic cells. Autophagy-related genes (ATG genes) are a group of genes that are involved in the regulation and execution of autophagy. In grouper, Atg5, Atg12, and Atg16 from orange-spotted grouper (*E. coioides*) were cloned and characterized (Li et al. 2019a, b, c). EcAtg5 and EcAtg16L1 mainly localized with a dot-like pattern in the cytoplasm, while EcAtg12 localized in the cytoplasm and nucleus in grouper spleen cells. Overexpression of EcAtg5, EcAtg12, and EcAtg16L1 significantly increased the replication of SGIV, as indicated by the increased CPE, viral gene transcription, and protein synthesis. All three Atg proteins decreased the expression levels of interferon related cytokines or effectors and pro-inflammatory factors, and inhibited the promoter activation of NFκB and IFN. Thus, aside from their function in autophagosome formation, Atg5, Atg12, and Atg16L1 regulated the production of type I interferons in response to SGIV infections.

Apoptosis-related genes Apoptosis is a highly regulated process of cell death critical for normal development and homeostasis in multicellular organisms, but it also plays a crucial role in response to virus infection (Verburg et al. 2022). Grouper FADD (EcFADD) was abundantly distributed in the cytoplasm and nucleus in grouper spleen (GS) cells. Overexpression of EcFADD inhibited SGIV infection and replication and SGIV-induced apoptosis. EcFADD promoted the activation of interferon-stimulated response element (ISRE) and type I interferon (IFN) genes in the antiviral IFN signaling pathway and inhibited the activation of apoptosis-related transcription factors p53 (Zhang et al. 2018).

Tumor necrosis factor (TNF) receptor type 1-associated DEATH domain protein (TRADD) is a TNFR1-associated signal transducer and an essential component of the TNFR1 complex that is involved in activating both apoptotic and nuclear factor

(NF)-κB pathways as an adaptor. Grouper TRADD (EcTRADD) exhibited a clear pattern of discrete and interconnecting cytoplasmic filaments resembling the death-effector filaments. After infection with SGIV, EcTRADD was transferred to the nucleus. Overexpression of EcTRADD inhibited replication of SGIV in vitro. EcTRADD induced the caspase-dependent apoptosis in control and infected cells. Additionally, EcTRADD significantly increased the promotor activation of NF-κB and p53 (Zhang et al. 2019a).

Receptor interacting protein (RIP) is an essential sensor of cellular stress, which may respond to apoptosis or cell survival and participate in antiviral pathways. Grouper (*E. coioides*) (EcRIP1) was distributed in the cytoplasm with point-like uniform and dot-like aggregation forms, while EcRIP2 existed in cytoplasmic filaments and dot aggregation patterns. Overexpression of EcRIP1 and EcRIP2 inhibited SGIV replication and positively regulated the expression levels of interferon (IFN) and IFN-stimulated genes and pro-inflammatory factors, respectively. EcRIP1 interacted with grouper tumor necrosis factor receptor type 1-associated DEATH domain protein (EcTRADD) to promote SGIV-induced apoptosis and interact with grouper Toll/interleukin-1 receptor (TIR) domain containing adapter inducing interferon-β (EcTRIF) to participate in Myeloid Differentiation Factor 88 (MyD88)-independent toll-like receptor (TLR) signaling. EcRIP1 may also interact with grouper tumor necrosis factor receptor-associated factors (TRAFs) as intracellular linker proteins and mediate the signaling of various downstream signaling pathways, including NF-κB and IFN (Zhang et al. 2020). The elevated expression levels of inflammatory cytokines induced by SGIV were markedly inhibited by EcRIP2 in a concentration-dependent manner. Moreover, EcRIP2 competed with EcASC in the binding of EcCaspase-1 in a dose-dependent manner. With increasing time after SGIV infection, EcCaspase-1 gradually combined with more EcRIP2 than EcASC (Zhang et al. 2023b). Thus, EcRIP1 and EcRIP2 may inhibit SGIV replication by interacting with multiple apoptosis related signaling molecules and regulating apoptosis during SGIV infection.

The apoptosis adaptor protein caspase and receptor interacting protein adaptor with death domain (CRADD) is a molecule that contains a death domain (DD)- and a caspase activation recruitment domain (CARD) and assists in inducing apoptosis and activating caspase-2. EcCRADD engages the pro-apoptotic protein EcRIP1 through the DD domain and activates caspase-2 through the CARD domain. This bridging allows EcCRADD to promote SGIV-induced apoptosis and synergistically inhibit SGIV replication in vitro. In contrast to EcCRADD, the overexpression of SGIV CARD protein VP48 significantly promotes SGIV replication and inhibits SGIV-induced apoptosis. Furthermore, VP48 may utilize caspase-2 as a scaffolding molecule to activate NF-κB and promote cell survival. Of note, EcCRADD competes with VP48 to bind with caspase-2, and inhibit VP48-induced NF-κB activity in a dose-dependent manner. Thus, EcCRADD inhibited virus replication via the caspase-2-activated p53 pathway and interfered with the VP48-activated NF-κB signaling pathway (Zhang et al. 2023c).

4.2 Host Metabolism Response Against SGIV Infection

During virus-host interactions, viral genes not only regulate immune related genes, but also manipulate different host metabolic pathways to ensure efficient replication by disrupting critical metabolic pathways and targeting master regulator proteins of metabolism. During SGIV infection, the expression levels of key enzymes in fatty acid synthesis in vivo and in vitro, including acetyl-Coenzyme A carboxylase alpha (ACC1), fatty acid synthase (FASN), medium-chain acyl-CoA dehydrogenase (MCAD), adipose triglyceride lipase (ATGL), lipoprotein lipase (LPL), and sterol regulatory element-binding protein-1 (SREBP1), were significantly increased. Moreover, SGIV infection induced the formation of lipid droplets (LDs), suggesting that SGIV altered de novo fatty acid synthesis in host cells. Using specific inhibitors, fatty acid synthesis and fatty acid β-oxidation were demonstrated to be essential for SGIV replication (Zheng et al. 2022b). In addition, the expression of both glucose transporters (GLUT1 and GLUT2) and the enzymes of glucose metabolism (hexokinase 2, HK2, and the pyruvate dehydrogenase complex, PDHX) were upregulated during SGIV infection in vivo and in vitro, suggesting that glycolysis might be involved in SGIV infection. Exogenous glucose supplementation promoted the expression of viral genes and infectious virion production, while glutamine had no effect on SGIV infection, indicating that glucose was required for SGIV replication. Consistently, pharmacological inhibition of glycolysis dramatically reduced the synthesis of SGIV MCP and virion production, and the promotion of glycolysis significantly increased SGIV replication. Furthermore, the knockdown of HK2, PDHX, or GLUT1 by siRNA decreased the transcription and protein synthesis of SGIV MCP and suppressed viral replication, indicating that those enzymes exerted essential roles in SGIV replication. Inhibition of mTOR activity in SGIV-infected cells effectively reduced the expression of key glycolytic enzymes, including HK2, PDHX, GLUT1, and GLUT2, and finally inhibited SGIV replication, suggesting that mTOR was involved in SGIV-induced glycolysis (Guo et al. 2022b).

In addition, metabolism related genes were also found to be crucial for SGIV replication. Grouper interferon-induced transmembrane protein 1 (EcIFITM1) encoded a cytoplasmic protein, which was partly colocalized with early endosomes, late endosomes, and lysosomes. Of note, the ectopic expression of EcIFITM1 significantly inhibited the replication of SGIV, which was demonstrated by reduced virus production and levels of viral gene transcription and protein expression. In contrast, the knockdown of EcIFITM1 using small interfering RNAs (siRNAs) promoted SGIV replication. The antiviral activity of EcIFITM1 might start at the step of viral entry into the host cells. Furthermore, the results of non-targeted lipometabolomics showed that EcIFITM1 overexpression induced lipid metabolism remodeling in vitro. All of the detected ceramides were significantly increased following EcIFITM1 overexpression, suggesting that EcIFITM1 may suppress SGIV entry by regulating the ceramide level in the lysosomal system. In addition, EcIFITM1 overexpression positively regulated both interferon-related molecules and ceramide synthesis-related genes, suggesting that EcIFITM1 might exert a bi-functional role, including immune regulation and lipid metabolism in response to SGIV infection (Zhang et al. 2021).

Cholesterol 25-hydroxylase (CH25H), is an ER-associated membrane protein that catalyzes the conversion of cholesterol to 25-hydroxycholesterol (25HC) to reduce cholesterol accumulation (Cyster et al. 2014). CH25H exerted antiviral activity through different mechanisms, such as blocking membrane fusion or activating an inflammatory cytokine response (Zang et al. 2020; Serquiña et al. 2021). Grouper CH25H (EcCH25H) contained a conserved fatty acid (FA) hydroxylase and ERG3 domains. Subcellular localization showed that EcCH25H and mutant EcCH25H-M were distributed in the cytoplasm and partly colocalized with the endoplasmic reticulum. SGIV replication was significantly decreased by EcCH25H overexpression, which was reflected in the reduced severity of cytopathic effect and a decrease in viral gene transcription but increased by knockdown of EcCH25H. Besides, the antiviral activity was dependent on its enzymatic activity. EcCH25H overexpression positively regulated IFN-related molecules and proinflammatory cytokines and increased IFN and ISRE promoter activities. Consistently, treatment with 25HC, the enzymatic product of CH25H, significantly inhibited replication of SGIV at the step of viral entry into the host cells. Thus, EcCH25H was hypothesized to exert antiviral activity against SGIV infection by regulating IFN signaling molecules and affecting viral entry via an effect on cholesterol (Zhang et al. 2019b).

5   Immuno-Biological Control of SGIV

5.1   Diagnosis Methods of SGIV

Accurate early diagnosis of viral pathogens will contribute greatly to the appropriate treatment, infection control, and prevention of further transmission of viral disease. For grouper iridoviral disease, several diagnostic methods have been developed to rapidly detect SGIV in recent years. Based on a monoclonal antibody specific to SGIV, a sensitive and accurate flow cytometry (FCM)-based method was developed to detect and quantitate SGIV after propagation in cell cultures (Qin et al. 2005). Using FCM, SGIV-infected cells could be distinguished from uninfected cells as early as 8 hr post infection (hpi), whereas the fluorescence of viral antigens was only able to be detected as early at 24 hpi by conventional immunofluorescence microscopy. The sensitivity of FCM was also comparable to that of PCR. Hence, FCM is a useful tool to detect SGIV from diseased fish and seawater samples after amplification in cells.

In addition, using a nonradioactive digoxigenin labeled DNA probe, an in situ hybridization method was established to complement histopathological methods. In SGIV-infected malabar grouper, strong hybridization signals were obtained from the kidney and spleen tissues, while intermediate intensity signals were observed in the intestine and liver tissues. The weakest signals were obtained from the stomach and gills (Huang et al. 2004).

Mao et al. developed a loop-mediated isothermal amplification (LAMP) for rapid, sensitive, and inexpensive detection of SGIV (Mao et al. 2008). The detection limit is about 0.02 fg of positive plasmid. A test of fish samples with suspected SGIV infection showed good concordance between LAMP and nested PCR. Moreover, the entire LAMP process requires only a water bath, making it simple and easy to operate in laboratory and field tests.

Aptamers are single-stranded oligonucleotides with high specificity and affinity to a wide range of targets, including toxins, antibiotics, proteins, viruses, and cells. Due to their secondary or tertiary spatial structures, aptamers specifically bind to targets and because of that are also called "chemical antibodies." Moreover, aptamers show many advantages over antibodies such as easy synthesis and modification, low cost, high stability, and the lack of immunogenicity, making them excellent components in a biosensor. For SGIV or SGIV-infected cells, a series of aptamers were selected by selective evolution of ligands by exponential enrichment. These aptamers formed stable stem-loop structures and specifically bound to SGIV with high affinity at the nanomolar level, exhibiting the potential to develop detection methods (Li et al. 2014, 2015).

A novel aptamer-based sandwich enzyme-linked apta-sorbent assay (ELASA) was developed for the rapid detection of SGIV (Li et al. 2016b). Apart from displaying good specificity, the detection limit of ELASA was as low as 5×10^4 SGIV-infected cells, with an incubation time as short as 1 min. In a validation study, 10 of 30 fish (33.3%) tested positive for SGIV-infection by PCR, while 33.3% of liver tissues, 23.3% of kidney tissues, and 26.7% of spleen tissues tested positive by the ELASA. Although ELASA assays appeared slightly less sensitive than PCR assays, they were much more convenient and less time consuming to perform than the latter. The procedure of ELASA mainly requires washing and incubation with completion in 1–2 h.

Meanwhile, a novel aptamer-based lateral flow biosensor (LFB) combined with strand displacement amplification (SDA) and colloidal gold nanoparticles was also developed to detect SGIV (Liu et al. 2021a). The LFB detected SGIV infection in a cell number-dependent manner with cell numbers as low as 5×10^4 /mL, with the complete process needing no more than 1.5 hr to complete. In groupers, positive detection of SGIV infection by LFB was 76.7%, which was slightly lower than that achieved by RT-PCR (86.6%). However, the LFB method did not need sophisticated manipulation and equipment compared to RT-PCR. Furthermore, the aptamers could be optimized to improve the detection sensitivity. Therefore, these aptamer-based methods have great application potentials in rapid and sensitive field detection of SGIV.

5.2   *Biological Control of SGIV*

Vaccination is one of the best methods to prevent and control viral disease. It is generally accepted that vaccination is a pivotal strategy that could lead to effective management of disease and significantly decrease economic losses in the fish industry. For example, fish immunized with formalin-inactivated vaccines exhibited good protective effects against the megalocytivirus, red seabream iridovirus (RSIV). Effective vaccination simultaneously triggered an early and efficient innate immune response and induced both specific humoral and cell-mediated immunity, while cell-mediated immunity was presumed to be mainly responsible for protecting the fish against infection (Caipang et al. 2006). Experimental, inactivated SGIV vaccines were prepared, and immunogenicity and protection against virus infection were investigated in orange-spotted grouper, *E. coioides* (Ou-yang et al. 2012a). Two kinds of vaccines, including β-propiolactone (BPL) inactivated virus (4 °C for 12 h) and formalin inactivated virus (4 °C for 12 d), were highly protective against SGIV challenge at 30-day post-vaccination and produced relative percentage survival rates of 91.7% and 100%, respectively. These effective vaccinations induced potent innate immune responses mediated by pro-inflammatory cytokines and type I interferon (IFN)-stimulated genes (ISGs). Moreover, effective vaccination also significantly up-regulated the expression of MHC class I genes and produced a substantial amount of specific serum antibody at 4 weeks post vaccination. The official clinical trials certificate of the SGIV-HN strain inactivated vaccine was approved by the Chinese Ministry of Agriculture and Rural Affairs in 2021 (approve No. 2021006). The clinical trials were carried out in Hanan, Guangdong, and Shandong provinces, and immunized groupers displayed more than 60% relative protection survival (RPS) in marine farming.

In addition to traditional inactivated virus vaccines, new generation vaccines, such as recombinant subunit vaccine and DNA vaccine, have gained considerable attention. A total of 162 potential open reading frames (ORFs) of SGIV were subjected to extensive sequence similarity searches, as well as motif, cellular location, and domain prediction. Among them, 13 DNA vaccine constructs were used to immunize fish, and the level of protection and survival was assessed after challenge with SGIV. Fish vaccinated with plasmid DNA encoding viral ORF072, ORF039, and ORF036 (designated as pcDNA-72, pcDNA-39, and pcDNA-36, respectively) exhibited 66.7%, 66.7%, and 58.3% relative percent survival rates, respectively, in comparison with the control fish. These three DNA vaccines induced innate immune responses, resulting in high levels of Mx expression relative to the fish vaccinated with the empty plasmid at 3 days post-vaccination. Furthermore, recombinant protein from ORF072 was also used to immunize another set of fish, and a similar protective effect was obtained (Ou-yang et al. 2012b).

Acknowledgements The authors would like to thank V. Gregory Chinchar and Kutti Subramaniam for their helpful comments in preparing this manuscript. We thank the following organizations for providing funds to support Open Access publishing of the second edition: University of Tennessee's (UT) Open Publishing Support Fund, UT School of Natural Resources, UT Center for Wildlife Health, Washington State University Libraries, University of Mississippi Medical Center, Gordon State College, Association of Reptile and Amphibian Veterinarians, and the Global Ranavirus Consortium.

References

Banks L, Pim D, Thomas M (2003) Viruses and the 26S proteasome: hacking into destruction. Trends Biochem Sci 28:452–459

Berthiaume L, Alain R, Robin J (1984) Morphology and ultrastructure of Lymphocystis disease virus, a fish iridovirus, grown in tissue culture. Virology 135(1):10–19

Cai J, Huang Y, Wei S, Huang X, Ye F, Fu J, Qin Q (2011) Characterization of p38 MAPKs from orange-spotted grouper, Epinephelus coioides involved in SGIV infection. Fish Shellfish Immunol 31(6):1129–1136

Caipang CM, Hirono I, Aoki T (2006) Immunogenicity, retention and protective effects of the protein derivatives of formalin-inactivated red seabream iridovirus (RSIV) vaccine in red seabream, Pagrus major. Fish Shellfish Immunol 20(4):597–609

Casorla-Pérez LA, López T, López S, Arias CF (2017) The ubiquitinproteasome system is necessary for the efficient replication of human astrovirus. J Virol 92:e1809–e1817

Chen LM, Wang F, Song W, Hew CL (2006) Temporal and differential gene expression of Singapore grouper iridovirus. J Gen Virol 87(Pt 10):2907–2915

Chi SC, Hu WW, Lo BJ (2010) Establishment and characterization of a continuous cell line (GF-1) derived from grouper, Epinephelus coioides (Hamilton): a cell line susceptible to grouper nervous necrosis virus (GNNV). J Fish Dis 22:173–182

Chinchar VG, Waltzek TB, Subramaniam -K (2017) Ranaviruses and other members of the family Iridoviridae: their place in the virosphere. Virology 511:259–271

Chua FHC, Ng ML, Ng KL, Loo JJ, Wee JY (1994) Investigation of outbreaks of a novel disease, "Sleepy Grouper Disease", affecting the brown-spotted grouper, Epinephelus tauvina. Forskal J Fish Diseas 17(4):417–427

Cyster JG, Dang EV, Reboldi A, Yi T (2014) 25-Hydroxycholesterols in innate and adaptive immunity. Nat Rev Immunol 14(11):731–743

Fensterl V, Sen GC (2015) Interferon-induced Ifit proteins: their role in viral pathogenesis. J Virol 89(5):2462–2468

Fu J, Huang Y, Cai J, Wei S, Ouyang Z, Ye F, Huang X, Qin Q (2014) Identification and characterization of Rab7 from orange-spotted grouper, Epinephelus coioides. Fish Shellfish Immunol 36(1):19–26

Gong J, Huang YH, Huang XH, Zhang R, Qin QW (2010) Nuclear-export-signal-dependent protein translocation of dUTPase encoded by Singapore grouper iridovirus. Arch Virol 155(7):1069–1076

Gong J, Huang Y, Huang X, Ouyang Z, Guo M, Qin Q (2011) Establishment and characterization of a new cell line derived from kidney of grouper, Epinephelus akaara (Temminck & Schlegel), susceptible to Singapore grouper iridovirus (SGIV). J Fish Dis 34:677–686

Guo CJ, Liu D, Wu YY, Yang XB, Yang LS, Mi S, Huang YX, Luo YW, Jia KT, Liu ZY, Chen WJ, Weng SP, Yu XQ, He JG (2011) Entry of tiger frog virus (an iridovirus) into HepG2 cells via a pH-dependent, atypical, caveola-mediated endocytosis pathway. J Virol 2011(85):6416–6426

Guo CJ, Wu YY, Yang LS, Yang XB, He J, Mi S, Jia KT, Weng SP, Yu XQ, He JG (2012a) Infectious spleen and kidney necrosis virus (a fish iridovirus) enters Mandarin fish fry cells via caveola-dependent endocytosis. J Virol 86:2621–2631

Guo M, Wei J, Huang X, Huang Y, Qin Q (2012b) Antiviral effects of β-defensin derived from orange-spotted grouper (Epinephelus coioides). Fish Shellfish Immunol 32(5):828–838

Guo C, Yan Y, Cui H, Huang X, Qin Q (2013) miR-homoHSV of Singapore grouper iridovirus (SGIV) inhibits expression of the SGIV pro-apoptotic factor LITAF and attenuates cell death. PLoS One 8(12):e83027

Guo CY, Huang YH, Wei SN, Ouyang ZL, Yan Y, Huang XH, Qin QW (2015) Establishment of a new cell line from the heart of giant grouper, Epinephelus lanceolatus (Bloch), and its application in toxicology and virus susceptibility. J Fish Dis 38(2):175–186

Guo M, Wei J, Zhou Y, Qin Q (2016a) Molecular clone and characterization of c-Jun N-terminal kinases 2 from orange-spotted grouper, Epinephelus coioides. Fish Shellfish Immunol 49:355–363

Guo M, Wei J, Zhou Y, Qin Q (2016b) MKK7 confers different activities to viral infection of Singapore grouper iridovirus (SGIV) and nervous necrosis virus (NNV) in grouper. Fish Shellfish Immunol 57:419–427

Guo M, Wei J, Huang X, Zhou Y, Yan Y, Qin Q (2016c) JNK1 derived from Orange-Spotted Grouper, Epinephelus coioides, involving in the evasion and infection of Singapore Grouper Iridovirus (SGIV). Front Microbiol 7:121

Guo M, Wei J, Zhou Y, Qin Q (2016d) c-Jun N-terminal kinases 3 (JNK3) from orange-spotted grouper, Epinephelus coioides, inhibiting the replication of Singapore grouper iridovirus (SGIV) and SGIV-induced apoptosis. Dev Comp Immunol 65:169–181

Guo X, Wang W, Zheng Q, Qin Q, Huang Y, Huang X (2022a) Comparative transcriptomic analysis reveals different host cell responses to Singapore grouper iridovirus and red-spotted grouper nervous necrosis virus. Fish Shellfish Immunol 128:136–147

Guo X, Zheng Q, Pan Z, Huang Y, Huang X, Qin Q (2022b) Singapore Grouper Iridovirus induces glucose metabolism in infected cells by activation of mammalian target of Rapamycin Signaling. Front Microbiol 13:827818

Guo M, Wei J, Zhou Y, Qin Q (2023) Antiviral immunity of grouper MAP kinase phosphatase 1 to Singapore grouper iridovirus infection. Dev Comp Immunol 143:104674

Heppell J, Berthiaume L (1992) Ultrastructure of lymphocystis disease virus (LDV) as compared to frog virus 3 (FV3) and chilo iridescent virus (CIV): effects of enzymatic digestions and detergent degradations. Arch Virol 125(1–4):215–226

Hu Y, Huang Y, Liu J, Zhang J, Qin Q, Huang X (2018) TBK1 from orange-spotted grouper exerts antiviral activity against fish viruses and regulates interferon response. Fish Shellfish Immunol 73:92–99

Huang C, Zhang X, Gin KY, Qin QW (2004) In situ hybridization of a marine fish virus, Singapore grouper iridovirus with a nucleic acid probe of major capsid protein. J Virol Methods 117(2):123–128

Huang X, Huang Y, Gong J, Yan Y, Qin Q (2008) Identification and characterization of a putative lipopolysaccharide-induced TNF-alpha factor (LITAF) homolog from Singapore grouper iridovirus. Biochem Biophys Res Commun 373(1):140–145

Huang XH, Huang YH, Sun JJ, Han X, Qin Q (2009) Characterization of two grouper Epinephelus akaara cell lines: application to studies of Singapore grouper iridovirus (SGIV) propagation and virus-host interaction. Aquaculture 292:172–179

Huang Y, Huang X, Yan Y, Cai J, Ouyang Z, Cui H, Wang P, Qin Q (2011a) Transcriptome analysis of orange-spotted grouper (Epinephelus coioides) spleen in response to Singapore grouper iridovirus. BMC Genomics 12:556

Huang X, Huang Y, Ouyang Z, Xu L, Yan Y, Cui H, Han X, Qin Q (2011b) Singapore grouper iridovirus, a large DNA virus, induces nonapoptotic cell death by a cell type dependent fashion and evokes ERK signaling. Apoptosis 16(8):831–845

Huang X, Huang Y, OuYang Z, Cai J, Yan Y, Qin Q (2011c) Roles of stress-activated protein kinases in the replication of Singapore grouper iridovirus and regulation of the inflammatory responses in grouper cells. J Gen Virol 92(Pt 6):1292–1301

Huang XH, Huang YH, Ouyang ZL, Qin QW (2011d) Establishment of a cell line from the brain of grouper (Epinephelus akaara) for cytotoxicity testing and virus pathogenesis. Aquaculture 311:65–73

Huang X, Gong J, Huang Y, Ouyang Z, Wang S, Chen X, Qin Q (2013) Characterization of an envelope gene VP19 from Singapore grouper iridovirus. Virol J 10:354

Huang YH, Huang XH, Ouyang ZL, Wei SN, Guo CY, Qin QW (2014) Development of a new cell line from the snout of giant grouper, Epinephelus lanceolatus (Bloch), and its application in iridovirus and nodavirus pathogenesis. Aquaculture 432:265–272

Huang Y, Ouyang Z, Wang W, Yu Y, Li P, Zhou S, Wei S, Wei J, Huang X, Qin Q (2015) Antiviral role of grouper STING against iridovirus infection. Fish Shellfish Immunol 47(1):157–167

Huang Y, Yang M, Yu Y, Yang Y, Zhou L, Huang X, Qin QW (2016a) Grouper TRIM13 exerts negative regulation of antiviral immune response against nodavirus. Fish Shellfish Immunol 55:106–115

Huang Y, Yu Y, Yang Y, Yang M, Zhou L, Huang X, Qin Q (2016b) Fish TRIM8 exerts antiviral roles through regulation of the proinflammatory factors and interferon signaling. Fish Shellfish Immunol 54:435–444

Huang Y, Yu Y, Yang Y, Yang M, Zhou L, Huang X, Qin Q (2016c) Antiviral function of grouper MDA5 against iridovirus and nodavirus. Fish Shellfish Immunol 54:188–196

Huang Y, Zhang J, Liu J, Hu Y, Ni S, Yang Y, Yu YP, Huang XH, Qin QW (2017) Fish TRIM35 negatively regulates the interferon signaling pathway in response to grouper nodavirus infection. Fish Shellfish Immunol 69:142–152

Huang XH, Wei SN, Ni SW, Huang YH, Qin QW (2018) Ubiquitin-Proteasome system is required for efficient replication of Singapore Grouper Iridovirus. Front Microbiol 9:2798

Ji I, Ishida T, Tsukamoto N, Kobayashi N, Naito A, Azuma S, Yamamoto T (2000) Tumor necrosis factor receptor-associated factor (TRAF) family: adapter proteins that mediate cytokine signaling. Exp Cell Res 254(1):14–24

Kato A, Hirohata Y, Arii J, Kawaguchi Y (2014) Phosphorylation of herpes simplex virus 1 dUTPase upregulated viral dUTPase activity to compensate for low cellular dUTPase activity for efficient viral replication. J Virol 88(14):7776–7785

Lai YS, Murali S, Ju HY, Wu MF, Guo IC, Chen SC, Fang K, Chang CY (2000) Two iridovirus-susceptible cell lines established fromkidney and liver of grouper, Epinephelus awoara (Temminck & Schlegel), and partial characterization of grouper iridovirus. J Fish Dis 23:379–388

Lai YS, John JA, Lin CH, Guo IC, Chen SC, Fang K, Chang CY (2003) Establishment of cell lines from a tropical grouper, Epinephelus awoara (Temminck & Schlegel), and their susceptibility to grouper irido- and nodaviruses. J Fish Dis 26:31–42

Lakra WS, Swaminathan TR, Joy KP (2011) Development, characterization, conservation and storage of fish cell lines: a review. Fish Physiol Biochem 37(1):1–20

Levraud JP, Jouneau L, Briolat V, Laghi V, Boudinot P (2019) IFN-stimulated genes in Zebrafish and humans define an ancient arsenal of antiviral immunity. J Immunol 203(12):3361–3373

Li P, Yan Y, Wei S, Wei J, Gao R, Huang X, Huang Y, Jiang G, Qin Q (2014) Isolation and characterization of a new class of DNA aptamers specific binding to Singapore grouper iridovirus (SGIV) with antiviral activities. Virus Res 188:146–154

Li P, Wei S, Zhou L, Yang M, Yu Y, Wei J, Jiang G, Qin Q (2015) Selection and characterization of novel DNA aptamers specifically recognized by Singapore grouper iridovirus-infected fish cells. J Gen Virol 96(11):3348–3359

Li P, Zhou L, Ni S, Xu M, Yu Y, Cai J, Wei S, Qin Q (2016a) Establishment and characterization of a novel cell line from the brain of golden pompano (Trachinotus ovatus). In Vitro Cell Dev Biol Anim 52(4):410–418

Li P, Zhou L, Wei J, Yu Y, Yang M, Wei S, Qin Q (2016b) Development and characterization of aptamer-based enzyme-linked apta-sorbent assay for the detection of Singapore grouper iridovirus infection. J Appl Microbiol 121(3):634–643

Li P, Zhou L, Wei S, Yang M, Ni S, Yu Y, Cai J, Qin Q (2017) Establishment and characterization of a cell line from the head kidney of golden pompano Trachinotus ovatus and its application in toxicology and virus susceptibility. J Fish Biol 90(5):1944–1959

Li C, Liu J, Zhang X, Wei S, Huang X, Huang Y, Wei J, Qin Q (2019a) Fish autophagy protein 5 exerts negative regulation on antiviral immune response against Iridovirus and Nodavirus. Front Immunol 10:517

Li C, Yu Y, Zhang X, Wei J, Qin Q (2019b) Grouper Atg12 negatively regulates the antiviral immune response against Singapore grouper iridovirus (SGIV) infection. Fish Shellfish Immunol 93:702–710

Li C, Wang L, Zhang X, Wei J, Qin Q (2019c) Molecular cloning, expression and functional analysis of Atg16L1 from orange-spotted grouper (Epinephelus coioides). Fish Shellfish Immunol 94:113–121

Li C, Wang L, Liu J, Yu Y, Huang Y, Huang X, Wei J, Qin Q (2020) Singapore Grouper Iridovirus (SGIV) inhibited autophagy for efficient viral replication. Front Microbiol 11:1446

Liang X, Liang J, Cao J, Liu S, Wang Q, Ning Y, Liang Z, Zheng J, Zhang Z, Luo J, Chen Y, Huang X, Huang Y, Qin Q, Zhou S (2023) Oral immunizations with Bacillus subtilis spores displaying VP19 protein provide protection against Singapore grouper iridovirus (SGIV) infection in grouper. Fish Shellfish Immunol 138:108860

Liu J, Huang Y, Huang X, Li C, Ni SW, Yu Y, Qin Q (2019) Grouper DDX41 exerts antiviral activity against fish iridovirus and nodavirus infection. Fish Shellfish Immunol 91:40–49

Liu J, Zhang X, Zheng J, Yu Y, Huang X, Wei J, Mukama O, Wang S, Qin Q (2021a) A lateral flow biosensor for rapid detection of Singapore grouper iridovirus (SGIV). Aquaculture 541(2021):736756

Liu ZT, Zhang X, Zhang Y, Qin QW, Huang XH, Huang YH (2021b) Establishment of a cell line from the head kidney of giant grouper (Epinephelus lanceolatus) and its susceptibility to fish viruses. Aquacult Rep 21:100899

Lv S, Zhang Y, Zheng J, Huang X, Huang Y, Qin Q (2019) Negative regulation of the interferon response by finTRIM82 in the orange spotted grouper. Fish Shellfish Immunol 88:391–402

Mao XL, Zhou S, Xu D, Gong J, Cui HC, Qin QW (2008) Rapid and sensitive detection of Singapore grouper iridovirus by loop-mediated isothermal amplification. J Appl Microbiol 105(2):389–397

OuYang ZL, Huang XH, Huang EY, Huang YH, Gong J, Sun JJ, Qin QW (2010) Establishment and characterization of a new marine fish cell line derived from red-spotted grouper Epinephelus akaara. J Fish Biol 77:1083–1095

Ou-yang Z, Wang P, Huang X, Cai J, Huang Y, Wei S, Ji H, Wei J, Zhou Y, Qin Q (2012a) Immunogenicity and protective effects of inactivated Singapore grouper iridovirus (SGIV) vaccines in orange-spotted grouper, Epinephelus coioides. Dev Comp Immunol 38(2):254–261

Ou-yang Z, Wang P, Huang Y, Huang X, Wan Q, Zhou S, Wei J, Zhou Y, Qin Q (2012b) Selection and identification of Singapore grouper iridovirus vaccine candidate antigens using bioinformatics and DNA vaccination. Vet Immunol Immunopathol 149(1–2):38–45

Pan Y, Wang S, Shan Y, Zhang D, Gao J, Zhang M, Liu S, Cai M, Xu H, Li G, Qin Q, Wang H (2015) Ultrafast tracking of a single Live Virion during the invagination of a cell membrane. Small 11(23):2782–2788

Parameswaran V, Ishaq Ahmed VP, Shukla R, Bhonde RR, Sahul Hameed AS (2007) Development and characterization of two newcell lines from milkfish (Chanos chanos) and grouper (Epinephelus coioides) for virus isolation. Mar Biotechnol (NY) 9:281–291

Pham PH, Lai YS, Lee FF, Bols NC, Chiou PP (2012) Differential viral propagation and induction of apoptosis by grouper iridovirus (GIV) in cell lines from three non-host species. Virus Res 167(1):16–25

Pintilie G, Chen DH, Tran BN, Jakana J, Wu J, Hew CL, Chiu W (2019) Segmentation and Comparative Modeling in an 8.6-Å Cryo-EM Map of the Singapore Grouper Iridovirus. Structure 27(10):1561–1569

Qin QW, Lam TJ, Sin YM, Shen H, Chang SF, Ngoh GH, Chen CL (2001) Electron microscopic observations of a marine fish iridovirus isolated from brown-spotted grouper, Epinephelus tauvina. J Virol Methods 98(1):17–24

Qin QW, Chang SF, Ngoh-Lim GH, Gibson-Kueh S, Shi C, Lam TJ (2003) Characterization of a novel ranavirus isolated from grouper Epinephelus tauvina. Dis Aquat Org 53(1):1–9

Qin QW, Gin KY, Lee LY, Gedaria AI, Zhang S (2005) Development of a flow cytometry based method for rapid and sensitive detection of a novel marine fish iridovirus in cell culture. J Virol Methods 125(1):49–54

Qin QW, Wu TH, Jia TL, Hegde A, Zhang RQ (2006) Development and characterization of a new tropical marine fish cell line from grouper, Epinephelus coioides susceptible to iridovirus and nodavirus. J Virol Methods 131(1):58–64

Serquiña AKP, Tagawa T, Oh D, Mahesh G, Ziegelbauer JM (2021) 25-hydroxycholesterol inhibits Kaposi's Sarcoma Herpesvirus and Epstein-Barr virus infections and activates inflammatory cytokine responses. MBio 12(6):e0290721

Smith VJ, Fernandes JM, Jones SJ, Kemp GD, Tatner MF (2000) Antibacterial proteins in rainbow trout, Oncorhynchus mykiss. Fish Shellfish Immunol 10(3):243–260

Song WJ, Qin QW, Qiu J, Huang CH, Wang F, Hew CL (2004) Functional genomics analysis of Singapore grouper iridovirus: complete sequence determination and proteomic analysis. J Virol 78(22):12576–12590

Song W, Lin Q, Joshi SB, Lim TK, Hew CL (2006) Proteomic studies of the Singapore grouper iridovirus. Mol Cell Proteomics 5(2):256–264

Sun M, Wu S, Zhang X, Zhang L, Kang S, Qin Q, Wei J (2021) Grouper TRAF5 exerts negative regulation on antiviral immune response against iridovirus. Fish Shellfish Immunol 115:7–13

Teng Y, Hou Z, Gong J, Liu H, Xie X, Zhang L, Chen X, Qin QW (2008) Whole-genome transcriptional profiles of a novel marine fish iridovirus, Singapore grouper iridovirus (SGIV) in virus-infected grouper spleen cell cultures and in orange-spotted grouper, Epinephulus coioides. Virology 377(1):39–48

Tran BN, Chen LM, Liu Y, Wu JL, Velázquez-Campoy A, Sivaraman J, Hew CL (2011) Novel histone H3 binding protein ORF158L from the Singapore grouper iridovirus. J Virol 85(17):9159–9166

Verburg SG, Lelievre RM, Westerveld MJ, Inkol JM, Sun YL, Workenhe ST (2022) Viral-mediated activation and inhibition of programmed cell death. PLoS Pathog 18(8):e1010718

Wan QJ, Gong J, Huang XH, Huang YH, Zhou S, Ou-Yang ZL, Cao JH, Ye LL, Qin QW (2010) Identification and characterization of a novel capsid protein encoded by Singapore grouper iridovirus ORF038L. Arch Virol 155(3):351–359

Wang F, Bi X, Chen LM, Hew CL (2008) ORF018R, a highly abundant virion protein from Singapore grouper iridovirus, is involved in serine/threonine phosphorylation and virion assembly. J Gen Virol 89(Pt 5):1169–1178

Wang L, Chong QY, Wu J (2013) A DNA-binding protein encoded by ORF008L of Singapore grouper iridovirus. Virus Res 176(1–2):37–44

Wang S, Huang X, Huang Y, Hao X, Xu H, Cai M, Wang H, Qin Q (2014) Entry of a novel marine DNA virus, Singapore grouper iridovirus, into host cells occurs via clathrin-mediated endocytosis and macropinocytosis in a pH-dependent manner. J Virol 88(22):13047–13063

Wang F, Liu Y, Zhu Y, Ngoc Tran B, Wu J, Leong HC (2015) Singapore Grouper Iridovirus ORF75R is a Scaffold protein essential for viral assembly. Sci Rep 5:13151

Wang S, Liu H, Zu X, Liu Y, Chen L, Zhu XQ, Zhang LK, Zhou Z, Xiao GF, Wang W (2016a) The ubiquitinproteasome system is essential for the productive entry of Japanese encephalitis virus. Virology 498:116–127

Wang W, Huang Y, Yu Y, Yang Y, Xu M, Chen X, Ni S, Qin Q, Huang X (2016b) Fish TRIM39 regulates cell cycle progression and exerts its antiviral function against iridovirus and nodavirus. Fish Shellfish Immunol 50:1–10

Wang L, Li Q, Ni S, Huang Y, Wei J, Liu J, Yu Y, Wang S, Qin Q (2019) The roles of grouper clathrin light chains in regulating the infection of a novel marine DNA virus, Singapore grouper iridovirus. Sci Rep 9(1):15647

Wang L, Li C, Zhang X, Yang M, Wei S, Huang Y, Qin Q, Wang S (2020) The small GTPase Rab5c exerts Bi-Function in Singapore Grouper Iridovirus infections and cellular responses in the grouper, Epinephelus coioides. Front Immunol 11:2133

Wang LQ, Zhang XY, Li JR, Yang M, Wang Q, Wei SN, Guan LF, Qin QW, Wang SW (2022) Rab20, a novel Rab small GTPase from orange-spotted grouper positively regulates host immune response against iridoviruses infection. Aquaculture 547:737534

Wang W, Zhang Y, Guo X, Xu W, Qin Q, Huang Y, Huang X (2023) Singapore grouper iridovirus infection counteracts poly I:C induced antiviral immune response in vitro. Fish Shellfish Immunol 135:108685

Wei S, Huang Y, Huang X, Qin Q (2015) Characterization of c-Jun from orange-spotted grouper, Epinephelus coioides involved in SGIV infection. Fish Shellfish Immunol 43(1):230–240

Wei S, Yu Y, Qin Q (2018) Establishment of a new fish cell line from the caudal fin of golden pompano Trachinotus ovatus and its susceptibility to iridovirus. J Fish Biol 92:1675–1686

Wei J, Zang S, Xu M, Zheng Q, Chen X, Qin Q (2017) TRAF6 is a critical factor in fish immune response to virus infection. Fish Shellfish Immunol 60:6–12

Wei J, Huang Y, Zhu W, Li C, Huang X, Qin Q (2019) Isolation and identification of Singapore grouper iridovirus Hainan strain (SGIV-HN) in China. Arch Virol 164:1869–1872.

Wen CM, Huang JY, Ciou JH, Kao YL, Cheng YH (2009) Immunochemical and molecular characterization of GBC4 as a tanycyte-like cell line derived from grouper brain. Comp Biochem Physiol A Mol Integr Physiol 153(2):191–201

White SH, Wimley WC, Selsted ME (1995) Structure, function, and membrane integration of defensins. Curr Opin Struct Biol 5(4):521–527

Wu JL, Chan R, Wenk MR, Hew CL (2010) Lipidomic study of intracellular Singapore grouper iridovirus. Virology 399(2):248–256

Wu G, Lin Q, Lim TK, Zhang Y, Aweya JJ, Zhu J, Yao D (2021) The interactome of Singapore grouper iridovirus protein ICP18 as revealed by proximity-dependent BioID approach. Virus Res 291:198218

Xia L, Cao J, Huang X, Qin Q (2009) Characterization of Singapore grouper iridovirus (SGIV) ORF086R, a putative homolog of ICP18 involved in cell growth control and virus replication. Arch Virol 154(9):1409–1416

Xia L, Liang H, Huang Y, Ou-Yang Z, Qin Q (2010) Identification and characterization of Singapore grouper iridovirus (SGIV) ORF162L, an immediate-early gene involved in cell growth control and viral replication. Virus Res 147(1):30–39

Xia LQ, Chen JL, Zhang HL, Cai J, Zhou S, Lu YS (2019) Identification of virion-associated transcriptional transactivator (VATT) of SGIV ICP46 promoter and their binding site on promoter. Virol J 16(1):110

Xu L, Liu M, Chen H, Zhang L, Xu Q, Zhan Z, Xu Z, Liu S, Wu S, Zhang X, Qin Q, Wei J (2023) Singapore grouper iridovirus VP122 targets grouper STING to evade the interferon immune response. 140:108990

Yan Y, Cui H, Jiang S, Huang Y, Huang X, Wei S, Xu W, Qin Q (2011) Identification of a novel marine fish virus, Singapore grouper iridovirus-encoded microRNAs expressed in grouper cells by Solexa sequencing. PLoS One 6(4):e19148

Yan Y, Cui H, Guo C, Li J, Huang X, Wei J, Qin Q (2013) An insulin-like growth factor homologue of Singapore grouper iridovirus modulates cell proliferation, apoptosis and enhances viral replication. J Gen Virol 94(Pt 12):2759–2770

Yan Y, Cui H, Guo C, Wei J, Huang Y, Li L, Qin Q (2014) Singapore grouper iridovirus-encoded semaphorin homologue (SGIV-sema) contributes to viral replication, cytoskeleton reorganization and inhibition of cellular immune responses. J Gen Virol 95(Pt 5):1144–1155

Yan Y, Guo C, Ni S, Wei J, Li P, Wei S, Cui H, Qin Q (2015) Singapore grouper iridovirus (SGIV) encoded SGIV-miR-13 attenuates viral infection via modulating major capsid protein expression. Virus Res 205:45–53

Yang Y, Huang Y, Yu Y, Yang M, Zhou S, Qin Q, Huang X (2016a) RING domain is essential for the antiviral activity of TRIM25 from orange spotted grouper. Fish Shellfish Immunol 55:304–314

Yang Y, Huang Y, Yu Y, Zhou S, Wang S, Yang M, Qin QW, Huang XH (2016b) Negative regulation of the innate antiviral immune response by TRIM62 from orange spotted grouper. Fish Shellfish Immunol 57:68–78

Yao D, Liu Y, Chen X, Lim TK, Wang L, Aweya JJ, Zhang Y, Lin Q (2019) In-depth proteomic profiling of the Singapore grouper iridovirus virion. Arch Virol 164(7):1889–1895

Yoneyama M, Kikuchi M, Matsumoto K, Imaizumi T, Miyagishi M, Taira K, Foy E, Loo YM, Gale M Jr, Akira S, Yonehara S, Kato A, Fujita T (2005) Shared and unique functions of the DExD/H-box helicases RIG-I, MDA5, and LGP2 in antiviral innate immunity. J Immunol 175(5):2851–2858

Yu Y, Huang X, Zhang J, Liu J, Hu Y, Yang Y, Cai J, Huang Y, Qin Q (2016a) Fish TRIM16L exerts negative regulation on antiviral immune response against grouper iridoviruses. Fish Shellfish Immunol 59:256–267

Yu Y, Huang Y, Wei S, Li P, Zhou L, Ni S, Huang X, Qin Q (2016b) A tumour necrosis factor receptor-like protein encoded by Singapore grouper iridovirus modulates cell proliferation, apoptosis and viral replication. J Gen Virol 97(3):756–766

Yu Y, Huang Y, Yang Y, Wang S, Yang M, Huang X, Qin Q (2016c) Negative regulation of the antiviral response by grouper LGP2 against fish viruses. Fish Shellfish Immunol 56:358–366

Yu Y, Wei S, Wang Z, Huang X, Huang Y, Cai J, Li C, Qin Q (2016d) Establishment of a new cell line from the snout tissue of golden pompano Trachinotus ovatus, and its application in virus susceptibility. J Fish Biol 88(6):2251–2262

Yu Y, Huang X, Liu J, Zhang J, Hu Y, Yang Y, Huang Y, Qin Q (2017a) Fish TRIM32 functions as a critical antiviral molecule against iridovirus and nodavirus. Fish Shellfish Immunol 60:33–43

Yu Y, Huang Y, Ni S, Zhou L, Liu J, Zhang J, Zhang X, Hu Y, Huang X, Qin Q (2017b) Singapore grouper iridovirus (SGIV) TNFR homolog VP51 functions as a virulence factor via modulating host inflammation response. Virology 511:280–289

Yu NT, Zheng XB, Liu ZX (2019) Protective immunity induced by DNA vaccine encoding viral membrane protein against SGIV infection in grouper. Fish Shellfish Immunol 92:649–654

Yu Y, Li C, Liu J, Zhu F, Wei S, Huang Y, Huang X, Qin Q (2020) Palmitic acid promotes virus replication in fish cell by modulating autophagy flux and TBK1-IRF3/7 pathway. Front Immunol 11:1764

Yuan Y, Hong Y (2016) Subcellular redistribution and sequential recruitment of macromolecular components during SGIV assembly. Protein Cell 7(9):651–661

Yuan Y, Wang Y, Liu Q, Zhu F, Hong Y (2016) Singapore grouper iridovirus protein VP088 is essential for viral infectivity. Sci Rep 6:31170

Zang R, Case JB, Yutuc E, Ma X, Shen S, Gomez Castro MF, Liu Z, Zeng Q, Zhao H, Son J, Rothlauf PW, Kreutzberger AJB, Hou G, Zhang H, Bose S, Wang X, Vahey MD, Mani K, Griffiths WJ, Kirchhausen T, Fremont DH, Guo H, Diwan A, Wang Y, Diamond MS, Whelan SPJ, Ding S (2020) Cholesterol 25-hydroxylase suppresses SARS-CoV-2 replication by blocking membrane fusion. Proc Natl Acad Sci USA 117(50):32105–32113

Zhao Z, Huang Y, Liu C, Zhu D, Gao S, Liu S, Peng R, Zhang Y, Huang X, Qi J, Wong CCL, Zhang X, Wang P, Qin Q, Gao GF (2023) Near-atomic architecture of Singapore grouper iridovirus and implications for giant virus assembly. Nat Commun 14(1):2050

Zhang Z, Bao M, Lu N, Weng L, Yuan B, Liu YJ (2013) The E3 ubiquitin ligase TRIM21 negatively regulates the innate immune response to intracellular double-stranded DNA. Nat Immunol 14(2):172–178

Zhang H, Zhou S, Xia L, Huang X, Huang Y, Cao J, Qin Q (2015) Characterization of the VP39 envelope protein from Singapore grouper iridovirus. Can J Microbiol 61(12):924–937

Zhang X, Zang S, Li C, Wei J, Qin Q (2018) Molecular cloning and characterization of FADD from the orange-spotted grouper (Epinephelus coioides). Fish Shellfish Immunol 74:517–529

Zhang X, Liu Z, Li C, Zhang Y, Wang L, Wei J, Qin Q (2019a) Grouper TRADD mediates innate antiviral immune responses and apoptosis induced by Singapore Grouper Iridovirus (SGIV) infection. Front Cell Infect Microbiol 9:329

Zhang Y, Wang L, Huang X, Wang S, Huang Y, Qin Q (2019b) Fish cholesterol 25-hydroxylase inhibits virus replication via regulating interferon immune response or affecting virus entry. Front Immunol 10:322

Zhang Y, Wang Y, Liu Z, Zheng J, Huang Y, Huang X, Qin Q (2019c) Grouper IFIT1 inhibits iridovirus and nodavirus infection by positively regulating interferon response. Fish Shellfish Immunol 94:81–89

Zhang X, Liu Z, Wu S, Sun M, Wei J, Qin Q (2020) Fish RIP1 mediates innate antiviral immune responses induced by SGIV and RGNNV infection. Front Immunol 11:1718

Zhang Y, Wang L, Zheng J, Huang L, Wang S, Huang X, Qin Q, Huang Y (2021) Grouper interferon-induced transmembrane protein 1 inhibits Iridovirus and Nodavirus replication by regulating virus entry and host lipid metabolism. Front Immunol 12:636806

Zhang Y, Gao X, Yang X, Wang Y, Wang W, Huang X, Qin Q, Huang Y (2022) Singapore Grouper Iridovirus VP131 drives degradation of STING-TBK1 pathway proteins and negatively regulates antiviral innate immunity. J Virol 96(20):e0068222

Zhang L, Zhang X, Liao J, Xu L, Kang S, Chen H, Sun M, Wu S, Xu Z, Wei S, Qin Q, Wei J (2023a) Grouper cGAS is a negative regulator of STING-mediated interferon response. Front Immunol 14:1092824

Zhang X, Wu S, Liu Z, Chen H, Liao J, Wei J, Qin Q (2023b) Grouper RIP2 inhibits Singapore grouper iridovirus infection by modulating ASC-caspase-1 interaction. Front Immunol 14:1185907

Zhang X, Wu ST, Liu ZT, Liao JM, Wei JG, Qin QW (2023c) A novel fish CRADD inhibits Singapore Grouper Iridovirus (SGIV) infection by mediating apoptosis and modulating SGIV CARD (VP48)-caspase-2 interaction. Aquaculture 576:739828

Zheng J, Zhi L, Wang W, Ni N, Huang Y, Qin Q, Huang X (2022a) Fish TRIM21 exhibits antiviral activity against grouper iridovirus and nodavirus infection. Fish Shellfish Immunol 127:956–964

Zheng Q, Huang Y, Wang L, Zhang Y, Guo X, Huang X, Qin Q (2022b) SGIV induced and exploited cellular De Novo fatty acid synthesis for virus entry and replication. Viruses 14(2):180

Zhi L, Wang W, Zheng J, Liu S, Zhou S, Qin Q, Huang Y, Huang X (2022) Grouper TRIM23 exerts antiviral activity against iridovirus and nodavirus. Front Immunol 13:985291

Zhou GZ, Li ZQ, Yuan XP, Zhang QY (2007) Establishment, characterization, and virus susceptibility of a new marine cell line from red spotted grouper (Epinephelus akaara). Mar Biotechnol (NY). 9:370–376

Zhou JG, Wei JG, Xu D, Cui HC, Yan Y, Ou-Yang ZL, Huang XH, Huang YH, Qin QW (2011a) Molecular cloning and characterization of two novel hepcidins from orange-spotted grouper, Epinephelus coioides. Fish Shellfish Immunol 30(2):559–568

Zhou S, Wan Q, Huang Y, Huang X, Cao J, Ye L, Lim TK, Lin Q, Qin Q (2011b) Proteomic analysis of Singapore grouper iridovirus envelope proteins and characterization of a novel envelope protein VP088. Proteomics 11(11):2236–2248

Zhou L, Li P, Liu J, Ni S, Yu Y, Yang M, Wei S. Qin Q (2017) Establishment and characterization of a mid-kidney cell line derived from golden pompano Trachinotus ovatus, a new cell model for virus pathogenesis and toxicology studies. In Vitro Cell Dev Biol Anim 53(4):320–327

Ranavirus Distribution and Host Range

Rachel E. Marschang, Jonathan I. Meddings, Thomas B. Waltzek, Paul Hick, Matthew C. Allender, Wytamma Wirth, and Amanda L. J. Duffus

1 Introduction

The genus *Ranavirus* is a group of globally emerging pathogens infecting fish, amphibians, and reptiles impacting both captive and wild animals globally (Table 1). Ranaviruses are pathogens capable of infecting multiple species at a site (e.g., Mao et al. 1999a; Duffus et al. 2008) and can be transmitted between taxonomic classes of ectothermic vertebrates (e.g., Brenes et al. 2014a, b). Ranaviruses are known to infect at least 262 species across 67 families of ectothermic vertebrates globally (Table 2, Fig. 1). Knowledge of the host range of various ranaviruses has been growing slowly but steadily, and the number of species now reported to have been

R. E. Marschang (✉)
Laboklin GmbH & Co. KG, Bad Kissingen, Germany

J. I. Meddings
Faculty of Medicine, Nursing and Health Sciences, Eastern Health Clinical School, Monash University, Melbourne, Victoria, Australia

T. B. Waltzek
College of Veterinary Medicine, Washington State University, Pullman, WA, USA

P. Hick
Elizabeth Macarthur Agricultural Institute, Department of Primary Industries, Menangle, NSW, Australia

M. C. Allender
Wildlife Epidemiology Lab, University of Illinois, Urbana, IL, USA

Brookfield Zoo Chicago, Brookfield, IL, USA

W. Wirth
Peter Doherty Institute for Infection and Immunity, University of Melbourne, Parkville, VIC, Australia

A. L. J. Duffus
Department of Natural Sciences, Gordon State College, Barnesville, GA, USA

© The Author(s) 2025
M. J. Gray, V. G. Chinchar (eds.), *Ranaviruses*,
https://doi.org/10.1007/978-3-031-64973-8_6

155

Table 1 Ranavirus species (old an recently approved binomial nomenclature) and named strains discussed in this chapter as well as the host classes in which they have been described and continents on which they have been found

Ranavirus species[a]	Named strains[b]	Host classes	Global range
Ambystoma tigrinum virus, Ranavirus ambystoma1	Ambystoma tigrinum virus (ATV) Orchinik ranavirus (ORV)	Amphibians	North America
Common midwife toad virus, Ranavirus alytes1	Andrias davidianus ranavirus (ADRV) Chinese giant salamander iridovirus (CGSIV) Chinese giant salamander virus (CGSV) Common midwife toad virus (CMTV) Pike-perch iridovirus (PPIV) White sturgeon ranavirus	Amphibians, fish, reptiles	Asia, Europe, North America
Epizootic hematopoietic necrosis virus, Ranavirus perca1	Epizootic hematopoietic necrosis virus (EHNV) European catfish virus (ECV) European sheatfish virus (ESV)	Fish, reptiles	Europe, Oceania
European North Atlantic ranavirus, Ranavirus gadus1	Cod iridovirus (CoIV) European North Atlantic ranavirus (ENARV) Lumpfish ranavirus Ranavirus maximus (Rmax)	Fish	Europe, North Atlantic
Frog virus 3, Ranavirus rana1	Bohle iridovirus (BIV) Frog virus 3 (FV3) German gecko ranavirus (GGRV) Rana grylio virus (RGV) Rana nigromaculata ranavirus (RNRV) Soft-shelled turtle iridovirus (STIV) Tiger frog virus (TFV) Wamena virus Zoo ranavirus	Amphibians, fish, reptiles	Africa, Asia, Europe, Oceania, North and South America
Santee-Cooper ranavirus, Ranavirus micropterus1	Doctor fish virus (DFV) Koi ranavirus Guppy virus 6 (GV6) Largemouth bass virus (LMBV) Mandarin fish ranavirus (MRV) Santee-Cooper ranavirus (SCRV)	Fish	Asia, North America
Singapore grouper iridovirus, Ranavirus epinephelus1	Grouper iridovirus (GIV) Singapore grouper iridovirus (SGIV)	Fish	Asia

(continued)

Table 1 (continued)

Ranavirus species[a]	Named strains[b]	Host classes	Global range
Taxonomically unassigned ranaviruses	Short-finned eel ranavirus (SERV)	Fish	Europe
Chimeric ranaviruses	Rana catesbeiana virus-Z2 (RCV-Z2) Tortoise ranavirus 1 and 2 (ToRV-1 and ToRV-2)	Amphibians, reptiles	Europe, North America

[a]Virus species names used in this text, followed by recently approved binomial species names
[b]Includes only strains mentioned in this chapter. Some individual strains have been given more than one name

Table 2 The taxonomic distribution of ranavirus cases among amphibian, fish, and reptilian hosts

Class	Family	No. species affected
Amphibians	Alytidae	1
	Ambystomatidae	10[a]
	Bufonidae	14[a]
	Centrolendae	3
	Craugastoridae	8
	Cryptobranchidae	2
	Dendrobatidae	5
	Dicroglossidae	2
	Eleutherodactylidae	1
	Hylidae	27
	Hynobiidae	1
	Leptodactylidae	4[a]
	Mantellidae	2
	Megophryidae	1
	Microhylidae	3
	Myobatrachidae	2
	Pipidae	2
	Plethodontidae	26
	Ptychadenidae	1[a]
	Ranidae	35[a]
	Rhacophoridae	1
	Salamandridae	15
	Scaphinopodidae	2
	Strabomantidae	8
	Telmatobiidae	1

(continued)

Table 2 (continued)

Class	Family	No. species affected
Fish	Acipenseridae	3
	Anguillidae	1
	Catostomidae	1
	Centrarchidae	9
	Channidae	1
	Cyprinidae	5
	Eleotridae	1
	Esocidae	2
	Gadidae	1
	Gasterosteidae	1
	Ictaluridae	2
	Labridae	1
	Latidae	1
	Lutjanidae	1
	Moronidae	3
	Nemacheilidae	1
	Percidae	2
	Poeciliidae	1
	Salmonidae	1
	Sciaenidae	1
	Scophthalmidae	1
	Serranidae	5
	Siluridae	1
	Sinipercidae	2
	Terapontidae	1
Reptiles	Agamidae	5
	Anguidae	2
	Anolidae	2
	Boidae	1[a]
	Chamaeleonidae	1
	Chelydridae	1
	Colubridae	1
	Emydidae	4
	Gekkonidae	1
	Iguanidae	1
	Kinosternidae	1
	Lacertidae	2
	Phrynosomatidae	1
	Pythonidae	4
	Testudinidae	8
	Trionychidae	1
	Varanidae	1

[a]Some reports only include the genera

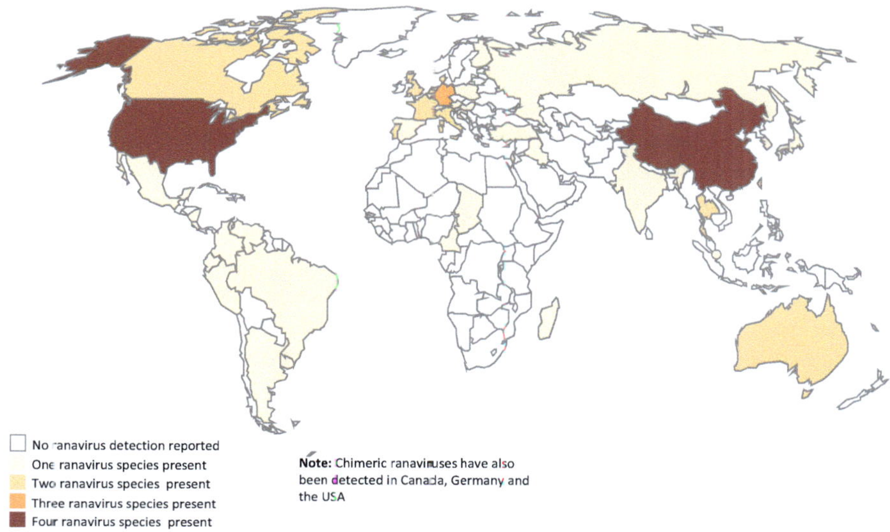

Fig. 1 Global distribution of vertebrate ranaviruses and heat map indicating the number of different ranavirus species found in each country. Shown are countries in which ranaviruses have been detected, but not a more detailed breakdown of the locations in which viruses have been found

naturally infected with a ranavirus has grown by 50% since the first edition of this chapter was published in 2015 (Duffus et al. 2015). Most of what is known about the epidemiology, geography, and host range of ranaviruses comes from investigations of obvious die-offs, sporadic surveillance efforts in small numbers of populations at one or two time points, and a few larger-scale surveillance efforts focused on a handful of species of economic importance or conservation interest (Grizzle and Brunner 2003; Gray et al. 2009b, 2024; Whittington et al. 2010; Miller et al. 2011). The known geographic and host range of ranaviruses likely remains an underestimate because of occasional misdiagnosis, subclinical infections that lack gross signs of infection, many host species being cryptic and difficult to detect, and a lack of awareness of ranaviruses as significant pathogens. Additionally, in many regions of the world, expertise in ranavirus diagnostics and the capacity for molecular screening for these viruses are simply not available. Screening for individual ranavirus species or strains in areas in which other strains and species are present may also contribute to underdetection as well as bias in interpretation of host and geographical range of individual ranavirus species.

There are currently seven species of ranavirus (Waltzek et al. 2024). This text uses the taxonomic terminology valid at the time of writing. The most recently approved binomial nomenclature of the recognized species in the genus are listed in Table 1, together with the names used in this text. Evaluation of worldwide reports shows individual virus species identified in some countries, while multiple species are found in others (Fig. 1). In many cases, the ranavirus detected has not been further characterized. By continent, only *Frog virus 3* has currently been identified on the African continent; *Common midwife toad virus*, *Frog virus 3*, *Santee-Cooper ranavirus*, and *Singapore grouper iridovirus* have been identified in Asia; *Common midwife toad virus*, *Frog virus 3*, *Epizootic hematopoietic necrosis virus*, and *European North Atlantic ranavirus* and a taxonomically unassigned ranavirus (short-finned eel ranavirus) have been identified in Europe; *Ambystoma tigrinum virus*, *Common midwife toad virus*, *Frog virus 3*, and *Santee-Cooper ranavirus* have been identified in North America; *Frog virus 3* and *Epizootic hematopoietic necrosis virus* have been identified in Oceania; and only *Frog virus 3* has been identified in South America (Table 3).

While the outcome of infection varies among hosts and strains of ranaviruses, it is clear these viruses have the potential to cause population declines and extinctions (Teacher et al. 2010; Earl and Gray 2014). They may present a significant threat to host species that are geographically isolated or exist at low abundance (Heard et al. 2013; Earl and Gray 2014), with rare and highly susceptible host species at greatest risk (Earl and Gray 2014). However, common species also can be affected. For example, populations of the common frog (*Rana temporaria*) have declined 80% on average in the UK, where ranavirus die-offs have reoccurred (Teacher et al. 2010). It is therefore important to understand the geographic extent, host range, and phylogenetic relationships of these emerging pathogens (Waltzek et al. 2024).

Transmission between host species and classes is well documented for individual isolates, especially in the species *Frog virus 3*, but more study is necessary to understand specific factors influencing host range and pathogenicity for these viruses. A comparison of geographic ranges known for specific ranaviruses over time indicates that some ranaviruses are clearly spreading, at least in part due to the pet, food, and angling industries (Schloegel et al. 2009; Kolby et al. 2014; Saucedo et al. 2017). This chapter presents an overview of ranaviruses found in each of the known ranavirus host classes, as well as information on the role of international trade in the global dissemination of ranaviruses and information on transmission studies demonstrating the ability of specific ranaviruses to infect hosts belonging to multiple classes. This chapter also provides an overview of the Global Ranavirus Reporting System (GRRS), a tool for the collection and sharing of data on species and locations in which ranaviruses are found.

2 Ranaviruses Infecting Amphibians

Since the 1990s, there has been a dramatic increase in the number of papers reporting ranaviruses in amphibians (Brunner et al. 2021; Wirth et al. 2021). While greater awareness and improved surveillance have no doubt impacted this trend, these increases are not solely a sampling artifact. We now understand that ranaviruses are globally distributed in amphibians and that mortality and morbidity events are happening in new places (Duffus et al. 2015; Chinchar et al. 2021). Indeed, ranaviruses threaten many different amphibian populations (e.g., Teacher et al. 2010; Earl and Gray 2014; Peace et al. 2019) and communities (e.g., Price et al. 2014; Rosa et al.

Table 3 Ranavirus species detected in individual countries

Continent	Country	Ranavirus species amphibians	Ranavirus species fish	Ranavirus species reptiles
Africa	Cameroon	Unknown		
	Chad	*FV3*		
	Madagascar	Unknown		
Asia	China	*CMTV, FV3*	*SCRV, CMTV, SGIV*	*FV3*
	India		*SCRV*	
	Iraq	*CMTV*		
	Israel	Unknown		
	Japan	*FV3*		*FV3*
	Singapore		*SGIV*	
	South Korea	*FV3*		
	Taiwan	*FV3*	*SGIV*	
	Thailand	*FV3*	*FV3, SCRV*	
	Turkey	Unknown		
Europe	Austria			*FV3*
	Belgium	*CMTV*		
	Croatia	Unknown		
	Denmark	*FV3*	*ENARV*	
	Finland		*CMTV*	
	France	*CMTV*	*EHNV*	
	Germany	*CMTV*	*EHNV*	*FV3, EHNV*
	Hungary	Unknown	*EHNV*	
	Italy	Unknown	*EHNV*, unassigned	
	Netherlands	*CMTV, FV3*		
	Poland			*CMTV*
	Portugal	*CMTV*		*FV3*
	Spain	*CMTV*		*CMTV*
	Switzerland	*CMTV*		*CMTV*
	Russia	Unknown		
	UK	*CMTV, FV3*		*FV3*
North America	Canada	*ATV, FV3*		*FV3*
	Costa Rica	Unknown		
	Nicaragua	Unknown		
	Mexico	*FV3*		
	USA	*ATV, FV3, CMTV*	*SCRV, FV3, CMTV*	*FV3*
Oceania	Australia	*FV3*	*EHNV*	*FV3*
South America	Argentina	Unknown		
	Brazil	*FV3*		
	Ecuador	Unknown		
	Peru	Unknown		
	Uruguay	Unknown		
	Venezuela	Unknown		

Abbreviations: *ATV Ambystoma tigrinum virus, CMTV Common midwife toad virus, EHNV Epizootic hematopoietic necrosis virus, ENARV European North Atlantic ranavirus, FV3 Frog virus 3, SCRV Santee-Cooper ranavirus, SGIV Singapore grouper iridovirus*

Fig. 2 Global distribution of amphibian ranaviruses. Shown are countries in which ranaviruses have been detected, but not a more detailed breakdown of the locations in which viruses have been found

2017; Bienentreu et al. 2022). In 2015, when the first edition of this book was published, ranaviruses had been reported in at least 105 species of amphibians from 18 families in 25 countries. By 2020, reports had increased to at least 145 species from 23 families in 32 countries (Chinchar et al. 2021), and they have increased further to at least 177 species from 25 families in 37 countries at the time of writing (Table 4, Fig. 2). As in 2015, these reported numbers are underestimates. In amphibians, underestimation of case numbers is exacerbated by the cryptic nature of many amphibian species, the fact that the gross clinical signs of ranavirosis are not necessarily specific to ranaviral disease, and difficulty observing morbidity and mortality events in amphibians due to their rapid decomposition and the short timespan during which these mortality events can occur (Brunner et al. 2015; Gray et al. 2024). Ranaviruses are known to infect anurans and caudates, but to date there have been no reports of ranavirus infection in caecilians. This could be due to their fossorial nature and the fact that less research is conducted on this group than on anurans and caudates.

There are currently only three recognized species of ranaviruses that are known to affect amphibians: *Frog virus 3*, *Ambystoma tigrinum virus*, and *Common midwife toad virus* (Table 1) (Chinchar et al. 2017a). There are also several chimeric ranaviruses (i.e., ranaviruses with genomes derived from two or more ranavirus species) that are known to affect amphibians, as well as some taxonomically unassigned viruses discussed in greater detail below. As discussed by Waltzek et al. (2024), declaring a specific ranavirus isolate a unique viral species is complex due to considerable sequence conservation that is often greater than 95% at the level of amino acids among isolates (e.g., a core set of genes; Eaton et al. 2007; Chinchar et al. 2017a; Waltzek et al. 2024).

Table 4 Distribution of ranavirus infections or mortality in wild and captive amphibians, including the location of origin for imported animals

Continent	Nation	Origin[a]	Family	Scientific name	I, M[b]	W, C[c]	Virus species[d]	References
Africa	Cameroon	–	Pipidae	*Xenopus longipes*	I	W		Docherty-Bone et al. (2013)
	Chad	–	Dicroglossidae	*Hoplobatrachus occipitalis*	I	W	*FV3*	Box et al. (2021)
			Ptychadenidae	*Ptychadena* spp.	I	W	*FV3*	Box et al. (2021)
	Madagascar	–	Mantellidae	*Mantidactylus cowanii*	I	W		Kolby et al. (2015)
				Mantidactylus mocquardi	I	W		Kolby et al. (2015)
Asia	China	–	Cryptobranchidae	*Andrias davidianus*	M	C	*CMTV*	Geng et al. (2011), Zhou et al. (2013), Ma et al. (2014)
			Ranidae	*Rana amurensis*	M	W	*FV3*	Zhu and Wang (2016)
		–		*Rana dybowskii*	I, M	W	*FV3*	Xu et al. (2010), Zhu and Wang (2016)
		–		*Rana grylio*	M	C	*FV3*	Zhang et al. (1996), Zhang et al. (2001)
		–		*Rana nigromaculata*	I, M	C	*FV3*	Mu et al. (2018), Yu et al. (2020)
		–		*Hoplobatrachus tigerinus* formerly *Rana tigrina*	M	C	*FV3*	Weng et al. (2002)
	Iraq	–	Salamandridae	*Neurergus crocatus*	M	W, C	*CMTV*	Stöhr et al. (2013c)
	Israel	–	Bufonidae	*Pseudepidalea viridis*	I	W		Milstein (2011)
	Japan	–	Hynobiidae	*Hynobius nebulosus*	M	C	*FV3*	Une et al. (2009a)
		–	Ranidae	*Lithobates catesbeianus*	M	W	*FV3*	Une et al. (2009b)
	South Korea	–	Hylidae	*Dryophytes japonicus* formerly *Hyla japonica*	I	W	*FV3*	Kwon et al. (2017), Roh et al. (2022)
			Microhylidae	*Kaloula borealis*	I	W	*FV3*	Park et al. (2017)
		–	Ranidae	*Lithobates catesbieanus*	I	W		Roh et al. (2022)
		–		*Pelophylax nigromaculatus*	I	W		Roh et al. (2022)
		–		*Rana dybowskii*	M	W	*FV3*	Park et al. (2021)
		–		*Rana huanrenensis*	M	W	*FV3*	Kwon et al. (2017)

(continued)

Table 4 (continued)

Continent	Nation	Origin[a]	Family	Scientific name	I, M[b]	W, C[c]	Virus species[d]	References
	Taiwan	–	Ranidae	*Lithobates catesbieanus*	M	W	*FV3*	Hsieh et al. (2021)
	Thailand	–	Dicroglossidae	*Fejervarya limnocharis*	M	C	*FV3*	Sriwanayos et al. (2020)
		–	Ranidae	*Hoplobatrachus tigerinus*	M	C	*FV3*	Kanchanakhan (1998), Sriwanayos et al. (2020)
		–		*Hoplobatrachus rugulosus*	M	C	*FV3*	Sriwanayos et al. (2020)
		–		*Pelophylax chosenicus*	M	C	*FV3*	Kim et al. (2009)
	Turkey	–	Ranidae	*Pelophylax caralitanus*	I	W		Erişmiş et al. (2019)
Australia	Australia	–	Hylidae	*Litoria caerulea*	M	W, C	*FV3*	Cullen and Owens (2002), Weir et al. (2012)
		–		*Litoria nannotis*	I	W		Wynne (2019)
		–		*Litoria lorica*	I	W		Wynne (2019)
		–		*Litoria splendida*	M	C	*FV3*	Weir et al. (2012)
		–	Myobatrachidae	*Limnodynastes ornatus*	M	C	*FV3*	Speare and Smith (1992)
		–		*Pseudophryne coriacea*	M	C	*FV3*	Cullen and Owens (2002)
Europe	Belgium	–	Bufonidae	*Bufo bufo*	I	W		Martel et al. (2012)
		–	Ranidae	*Lithobates catesbieanus*	I	W	*CMTV*	Sharifian-Fard et al. (2011)
		CN[e]	Salamandridae	*Tylototriton kweichowensis*	M	C		Pasmans et al. (2008)
	Croatia	–	Ranidae	*Pelophylax esculenta*	M	W		Fijan et al. (1991)
	Denmark	–	Ranidae	*Pelophylax esculenta*	M	W	*FV3*	Ariel et al. (2009)
	France	–	Ranidae	*Rana temporaria*	M	W	*CMTV*	Miaud et al. (2016)
	Germany	–	Ranidae	*Pelophylax esculenta*	M	W	*CMTV*	Stöhr et al. (2013a)
	Hungary	–	Bufonidae	*Bufo bufo*	I	W		Vörös et al. (2020)
		–	Ranidae	*Pelophylax ridibundus*	I	W		Vörös et al. (2020)
		–	Salamandridae	*Salamandra salamandra*	I	W		Vörös et al. (2020)

Country		Family	Species				Reference
	–	Salamandridae	*Triturus carnifex*	I	W		Vörös et al. (2020)
	–	Salamandridae	*Triturus dobrogicus*	I	W		Vörös et al. (2020)
Italy	–	Ranidae	*Pelophylax esculenta*	–	–		Ariel et al. (2010)
Netherlands	–	Dendrobatidae	*Dendrobates auratus*	M	C	*CMTV*	Kik et al. (2012)
		Dendrobatidae	*Oophaga pumilio*	I	C	*FV3*	Saucedo et al. (2017)
			Phyllobates bicolor	M	C	*CMTV*	Kik et al. (2012)
			Phyllobates vittatus	M	C	*CMTV*	Kik et al. (2012)
	–	Ranidae	*Pelophylax* spp.	I, M	W	*CMTV*	Kik et al. (2011), Saucedo et al. (2018)
	–	Salamandridae	*Lissotriton vulgaris*	I, M	W	*CMTV*	Kik et al. (2011), Saucedo et al. (2018)
		Salamandridae	*Triturus cristatus*	I	W	*CMTV*	Saucedo et al. (2018)
Portugal	–	Alytidae	*Alytes obstetricans*	I	W	*CMTV*	Rosa et al. (2017)
		Bufonidae	*Bufo spinosus*	I	W	*CMTV*	Rosa et al. (2017)
		Hylidae	*Hyla molleri*	I	W	*CMTV*	Rosa et al. (2017)
		Pipidae	*Xenopus laevis*	I	W		Coutinho (2019)
		Ranidae	*Pelophylax perezi*	I	W	*CMTV*	Rosa et al. (2017), Coutinho (2019)
	–	Salamandridae	*Lissotriton boscai*	I	W	*CMTV*	Rosa et al. (2017)
			Salamandra salamandra	I	W	*CMTV*	Rosa et al. (2017)
	–		*Triturus marmoratus*	I, M	W	*CMTV*	Alves de Matos et al. (2008), Rosa et al. (2017)
	–		*Triturus boscai*	M	W		Alves de Matos et al. (2008)
Spain	–	Alytidae	*Alytes obstetricans*	M	W	*CMTV*	Balseiro et al. (2009)
	–	Salamandridae	*Ichthyosaura alpestris*	I, M	W	*CMTV*	Balseiro et al. (2010), Martinez-Silvestre et al. (2017)

(continued)

Table 4 (continued)

Continent	Nation	Origin[a]	Family	Scientific name	I, M[b]	W, C[c]	Virus species[d]	References
	Switzerland	CZ, D, PL, SE, SK	Ranidae	*Pelophylax* kl. *esculentus*	M	C	*CMTV*	Stöhr et al. (2013a)
	Russia	–	Bufonidae	*Bufo bufo*	I	W		Reshetnikov et al. (2014), Lisachov et al. (2022)
	UK	–	Alytidae	*Alytes obstetricans*	M	W	*FV3*	Duffus et al. (2014)
		–	Bufonidae	*Bufo bufo*	M	W	*FV3, CMTV*	Hyatt et al. (2000), Duffus et al. (2014), Price et al. (2017)
		–	Ranidae	*Rana temporaria*	M	W	*FV3, CMTV*	Cunningham et al. (1993, 1996), Drury et al. (1995), Teacher et al. (2010), Duffus et al. (2013), Price et al. (2017)
		–	Salamandridae	*Lissotriton vulgaris*	I	W	*FV3, CMTV*	Duffus et al. (2014), Price et al. (2017)
North America	Canada	–	Ambystomatidae	*Ambystoma mavortium*	M	W	*ATV*	Bollinger et al. (1999), Schock et al. (2008)
		–		*Ambystoma* spp.	I	W	*FV3*	Duffus et al. (2008)
			Bufonidae	*Anaxyrus hemiophrys*	I, M	W	*FV3*	Forzán et al. (2019), Grant et al. (2019)
		–	Hylidae	*Hyla versicolor*	I	W	*FV3*	Duffus et al. (2008)
		–		*Pseudacris crucifer*	M	W		Miller et al. (2011)
		–		*Pseudacris* spp.	I	W	*FV3*	Duffus et al. (2008)
				Pseudacris maculata	I, M	W	*FV3*	Forzán et al. (2019), Grant et al. (2019)
		–	Ranidae	*Lithobates clamitans*	I, M	W	*FV3, chimeric*	St Amour et al. (2008), Forzán and Wood (2013), Grant et al. (2019)
		–		*Lithobates pipiens*	I, M	W, C	*FV3*	Greer et al. (2005), Schock et al. (2008), Echaubard et al. (2010), Paetow et al. (2011), Grant et al. (2019)

				I, M	W	FV3	
	–		Lithobates sylvaticus	I, M	W	FV3	Greer et al. (2005), Duffus et al. (2008), Schock et al. (2008), Schock et al. (2010), Grant et al. (2019), Forzán et al. (2019)
	–		Rana pretiosa	M	C	FV3	Schock et al. (2008)
	–	Salamandridae	Notophthalmus viridescens	I	W	FV3	Duffus et al. (2008)
Costa Rica	–	Bufonidae	Rhaebo haematiticus	I	W		Whitfield et al. (2013)
			Rhinella horribilis	I	W		Whitfield et al. (2021)
	–		Rhinella marina/Bufo marinus	I	W, C		Speare et al. (1991), Whitfield et al. (2013)
		Centrolenidae	Espadarana prosoblepon	I	W		Whitfield et al. (2021)
			Hyalinobatrachium colymbiphyllum	I	W		Whitfield et al. (2021)
			Teratohyla spinosa	I	W		Whitfield et al. (2013)
		Craugastoridae	Craugastor bransfordii	I	W		Whitfield et al. (2013), Whitfield et al. (2021)
			Craugastor crassidigitus	I	W		Whitfield et al. (2021)
			Craugastor fitzingeri	I	W		Whitfield et al. (2013), Whitfield et al. (2021)
			Craugastor megacephalus	I	W		Whitfield et al. (2013)
			Craugastor punctariolus	I	W		Puschendorf et al. (2019)
			Craugastor podiciferus	I	W		Whitfield et al. (2021)
			Craugastor stejnegerianus	I	W		Whitfield et al. (2021)
			Craugastor taurus	I	W		Whitfield et al. (2021)
	–	Dendrobatidae	Oophaga pumilio	I	W		Whitfield et al. (2013)
	–	Hylidae	Agalychnis lemur	I	W		Whitfield et al. (2021)
			Dendropsophus phlebodes	I	W		Whitfield et al. (2021)

(continued)

Table 4 (continued)

Continent	Nation	Origin[a]	Family	Scientific name	I, M[b]	W, C[c]	Virus species[d]	References
				Duellmanohyla legleri	I	W		Whitfield et al. (2021)
				Scinax elaeochroa	I	W		Whitfield et al. (2013), Whitfield et al. (2021)
		–		*Smilisca baudinii*	I	W		Whitfield et al. (2013)
			Ranidae	*Lithobates forreri*	I	W		Whitfield et al. (2021)
				Lithobates vibicarius	I	W		Whitfield et al. (2021)
				Lithobates warszewitschii	I	W		Whitfield et al. (2021)
	Nicaragua	–	Hylidae	*Agalychnis callidryas*	M	W		Stark et al. (2014)
		–	Ranidae	*Lithobates forrei*	M	W		Stark et al. (2014)
	Mexico	–	Ranidae	*Lithobates catesbeianus*	I, M	C	*FV3*	Saucedo et al. (2019)
	USA	–	Ambystomatidae	*Ambystoma jeffersonianum*	M	W		Green and Ip (2011)
				Ambystoma annulatum	I	W		Watters et al. (2018)
		–		*Ambystoma macrodactylum*	M	W		Driskell et al. (2009), Watters et al. (2018), Davis et al. (2019), Smith et al. (2019)
		–		*Ambystoma maculatum*	I, M	W		Green et al. (2002), Petranka et al. (2003), Gahl and Calhoun (2010), Todd-Thompson (2010), Brunner et al. (2011), O'Bryan et al. (2012), Homan et al. (2013), Olori et al. (2018), Watters et al. (2018), Mosher et al. (2019), Millikin et al. (2023)
		–		*Ambystoma mavortium*	I, M	W	*ATV*	Jancovich et al. (1997), Docherty et al. (2003), Jancovich et al. (2005), Picco and Collins (2008), Greer et al. (2009), Davis et al. (2020), Davis and Kerby (2016), Patla et al. (2016)

−		*Ambystoma opacum*	I, M	W		Todd-Thompson (2010), Davis et al. (2019)
		Ambystoma talpoideum	I	W		O'Bryan et al. (2012)
		Ambystoma texanum	I	W		Watters et al. (2018), Davis et al. (2019)
−		*Ambystoma tigrinum*	I, M	W, C		Green et al. (2002), Hoverman et al. (2012b)
−	Bufonidae	*Anaxyrus americanus*	I	W		Hoverman et al. (2012a), Olori et al. (2018), Watters et al. (2018), Davis et al. (2019), Smith et al. (2019)
−		*Anaxyrus boreas boreas*	I, M	C, W		Cheng et al. (2014), Tornabene et al. (2018), Patla et al. (2016)
		Anaxyrus fowleri	M	W		Monson-Collar et al. (2013)
		Anaxyrus woodhousii	I	W		Watters et al. (2018), Davis et al. (2019), Smith et al. (2019), Davis and Kerby (2016)
		Melanophryniscus stelzneri	I	C		Cheng et al. (2014)
−	Cryptobranchidae	*Cryptobranchus alleganiensis alleganiensis*	I	W		Souza et al. (2012), Hardman et al. (2020)
−	Dendrobatidae	*Dendrobates auratus*	I	C		Miller et al. (2008)
−		*Phyllobates terribilis*	I	C		Miller et al. (2008)
	Eleutherodactylidae	*Eleutherodactylus planirostris*	I	W		Rivera et al. 2019
	Hylidae	*Acris blanchardi*	I	W		Watters et al. (2018), Davis et al. (2019), Smith et al. (2019), Davis and Kerby (2016)
−		*Acris crepitans*	I	W		Hoverman et al. (2012a), Adamovicz et al. (2018)

(continued)

Table 4 (continued)

Continent	Nation	Origin[a]	Family	Scientific name	I, M[b]	W, C[c]	Virus species[d]	References
		–		*Hyla chrysoscelis*	I, M	W, C		Driskell et al. (2009), Watters et al. (2018), Davis et al. (2019), Smith et al. (2019), Davis and Kerby (2016)
		–		*Hyla chrysoscelis/Hyla versicolor* complex	I	W		O'Bryan et al. (2012)
		–		*Hyla cinerea*	M	W		Green and Converse (2005), Isidoro-Ayza et al. (2017), Davis et al. (2019), Smith et al. (2019), Rivera et al. (2019)
				Hyla squirella	I	W		Rivera et al. (2019)
				Hyliola regilla	I	W		Tornabene et al. (2018)
				Osteopilus septentrionalis	I	W		Galt et al. (2021)
		–		*Pseudacris clarkii*	M	W		Torrence et al. (2010), Isidoro-Ayza et al. (2017), Olori et al. (2018), Watters et al. (2018), Davis et al. (2019)
		–		*Pseudacris crucifer*	M	W		Green et al. (2002), Gahl and Calhoun (2010), Todd-Thompson (2010)
		–		*Pseudacris feriarum*	I, M	W		Todd-Thompson (2010), Hoverman et al. (2012a)
				Pseudacris fouquettei	I	W		Davis et al. (2019), Smith et al. (2019)
				Pseudacris maculata/triseriata	I	W		Davis and Kerby (2016), Talbott et al. (2018)
		–		*Pseudacris regilla*	M	W		Green and Ip (2011)
		–		*Pseudacris sierra*	M	W		Russell et al. (2011)
				Pseudacris maculata	M	W		Patla et al. (2016)
			Microhylidae	*Gastrophryne carolinensis*	I	W		Watters et al. (2018), Davis et al. (2019)
				Gastrophryne olivacea	I	W		Davis et al. (2019), Smith et al. (2019)
			Megophryidae	*Megophrys nasuta*	I	C		Cheng et al. (2014)

		Plethodontidae	*Aneides aeneus*	I	W		Blackburn et al. (2015), Newman et al. (2019)
	–		*Desmognathus conanti*	I	W		Gray et al. (2009a), Sutton et al. (2015)
	–		*Desmognathus folkertsi*	I	W		Rothermel et al. (2013)
	–		*Desmognathus fuscus*	I	W		Davidson and Chambers (2011), Hamed et al. (2013)
	–		*Desmognathus imitator*	I	W		Gray et al. (2009b), Sutton et al. (2015)
	–		*Desmognathus marmoratus*	I	W		Rothermel et al. (2013), Sutton et al. (2015)
	–		*Desmognathus monticola*	I	W		Gray et al. (2009b), Davidson and Chambers (2011), Hamed et al. (2013), Rothermel et al. (2013), Sutton et al. (2015)
	–		*Desmognathus ocoee*	I	W		Gray et al. (2009b), Rothermel et al. (2013), Sutton et al. (2015)
	–		*Desmognathus orestes*	I	W		Hamed et al. (2013)
	–		*Desmognathus organi*	I	W		Hamed et al. (2013)
	–		*Desmognathus quadramaculatus*	I	W		Gray et al. (2009b), Davidson and Chambers (2011), Hamed et al. (2013), Rothermel et al. (2013), Sutton et al. (2015)
	–		*Desmognathus santeetlah*	I	W		Gray et al. (2009b), Sutton et al. (2015)
	–		*Desmognathus wrighti*	I	W		Gray et al. (2009b), Sutton et al. (2015)
	–		*Eurycea cirrigera*	I	W		Davidson and Chambers (2011)
			Eurycea bislineata	I	W		Adamovicz et al. (2018), Olori et al. (2018)
	–		*Eurycea longicauda*	I	W		Davidson and Chambers (2011), Watters et al. (2018)

(continued)

Table 4 (continued)

Continent	Nation	Origin[a]	Family	Scientific name	I, M[b]	W, C[c]	Virus species[d]	References
		–		*Eurycea lucifuga*	I	W		Davidson and Chambers (2011), Watters et al. (2018)
				Eurycea multiplicata	I	W		Watters et al. (2018)
		–		*Eurycea wilderae*	I	W		Gray et al. (2009b), Sutton et al. (2015)
				Eurycea tynerensis	I	W		Watters et al. (2018)
		–		*Gyrinophilus porphyriticus*	I	W		Gray et al. (2009b)
				Plethodon albagula	I	W		Watters et al. (2018)
		–		*Plethodon glutinosus* complex	I	W		Davidson and Chambers (2011)
		–		*Plethodon jordani*	I	W		Gray et al. (2009b), Sutton et al. (2015)
		–		*Plethodon montanus*	I	W		Hamed et al. (2013)
		–		*Plethodon welleri*	I	W		Hamed et al. (2013)
		–	Ranidae	*Lithobates blairi*	M	W		Green and Ip (2011), Davis and Kerby (2016)
				Lithobates capito	M	W	*FV3*	Hartmann et al. (2022)
		–		*Lithobates catesbeianus*	I, M	W, C	*FV3, CMTV,* chimeric	Wolf et al. (1969), Green et al. (2002), Gray et al. (2007), Majji et al. (2006), Miller et al. (2007, 2009), Gahl and Calhoun (2010), Davidson and Chambers (2011), Homan et al. (2013), Landsberg et al. (2013), Olori et al. (2018), Isidoro-Ayza et al. (2017), Davis et al. (2019), Watters et al. (2018), Tornabene et al. (2018), Brunner et al. (2019), Davis and Kerby (2016), Claytor et al. (2017)

	–		*Lithobates clamitans*	I, M	W	Green et al. (2002), Gahl and Calhoun (2010), Gray et al. (2007), Miller et al. (2009), Homan et al. (2013), Monson-Collar et al. (2013), Olori et al. (2018), Youker-Smith et al. (2018), Davis et al. (2019), Isidoro-Ayza et al. (2017), Adamovicz et al. (2018), Watters et al. (2018), Winzeler et al. (2016)
			Lithobates clamitans melanota	I	W	Julian et al. (2019)
	–		*Lithobates palustris*	I, M	W	Green et al. (2002), Hoverman et al. (2012a), Davidson and Chambers (2011), Isidoro-Ayza et al. (2017), Watters et al. (2018)
	–		*Lithobates pipiens*	I, M	W	Granoff et al. (1965), Clark et al. (1968), Green et al. (2002), Uyehara et al. (2010), Olori et al. (2018), Davis and Kerby (2016)
	–		*Lithobates septentrionalis*	M	W	Green et al. (2002)
	–		*Lithobates sphenocephalus*	I, M	W	Green and Ip (2011), Hoverman et al. (2012a), O'Bryan et al. (2012), Landsberg et al. (2013), Monson-Collar et al. (2013), Smalling et al. (2018), Smith et al. (2019)

(continued)

Table 4 (continued)

Continent	Nation	Origin[a]	Family	Scientific name	I, M[b]	W, C[c]	Virus species[d]	References
		–		*Lithobates sylvaticus*	I, M	W		Green et al. (2002), Petranka et al. (2003), Harp and Petranka (2006), Gahl and Calhoun (2010), Todd-Thompson (2010), Uyehara et al. (2010), Brunner et al. (2011), Kolozsvary and Brunner (2016), Olori et al. (2018), Youker-Smith et al. (2018), Youker-Smith et al. (2018), Talbott et al. (2018), Smalling et al. (2018), Mosher et al. (2019), Isidoro-Ayza et al. (2017), Hall et al. (2018, 2020), Kirschman et al. (2017), O'Connor et al. (2016)
		–		*Pyxicephalus adspersus*	M	C		Miller et al. (2007)
		–		*Rana aurora*	M	W		Mao et al. (1999a)
		–		*Rana draytonii*	M	W		Green and Ip (2011)
		–		*Rana heckscheri*	M	W		Green and Ip (2011)
		–		*Rana luteiventris*	M	W		Converse and Green (2005), Green and Converse (2005), Russell et al. (2011), Patla et al. (2016)
		–		*Rana mucosa*	I, M	W		Converse and Green (2005), Smith et al. (2017)
		–		*Rana sierra*	M	W		Smith et al. (2017)
		–	Rhacophoridae	*Rhacophorus dennysi*	M	C		Miller et al. (2008)
			Salamandridae	*Notophthalmus viridescens*	I, M	W		Granoff et al. (1965), Green et al. (2002), Glenney et al. (2010), Richter et al. (2013), Watters et al. (2018), Davis et al. (2019)
				Notophthalmus perstriatus	M	W	*FV3*	Hartmann et al. (2022)

		–		*Taricha torosa*	I	W		Tornabene et al. (2018)
			Scaphiopodidae	*Scaphiopus holbrookii*	I, M	W		Kirschman et al. (2017)
				Spea bombifrons	I	W		Davis and Kerby (2016)
South America	Argentina	–	Leptodactylidae	*Atelognathus patagonicus*	M	W		Fox et al. (2006)
	Brazil	–	Ranidae	*Lithobates catesbeianus*	I, M	C	*FV3*	Mazzoni et al. (2009), Ruggeri et al. (2019), Candido et al. (2019)
	Columbia		Bufonidae	*Osornophryne* sp.	I	W		Flechas et al. (2023)
			Strabomantidae	*Pristimantis bogotensis*	I	W		Flechas et al. (2023)
				Pristimantis elegans	I	W		Flechas et al. (2023)
			Leptodactylidae	*Leptodactylus validus*	I	W		Flechas et al. (2023)
				Leptodactylus fragilis	I	W		Flechas et al. (2023)
			Ranidae	*Lithobates catesbeianus*	I	W		Flechas et al. (2023)
	Ecuador		Strabomantidae	*Pristimantis orestes*	I	W		Urgiles et al. (2021)
				Pristimantis phoxocephalus	I	W		Urgiles et al. (2021)
	Peru		Bufonidae	*Rhinella manu*	I	W		Warne et al. (2016)
			Hylidae	*Hypsiboas gladiator*	I	W		Warne et al. (2016)
			Strabomantidae	*Pristimantis lindae*	I	W		Warne et al. (2016)
				Pristimantis pharangobates	I	W		Warne et al. (2016)
				Pristimantis platydactylus	I	W		Warne et al. (2016)

(continued)

Table 4 (continued)

Continent	Nation	Origin[a]	Family	Scientific name	I, M[b]	W, C[c]	Virus species[d]	References
				Pristimantis toftae	I	W		Warne et al. (2016)
			Telmatobiidae	*Telmatobius marmoratus*	I	C (but wild caught)		Warne et al. (2016)
	Uruguay	–	Ranidae	*Lithobates catesbeianus*	I	C		Galli et al. (2006)
	Venezuela	–	Bufonidae	*Rhinella marina/Bufo marinus*	I	W		Zupanovic et al. (1998a)
		–	Leptodactylidae	*Leptodactylus* spp.	I	W		Zupanovic et al. (1998b)

List includes only cases in which virus was detected by culture or molecular methods in naturally infected hosts

[a]Provided when available

[b]*I* infection with no gross signs of ranaviral disease, *M* mortality due to ranaviral disease

[c]*W* wild population, *C* captive animals including zoological and ranaculture facilities, does not include controlled virus challenge studies

[d]Virus species identified in a given host species and location based at least on sequencing of PCR products. *ATV Ambystoma tigrinum virus, CMTV Common midwife toad virus, FV3 Frog virus 3*, – insufficient data available

[e]It is thought that these animals were imported from China

2.1   Frog virus 3 (**Ranavirus rana1**)

Frog virus 3 (FV3) infections and outbreaks have been reported in an increasing number of amphibian species around the globe since their first isolation from leopard frogs in the USA (Granoff et al. 1965). Animals infected with FV3 strains may be asymptomatic, have clinical signs of disease, and/or may die because of infection (Miller et al. 2024). While FV3 strains cause many infections in amphibian populations worldwide, the name is still somewhat of a misnomer as viruses in this species also infect fish and reptile species (see Sects. 3 and 4).

Outbreaks of FV3 have occurred across large geographical areas of North America in many different amphibian species. For example, FV3 is widely distributed across central and northwest Canada (Grant et al. 2019). However, the full extent of FV3's range in North America is unknown since many studies still do not report the virus species found. Recently, the first outbreak of FV3 in a ranaculture facility was reported in Mexico (Saucedo et al. 2019). A risk analysis for the threat of the virus spreading to other amphibians outside of the facility indicated a high risk. The authors concluded that containment of the animals in the facility was important and recommended surveillance of amphibians outside of the facility (Saucedo et al. 2019).

Reports of infections with FV3 strains from Central and South America are also increasing in number. This is likely due to increased awareness and enhanced surveillance for ranaviral infection/disease in these areas. For example, in Chile, the introduction of *Xenopus laevis* has been linked to FV3's range in that country (Peñafiel-Ricaurte et al. 2023). Cases of FV3 in South America have typically been associated with ranaculture of American bullfrogs (*Lithobates catesbeianus*; Mazzoni et al. 2009; Candido et al. 2019). However, there are an increasing number of examples of infection of wild, native amphibians in Central and South America (e.g., Ruggeri et al. 2019; Flechas et al. 2023).

In Europe, FV3 outbreaks have been occurring since the late 1980s/early 1990s. The first documented cases were in the southeastern portion of the UK in common frogs (Cunningham et al. 1993, 1996; Drury et al. 1995), followed by infections in common toads (*Bufo bufo*; Hyatt et al. 2000; Cunningham et al. 2007). FV3 infections have now been documented in several other native and introduced species of amphibians in the UK (Duffus et al. 2013, 2014; Price et al. 2017). In continental Europe, FV3 strains are also present, but less prevalent (e.g., Ariel et al. 2009; Stöhr et al. 2013c; Price et al. 2014).

In Asia, FV3 appears to be the dominant species of ranavirus in amphibians (Herath et al. 2021). Reports of ranaviruses were once sparse in Asia, but with improved surveillance and increased awareness of the risks these viruses pose, the picture of the distribution and host range of ranaviruses in this region is now better defined. There are reports of FV3 in captive and wild amphibians and native and introduced species across Asia (e.g., Une et al. 2009a, b, c; Roh et al. 2022). Ranaviruses in Asia are a major conservation concern because of their ability to infect endemic species that are often threatened or endangered. For example, an outbreak occurred in a captive population of Japanese clouded salamanders (*Hynobius nebulosus*), a vulnerable species, and within 20 days the entire colony was destroyed (Une et al. 2009c). The appearance of FV3 ranaviruses in Asia has been associated with outbreaks first observed in American bullfrogs (e.g., in Japan, Une et al. 2009a, b). In Taiwan, between 2007 and 2012, there were several mortality events of American bullfrog tadpoles in ranaculture facilities (Hsieh et al. 2021). This was linked to an outbreak of FV3, the first report of this virus in Taiwan (Hsieh et al. 2021). However, to our knowledge no surveillance outside of the ranaculture facilities has been performed in Taiwan. In South Korea, the prevalence of ranavirus in two native and one introduced species was studied (Roh et al. 2022). Of the three species examined, the introduced American bullfrog had the highest prevalence of FV3 in both the tadpole and adult stage (Roh et al. 2022), suggesting they are possible vectors of infection across the region. Several other ranaviruses in the *Frog virus 3* species have been isolated from amphibians and named. Rana grylio virus (RGV) was isolated in the mid-1990s in China (Zhang et al. 1996) and appears to belong in this species (Lei et al. 2012). Tiger frog virus (TFV) was isolated in 2000 from Chinese amphibians involved in a mass mortality event at a ranaculture facility (Weng et al. 2002). Genetically related viruses have also been found in diseased fish (marbled sleeper goby (*Oxyeleotris marmorata*), guppy (*Poecilia reticulata*), goldfish (*Carassius auratus*)) and amphibians (tiger frog (*Hoplobatrachus tigerinus*), Asian grass frog (*Fejervarya limnocharis*), East Asian bullfrog (*Hoplobatrachus rugulosus*)) in Thailand (Sriwanayos et al. 2020). Both ranaviruses may be a threat to aquaculture facilities in Asia. Phylogenetic analysis of the complete genome of Rana nigromaculata ranavirus (RNRV) revealed it belongs to the species *Frog virus 3* and is most closely related to RGV and soft-shelled turtle iridovirus (STIV) (Yu et al. 2020), while TFV forms a separate clade with isolates from China and Thailand, along with Wamena virus from a python in Australia (Sriwanayos et al. 2020). Chinese researchers produce an equivalent number of research outputs when compared to their North American counterparts (Wirth et al. 2021). However, significantly less ranaviral disease events have been reported in the GRRS, suggesting underreporting in this area (Brunner et al. 2021; Wirth et al. 2021).

In Africa, studies to date have found multiple species of amphibians affected by ranaviruses. Box et al. (2021) screened 160 individuals from five genera for the presence of ranaviruses and found *Hoplobatrachus occipitalis* and *Ptychadena* spp. infected with FV3 in Chad. A suspected FV3 strain has also been reported by Docherty-Bone et al. (2013) in *Xenopus longipes*, a species endemic to Lake Oku, Cameroon. The primers used for viral DNA detection were those developed for the major capsid protein of FV3, but the PCR products were of poor quality and could not be sequenced (Docherty-Bone et al. 2013). Other reports of ranavirus in wild amphibians come from Madagascar (Kolby et al. 2015).

In Australia, Bohle iridovirus (BIV) was first described in the early 1990s when it was isolated from recently metamorphosed ornate burrowing frogs (*Limnodynastes ornatus*) that had been raised in captivity and suddenly died (Speare and Smith 1992). Subsequently, BIV was shown to be pathogenic in additional species of Australian anurans and involved in mortality events in captive and wild settings

(Cullen et al. 1995; Cullen and Owens 2002). This strain of FV3 is also known to be transmissible to reptiles (e.g., Ariel et al. 2015; Wirth et al. 2019; see Sects. 4.2 and 6). In 2010, a similar strain was isolated from boreal toads (*Anaxyrus boreas boreas*) that experienced a mass mortality in an Iowa aquarium (Cheng et al. 2014). Originally named zoo ranavirus, after full sequencing it was determined to be a strain of FV3 (Subramaniam et al. 2019). However, it is still unclear if this is a novel FV3 isolate from North America or one from Australia that was transmitted from another captive animal.

2.2 *Ambystoma tigrinum virus* (Ranavirus ambystoma1)

Ambystoma tigrinum virus (ATV) was first described in Sonoran tiger salamander larvae (*Ambystoma mavortium* [*tigrinum*] *stebbinsi*) collected from the San Rafael Valley in Arizona USA in 1995 (Jancovich et al. 1997). The virus was isolated from a population containing both apparently healthy and visibly diseased salamander larvae (Jancovich et al. 1997). Healthy salamander larvae were infected with this virus via the water and by the feeding of infected tissues to healthy salamander larvae in the laboratory (Jancovich et al. 1997). After fulfilling Koch's postulates, it was determined that ATV was the causative agent of the disease seen in the Sonoran tiger salamander larvae and the likely cause of the recurrent epizootics that were first described in 1985 (Collins et al. 1988). A similar virus was also detected in tiger salamanders (*Ambystoma mavortium* [*tigrinum*] *diaboli*) in multiple ponds in Saskatchewan, Canada, in 1997 (Bollinger et al. 1999).

In the wild, ATV appears to be restricted to western North America (Jancovich et al. 2005; Ridenhour and Storfer 2008) and has been isolated from western tiger salamanders (*Ambystoma mavortium*) as far east as Manitoba, Canada (Schock et al. 2009). Phylogeographic studies have suggested local range expansion and long-distance colonization events, which may be attributed to anthropogenic spread (Jancovich et al. 2005). ATV has been found in tiger salamander larvae sold commercially as fish bait (Picco and Collins 2008), providing an anthropogenic explanation for range expansion.

The intra-year variation of ATV infection patterns among ponds is lower compared to the inter-year variation (Greer et al. 2009). In natural populations of tiger salamanders located on the Kaibab Plateau in the Northern Kaibab National Forest in Arizona, outbreaks of ATV appear to be synchronous (Greer et al. 2009). Of interest is that despite 4 years of observation, no visible signs of disease were seen, even in cases where the infection rate in the pond was greater than 50% (Greer et al. 2009). It is thought that the lack of observed morbidity and mortality in these populations is due to co-evolution between ATV and the host (Greer et al. 2009). This may be the case as there is evidence of local adaptation in ATV strains isolated from the southwestern USA (Ridenhour and Storfer 2008).

Although natural infections have only been described in western tiger salamanders so far, infection studies have shown that ATV can infect and cause disease and mortality in other species. Long-toe salamanders (*Ambystoma macrodactylum*) can be infected, although with lower than expected infection rates (Forson and Storfer 2006). In the lab, the endangered California tiger salamander (*Ambystoma californiense*) is susceptible to ATV and experienced mortality associated with infection (Picco et al. 2007). A transmission study using an ATV strain in western tiger salamanders, wood frogs (*Lithobates* [*Rana*] *sylvatica*), northern leopard frogs (*Lithobates* [*Rana*] *pipiens*), and Pacific tree frogs (*Pseudacris* [*Hyla*] *regilla*) showed that the virus was able to infect all four species and led to mortalities in all but the northern leopard frogs (Schock et al. 2009), although infection and mortality rates were lower in species other than western tiger salamanders. The lack of reports of natural infections in species other than western tiger salamanders may be due to differences in susceptibility of other amphibian species to this virus species, but other factors may also play a role, including testing bias, as methods capable of detecting ATV are predominantly used in areas and species in which this virus has been previously found. Without sequence data it is difficult to designate observed ranaviruses at the species level; therefore, caution must be taken when examining ranaviral species distributions.

2.3   *Common midwife toad virus* (Ranavirus alytes1)

Common midwife toad virus (CMTV) is now known to be the most widespread ranavirus in continental Europe, but it is also found in the UK (e.g., Price et al. 2017) and in Asia (primarily in fish; see Sect. 3.4) and has been detected in disease outbreaks in the USA (Claytor et al. 2017). *Common midwife toad virus* has only recently been given species status (Chinchar et al. 2017a), but it has been long known that it was evolutionarily distinct (Mavian et al. 2012). CMTV was first isolated from a mass mortality event in northern Spain in common midwife toad (*Alytes obstetricans*) tadpoles in 2007 (Balseiro et al. 2009). In 2008, a second mortality event in the same region of Spain affected both common midwife toad tadpoles and juvenile alpine news (*Ichthyosaura* [*Mesotriton*] *alpestris*, Balseiro et al. 2010). CMTV infections have now been documented in both invasive (e.g., American bullfrogs, Sharifian-Fard et al. 2011), and native (e.g., common frogs, Miaud et al. 2016) species on the continent. They have also been documented in both wild and captive animals (e.g., Kik et al. 2011, 2012; Spitzen-van der Sluijs et al. 2016).

CMTV affects a substantial number of amphibian species and has been documented as the cause of collapse of amphibian communities in Spain (Price et al. 2014; Rosa et al. 2017). Since CMTV affects multiple species in the same amphibian community, they may serve as agents of dispersal for the virus (von Essen et al. 2020). CMTV is thought to be native to the Iberian Peninsula and from there to have spread across continental Europe primarily driven by climate change and to Asia via translocation (Thumsová et al. 2022).

Chinese giant salamander virus (CGSV), sometimes referred to as Chinese giant salamander iridovirus (CGSIV) or Andrias davidianus ranavirus (ADRV), has been isolated from the endangered Chinese giant salamander (*Andrias davidianus*, Geng et al. 2011; Chen et al. 2013; Ma et al. 2014; Cunningham et al. 2016; Du et al. 2016). These viruses were isolated from captive populations and have been associated with high morbidity and mortality (Geng et al. 2011; Chen et al. 2013; Ma et al. 2014; Cunningham et al. 2016). Phylogenetic analyses group these viruses in the *Common midwife toad virus* species ranaviruses (Chen et al. 2013; Stöhr et al. 2015; Herath et al. 2021). These viruses have been documented in Chinese giant salamander farms across China (Herath et al. 2021). The emergence of this salamander ranavirus in China is a serious conservation threat (Cunningham et al. 2016).

2.4 *Chimeric Ranaviruses*

In recent years, novel recombinant ranaviruses with increased virulence have been identified (Claytor et al. 2017; Vilaça et al. 2019). Species that are susceptible to multiple ranaviruses, such as the American bullfrog, may act as mixing vessels for the emergence of these chimeric ranaviruses (Claytor et al. 2017). Co-housing diverse species may increase opportunities for recombination among divergent ranaviruses (Price 2015). These chimeric strains may spread from the animal trade and production facilities to wild animals, putting affected populations at risk of local extinction (Vilaça et al. 2019).

The known geographic distribution of chimeric ranaviruses in amphibians is currently limited to North and South America (Claytor et al. 2017; Vilaça et al. 2019). However, the number of recombinant viruses is likely to be underreported as recombination detection typically requires full genome sequencing (Grant et al. 2019; Vilaça et al. 2019).

Rana catesbeiana virus-Z2 (RCV-Z2), a CMTV-FV3 recombinant, emerged in a ranaculture facility in North America following a CMTV outbreak at the same facility 8 years earlier (Claytor et al. 2017). This chimeric virus has one of the highest basic reproduction numbers reported for any amphibian pathogen and possesses the potential to decimate populations through multiple transmission pathways (Hoverman et al. 2011a, b; Peace et al. 2019). The increased virulence of this virus is believed to result from the acquisition of CMTV genes by an FV3 virus.

The emergence of chimeric ranaviruses like RCV-Z2 highlights the potential for ranaviruses to evolve virulence rapidly and potentially adapt to new host species. Understanding the mechanisms of genetic recombination in ranaviruses is important for predicting the emergence of new strains and developing effective management strategies. Continued research into the molecular biology and ecology of ranaviruses, including chimeric ranaviruses, is necessary to better understand their evolution and spread and to mitigate their impact on wildlife populations (Chinchar et al. 2017b).

3 Ranaviruses Infecting Fish

Ranaviruses can cause severe systemic diseases in finfish in both marine and freshwater environments (Whittington et al. 2010). *Epizootic hematopoietic necrosis virus* was the first ranavirus associated with fish die-offs and was isolated in 1985 in Australia (Langdon et al. 1986b). A distinct strain of epizootic hematopoietic necrosis virus (EHNV), European catfish virus (ECV), was detected soon after in Europe (Ahne et al. 1989). The Santee-Cooper ranavirus (SCRV), known informally as largemouth bass virus (LMBV), in the species *Santee-Cooper ranavirus*, was associated with wild fish epizootics in the USA (Plumb et al. 1996) and is now an emerging pathogen of fish in Asia (Zhang et al. 2020). Although typically associated with morbidity in amphibians and reptiles, FV3 also has been isolated from a moribund three-spined stickleback (*Gasterosteus aculeatus*) during a sympatric epizootic in the northern red-legged frog (*Rana aurora*; Mao et al. 1999a; Conrad et al. 2021). In Finland, the pike-perch iridovirus (PPIV), a strain of *Common midwife toad virus*, was isolated from an apparently healthy pike-perch (*Sander lucioperca*) (Tapiovaara et al. 1998; Holopainen et al. 2016). The *European North Atlantic ranavirus* includes ranavirus strains isolated from European marine fishes including Atlantic cod (*Gadus morhua*) in 1979 (Jensen et al. 1979), turbot (*Scophthalmus maximus*) in 1999 (Ariel et al. 2010), and lumpfish (*Cyclopterus lumpus*) from 2014 to 2016 (Stagg et al. 2020). There are four species of ranavirus recognized by the ICTV that primarily infect fish: EHNV is known to occur naturally in Australia and Europe; SCRV was initially detected in North American fishes and is now emerging in Asia; European North Atlantic ranavirus (ENARV) occurs in European marine fishes, and the most divergent species in the genus, *Singapore grouper iridovirus* (Singapore grouper iridovirus, SGIV), is a pathogen primarily of grouper species in mariculture in Asia.

While EHNV has impacted aquaculture in Australia and Europe (Whittington et al. 2010), BIV, a strain of *Frog virus 3*, appears to be restricted to a single outbreak in hatchery-reared Nile tilapia fry (*Oreochromis niloticus*) in Australia (Ariel and Owens 1997). Strains of SCRV and FV3 have been repeatedly detected among hatchery-reared freshwater fishes in North America and Asia (Woodland et al. 2002b; Prasankok et al. 2005; Deng et al. 2011; George et al. 2015; Chinchar and Waltzek 2014; Waltzek et al. 2014; Sriwanayos et al. 2020). SGIV and the related strain, grouper iridovirus (GIV), have negatively impacted grouper mariculture in some Asian locations since the 1990s (Chua et al. 1994; Murali et al. 2002; Qin et al. 2003) and are emerging in China (Ma et al. 2016). The reasons for the emergence of ranaviruses as pathogens of finfish within both natural and managed populations are not fully understood; however, international trade is a contributing factor (see Sect. 5). Overviews of ranavirus distribution in fish are provided in Table 5 and Fig. 3.

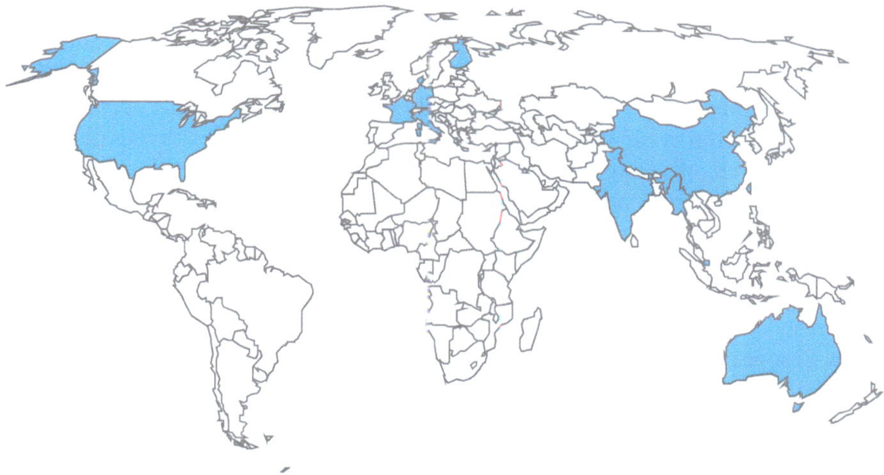

Fig. 3 Global distribution of fish ranaviruses. Shown are countries in which ranaviruses have been detected, but not a more detailed breakdown of the locations in which viruses have been found

3.1 *Epizootic hematopoietic necrosis virus* (**Ranavirus perca1**)

EHNV was the first ranavirus to be associated with systemic infection and mass mortality in any vertebrate species. It was identified as the cause of epizootic mortality of redfin perch, an introduced species in Australia now called European perch (*Perca fluviatilis*) and rainbow trout (*Oncorhynchus mykiss*) in Australia in 1985 (Langdon et al. 1986b, 1988; Langdon and Humphrey 1987). The source of the outbreak was not determined with no previous evidence of infection in salmonid fish or mass mortality of European perch (Langdon et al. 1986a; Whittington et al. 1996). To date, there have been no known cases of EHNV mortality in species other than European perch, despite the fact that at least 14 additional species are known to be susceptible to experimental challenges (Whittington et al. 2010; Becker et al. 2013). EHNV is endemic in the upper Murrumbidgee River catchment, and disease in European perch continues to occur in sporadic outbreaks in south-eastern Australia (Becker et al. 2019). The survey did not find any evidence of infection of native fishes. Natural epizootics in European perch occur most often in summer, and there is evidence of a positive relationship between EHNV pathogenicity and water temperature. European perch are not susceptible to EHNV below 10 °C, and incubation periods for the virus are shorter at higher temperatures (Whittington and Reddacliff 1995). In rainbow trout, EHNV outbreaks have occurred between 11 and 20 °C (Whittington and Reddacliff 1995; Whittington et al. 1994, 1999). Temperature-dependent pathogenicity may be related to viral replication rates (Ariel et al. 2009).

The impact of EHNV on aquaculture has been limited to farmed rainbow trout in south-eastern Australia. European perch are highly susceptible to EHNV, while rainbow trout are relatively resistant (Whittington and Reddacliff 1995). In affected trout farms, EHNV tends to occur in only a small proportion of fish (Whittington et al. 1994, 1999), with total mortality generally ≤4% across all age classes. While few fish become infected, the mortality rate of infected individuals appears to be high (Whittington et al. 1994, 1999). In contrast, EHNV causes severe disease in European perch, affecting high proportions of populations of fingerlings and juveniles in endemic areas and also naïve adults that enter new areas (Langdon et al. 1986b; Langdon and Humphrey 1987; Whittington et al. 1996). There is some evidence, based on virus isolation and serology, that both European perch and rainbow trout are capable of living with subclinical infections of EHNV, making them possible reservoirs for the pathogen (Whittington et al. 2010).

In European perch, there was an initial progressive spread of EHNV into river systems, possibly due to natural fish migration, fish releases, and avifauna (Whittington et al. 1996). Molecular epidemiology indicates that the virus has been very stable over time and in different locations (Hick et al. 2017). Waterborne infection and ingestion of infected fish are transmission routes of EHNV between susceptible hosts within a population, but longer distance spread is likely a result of human activity, particularly by movement of infected trout fingerlings in aquaculture (Langdon et al. 1988; Whittington et al. 1994, 1999).

The first transmission studies with EHNV were conducted by Langdon (1989) who identified a wide range of susceptible hosts, a factor that contributed to the listing of EHNV by the World Organisation for Animal Health (WOAH). Although there are recognized deficiencies in laboratory challenge models to determine the susceptibility of host fish to virus isolates under natural conditions, there is the potential for an increase in the geographical and host range of EHNV. For example, European perch and rainbow trout of Polish origin were susceptible (Borzym and Maj-Paluch 2015). In three separate challenge studies, black bullhead (*Ameiurus melas*), northern pike (*Esox lucius*), and pike-perch (*Sander lucioperca*) experienced significant mortality following bath exposure to EHNV (Jensen et al. 2009, 2011a; Gobbo et al. 2010). On the other hand, goldfish, common carp (*Cyprinus carpio*), and European sheatfish (*Silurus glanis*) did not experience significant mortality following bath exposure to EHNV (Jensen et al. 2011b; Leimbach et al. 2014). Like other ranaviruses, the outcome of EHNV infection may depend on various viral, host, and environmental factors including virus concentration and route of delivery, viral strain, host genetics, host density and age, and water temperature (Brunner et al. 2015). For example, lower mortality was observed when European redfin perch stocks were challenged with EHNV, but it was these individuals that caused transmission of the pathogen to and extensive mortality in Australian European perch that were cohoused with them (Ariel and Jensen 2009). There is evidence of European perch from endemic areas having reduced mortality when exposed to EHNV compared to European perch sourced from EHNV-free regions of Australia (Becker et al. 2016).

ECV, also referred to as European sheatfish virus (ESV), prior to its formal classification, is now recognized within the species *Epizootic hematopoietic necrosis virus* (Chinchar et al. 2017a). This pathogen has triggered disease in cultivated sheatfish in Germany (Ahne et al. 1989, 1991) and wild black bullheads in France and Italy (Pozet et al. 1992; Bovo et al. 1993; Bigarré et al. 2008). Evidently, the virus is endemic in some locations (e.g., Lake le Bourget and Lake Apremont in France; Bigarré et al. 2008). In Italy, the disease occurs in both farmed and wild black bullhead and in farmed brown bullhead (*Ameiurus nebulosus*) and affects production of these species (Ariel et al. 2010). An ECV outbreak was detected in brown bullheads in Hungary in 2008 (Juhász et al. 2013), and the isolate caused disease in juvenile Wels catfish (*Silurus glanis*) (Abonyi et al. 2022).

The host range, geographic distribution, and diversity of ECV strains in Europe are not fully understood. The virus is readily transmitted to catfish via a range of challenge methods including bath exposure, cohabitation, and intramuscular injection (Ahne et al. 1990; Pozet et al. 1992) and results in only a small proportion of exposed catfish surviving (Pozet et al. 1992). Interestingly, Gobbo et al. (2010) found different patterns of susceptibility based on closely related ranavirus strains, as black bullheads were susceptible to ECV, but not to the ESV isolate used in this study. More recent experiments have demonstrated variable pathogenicity of different isolates of ECV and a strong impact of water temperature on disease outcome, with mortality varying between 8% and 10% among challenged sheatfish (Leimbach et al. 2014). In three separate experimental challenge studies, black bullhead, pike, and sheatfish experienced significant mortality following bath exposure to strains of ECV (Jensen et al. 2009; Gobbo et al. 2010; Leimbach et al. 2014). Goldfish, common carp, and pike-perch did not experience significant mortality following bath exposure to ECV strains (Jensen et al. 2011a, b).

3.2 *Santee-Cooper ranavirus* (Ranavirus micropterus1)

The virus species *Santee-Cooper ranavirus* includes a number of viruses with an emerging geographical and host range. An iridovirus was identified in largemouth bass (*Micropterus salmoides*) during a disease outbreak that occurred in the Santee-Cooper Reservoir, South Carolina, USA, in 1995 (Plumb et al. 1996). Subsequent genetic analyses confirmed LMBV to be a unique member of the genus *Ranavirus* (Mao et al. 1997, 1999b) and nearly identical to doctor fish virus (DFV) and guppy virus 6 (GV6), which had previously been isolated from imported ornamental fishes originating in Southeast Asia (Hedrick and McDowell 1995). The designation as SCRV based on the location where the virus was isolated was disputed by Grizzle et al. (2002) citing the previous isolation of LMBV from Lake Weir, Florida, USA, in 1991 (Francis-Floyd 1992). The aforementioned studies and more recent phylogenetic analyses support LMBV, GV6, and DFV as strains of the same species that is formally known by the ICTV as the *Santee-Cooper ranavirus* (Holopainen et al. 2009). Based on the genetic sequence as well as epidemiological and pathobiological characteristics, *Santee-Cooper ranavirus* is a divergent species within the genus *Ranavirus* (Hyatt et al. 2000; Whittington et al. 2010; Jancovich et al. 2015).

Disease outbreaks attributed to SCRV have been repeatedly reported among wild populations of North American largemouth bass (Grizzle and Brunner 2003; Plumb and Hanson 2011), and these have caused disease and die-offs of young-of-year smallmouth bass (*Micropterus dolomieu*) in Pennsylvania, USA (Boonthai et al. 2018). Although SCRV virulence appears variable in natural and experimental settings, typical outbreaks involve adult fish observed during summer at the surface with buoyancy or equilibrium problems. Experimental infections of SCRV in largemouth bass and striped bass (*Morone saxatilis*) revealed a direct correlation between virus titer and mortality; however, striped bass experienced lower overall cumulative mortality (Plumb and Zilberg 1999b; Zilberg et al. 2000). Experimental transmission of SCRV in largemouth bass via oral administration resulted in infection of the skin and internal organs (e.g., swim bladder) without mortality (Woodland et al. 2002a). In the USA, SCRV has also been identified from a wide range of wild freshwater fishes (6 families and 17 species) in 31 states ranging as far south as Florida to as far west as Arizona and to the northern states of Wisconsin, Michigan, New York, Vermont, and Delaware (Goldberg 2002; Woodland et al. 2002b; Groocock et al. 2008; USFWS 2011; Iwanowicz et al. 2013; Table 5). An SCRV strain was isolated from the exotic Northern snakehead (*Channa argus*) introduced into the Chesapeake Bay watershed (Iwanowicz et al. 2013). The DFV and GV6 strains of SCRV were shown experimentally to infect and induce low mortality in rainbow trout and chinook salmon (*Oncorhynchus tshawytscha*), but not channel catfish (*Ictalurus punctatus*) (Hedrick and McDowell 1995).

Largemouth bass aquaculture in China has expanded since introduction of the species in the 1980s with ranaviruses proving to be a disease threat to this important industry (Zhao et al. 2020). A virus closely related to SCRV first caused disease of largemouth bass in Guangdong in 2006, and a disease syndrome featuring skin ulceration was also attributed to a strain of SCRV in the same province in 2008 (Deng et al. 2011). Ulcerative disease caused by SCRV has continued to impact aquaculture of largemouth bass (Jin et al. 2022; Zhao et al. 2020), and strain variation within the species may impact disease outcomes (Liu et al. 2023b). An ulcerative disease of farmed Barcoo grunter (*Scortum barcoo*) in Thailand in 2013 was also attributed to an SCRV (Kayansamruaj et al. 2017). Little or no mortality was observed for an SCRV isolated from largemouth bass when tested experimentally in seven other species including koi (*Cyprinus carpio*; Table 5). Contrary to these findings, a mass mortality event of farmed koi in southern India was attributed to koi ranavirus, a strain of SCRV (George et al. 2015). There is evidence of a similar virus causing disease in the freshwater fish *Osteobrama belangeri* in Manipur, India (John et al. 2016). SCRV infection in the ornamental marine species similar to damselfish (*Pomacentrus similis*) was also reported from India. The isolate had homology to koi ranavirus and was shown to be virulent for barramundi (*Lates calcarifer*), an important aquaculture species (Sivasankar et al. 2017).

Until 2013, disease in China caused by viruses related to SCRV was limited to largemouth bass. Mandarin fish ranavirus (MRV) was first detected in 2013 as the cause of disease in another important freshwater farmed fish in China, mandarin fish

Table 5 Distribution of ranavirus infections or mortality in wild and captive fish, including the location of origin for imported animals

Continent	Nation	Origin[a]	State/province[a]	Family	Scientific name	I, M[b]	W, C[c]	Virus species[d]	References
Asia	China	–	Guangdong, Sichuan	Centrarchidae	*Micropterus salmoides*	M	C	*SCRV*	Deng et al. (2011), Jin et al. (2022), Zhao et al. (2020)
	China		Sichuan	Nemacheilidae	*Triplophysa siluroides*	M	C	*CMTV*	Deng et al. (2020)
	China	–	Sichuan	Cyprinidae	*Percocypris pingi*	M	C	*CMTV*	Liu et al. (2023a)
	China		Guangdong	Sinipercidae	*Siniperca chuatsi, Siniperca scherzeri × Siniperca chuatsi* (hybrid)	M	C	*SCRV*	Dong et al. (2017), Zhang et al. (2020)
	China		Hainan, Guangxi, Hainan	Serranidae	*Epinephelus coioides, Epinephelus fuscoguttatus × Epinephelus lanceolatus* (hybrid)	M	C	*SGIV*	Ma et al. (2016), Wei et al. (2019), Xiao et al. (2019)
	India			Cyprinidae	*Cyprinus carpio*	M	C	*SCRV*	George et al. (2015)
	Singapore	–	–	Serranidae	*Epinephelus tauvina*	M	C	*SGIV*	Chua et al. (1994), Qin et al. (2003)
	Taiwan	–	–	Centrarchidae	*Micropterus salmoides*	–	C	*SGIV*	Huang et al. (2011)
		–	–	Latidae	*Lates calcarifer*	–	C	*SGIV*	Huang et al. (2011)
		–	–	Lutjanidae	*Lutjanus erythropterus*	–	C	*SGIV*	Huang et al. (2011)
		–	–	Serranidae	*Epinephelus awoara*	M	C	*SGIV*	Murali et al. (2002), Huang et al. (2011)
		–	–		*Epinephelus coioides*	–	C	*SGIV*	Huang et al. (2011), Chuang et al. (2022)
		–	–		*Epinephelus lanceolatus*	–	C	*SGIV*	Huang et al. (2011)
	Thailand	–	Nakhon Pathom	Eleotridae	*Oxyeleotris marmoratus*	M	C	*FV3*	Prasankok et al. (2005)
	Thailand	–	Samut Sakhon	Poeciliidae	*Poecilia reticulata*	I	C	*FV3*	Sriwanayos et al. (2020)
	Thailand	–	Bangkok	Cyprinidae	*Carassius auratus*	I	C	*FV3*	Sriwanayos et al. (2020)
	Thailand	–	Ayutthaya, Phetchaburi	Terapontidae	*Scortum barcoo*	M	C	*SCRV*	Kayansamruaj et al. (2017)

(conitnued)

Table 5 (continued)

Continent	Nation	Origin[a]	State/province[a]	Family	Scientific name	I, M[b]	W, C[c]	Virus species[d]	References
Europe	Denmark	–	–	Gadidae	*Gadus morhua*	I	W	*ENARV*	Ariel et al. (2010)
		–	–	Scophthalmidae	*Scophthalmus maximus*	M	C	*ENARV*	Ariel et al. (2010)
	Finland	–	–	Percidae	*Sander lucioperca*	I	W	*CMTV*	Tapiovaara et al. (1998)
	France	–	–	Ictaluridae	*Ictalurus melas*	M	W	*EHNV*	Pozet et al. (1992), Bigarré et al. (2008)
	Germany	–	–	Siluridae	*Silurus glanis*	M	C	*EHNV*	Ahne et al. (1989, 1991)
	Hungary	–	–	Ictaluridae	*Ameiurus nebulosus*	M	W	*EHNV*	Juhász et al. (2013)
	Italy	–	–	Ictaluridae	*Ictalurus melas*	M	W	*EHNV*	Bovo et al. (1993)
	Italy	New Zealand	–	Anguillidae	*Anguilla australis*	–	W	unassigned	Bovo et al. (1999)

North America	USA	Asia	CA	Labridae	*Labroides dimidiatus*	–	–	*SCRV*	Hedrick and McDowell (1995)
		Asia	VA	Channidae	*Channa argus*	I	W	*SCRV*	Iwanowicz et al. (2013)
		–	MN	Centrarchidae	*Ambloplites rupestris*	–	–	*SCRV*	USFWS (2011)
		–	MD, TN	Centrarchidae	*Lepomis auritus*	–	–	*SCRV*	USFWS (2011)
		–	GA, IL, MN, OH	Centrarchidae	*Lepomis macrochirus*	–	–	*SCRV*	Woodland et al. (2002a), USFWS (2011)
		–	AR	Centrarchidae	*Lepomis megalotis*	–	–	*SCRV*	USFWS (2011)
		–	AZ, KY, MI, MN, NY, OH, OK, PA, WI, WV	Centrarchidae	*Micropterus dolomieu*	M	W		Groocock et al. (2008), USFWS (2011)
		–	FL	Centrarchidae	*Micropterus notius*	–	–	*SCRV*	USFWS (2011)
		–	AL, AR, GA, KY, LA, MI, OK, TN	Centrarchidae	*Micropterus punctulatus*	–	–	*SCRV*	USFWS (2011)
		–	AL, AR, AZ, CT, DE, FL, GA, IA, IL, KS, KY, LA, MD, MI, MN, MO, MS, NC, NJ, NY, OH, OK, PA, SC, TN, TX, VA, VT, WI	Centrarchidae	*Micropterus salmoides*	I, M	W, C		Grant et al. (2005), Groocock et al. (2008), Grizzle et al. (2002), Hanson et al. (2001), Mao et al. (1999b), Neal et al. (2009), Plumb et al. (1996, 1999), Southard et al. (2009), USFWS (2011), Woodland et al. (2002b)
		–	WI	Centrarchidae	*Pomoxis nigromaculatus*	–	–	*SCRV*	USFWS (2011)
		–	KY	Catostomidae	*Minytrema melanops*	–	–	*SCRV*	USFWS (2011)
		–	NC	Cyprinidae	*Nocomis leptocephalus*	–	–	*SCRV*	USFWS (2011)
		–	AR	Cyprinidae	*Pimephales promelas*	–	C	*FV3*	Waltzek et al. (2014)

(conitnued)

Table 5 (continued)

Continent	Nation	Origin[a]	State/province[a]	Family	Scientific name	I, M[b]	W, C[c]	Virus species[d]	References
		–	IL	Esocidae	*Esox masquinongy*	–	–	*SCRV*	USFWS (2011)
		–	OH	Esocidae	*Esox niger*	–	C	*FV3*	Waltzek et al. (2014)
		–	–	Esocidae	*Esox niger*	–	–	*SCRV*	Goldberg (2002)
		–	SC	Moronidae	*Morone americana*	–	–	*SCRV*	USFWS (2011)
		–	KS	Moronidae	*Morone chrysops*	–	–	*SCRV*	USFWS (2011)
		–	AZ	Moronidae	*Morone mississippiensis*	–	–	*SCRV*	USFWS (2011)
		–	MN, OH	Sciaenidae	*Aplodinotus grunniens*	–	–	*SCRV*	USFWS (2011)
		Asia	CA	Poeciliidae	*Poecilia reticulata*	–	–	*SCRV*	Hedrick and McDowell (1995)
		–	CA	Gasterosteidae	*Gasterosteus aculeatus*	M	W	*FV3*	Mao et al. (1999a)
		–	GA	Acipenseridae	*Acipenser gueldenstaedtii*	M	C	*FV3*	Waltzek et al. (2014)
		–	CA	Acipenseridae	*Acipenser transmontanus*	M	C	*CMTV*	Waltzek et al. (2014)
		–	MO	Acipenseridae	*Scaphirhynchus albus*	M	C	*FV3*	Waltzek et al. (2014)
Oceania	Australia	–	NSW, SA, Vic	Percidae	*Perca fluviatilis*	M	W	*EHNV*	Whittington et al. (2010)
		–	NSW, SA, Vic	Salmonidae	*Oncorhynchus mykiss*	M	C	*EHNV*	Whittington et al. (2010)

[a]Provided when available

[b]*I* infection with no gross signs of ranaviral disease, *M* mortality due to ranaviral disease

[c]*W* wild population, *C* captive animals including zoological and ranaculture facilities, does not include controlled virus challenge studies

[d]Virus species identified in a given host species and location based at least on sequencing of PCR products. *CMTV Common midwife toad virus, EHNV Epizootic hematopoietic necrosis virus, ENARV European North Atlantic ranavirus, FV3 Frog virus 3, SCRV Santee-Cooper ranavirus, SGIV Singapore grouper iridovirus,* – insufficient data available

(*Siniperca chuatsi*), and was identified as a strain within the species *Santee-Cooper ranavirus* (Dong et al. 2017). The MRV member of *Santee-Cooper ranavirus* has continued to impact mandarin fish production in Guangdong with a varied disease presentation sometimes including ascites prior to mortality (Zhang et al. 2020).

It seems likely that SCRV has been disseminated across the USA and globally through the unrestricted movement of live fish and their products associated with the ornamental (Hedrick and McDowell 1995; Deng et al. 2011; George et al. 2015), food (Plumb and Zilberg 1999a), and angling industries (Grant et al. 2005; Schramm Jr and Davis 2006). For example, in the USA, largemouth bass angling tournaments may contribute to the spread of SCRV to naïve fish by placing infected and uninfected fish in close proximity; however, the stress associated with angling has not been shown to greatly increase SCRV-associated mortality (Grant et al. 2005; Schramm Jr and Davis 2006). Given that SCRV remains infectious in frozen tissues, the import/export of frozen fish tissues may represent another mechanism by which the virus can be spread (Plumb and Zilberg 1999a). Future concerted surveillance efforts are needed to confirm the risk that the aforementioned industries play in the global dissemination of SCRV.

3.3   *Frog virus 3* (**Ranavirus rana1**)

Only a single case of FV3 infection has been reported in wild fish (Mao et al. 1999a; Conrad et al. 2021). An FV3 strain was recovered from a single moribund three-spined stickleback (*Gasterosteus aculeatus*) that was coinfected with myxozoan parasites, obscuring the role of the virus in the disease (Mao et al. 1999a). Several cases of piscine infection with FV3 have been reported among managed stocks including cultured marbled sleeper goby, goldfish, and guppy in Thailand (Prasankok et al. 2005; Sriwanayos et al. 2020). Furthermore, FV3 outbreaks have impeded efforts to restore populations of the critically endangered pallid sturgeon (*Scaphirhynchus albus*) in the Missouri River Basin of the USA (Waltzek et al. 2014; Stilwell et al. 2022). High-mortality epizootics were reported among young-of-the-year pallid sturgeon in 2001, 2009, 2013, and 2015 at the Blind Pony Hatchery in Sweet Springs, Missouri, USA (Chinchar and Waltzek 2014; Waltzek et al. 2014; Stilwell et al. 2022). Experimental transmission of the 2009 isolate recreated the same high-mortality disease in naïve juvenile pallid sturgeon following bath exposure (Waltzek et al. 2014; Stilwell et al. 2022). The severity of disease and mortality rate in virus-exposed pallid sturgeon are significantly greater at a water temperature of 23 °C, compared to 17 °C. Furthermore, an FV3 strain isolated from moribund hatchery-reared Russian sturgeon (*Acipenser gueldenstaedtii*) was found to be lethal to both Russian and lake (*A. fulvescens*) sturgeon following intraperitoneal injection (Waltzek et al. 2014).

Experimental transmission studies using other FV3 strains isolated from a diversity of ectothermic vertebrate classes have been shown to infect black bullhead, northern pike, pike-perch, mosquito fish (*Gambusia affinis*), and bluegill (*Lepomis macrochirus*), although little or no mortality was observed in these species (Gobbo et al. 2010; Jensen et al. 2009, 2011a, b; Brenes et al. 2014a). Similarly, recent North American fish health surveys resulted in the isolation of FV3 from healthy appearing fathead minnow (*Pimephales promelas*) and northern pike (Waltzek et al. 2014). Although preliminary, these data suggest that imperiled sturgeon may be predisposed to infections with FV3 strains, whereas other fishes may simply act as viral carriers or dead-end hosts. Future studies are needed to explore the importance of FV3 strains across a wider range of fish species, as well as the potential role of aquaculture in the global dissemination of these important pathogens.

Experimental transmission studies with BIV, an Australian FV3 strain (Chinchar et al. 2017a), demonstrated its pathogenicity to barramundi, a popular sport fish in Australia (Moody and Owens 1994). On just one occasion, BIV may have been associated with high mortality in hatchery-reared Nile tilapia fry in Australia (Ariel and Owens 1997). Although the authors did not genetically characterize the iridovirus, feeding moribund tilapia fry to barramundi fingerlings reproduced disease similar to what had been reported following challenge studies of barramundi to BIV (Moody and Owens 1994).

3.4 *Common midwife toad virus* (Ranavirus alytes1)

CMTV primarily infects amphibian species across Europe (Balseiro et al. 2009, Sect. 2.3). In Finland, PPIV, a CMTV strain, was isolated in 1995 from apparently healthy pike-perch (*Sander lucioperca*) fingerlings during routine disease surveillance (Tapiovaara et al. 1998; Holopainen et al. 2016). Experimental PPIV infection trials did not generate disease in either pike-perch or rainbow trout; however, the virus was recovered from survivors in both studies, suggesting these hosts might serve as carriers (Jensen et al. 2011a). Other studies have demonstrated that PPIV can cause lethal infections in northern pike fry (Jensen et al. 2009). The white sturgeon ranavirus is a strain of CMTV that was isolated from juvenile white sturgeon (*Acipenser transmontanus*) on a California, USA, farm during an unusual mortality event in 1998 (Waltzek et al. 2014). A CMTV strain was isolated from diseased catfish-like loach (*Triplophysa siluroides*) cultured in Sichuan Province, China (Deng et al. 2020). Following exposure of catfish-like loach to the CMTV strain, similar clinical signs were observed, and the fish experienced significant mortality. A related CMTV strain was isolated from diseased *Percocypris pingi* on a farm in Sichuan Province, China, in 2021 (Liu et al. 2023a). Experimental challenge studies reproduced the lethal disease that had been observed on the farm.

3.5 *Singapore grouper iridovirus* (**Ranavirus epinephelus1**)

The species *Singapore grouper iridovirus* is the most divergent of the ranaviruses (Chinchar et al. 2017a) and is identified as a pathogen of marine fish. SGIV has become increasingly widespread throughout Southeast Asia and China where it causes disease and substantial economic losses in marine aquaculture. Infection can cause high mortality, particularly of grouper (*Epinephelus* spp.) with splenomegaly, hemorrhage, and necrosis evident on histopathology (Qin et al. 2003). The species *Singapore grouper iridovirus* includes viruses that were referred to as grouper iridovirus (GIV). These need to be distinguished from megalocytiviruses of the species *Infectious spleen and kidney necrosis virus* which are also sometimes referred to as grouper iridoviruses, but are in a different genus and cause a different pathology (Ma et al. 2016).

The first description of "sleepy grouper disease" was a mass mortality of farmed brown-spotted grouper (*E. tauvina*) in Singapore in 1994 (Chua et al. 1994). Later, the causative virus was isolated and characterized as SGIV (Qin et al. 2001; Song et al. 2004). It was recognized in recurrent disease outbreaks in Singapore (Chang et al. 2002). Meanwhile, disease detected around the same time in farmed yellow grouper (*E. awoara*) in Taiwan was caused by infection with GIV, which was very similar to SGIV (Murali et al. 2002; Tsai et al. 2005). The movement of grouper fry for farming may have resulted in the spread of this pathogen between these locations (Qin et al. 2003). Surveys of iridoviruses in marine fish in Taiwan found SGIV more frequently between 2000 and 2005, followed by a shift to a predominance of infectious spleen and kidney necrosis virus (ISKNV) by 2009 (Huang et al. 2011). This study identified SGIV in non-grouper marine species and the first detection in freshwater (*Micropterus salmoides*) and catadromous (*Lates calcarifer*) fishes.

SGIV was first isolated in mainland China in 2013 from an outbreak of disease in farmed hybrid grouper (*E. fuscoguttatus* × *E. lanceolatus*) in Hainan province (Ma et al. 2016). This signalled the onset of severe economic impacts as this pathogen affected important aquaculture regions and species. Reports of SGIV include detection in orange-spotted grouper (*E. coioides*) in Hainan province (Wei et al. 2019) and in hybrid grouper in Guangxi province (Xiao et al. 2019).

While SGIV has been frequently identified as a disease agent of *Epinephelus* spp. at aquaculture farms, there is experimental evidence for a broader host range including the susceptibility of barramundi to infection and disease (Wang et al. 2017). The freshwater ornamental fish, dwarf gourami (*Trichogaster lalius*), and swordtail (*Xiphophorus hellerii*) were susceptible to experimental infection with SGIV, demonstrating the potential for cross-infection of SGIV between freshwater and marine fish (Chuang et al. 2022). Biosecurity considerations are important to protect susceptible species present in regions that are currently unaffected by SGIV. This will benefit from validated diagnostic tests and recognition of the disease risks, similar to the case of red sea bream iridoviral disease caused by megalocytiviruses which are notifiably emerging as global fish pathogens (WOAH 2021).

3.6   European North Atlantic ranavirus (Ranavirus gadus1)

The *European North Atlantic ranavirus* includes ranavirus strains isolated from European marine fishes including Atlantic cod (Jensen et al. 1979), turbot (Ariel et al. 2010), and lumpfish (Stagg et al. 2020). The cod iridovirus (CoIV) was isolated from wild Atlantic cod in Danish coastal waters in 1979 during an investigation of cutaneous ulcerative disease (Jensen et al. 1979). However, results of infection trials could not definitively link the CoIV to the cod ulcerative disease (Jensen and Larsen 1982). Ranavirus maximus (Rmax) was isolated from clinically healthy turbot fry from an aquaculture facility during a screening for export certification in 1999 (Ariel et al. 2010). The lumpfish ranavirus was isolated from lumpfish at multiple locations in the North Atlantic including the Faroe Islands in 2014, Iceland in 2015, and Scotland and Ireland in 2016 (Stagg et al. 2020). Complete genome sequencing and phylogenetic analysis based on the concatenated 26 iridovirus core genes supported the lumpfish ranavirus isolates and the closely related CoIV and Rmax isolates as a distinct clade among ranaviruses (Stagg et al. 2020). Scholz et al. (2022) demonstrated that the lumpfish ranavirus can be horizontally transmitted and that some strains can reduce survival of lumpfish juveniles following IP injection associated with consistent pathological changes.

3.7   Taxonomically Unassigned Ranaviruses That Infect Fish

The short-finned eel ranavirus (SERV) was isolated in 1999 during routine screening of live short-finned eel (*Anguilla australis*) imported into Italy from New Zealand (Jensen et al. 2009). Bath challenges with SERV resulted in significant mortality in northern pike fry (Jensen et al. 2009) versus no appreciable disease in juvenile black bullheads (Gobbo et al. 2010). Phylogenetic analyses based on the concatenated nucleotide sequences of the 26 iridovirus core genes revealed that SERV forms a distinct branch at the base of the ranavirus tree near other fish ranaviruses (Subramaniam et al. 2016), suggesting that SERV likely represents a new ranavirus species.

4   Ranaviruses Infecting Reptiles

As for ranaviruses in other host classes, it is likely that ranaviruses in reptiles remain underreported. However, there has been an increase in testing for ranaviruses and reports of ranaviral disease in reptiles over the past two decades. This has corresponded with a heightened awareness and improved surveillance of ranaviruses as pathogens of reptiles in general and particularly in chelonians (Wirth et al. 2018;

Adamovicz et al. 2020; Marschang and Hyndman 2022). There are multiple reports of single cases and outbreaks in reptiles (Table 6). In North America, many of these involve box turtles (*Terrapene* spp.) (De Voe et al. 2004; Allender et al. 2006; Johnson et al. 2008, 2010; Ruder et al. 2010; Allender 2012; Kimble et al. 2014; Sim et al. 2016; Perpiñán et al. 2016; Agha et al. 2017; Adamovicz et al. 2018). More recently, other aquatic turtles and terrestrial tortoises worldwide have been observed with ranavirus infections associated with mortality events or during routine surveys (Benetka et al. 2007; Blahak and Uhlenbrok 2010; Stöhr et al. 2015; Winzeler et al. 2015; McKenzie et al. 2019; Carstairs et al. 2020; Cozad et al. 2020; von Essen et al. 2020) caused by a variety of ranavirus species (Table 6). Characterization of these viruses has most often been based on partial MCP gene sequences, but additional sequence data are becoming available to help understand relationships between the ranaviruses found in reptiles. The majority of ranaviruses detected in reptiles so far have been FV3 strains (Huang et al. 2009; Allender et al. 2011; Adamovicz et al. 2018; Stöhr et al. 2015). In North America, only FV3 has been detected in reptiles. In addition, EHNV and CMTV viruses have been detected in several reptile species in Europe (Marschang et al. 2013; Price et al. 2014; Price 2015; Stöhr et al. 2015). Although ranavirus detection in chelonians has been reported most frequently, detection of these viruses in lizards and snakes has been increasing, mostly from captive animals (Stöhr et al. 2013b, 2015; Behncke et al. 2013; Tamukai et al. 2016), but reports from wild squamates are also available (Alves de Matos et al. 2011; Price et al. 2014, 2017; Goodman et al. 2018). Another aspect of increased surveillance and reporting of ranaviral infections in reptiles is the increased finding of co-occurring infections with various other viruses as well as parasites and bacteria in ranavirus-infected reptiles, making an exact assessment of the pathogenicity of ranaviruses in some cases difficult. Overviews of ranavirus detections and distribution in reptiles are provided in Table 6 and Fig. 4.

4.1 Historical Reports of Ranaviruses in Reptiles

In the 1980s, two cases of iridovirus infections in tortoises were described in Switzerland (Heldstab and Bestetti 1982; Müller et al. 1988). Due to their clinical, histological, and electron microscopical findings, these animals are believed to have been infected with a ranavirus and are therefore likely the first documented cases of ranaviral infection and disease in reptiles. This was followed by a small number of proven detections of ranavirus infection in reptiles that were documented in captive and wild chelonians in the late 1990s (Table 6). Since then, most reports of ranaviruses in reptiles have included at least some genetic data to help identify the virus involved.

Table 6 Distribution of ranavirus infections or mortality in wild and captive reptiles, including the location of origin for imported animals

Continent	Nation	Origin[a]	Family	Scientific name	I, M[b]	W, C[c]	Virus species[d]	References
Asia	China	–	Trionychidae	*Pelodiscus (Trionyx) sinensis*	M	C	*FV3*	Chen et al. (1999)
	Japan	–	Agamidae	*Pogona vitticeps*	M	C	*FV3*	Tamukai et al. (2016)
Europe	Austria	Ethiopia	Testudinidae	*Stigmochelys (Geochelone) pardalis*	M	C	*FV3*	Benetka et al. (2007)
	Germany	Asia via FL	Agamidae	*Japalura splendida*	M	C/W	*FV3*	Behncke et al. (2013)
		–		*Pogona vitticeps*	I, M	C	*FV3/CMTV* chimeric	Stöhr et al. (2013b), Marschang et al. (2013)
		Asia	Anguidae	*Dopasia gracilis*	M	C/W	*FV3*	Stöhr et al. (2013b)
		–	Boidae	*Eunectes* sp.	I	C	–	Marschang et al. (2013)
		USA	Anolidae	*Anolis sagrei*	M	C/W	*FV3*	Stöhr et al. (2013b)
		USA		*Anolis carolinensis*	M	C/W	*FV3*	Stöhr et al. (2013b)
		–	Emydidae	*Emys orbicularis*	I	C	–	Stöhr et al. (2013d)
		–		*Trachemys scripta elegans*	I	C	–	Marschang et al. (2013)
		–	Gekkonidae	*Uroplatus fimbriatus*	M	C	*FV3*	Marschang et al. (2005)
		–	Iguanidae	*Iguana iguana*	M	C	*EHNV*	Stöhr et al. (2013b)
		–	Testudinidae	*Testudo graeca*	I	C	–	Marschang et al. (2013)
		–		*Testudo hermanni*	M	C	–	Blahak and Uhlenbrok (2010)
		–		*Testudo horsfieldii*	I	C	–	Marschang et al. (2013)
		–		*Testudo kleinmanni*	M	C	*FV3/CMTV* chimeric	Blahak and Uhlenbrok (2010)
		–		*Testudo marginata*	M	C	*FV3/CMTV* chimeric	Blahak and Uhlenbrok (2010)
		–		*Stigmochelys pardalis*	I	C	–	Marschang et al. (2013)
		ID	Pythonidae	*Python brongersmai*	M	C/W	*FV3*	Stöhr et al. (2015)
		–		*Python regius*	I	C	–	Marschang et al. (2013)
		–		*Python molurus*	I	C	–	Marschang et al. (2013)

		–	Varanidae	*Varanus macraei*	I	C	–	Marschang et al. (2013)
	Poland	–	Emydidae	*Trachemys scripta elegans*	M	W	*CMTV*	Borzym et al. (2020)
	Portugal	–	Lacertidae	*Lacerta monticola*	I	W	*FV3*	Alves de Matos et al. (2011)
	Spain	–	Colubridae	*Natrix maura*	M	W	*CMTV*	Price et al. (2014)
	Switzerland	Yugosl-avia	Testudinidae	*Testudo hermanni*	M	C/W	–	Müller et al. (1988)
		–		*Testudo hermanni*	M	C	*CMTV*	Heldstab and Bestetti (1982), Marschang et al. (1999)
	United Kingdom	–	Lacertidae	*Lacerta agilis*	I	W	–	Marschang et al. (2013)
		–	Testudinidae	*Testudo hermanni*	M	C	–	Marschang et al. (2013)
		–	Anguidae	*Anguis fragilis*	M	W	*FV3*	Price et al. (2017)
North America	USA	–	Phrynosomatidae	*Sceloporus undulatus*	I	W	–	Goodman et al. (2018)
		–	Emydidae	*Chrysemys picta*	I, M	W	–	Goodman et al. (2013)
		–		*Terrapene carolina bauri*	M	C,W	*FV3*	Johnson et al. (2008)
		–		*Terrapene carolina carolina*	I, M	C,W	*FV3*	Mao et al. (1997), De Voe et al. (2004), Allender et al. (2006), Johnson et al. (2008), Ruder et al. (2010), Farnsworth and Seigel (2013), Kimble et al. (2014), Sim et al. (2016), Perpiñán et al. (2016), Agha et al. (2017), Adamovicz et al. (2018)
			Kinosternidae	*Kinosternon subrubrum*	D	W	*FV3*	Winzeler et al. (2015)
		–	Testudinidae	*Geochelone platynota*	M	C	*FV3*	Johnson et al. (2008)
		–		*Gopherus polyphemus*	M	W	*FV3*	Westhouse et al. (1996), Johnson et al. (2008), Cozad et al. (2020)
		–		*Testudo horsfieldii*	–	C	*FV3*	Mao et al. (1997)
			Chamaeleonidae	*Trioceros melleri*	I, M	C	*FV3*	Peiffer et al. (2019)
	Canada		Chelydridae	*Chelydra serpentina*	I, M	W	*FV3*	McKenzie et al. (2019), Carstairs et al. (2020)
			Emydidae	*Chrysemys picta*	I	W	*FV3*	Carstairs et al. (2020)

(continued)

Table 6 (continued)

Continent	Nation	Origin[a]	Family	Scientific name	I, M[b]	W, C[c]	Virus species[d]	References
Oceania	Australia	Indonesia	Pythonidae	*Morelia* (*Chondropython*) *viridis*	M	C/W	*FV3*	Hyatt et al. (2002)
		–	Agamidae	*Intellagama lesueurii lesueurii*	I	C/W	*FV3*	Maclaine et al. (2019)
		–	Agamidae	*Pogona vitticeps*	I	C	*FV3*	Maclaine et al. (2019)
		–	Agamidae	*Hypsilurus boydii*	I	C	*FV3*	Maclaine et al. (2019)
		–	Agamidae	*Diporiphora nobbi*	I	W	*FV3*	Maclaine et al. (2019)

List includes cases in which virus was detected by electron microscopy, culture, or molecular methods in naturally infected hosts

[a]Provided when available

[b]*I* infection with no gross signs of ranaviral disease, *M* mortality due to ranaviral disease

[c]*W* wild population, *C* captive animals including zoological and ranaculture facilities, does not include controlled virus challenge studies

[d]Virus species identified in a given host species and location based at least on sequencing of PCR products. *CMTV Common midwife toad virus, EHNV Epizootic hematopoietic necrosis virus, FV3 Frog virus 3,* – insufficient data available

Fig. 4 Global distribution of reptilian ranaviruses. Shown are countries in which ranaviruses have been detected, but not a more detailed breakdown of the locations in which viruses have been found

4.2 Frog virus 3 (*Ranavirus rana1*)

The first cases of ranaviral infection in reptiles with available viral genome data were from a box turtle (*Terrapene c. carolina*) and a Horsfield's tortoise (*Testudo horsfieldii*) from North America. Both appeared to be FV3 strains based on partial MCP gene sequences and restriction endonuclease analysis (Mao et al. 1997). The Horsfield's tortoise had died with swelling of the eyelids, rhinitis, and dyspnea, while diphtheroid necrotizing glossitis, pharyngo-esophagitis, and tracheitis were noted in the box turtle (Marschang et al. 2021). Since then, the number of reports and cases in chelonians worldwide, especially box turtles (*Terrapene* sp.) in the USA, has increased continuously (De Voe et al. 2004; Allender et al. 2006; Johnson et al. 2008; Allender 2012; Winzeler et al. 2015; McKenzie et al. 2019; Carstairs et al. 2020; Cozad et al. 2020; von Essen et al. 2020; Table 6).

Adult chelonians have been more commonly reported to develop FV3 disease than juveniles (Johnson 2006), and surveillance in eastern box turtles demonstrated that juveniles are more likely to be FV3 positive using qPCR (Allender 2012). Challenge studies in juvenile turtles demonstrate a higher mortality rate and shorter median survival time than adults of the same or similar species (Allender et al. 2018). Therefore, it is likely that susceptibility of chelonians to ranaviruses differs among developmental stages similar to amphibians (Haislip et al. 2011). Some outbreaks in box turtles have involved translocation events that congregate many individuals, resulting in high infection prevalence and death (Belzer and Seibert 2011; Farnsworth and Seigel 2013; Kimble et al. 2014, 2017). Several surveys of emydid turtles in the USA have been performed, and ranavirus prevalence has not been reported above 5% in a population without abnormal mortality events (Allender et al. 2013a; Archer et al. 2017; Winzeler et al. 2018; Carstairs 2019; Winter et al. 2020). A single study of wild Eastern painted turtles (*Chrysemys picta picta*), an aquatic species, in Virginia, USA, reported infection prevalence of 4.8–31.6% in different ponds, with no apparent disease (Goodman et al. 2013).

Some studies have indicated that increased ranaviral mortality in chelonians may correlate with increased exposure to infected sympatric amphibians, possibly via predation on infected amphibians, exposure to water containing ranavirus shed by amphibians, or hematophagous insects (Belzer and Seibert 2011; Kimble et al. 2014). There are also multiple studies screening chelonian samples for ranaviruses with no virus detection reported (Winzeler et al. 2018; Kolesnik et al. 2017; Carstairs 2019; Winter et al. 2020; Calle et al. 2021; Schönbächler et al. 2022). However, a study in Canada following up on negative results screening wild turtles in an area in which ranavirus was known to occur in amphibians (Carstairs 2019) was able to detect ranaviruses in 7 of 46 turtles tested (in painted turtles and snapping turtles (*Chelydra serpentina*)), indicating that both sample material and virus detection method can strongly influence detection rates (Carstairs et al. 2020).

Koch's postulates have been fulfilled for FV3 ranaviral disease in chelonians, including box turtles, red-eared sliders (*Trachemys scripta elegans*), river cooters (*Pseudemys concinna concinna*), Mississippi map turtles (*Graptemys pseudogeographica kohni*), and false map turtles (*Graptemys pseudogeographica*) (Johnson et al. 2007; Allender et al. 2013b, 2018; Hausmann et al. 2015). Experimental challenge with FV3 isolates from either Burmese star tortoises (*Geochelone platynota*) or eastern box turtles has resulted in high mortality in red-eared sliders (Johnson et al. 2007; Allender et al. 2013b, 2018). Characteristic clinical signs of nasal discharge and oral plaques were seen but were inconsistent among individuals. Mortality rate and the presence of clinical signs were observed to be significantly greater in adult turtles exposed at 22 °C compared to 28 °C, with corresponding increased viral copy number and shorter median survival time at lower temperatures (Allender et al. 2013b). Conversely, in four species of aquatic turtles, juvenile turtles housed at higher temperatures had a shorter median survival time than those turtles housed at lower temperatures, but mortality in all groups was 100% (Allender et al. 2018). In a challenge study of previously infected eastern box turtles, only 14% of inoculated turtles died upon re-infection despite all turtles shedding viral DNA (Hausmann et al. 2015), indicating that immunity may be conferred from previous infection. Unfortunately, previously exposed turtles also shed viral DNA intermittently for years in the absence of inoculation, making these animals likely carriers of disease (Allender, personal communication).

STIV, which causes a severe systemic disease in farmed soft-shelled turtles (*Pelodiscus* [*Trionyx*] *sinensis*) in China, was the first reptilian ranavirus to be fully sequenced, demonstrating that it is a strain of FV3 (Huang et al. 2009). The disease in soft-shelled turtles is referred to as "red neck disease" and was first reported in 1998.

BIV was first reported in ornate burrowing frogs (*Platyplectrum ornatum*) in Australia in 1992 (Speare and Smith 1992) and has since been identified in other frog (Weir et al. 2012; Wynne 2019), snake (Hyatt et al. 2002), and lizard (Maclaine et al. 2020) species across northern Australia. Originally designated as a second ranavirus species, BIV was later reclassified based on molecular data as a strain of FV3 (Chinchar et al. 2017a). In Australia, FV3 viruses have been found in the continental north (EHNV in the south). Importantly, however, the vastness of Australia's geography, coupled with the limited number of research teams investigating ranaviruses there, suggests a possible wider geographic spread of these viruses than reported to date (Ariel et al. 2017; Wirth 2020).

Experimental challenge studies have shown at least two species of Australian freshwater turtle (*Myuchelys latisternum* and *Emydura macquarii krefftii*) are susceptible to FV3 (Ariel et al. 2015, 2017); however, reports of ranaviral infection in wild freshwater turtles in Australia are lacking. One reason for this could be infrequent occurrence in wild populations. A molecular survey of five species of freshwater turtles ($n = 379$) from five locations across North Queensland returned no positive result for qPCR assay for ranaviral DNA in turtle blood (Wirth et al. 2024). This lack of ranaviral infection reports in Australian freshwater turtles could also be the result of poor surveillance (Wirth et al. 2024). For example, a previous attempt to establish an enzyme-linked immunosorbent assay (ELISA) for BIV to conduct a sero-survey of freshwater turtles in Queensland failed due to an inability to develop a seropositive control, with two wild-caught adult male freshwater turtles (*Emydura macquarii krefftii*) failing to seroconvert despite multiple intra-muscular inoculations (Meddings 2011). This seroconversion failure was potentially due to seasonal variation in steroid hormones causing thymic involution (Leceta and Zapata 1985, 1986), which underscores the importance of considering seasonal variation in immune responses when conducting sero-surveys of reptiles (Meddings 2011). In a subsequent experimental challenge trial, three of six BIV-injected adult *Emydura macquarii krefftii* failed to sero-convert (Ariel et al. 2017), suggesting individual variation in immune responses is another factor to consider when developing seropositive controls. A sero-survey of Northern Queensland reptile fauna (freshwater turtles *Myuchelys latisternum* and *Emydura macquarii krefftii*, freshwater crocodiles *Crocodylus johnstoni*, and snakes *Boiga irregularis*, *Dendrelaphis punctulatus*, *Tropidonophis mairii*, *Liasis childreni*, and *L. fuscus*) sampled in 1994 and 1995 detected past exposure to BIV (Ariel et al. 2017). The wild-caught turtles, crocodiles, and snakes had anti-BIV antibodies ranging from 15% to 59% of the population surveyed according to species, location, and year sampled (Ariel et al. 2017). The difference between results from serological and molecular surveys of Australian turtles may be explained by adult turtles rapidly clearing ranaviral infection while producing a lasting antibody response (Wirth et al. 2024). Ranaviral DNA can be found in the blood of BIV-infected turtle hatchlings weeks after infection; however, freshwater turtle hatchlings are notoriously hard to capture in the wild, which limits the ability to detect active ranaviral infections in Australian turtle hatchlings (Wirth et al. 2024).

Until recently, ranaviruses were only rarely reported in squamate reptiles (snakes and lizards). The first report of ranaviruses in snakes was in a group of ten juvenile green tree pythons (*Morelia* [*Chondropython*] *viridis*) imported into Australia from Papua New Guinea with oral and hepatic lesions, with an FV3 strain isolated from pooled necropsy tissues (Hyatt et al. 2002). The virus was named Wamena virus and appears most closely related to TFV strains circulating among fish and amphibian aquaculture facilities in Asia (Sriwanayos et al. 2020). FV3 was also isolated from several organs of a red blood python (*Python brongersmai*) with similar pathology imported into Germany from Indonesia, with the isolated ranavirus also most closely related to TFV (Stöhr et al. 2015). In lizards, a ranavirus detected by virus isolation and later sequenced was found in a captive leaf-tailed gecko (*Uroplatus fimbriatus*) that had died with glossitis and hepatic necrosis in Germany (Marschang et al. 2005). That virus was later named German gecko ranavirus (GGRV) and found to be closely related to BIV based on full genome sequencing (Stöhr et al. 2015). In Portugal, an FV3 strain was isolated from a wild-caught Iberian mountain lizard (*Iberolacerta* [*Lacerta*] *monticola*) that did not show any clinical signs of disease, and a co-infection with erythrocytic necrosis virus was also found (Alves de Matos et al. 2011). In a study describing virological screening of samples from lizards (Stöhr et al. 2013b), FV3 strains were detected in brown anoles (*Anolis sagrei*), Asian glass lizards (*Dopasia gracilis*), and green anoles (*Anolis carolinensis*). All of the infected lizards had skin lesions. Sequencing part of the MCP gene of each virus showed that the five detected viruses were distinct from one another, and the obtained sequences were 98.4–100% identical to the corresponding portion of the FV3 genome. Further characterization based on larger portions of the viral genomes also clustered them with members of the *Frog virus 3* species (Stöhr et al. 2015). A ranavirus was also detected in green striped tree dragons (*Diploderma splendidum* [*Japalura splendida*]) imported from southwestern China via Florida into Germany during a mass mortality event. The ranavirus was closely related to FV3 (Behncke et al. 2013; Stöhr et al. 2015). A project screening archived tissue samples from amphibians and reptiles in the UK found an FV3 virus in a single sample from a slow worm (*Anguis fragilis*) (Price et al. 2017). The sample had been collected in 1993 along with samples from several common frogs and an unidentified newt during a disease incident. Ranaviruses were detected in tissues from all of the animals sampled in conjunction with that incident. In Japan, a ranavirus was detected in affected skin in a group of captive bearded dragons with brown crustaceous skin lesions (Tamukai et al. 2016). *Austwickia chelonae* was also identified in the skin lesions. In a study in which reptiles in Virginia, USA, were screened, ranaviral DNA was detected in 36.1% of the tail samples tested from wild-caught Eastern fence lizards (*Sceloporus undulatus*) (Goodman et al. 2018). No disease was noted in the animals. Eastern box turtles caught during the same study also tested positive with no signs of disease.

Transmission studies have shown Australian lizard species, such as eastern water dragon (*Intellagama lesueurii lesueurii*), to be susceptible to infection with BIV (Maclaine et al. 2018, 2019). While no epizootics in Australian lizards have been linked to ranaviral disease, ranaviral DNA has been detected in several Australian Agamidae species including wild eastern water dragons, nobbi dragons (*Diporiphora nobbi*), a Boyd's forest dragon (*Lophosaurus* [*Hypsilurus*] *boydii*), captive central bearded dragons (*Pogona vitticeps*), and a captive eastern water dragon (Maclaine 2019; Maclaine et al. 2020). The samples from captive lizards in Australia were obtained from collections with previous mortality events from unknown causes (Maclaine 2019). Australia is home to over 800 species of lizard including several species of (potentially susceptible) leaf tail gecko with limited geographical distributions (Marschang et al. 2005; Maclaine et al. 2020; https://arod.com.au/arod/reptilia/Squamata). The overlap of FV3 and potentially susceptible species is a conservation concern that deserves greater attention.

4.3   CMTV, EHNV, and Chimeric Viruses

While the majority of ranaviruses found in reptiles to date have been strains of FV3, there have been several exceptions. In some cases, extensive sequencing of the genome was necessary in order to understand the relationships between these viruses and other ranaviruses. One of the first cases in which ranaviruses were detected in chelonians in Europe was in a group of juvenile Hermann's tortoises (*Testudo hermanni*) in Switzerland (Marschang et al. 1999). Viruses were detected by virus isolation and PCR and sequencing in two juvenile diseased tortoises. All seven animals in the affected group died with similar signs. The associated virus was first described as FV3-like but has since been shown to be more closely related to CMTV (Stöhr et al. 2015). A ranavirus closely related to CMTV was found in the esophageal tissues of a wild viperine snake (*Natrix maura*) during a CMTV outbreak among amphibians in Spain (Price et al. 2014). The snake was ingesting diseased amphibians, and ulcerative lesions were found in the esophagus. In Poland, a ranavirus identified as a CMTV based on full genome sequencing was isolated from an invasive free-ranging red-eared slider with respiratory disease (Borzym et al. 2020). Splenomegaly and hepatic and tracheal necrosis were detected in the turtle.

Ranaviruses have also been detected in association with mortality events in captive Hermann's tortoises, Egyptian tortoises (*T. kleinmanni*), and marginated tortoises (*T. marginata*) in Germany. Affected animals developed stomatitis as well as splenic necrosis, enteritis, hepatitis, pancreatitis, and dermatitis in some cases (Uhlenbrok 2010). Sequence analysis of the genomes of the viruses associated with these outbreaks (tortoise ranavirus 1 and 2, ToRV-1, and ToRV-2) clusters them closely with FV3, while their genomic arrangement resembles that of CMTV (Stöhr et al. 2015; Price 2015). A similar virus was also detected in a captive central bearded dragon with skin lesions in Germany (Stöhr et al. 2013b).

In a study describing virological screening of samples from lizards (Stöhr et al. 2013b), a ranaviral infection was detected in green iguanas (*Iguana iguana*) with skin lesions (Stöhr et al. 2013b). Sequencing of part of the MCP gene showed that it was 100% identical to ECV, a strain of EHNV.

5 Impact of Trade

Ranaviruses are classified as emerging pathogens because their geographic and host ranges appear to be expanding (Daszak et al. 1999). It is becoming evident that ranaviruses are frequently spread through the regional and international trade of animals. For example, barred tiger salamander (*Ambystoma mavortium*) larvae are sold as fishing bait in the southwestern USA, and as many as 100% have been shown to be infected with ATV (Picco and Collins 2008). Amphibian ranaviruses have been found in animals that are traded over international borders for a variety of reasons, including human consumption and pet trade (Schloegel et al. 2009; Kolby et al. 2014; Saucedo et al. 2017). Schloegel et al. (2009) found 8.5% of amphibians imported into the USA at three major port cities were infected with ranaviruses. Similarly, Kolby et al. (2014) found over 50% of amphibians exported via Hong Kong International Airport were infected with ranaviruses. Reptiles infected with ranaviruses also have been discovered in internationally traded animals (Hyatt et al. 2002; Stöhr et al. 2013b). Internationally traded ornamental fishes have been shown to harbor ranaviruses (Hedrick and McDowell 1995), with the repeated detection of the same finfish ranaviruses (e.g., SCRV) around the globe suggesting that the international movement of live animals and their products likely plays an important role in the occurrence of these epizootics (Hedrick and McDowell 1995; Plumb and Zilberg 1999a; Grant et al. 2005; Schramm Jr and Davis 2006; Deng et al. 2011; George et al. 2015). Similarly, trade in farmed fish is believed to have fueled the spread of SGIV in Asia (Qin et al. 2003).

International trade and invasive species have contributed to the spread, diversity, and accumulation of ranaviruses. The African clawed frog (*Xenopus laevis*) is an invasive species in Chile and is thought to be a reservoir of infection due to its resistance to ranaviruses, with concerns this invasive species may contribute to the spread of ranavirus and subsequent decline of native amphibian species (Peñafiel-Ricaurte et al. 2023). Genetic diversity of ranaviruses is thought to result from the transportation of animals in close proximity promoting co-infections and interclass transmission (Duffus et al. 2015; Stöhr et al. 2015), with exposure to genetically diverse viruses from geographically distant locations a probable driver of naturally recombinant virus (Price 2015). For example, a ranavirus isolated from an Egyptian tortoise (Stöhr et al. 2015) was identified as a likely mosaic of two parent ranaviruses, as it contained genes clustering to both FV3 and CMTV clades (Price 2015). As for viral accumulation, a growing body of evidence points to this being driven by the pet trade, with a broad range of viral groups found to co-occur in pet reptiles, including ranaviruses with adenoviruses, other iridoviruses, ferlaviruses, and parvoviruses among others (Abbas et al. 2011; Bak et al. 2018; Ball et al. 2014; Behncke et al. 2013; Penzes et al. 2015).

Of interest is that within the legal international trade in reptiles, the largest numbers are traded through Europe, including both captive-bred and wild-caught animals (Bush et al. 2013). The majority of reptile infections with genetically diverse ranaviruses have also been reported in Europe, often with a connection to the pet trade (Stöhr et al. 2013b, 2015). The role of illegal trade in exotic pets for the epidemiology of ranaviral infections in reptiles has not been studied, although there is some indication that this has played a role in ranaviral outbreaks in pet reptiles as well (S. Blahak, CVUA-OWL, personal communication). Wild-caught and farmed reptiles which are globally transferred via the pet trade are often in contact with numerous different animal species (reptiles and amphibians) and are not regularly tested for the presence of infections.

6   Interclass Transmission of Ranaviruses

Interclass transmission of ranaviruses refers to their ability to infect a wide variety of hosts across three classes of ectothermic vertebrates (Amphibia, Reptilia, and Osteichthyes), with viral amplification in highly susceptible hosts and persistence in hosts with low susceptibility creating the ideal conditions for transmission (Paull et al. 2012). Evidence for interclass transmission has been found in field studies detecting ranaviruses in sympatric species in multiple classes (e.g., Currylow et al. 2014; Price et al. 2014, 2017). Genome studies also provide evidence for the role of interclass transmission in the evolution of ranaviruses (e.g., Jancovich et al. 2010). Experimental studies of interclass transmission of ranaviruses have used both injection and more natural routes of exposure, such as baths, to demonstrate the ability of specific ranaviruses to infect hosts from multiple classes (Table 7).

Moody and Owens (1994) provided the first evidence for interclass transmission of ranaviruses: Barramundi exposed to FV3 isolated from amphibians either by bath exposure or direct injection developed disease and experienced 100% mortality. In addition, Bayley et al. (2013) showed that a CMTV isolate originally derived from fish (PPIV) can cause mortality in common frog tadpoles. Tadpoles were also exposed to five other fish-derived ranavirus isolates via water bath, but only PPIV caused mortality. This was the first case where exposure to a fish-derived isolate caused death in an amphibian. Subsequent experiments by Brenes et al. (2014a, b) have shown transmission of FV3 isolates between fish, amphibians, and reptiles. In one experiment, FV3 originally isolated from a moribund pallid sturgeon was transmitted via the water from infected individuals to amphibians (Cope's grey treefrog, *Hyla chrysoscelis*), fish (mosquito fish, *Gambusia affinis*), and reptiles (red-eared sliders). In a similar experiment, three chelonians (Florida soft-shelled turtle, *Apalone ferox*; eastern river cooters; Mississippi map turtles) were assessed for susceptibility to the same FV3 isolate from pallid sturgeon, an FV3 isolate from an eastern box turtle (*Terrapene carolina carolina*), and a third FV3 isolate from an American bullfrog, from the USA (Brenes et al. 2014b). While no disease or mortality was observed in these experiments, infections were documented in soft-shelled turtles that were exposed to the fish isolate and those that were exposed to the turtle isolate (Brenes et al. 2014b). Infections were also observed in the Mississippi map turtles that were exposed to the turtle isolate (Brenes et al. 2014b).

Table 7 Ranavirus transmission studies testing interclass transmission

Original isolates				Transmission studies			
Continent	Region	Virus species and strain	Species isolated from	Source	Species infected	Transmission route	I,D,M[a]	References
Oceania	Australia, North Queensland	EHNV	European perch (*Perca fluviatilis*)	Langdon et al. (1986b)	European common frog tadpoles and post-metamorphs (*Rana temporaria*)	Bath immersion for 1 hour	M	Bayley et al. (2013)
		FV3 (BIV)	Ornate burrowing frog (*Limnodynastes ornatus*)	Speare and Smith (1992)	Barramundi (*Lates calcarifer*)	Bath exposure or intraperitoneal and intramuscular injection	I, D, M	Moody and Owens (1994)
					Juvenile short-necked turtles (*Emydura macquarii krefftii*) and saw-shelled turtles (*Myuchelys* (*Elseya*) *latisternum*)	Intraperitoneal injection	D, M	Ariel (1997)
					Adult short-necked turtles (*Emydura macquarii krefftii*) and saw-shelled turtles (*Myuchelys* (*Elseya*) *latisternum*)	Intraperitoneal injection or co-habitation	I	Ariel (1997)
					Juvenile crocodiles (*Crocodylus johnstoni*)	Intraperitoneal injection or co-habitation	–	Ariel (1997)
					Brown tree snakes, *Boiga irregularis*, common green tree snakes, *Dendrelaphis punctulatus*, and keelback snakes, *Tropidonophis* (*Amphiesma*) *mairii*)	Injection, co-habitation, or feeding of infected frogs	–	Ariel (1997)
Europe	Finland	CMTV (PPIV)	Pike-perch (*Stizostedion lucioperca*)	Tapiovaara et al. (1998)	European common frog tadpoles and post-metamorphs (*Rana temporaria*)	Bath immersion for 1 hour	M	Bayley et al. (2013)
	Italy	EHNV (ECV)	Black bullhead (*Ameiurus melas*)	Bovo et al. (1993)	European common frog tadpoles and post-metamorphs (*Rana temporaria*)	Bath immersion for 1 hour	M	Bayley et al. (2013)
	Italy (fish imported from New Zealand)	SERV	Short-finned eel (*Anguilla australis*)	Bovo et al. (1999)	European common frog tadpoles and post-metamorphs (*Rana temporaria*)	Bath immersion for 1 hour	M	Bayley et al. (2013)

North America	USA—Arizona	ATV (ORV)	Tiger salamander (*Ambystoma tigrinum*)	Picco and Collins (2008)	Largemouth. Bass (*Micropterus salmoides*)	Intraperitoneal injection	I	Picco et al. (2010)
	USA—California (fish imported from Asia)	SCRV (DFV)	Doctor fish (*Labroides dimidiatus*)	Hedrick and McDowell (1995)	European common frog tadpoles and post-metamorphs (*Rana temporaria*)	Bath immersion for 1 hour	M	Bayley et al. (2013)
	USA—California (fish imported from Asia)	SCRV (GV6)	Guppy (*Poecilia reticulata*)	Hedrick and McDowell (1995)	European common frog tadpoles and post-metamorphs (*Rana temporaria*)	Bath immersion for 1 hour	–	Bayley et al. (2013)
	USA—Georgia	FV3	American bullfrog (*Lithobates catesbeianus*)	Miller et al. (2007)	Nile tilapia (*Oreochromis niloticus*), channel catfish (*Ictalurus punctatus*), Western mosquitofish (*Gambusia affinis*), bluegill (*Lepomis macrochirus*), and fathead minnow (*Pimephales promelas*) fingerlings.	Bath immersion for 3 days	M	Brenes et al. (2014b)
					Florida soft-shelled turtle (*Apalone ferox*), eastern river cooter (*Pseudemys concinna*), and Mississippi map turtle (*Graptemys pseudogeographica kohni*).	Bath immersion for 3 days	I	Brenes et al. (2014b)

(continued)

Table 7 (continued)

Original isolates					Transmission studies			
Continent	Region	Virus species and strain	Species isolated from	Source	Species infected	Transmission route	I,D,M[a]	References
	USA—Kentucky	FV3	Eastern box turtle (*Terrapene carolina carolina*)	Ruder et al. (2010)	Nile tilapia (*Oreochromis niloticus*), channel catfish (*Ictalurus punctatus*), Western mosquito fish (*Gambusia affinis*), bluegill (*Lepomis macrochirus*), and fathead minnow (*Pimephales promelas*) fingerlings	Bath immersion for 3 days	M	Brenes et al. (2014b)
	USA—Missouri	FV3	Pallid sturgeon (*Scaphirhynchus albus*)	Waltzek et al. (2014)	Red-eared slider (*Trachemys scripta elegans*)	Bath immersion for 3 days, and co-habitation with exposed *Hyla chrysoscelis* tadpoles	I	Brenes et al. (2014a)
					Cope's gray treefrog tadpoles (*Hyla chrysoscelis*)	Bath immersion for 3 days, and co-habitation with exposed *Gambusia affinis, Trachemys scripta elegans, Hyla chrysoscelis* tadpoles	M	Brenes et al. (2014a)
					Florida soft-shelled turtle (*Apalone ferox*), eastern river cooter (*Pseudemys concinna*), and Mississippi map turtle (*Graptemys pseudogeographica kohni*)	Bath immersion for 3 days	I	Brenes et al. (2014b)

[a]*I* infection with no signs of ranaviral disease, *D* Gross and/or histological ranaviral disease reported, *M* mortality due to ranaviral disease, – no ranaviral infection or disease reported

A pathogenicity trial that used BIV, the strain of FV3 originally isolated from an ornate burrowing frog by Speare and Smith (1992), found it could infect and cause mortality in juvenile short-necked turtles (*Emydura macquarii krefftii*) and saw shelled turtles (*Myuchelys* [*Elseya*] *latisternum*), but infection was not detected in adult turtles of the same species (Ariel 1997; Ariel et al. 2015). In addition, BIV did not induce disease in brown tree snakes (*Boiga irregularis*), common green tree snakes (*Dendrelaphis punctulatus*), keelback snakes (*Tropidonophis* [*Amphiesma*] *mairii*), or juvenile crocodiles (*Crocodylus johnstoni*). Notably, BIV was reisolated from one of the brown tree snakes 4 weeks after inoculation. Isolation of the inoculated virus so long after initial subclinical infection suggested this species may be a viable reservoir (Ariel 1997; Ariel et al. 2015).

Transmission studies with ATV have been carried out with amphibians and fish. Although originally thought to be restricted to urodeles (Jancovich et al. 2001), ATV was later shown to be pathogenic to anurans (Schock et al. 2008). Experimental infection of largemouth bass was successful, but inoculated animals experienced no mortality or disease (Picco et al. 2010). There is also evidence that multiple FV3 and ATV strains may circulate in ponds and infect both urodeles and anurans (Schock et al. 2008). To our knowledge, no experimental infections of ATV in reptiles have been attempted to date.

Our understanding of the role of different host classes in the epidemiology of ranaviruses is limited, although as the results mentioned above demonstrate, reptiles, fish, and amphibians may act as reservoirs for FV3 for other taxa. Several field studies have indicated that ranavirus infections in one group of hosts can affect the health and survival of sympatric ectothermic vertebrates, but the role of various hosts as long-term carriers of virus and in the dynamics of transmission remains unknown. The fact that multiple studies have shown different viruses may have vastly different effects on various hosts is also important for the assessment of infection status in clinically healthy animals, both in the wild and in captivity, especially in trade. Healthy infected animals in which ranaviruses are not suspected could be a source of infection via direct contact or environmental contamination for other susceptible species of other animal classes. All these findings underline the need to reassess our understanding of ranaviruses as multi-species pathogens, not only pathogens of specific groups of animals.

7 The Global Ranavirus Reporting System

The effort to create a Global Ranavirus Reporting System (GRRS; *https://brunner-lab.shinyapps.io/GRRS_Interactive/*) began in 2011 (Duffus and Olson 2011). The GRRS is an online and open access database, which is designed to track the occurrence of ranavirus infections around the globe (Brunner et al. 2021). The portal tracks both presence and absence data for all taxa affected by ranaviruses (Brunner et al. 2021). The GRRS database has multiple layers that can be viewed on an interactive map or by downloading the information from the database (Fig. 5; Brunner

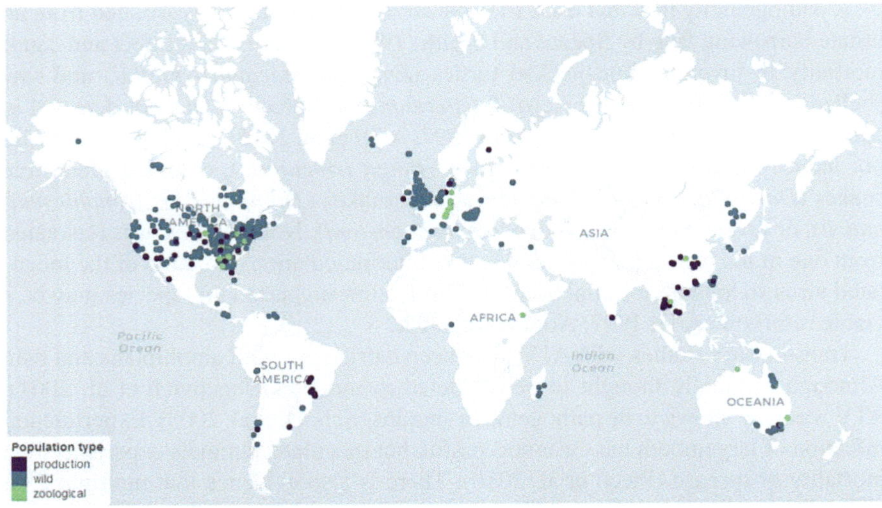

Fig. 5 Interactive map produced by the Global Ranavirus Reporting System. The inset is the color key that identifies the type of population that the sample comes from

et al. 2021). The layers of the database include but are not limited to species, life history stage, type of population (i.e., wild or captive), production (e.g., from aquaculture or ranaculture facilities) or zoological collection (this includes animals in the pet trade, private collections, and larger zoological collections), number of samples that were screened for infection, confirmed disease (if applicable), and diagnostic methods used (Fig. 6; Brunner et al. 2021). The number of animals screened for infection can be very different from those with confirmed disease. This is partially dependent on the diagnostic methods and definition of morbidity used. There are some difficulties with the specificity and sensitivity of diagnostic methods used for ranavirus detection (Miller et al. 2024). This can make comparison of data between publications and within the GRRS challenging.

8   Summary and Conclusions

Ranavirus infections in amphibians, fish, and reptiles are widespread and affect a diverse suite of species within these vertebrate classes. The species affected include some that are economically important (e.g., rainbow trout, soft-shelled turtle, bullfrogs) but also several that are of conservation concern (e.g., Chinese giant salamander, box turtles, dusky gopher frog). The economic and conservation risk of these multi-species pathogens is dependent upon many factors, including characteristics of the host species. One thing is certain—some host species are highly susceptible to specific ranaviruses, and these species are most likely to be affected during outbreaks. Thus, understanding host susceptibility to different ranaviruses is key to

Fig. 6 A zoom in of the interactive map with one of the data points "open." You can find the name of the event, the taxonomic information of the animal affected, the number of animals involved in the mortality/morbidity event, the number infected/number tested, and the number of diseased animals in that specific event

quantifying risk. Host-pathogen interactions between ranaviruses and amphibians are the best characterized. Research in this area needs to continue, but there also needs to be greater attention on the role of reptiles and fish in epizootic events.

All three classes of vertebrate hosts are intensively farmed in different regions of the world. The conditions of captive culture facilities that often maintain high densities of genetically similar individuals may be conducive to repeated outbreaks of ranaviral disease (Pearman and Garner 2005). Additionally, conditions that favor transmission can lead to increased virulence, according to the virulence trade-off hypothesis (Alizon et al. 2009). Thus, captive facilities with reoccurring ranavirus outbreaks may facilitate evolution of ranavirus types that are more virulent than wild types (Brunner et al. 2024).

The commercial trade of ranavirus hosts is likely a significant factor facilitating the global distribution of ranaviruses, as well as interclass transmission of the pathogen. If novel strains of ranaviruses are introduced into naïve populations, experimental evidence suggests that there could be devastating effects (e.g., Pearman et al. 2004; Storfer et al. 2007; Hoverman et al. 2011a). With the trade in animals being truly global and ranaviruses accompanying them (e.g., Schloegel et al. 2009; Kolby et al. 2014), it is important to understand what ranaviruses are being transported and where they end up.

The WOAH lists EHNV and ranaviruses that infect amphibians as "reportable infections of wildlife" (WOAH 2022). This designation requires countries that have agreed to WOAH policies to screen a sample of ranavirus hosts that are crossing international borders for the presence of ranaviruses (Schloegel et al. 2010). However, few countries have yet taken steps to implement import policies that require declaration of ranavirus-free animals. Moreover, infection of fish by ranaviruses (other than EHNV) and reptiles (for any ranavirus species) is not included in the WOAH regulations.

Despite progress since the first edition of this book in 2015, we still do not fully understand the distribution and host range of ranaviruses. Continued surveillance of wild and captive populations as well as commercially traded animals combined with the characterization of ranavirus strains is necessary to fully understand the distribution and host diversity of these viruses. Therefore, when ranavirus studies are undertaken, sufficient funding should be obtained to at least partially characterize the virus if detected. This approach will require researchers to work collaboratively in interdisciplinary groups.

Acknowledgments We would like to thank David Lesbarreres and Danna Schock for helpful comments in the preparation of this manuscript. We thank the following organizations for providing funds to support Open Access publishing of the second edition: University of Tennessee's (UT) Open Publishing Support Fund, UT School of Natural Resources, UT Center for Wildlife Health, Washington State University Libraries, University of Mississippi Medical Center, Gordon State College, the Association of Reptile and Amphibian Veterinarians, and the Global Ranavirus Consortium.

References

Abbas MD, Marschang RE, Schmidt V et al (2011) A unique novel reptilian paramyxovirus, four atadenovirus types and a reovirus identified in a concurrent infection of a corn snake (*Pantherophis guttatus*) collection in Germany. Vet Microbiol 150:70–79

Abonyi F, Varga Á, Sellyei B et al (2022) Juvenile Wels catfish (*Silurus glanis*) display age-related mortality to European catfish virus (ECV) under experimental conditions. Viruses 14(8):1832. https://www.mdpi.com/1999-4915/14/8/1832

Adamovicz L, Allender MC, Archer G, a. (2018) Investigation of multiple mortality events in eastern box turtles (*Terrapene carolina carolina*). PLoS One 13:e0195617

Adamovicz L, Allender MC, Gibbons PM (2020) Emerging infectious diseases of chelonians: an update. Vet Clin North Am Exot Anim Pract 23(2):263–283

Agha M, Price SJ, Nowakowski AJ et al (2017) Mass mortality of eastern box turtles with upper respiratory disease following atypical cold weather. Dis Aquat Org 124(2):91–100

Ahne W, Schlotfeldt HJ, Thomsen I (1989) Fish viruses: isolation of an icosahedral cytoplasmic deoxyribovirus from sheatfish (*Silurus glanis*). J Veterinary Med Ser B 36:333–336

Ahne W, Ogawa M, Schlotfeldt HJ (1990) Fish viruses: transmission and pathogenicity of an icosahedral cytoplasmic deoxyribovirus isolated from sheatfish (*Silurus glanis*). J Veterinary Med Ser B 37:187–190

Ahne W, Schlotfeldt HJ, Ogawa M (1991) Iridovirus infection of adult sheatfish (*Silurus glanis*). Bull Eur Assoc Fish Pathol 11(3):97–98

Alizon S, Hurford A, Mideo N et al (2009) Virulence evolution and the trade-off hypothesis: history, current state of affairs and the future. J Evol Biol 22:245–259

Allender MC (2012) Characterizing the epidemiology of ranavirus in North American chelonians: diagnosis, surveillance, pathogenesis, and treatment. Department of Veterinary Clinical Medicine. Ph.D. thesis, University of Illinois at Urbana-Champaign, Urbana, p 219

Allender MC, Fry MM, Irizarry AR et al (2006) Intracytoplasmic inclusions in circulating leukocytes from an eastern box turtle (*Terrapene carolina carolina*) with iridoviral infection. J Wildl Dis 42:677–684

Allender MC, Abd-Eldaim M, Schumacher J et al (2011) PCR prevalence of ranavirus in free-ranging eastern box turtles (*Terrapene carolina carolina*) at rehabilitation centers in three southeastern US states. J Wildl Dis 47:759–764

Allender MC, Mitchell MA, McRuer D et al (2013a) Prevalence, clinical signs, and natural history characteristics of frog virus 3-like infections in eastern box turtles (*Terrapene carolina carolina*). Herpetol Conserv Biol 8:308–320

Allender MC, Mitchell MA, Torres T et al (2013b) Pathogenicity of frog virus 3-like virus in red-eared slider turtles (*Trachemys scripta elegans*) at two environmental temperatures. J Comp Pathol 149(2–3):356–367

Allender MC, Barthel AC, Rayl JM, Terio KA (2018) Experimental transmission of Frog Virus 3-like ranavirus in juvenile chelonians at two temperatures. J Zoo Wildl Med 54(4):716–725

Alves de Matos AP, Caeiro MF, Marschang RE et al (2008) Adaptation of ranaviruses from Peneda-Gerês National Park (Portugal) to cell cultures and their characterizations. Microsc Microanal 14:139–140

Alves de Matos AP, Caeiro MF, Papp R et al (2011) New viruses from Lacerta monticola (Serra da Estrela, Portugal): further evidence for a new group of nucleo-cytoplasmic large deoxyriboviruses. Microsc Microanal 17:101–108

Archer GA, Phillips CA, Adamovicz L et al (2017) Detection of copathogens in free-ranging eastern box turtles (*Terrapene carolina carolina*) in Illinois and Tennessee. J Zoo Wildl Med 48(4):1127–1134

Ariel E (1997) Pathology and serological aspects of Bohle iridovirus infections in six selected water-associated reptiles in North Queensland, Department of Microbiology and Immunology. Ph.D. thesis, James Cook University, North Queensland, p 168

Ariel E, Jensen BB (2009) Challenge studies of European stocks of redfin perch, *Perca fluviatilis* L., and rainbow trout, *Oncorhynchus mykiss* (Walbaum), with epizootic haematopoietic necrosis virus. J Fish Dis 32(12):1017–1025

Ariel E, Owens L (1997) Epizootic mortalities in tilapia *Oreochromis mossambicus*. Dis Aquat Org 29:1–6

Ariel E, Kielgas J, Svart HE et al (2009) Ranavirus in wild edible frogs, *Pelophylax* kl. *esculentus* in Denmark. Dis Aquat Org 85:7–14

Ariel E, Holopainen R, Olesen NJ et al (2010) Comparative study of ranavirus isolates from cod (*Gadus morhua*) and turbot (*Psetta maxima*) with reference to other ranaviruses. Arch Virol 155:1261–1271

Ariel E, Wirth W, Burgess G et al (2015) Pathogenicity in six Australian reptile species following experimental inoculation with Bohle iridovirus. Dis Aquat Org 115:203–212

Ariel E, Elliott E, Meddings JI et al (2017) Serological survey of Australian native reptiles for exposure to ranavirus. Dis Aquat Org 126:173–183

Bak EJ, Jho Y, Woo GH (2018) Detection and phylogenetic analysis of a new adenoviral polymerase gene in reptiles in Korea. Arch Virol 163:1663–1669

Ball I, Behncke H, Schmidt V et al (2014) Partial characterization of new adenoviruses found in lizards. J Zoo Wildl Med 45:287–297

Balseiro A, Dalton KP, Cerrol A et al (2009) Pathology, isolation and molecular characterization of a ranavirus from the common midwife toad *Alytes obstetricans* on the Iberian Peninsula. Dis Aquat Org 84:95–104

Balseiro A, Dalton KP, Cerro A et al (2010) Outbreak of common midwife toad virus in alpine newts (*Mesotriton alpestris cyreni*) and common midwife toads (*Alytes obstetricans*) in northern Spain: comparative pathological study of an emerging ranavirus. Vet J 186:256–258

Bayley AE, Hill BJ, Feist SW (2013) Susceptibility of the European common frog *Rana temporaria* to a panel of ranavirus isolates from fish and amphibian hosts. Dis Aquat Org 103:171–183

Becker JA, Tweedie A, Gilligan D et al (2013) Experimental infection of Australian freshwater fish with epizootic haematopoietic necrosis virus (EHNV). J Aquat Anim Health 25:66–76

Becker JA, Tweedie A, Gilligan D et al (2016) Susceptibility of Australian Redfin Perch *Perca fluviatilis* experimentally challenged with *Epizootic Hematopoietic Necrosis Virus* (EHNV). J Aquat Anim Health 28(2):122–130

Becker JA, Gilligan D, Asmus M et al (2019) Geographic distribution of *Epizootic haematopoietic necrosis virus* (EHNV) in freshwater fish in south eastern Australia: lost opportunity for a notifiable pathogen to expand its geographic range. Viruses 11(4):315

Behncke H, Stöhr AC, Heckers KO et al (2013) Mass mortality in green striped tree dragons (*Japalura splendida*) associated with multiple viral infections. Vet Rec 173:248

Belzer WR, Seibert S (2011) A natural history of Ranavirus in an eastern box turtle population. Turtle Tortoise Newsl 15:18–25

Benetka V, Grabensteiner E, Gumpenberger M et al (2007) First report of an iridovirus (Genus *Ranavirus*) infection in a Leopard tortoise (*Geochelone pardalis pardalis*). Wien Tierarztl Monatsschr 94:243–248

Bienentreu JF, Schock DM, Greer AL et al (2022) Ranavirus amplification in low-diversity amphibian communities. Front Vet Sci 9:755426

Bigarré L, Cabon J, Baud M et al (2008) Ranaviruses associated with high mortalities in catfish in France. Bull Eur Assoc Fish Pathol 28:163–168

Blackburn ME, Wayland JA, Smith WH et al (2015) First report of Ranavirus and *Batrachochytrium dendrobatidis* in green salamanders (*Aneides aeneus*) from Virginia, USA. Herpetol Rev 46(3):357–361

Blahak S, Uhlenbrok C (2010) Ranavirus infections in European terrestrial tortoises in Germany. In: Öfner S, Weinzierl F (eds) Proceedings of the 1st international conference on reptile and amphibian medicine, Munich, Germany, 4–7 March, pp 17–23

Bollinger RK, Mao J, Schock DM et al (1999) Pathology, isolation, and preliminary molecular characterization of a novel iridovirus from tiger salamanders in Saskatchewan. J Wildl Dis 35:413–429

Boonthai T, Loch TP, Yamashita CJ et al (2018) Laboratory investigation into the role of largemouth bass virus (Ranavirus, Iridoviridae) in smallmouth bass mortality events in Pennsylvania rivers. BMC Vet Res 14(1):62

Borzym E, Maj-Paluch J (2015) Experimental infection with *Epizootic haematopoietic necrosis virus* (EHNV) of rainbow trout (*Oncorhynchus mykiss* Walbaum) and European perch (*Perca fluviatilis* L.). J Vet Res 59(4):473–478

Borzym E, Stachnik M, Reichert M, Rzeżutka A, Jasik A, Waltzek TB, Subramaniam K (2020) Genome sequence of a ranavirus isolated from a red-eared slider (*Trachemys scripta elegans*) in Poland. Microbiol Resour Announc 9(47):e00781–e00720

Bovo G, Comuzzi M, De Mas S et al (1993) Isolation of an irido-like viral agent from breeding cat fish (*Ictalurus melas*). Boll Soc Ital Patol Ittica 11:3–10

Bovo G, Giacometti P, Montesi F et al (1999) Isolation of an irido-like agent from New Zealand eel. In: Proceedings of the European Association of Fish Pathologists, Rhodes, Greece, p 53

Box EK, Cleveland CA, Subramaniam K et al (2021) Molecular confirmation of ranavirus infection in amphibians from Chad, Africa. Front Vet Sci 2021:1034

Brenes R, Gray MJ, Waltzek TB et al (2014a) Transmission of ranavirus between ectothermic vertebrate hosts. PLoS One 9:e92476

Brenes R, Miller DL, Waltzek TB et al (2014b) Susceptibility of fish and turtles to three ranaviruses isolated from different ectothermic vertebrate classes. J Aquat Anim Health 26:118–126

Brunner JL, Barnet KE, Gosier CJ et al (2011) Ranavirus infection in die-offs of vernal pool amphibians in New York, USA. Herpetol Rev 42:76–79

Brunner JL, Storfer A, Gray MJ, Hoverman JT (2015) Ranavirus ecology and evolution: from epidemiology to extinction. In: Gray MJ, Chinchar VG (eds) Ranaviruses: lethal pathogens of ectothermic vertebrates. Springer Online https://link.springer.com/book/10.1007/978-3-319-13755-1

Brunner JL, Olson AD, Rice JG et al (2019) Ranavirus infection dynamics and shedding in American bullfrogs: consequences for spread and detection in trade. Dis Aquat Org 135(2):135–150

Brunner JL, Olson DH, Gray MJ et al (2021) Global patterns of ranavirus detections. Facets 6(1):912–924

Brunner JL, Storfer A, LeSage EH et al (2024) Ranavirus exology: from individual infections to population epidemiology to community impacts. In: Gray MJ, Chinchar VG (eds) Ranaviruses: lethal pathogens of ectothermic vertebrates. 2nd edn. Springer Online. In press

Bush ER, Baker SE, MacDonald DW (2013) Global trade in exotic pets 2006–2012. Conserv Biol 28:663–676

Calle PP, Raphael BL, Lwin T et al (2021) Burmese roofed turtle (*Batagur trivittata*) disease screening in Myanmar. J Zoo Wildl Med 52(4):1270–1274

Candido M, Tavares LS, Alencar ALF et al (2019) Genome analysis of Ranavirus frog virus 3 isolated from American Bullfrog (*Lithobates catesbeianus*) in South America. Sci Rep 9(1). https://doi.org/10.1038/s41598-019-53626-z

Carstairs SJ (2019) Evidence for low prevalence of ranaviruses in Ontario, Canada's freshwater turtle population. PeerJ 7:e6987. https://doi.org/10.7717/peerj.6987

Carstairs SJ, Kyle CJ, Vilaça ST (2020) High prevalence of subclinical frog virus 3 infection in freshwater turtles of Ontario, Canada. Virology 543:76–83

Chang SF, Ngoh-Lim GH, Kueh LFS et al (2002) Initial investigations into two viruses isolated from marine food fish in Singapore. Vet Rec 150(1):15–16

Chen ZX, Zheng JC, Jiang YL (1999) A new iridovirus isolated from soft-shelled turtle. Virus Res 63:147–151

Chen Z, Gui X, Gao C et al (2013) Genome architecture changes and major gene variations of *Andrias davidianus* ranavirus (ADRV). Vet Res 44:101

Cheng K, Jones MEB, Jancovich JK et al (2014) Isolation of a Bohle iridovirus-like agent from boreal toads housed within a cosmopolitan aquarium collection. Dis Aquat Org 111(2):139–152

Chinchar VG, Waltzek TB (2014) Ranaviruses: not just for frogs. PLoS Path 10:e1003850

Chinchar VG, Hick P, Ince IA et al (2017a) ICTV virus taxonomy profile: *Iridoviridae*. J Gen Virol 98(5):890–891

Chinchar VG, Waltzek TB, Subramaniam K (2017b) Ranaviruses and other members of the family Iridoviridae: their place in the virosphere. Virology 511:259–271

Chinchar VG, Duffus ALJ, Brunner JL (2021) Ecology of viruses infecting ectothermic vertebrates: the impact of ranavirus infections on amphibians. In: Hurst C (ed) Studies in viral ecology, 2nd edn. Wiley, Sussex

Chua FHC, Ng ML, Ng KL et al (1994) Investigation of outbreaks of a novel disease, 'Sleepy Grouper Disease', affecting the brown-spotted grouper, *Epinephelus tauvina* Forskal. J Fish Dis 17:417–427

Chuang HC, Chu TW, Cheng AC et al (2022) Iridovirus isolated from marine giant sea perch causes infection in freshwater ornamental fish. Aquaculture 548:737588

Clark HF, Brennan JC, Zeigel RF et al (1968) Isolation and characterization of viruses from the kidneys of *Rana pipiens* with renal adenocarcinoma before and after passage in the red eft (*Triturus viridescens*). J Virol 2:629–640

Claytor SC, Subramaniam K, Landrau-Giovannetti N et al (2017) Ranavirus phylogenomics: signatures of recombination and inversions among bullfrog ranaculture isolates. Virology 511:330–343

Collins JP, Jones TR, Berna HJ (1988) Conserving genetically distinct populations: the case of the Huachuca tiger salamander (*Ambystoma tigrinum stebbinsi* Lowe). In: Proceedings of the symposium on management of amphibians, reptiles and small mammals in North America, Flagstaff, Arizona, 19–21 July

Conrad CRK, Subramaniam K, Chinchar VG, Waltzek TB (2021) Genomic sequencing of ranavirus isolates from a three-spined stickleback (*Gasterosteus aculeatus*) and a red-legged frog (*Rana aurora*). Microbiol Resour Announc 10(46):e0090221

Converse KA, Green DE (2005) Diseases of tadpoles. In: Majumadar SK, Huffman JE, Brenner FJ et al (eds) Wildlife diseases: landscape epidemiology, spatial distribution and utilization of remote sensing techniques. Pennsylvania Academy of Science, Easton, pp 72–88

Coutinho CD (2019) The role of the invasive African clawed frog as vector of Ranavirus for native fishes and amphibians. Masters dissertation, Universidade de Lisboa. Available from: http://hdl.handle.net/10451/41439

Cozad RA, Norton TM, Aresco MJ et al (2020) Pathogen surveillance and detection of ranavirus (*Frog virus 3*) in translocated gopher tortoises (*Gopherus polyphemus*). J Wildl Dis 56(3):679–683

Cullen BR, Owens L (2002) Experimental challenge and clinical cases of Bohle iridovirus (BIV) in native Australian anurans. Dis Aquat Org 49:83–92

Cullen BR, Owens L, Whittington RJ (1995) Experimental infection of Australian anurans (*Limnodynastes terraereginae* and *Litoria latopalmata*) with Bohle iridovirus. Dis Aquat Org 23:8392

Cunningham AA, Langton TES, Bennett PM et al (1993) Unusual mortality associated with poxvirus-like particles in frogs (*Rana temporaria*). Vet Rec 133:141–142

Cunningham AA, Langton TES, Bennet PM et al (1996) Pathological and microbial findings from incidents of unusual mortality on the common frog (*Rana temporaria*). Philos Trans R Soc Lon B Biol Sci 315:1539–1557

Cunningham AA, Hyatt AD, Russell P et al (2007) Experimental transmission of a ranavirus disease of common toads (*Bufo bufo*) to common frogs (*Rana temporaria*). Epidemiol Infect 135:1213–1216

Cunningham AA, Turvey ST, Zhou F et al (2016) Development of the Chinese giant salamander *Andrias davidianus* farming industry in Shaanxi Province, China: conservation threats and opportunities. Oryx 50(2):265–273

Currylow AF, Johnson AJ, Williams RN (2014) Evidence of ranavirus infections among sympatric larval amphibians and box turtles. J Herpetol 48:117–121

Daszak P, Berger L, Cunningham AA et al (1999) Emerging Infectious diseases and amphibian population declines. Emerg Infect Dis 6:735–748

Davidson SRA, Chambers DL (2011) Ranavirus prevalence in amphibian populations of Wise County, Virginia, USA. Herpetol Rev 42:214–215

Davis DR, Kerby JL (2016) First detection of ranavirus in amphibians from Nebraska, USA. Herpetol Rev 47(1):46–50

Davis DR, Farkas JK, Kruisselbrink TR et al (2019) Prevalence and distribution of ranavirus in amphibians from Southeastern Oklahoma, USA. Herpetol Conserv Biol 14(2)

Davis DR, Ferguson KJ, Schwarz MS, Kerby JL (2020) Effects of agricultural pollutants on stress hormones and viral infection in larval salamanders. Wetlands 40:577–586

De Voe RK, Geissler K, Elmore S et al (2004) Ranavirus-associated morbidity and mortality in a group of captive eastern box turtles (*Terrapene carolina carolina*). J Zoo Wildl Med 35:534–543

Deng G, Li S, Xie J et al (2011) Characterization of a ranavirus isolated from cultured largemouth bass (*Micropterus salmoides*) in China. Aquaculture 312:198–204

Deng L, Geng Y, Zhao R et al (2020) CMTV-like ranavirus infection associated with high mortality in captive catfish-like loach, *Triplophysa siluorides,* in China. Transbound Emerg Dis 67(3):1330–1335. https://doi.org/10.1111/tbed.13473

Docherty DE, Meteyer CU, Wang J et al (2003) Diagnostic and molecular evaluation of three iridovirus associated salamander mortality events. J Wildl Dis 39:556–566

Docherty-Bone TM, Ndifon RK, Nyingchia ON et al (2013) Morbidity and mortality of the critically endangered Lake Oku clawed frog (*Xenopus longipes*). Endanger Species Res 21:115–128

Dong C, Wang Z, Weng S, He J (2017) Occurrence of a lethal ranavirus in hybrid mandarin (Siniperca scherzeri×Siniperca chuatsi) in Guangdong, South China. Vet Microbiol 203:28–33

Driskell EA, Miller DL, Swist SL et al (2009) PCR detection of ranavirus in adult anurans from the Louisville Zoological Gardens. J Zoo Wildl Med 40:559–563

Drury SEN, Gough RE, Cunningham AA et al (1995) Isolation of an iridovirus-like agent from common frogs (*Rana temporaria*). Vet Rec 137:72–73

Du J, Wang L, Wang Y et al (2016) Autophagy and apoptosis induced by Chinese giant salamander (*Andrias davidianus*) iridovirus (CGSIV). Vet Microbiol 195:87–95

Duffus A, Olson D (2011) The establishment of a global Ranavirus reporting system. Froglog 96:37

Duffus ALJ, Pauli PD, Wozney K et al (2008) Frog virus 3-like infections in aquatic amphibian communities. J Wildl Dis 44:109–120

Duffus ALJ, Nichols RA, Garner TWJ (2013) Investigations into the life history stages of the common frog (*Rana temporaria*) affected by an amphibian ranavirus in the United Kingdom. Herpetol Rev 44:460–463

Duffus ALJ, Nichols RA, Garner TWJ (2014) Detection of a frog virus 3–like ranavirus in native and introduced amphibians in the United Kingdom in 2007 and 2008. Herpetol Rev 45:608–610

Duffus ALJ, Waltzek TB, Stöhr AC et al (2015) Distribution and host range of ranaviruses. In: Gray MJ, Chinchar VG (eds) Ranaviruses: lethal pathogens of ectothermic vertebrates. Springer Online. https://link.springer.com/book/10.1007/978-3-319-13755-1

Earl JE, Gray MJ (2014) Introduction of ranavirus to isolated wood frog populations could cause local extinction. EcoHealth. https://doi.org/10.1007/s10393-014-0950-y

Eaton HE, Metcalf J, Penny E et al (2007) Comparative genomic analysis of the family Iridoviridae: re-annotating and defining the core set of iridovirus genes. Virol J 4:1–17

Echaubard P, Little K, Pauli B, Lesbarrères D (2010) Context-dependent effects of ranaviral infection on northern leopard frog life history traits. PLoS One 5:e13723

Erişmiş UC, Yoldaş T, Cevdet UĞ (2019) Investigation of prevalence of co-infection by *Batrachochytrium dendrobatidis* and Ranavirus in endemic Beyşehir frog (*Pelophylax caralitanus*). Acta Aquat Turc 15(2):239–246

Farnsworth SD, Seigel RA (2013) Responses, movements, and survival of relocated box turtles during construction of the intercounty connector highway in Maryland. Transport Res Rec 2362:1–8. https://doi.org/10.3141/2362-01

Fijan N, Matasin Z, Petrinec Z et al (1991) Isolation of an iridovirus-like agent from the green frog (*Rana esculenta* L.). Vet Arch 61:151–158

Flechas SV, Urbina J, Crawford AJ, Gutiérrez K et al (2023) First evidence of ranavirus in native and invasive amphibians in Colombia. Dis Aquat Org 153:51–58

Forson D, Storfer A (2006) Effects of atrazine and iridovirus infection on survival and life-history traits of the long-toed salamander (*Ambystoma macrodactylum*). Environ Toxicol Chem 25(1):168–173

Forzán M, Wood J (2013) Low detection of ranavirus DNA in wild postmetamorphic green frogs, *Rana* (*Lithobates*) *clamitans*, despite previous or concurrent tadpole mortality. J Wildl Dis 49:879–886

Forzán MJ, Bienentreu J, Schock DM, Lesbarrères D (2019) Multi-tool diagnosis of an outbreak of ranavirosis in amphibian tadpoles in the Canadian boreal forest. Dis Aquat Org 135(1):33–41

Fox SF, Greer AL, Tores-Cervantes R et al (2006) First case of ranavirus-associated morbidity and mortality in natural populations of the South American frog, *Atelognathus patagonicus*. Dis Aquat Org 72:87–92

Francis-Floyd R (1992) Comparative hematology for largemouth bass *(Micropterus salmoides)* and black crappie *(Pomoxis nigromaculatus)* from Lake Weir, Lake Holy, and Newman's Lake. Final report. Florida Freshwater Game and Fish Commission, Tallahassee, FL

Gahl MK, Calhoun AJK (2010) The role of multiple stressors in ranavirus-caused amphibian mortalities in Acadia National Park wetlands. Can J Zool 88:108–118

Galli L, Pereira A, Márquez A et al (2006) Ranavirus detection by PCR in cultured tadpoles *(Rana catesbeiana* Shaw, 1802) from South America. Aquaculture 257:78–82

Galt N, Atkinson MS, Glorioso B et al (2021) Widespread ranavirus and perkinsea infections in Cuban treefrogs *(Osteopilus septentrionalis)* invading New Orleans, USA. Herpetol Conserv Biol 16(1):17–29

Geng Y, Wang KY, Zhou ZY et al (2011) First report of a ranavirus associated with morbidity and mortality in farmed Chinese giant salamanders. J Comp Pathol 145:95–112

George MR, John KR, Mansoor MM et al (2015) Isolation and characterization of a ranavirus from koi, *Cyprinus carpio* L., experiencing mass mortalities in India. J Fish Dis 38(4):389–403. https://doi.org/10.1111/jfd.12246

Glenney JW, Julian JT, Quartz WM (2010) Preliminary amphibian health survey in the Delaware Water Gap National Recreation Area. J Aquat Anim Health 22:102–114

Gobbo F, Cappellozza E, Pastore MR et al (2010) Susceptibility of black bullhead *Ameiurus melas* to a panel of ranavirus isolates. Dis Aquat Org 90:167–174

Goldberg TL (2002) Largemouth bass virus: an emerging problem for warmwater fisheries? In: Philipp DP, Ridgway MS (eds) Black bass: ecology, conservation and management. American Fisheries Society Symposium, Bethesda, MD

Goodman RM, Miller DL, Ararso YT (2013) Prevalence of ranavirus in Virginia turtles as detected by tail-clip sampling versus oral-cloacal swabbing. Northeast Nat 20:325–332

Goodman RM, Hargadon KM, Carter ED (2018) Detection of ranavirus in Eastern fence lizards and Eastern box turtles in Central Virginia. Northeast Nat 25:391–398

Granoff A, Came PE, Rafferty KA (1965) The isolation and properties of viruses from *Rana pipiens*: their possible relationship to the renal adenocarcinoma of the leopard frog. Ann N Y Acad Sci 126:237–255

Grant EC, Inendino KR, Love WJ et al (2005) Effects of practices related to catch-and-release angling on mortality and viral transmission in juvenile largemouth bass infected with largemouth bass virus. J Aquat Anim Health 17:315–322

Grant SA, Bienentreu JF, Vilaça ST et al (2019) Low intraspecific variation of Frog virus 3 with evidence for novel FV3-like isolates in central and northwestern Canada. Dis Aquat Org 134(1):1–13

Gray MJ, Miller DL, Schmuster AC et al (2007) Frog virus 3 prevalence in tadpole populations at cattle-access and non-access wetlands in Tennessee, USA. Dis Aquat Org 77:97–103

Gray MJ, Miller DL, Hoverman JT (2009a) Ecology and pathology of amphibian ranaviruses. Dis Aquat Org 87:243–266

Gray MJ, Miller DL, Hoverman JT (2009b) First report of ranavirus infecting lungless salamanders. Herpetol Rev 40:316–319

Gray MJ, Brunner JL, Earl JE et al (2024) Design and analysis of ranavirus studies: insights into planning surveillance, modeling host-pathogen dynamics, and performing risk analyses. In: Gray MJ, Chinchar VG (eds) Ranaviruses: lethal pathogens of ectothermic vertebrates, 2nd edn. Springer Online. In press

Green DE, Converse KA (2005) Diseases of frogs and toads. In: Majumadar SK, Huffman JE, Brenner FJ et al (eds) Wildlife diseases: landscape epidemiology, spatial distribution and utilization of remote sensing techniques. Pennsylvania Academy of Science, pp 89–117

Green DE, Ip HS (2011) US Geological Survey National Wildlife Health Center, Madison, Wisconsin, USA. Unpublished work

Green DE, Converse KA, Schrader AK (2002) Epizootiology of sixty-four amphibian morbidity and mortality events in the USA, 1996–2001. Ann N Y Acad Sci 969:323–339

Greer AL, Berrill M, Wilson PJ (2005) Five amphibian mortality events associated with ranavirus infection in south central Ontario, Canada. Dis Aquat Org 67:9–14

Greer AL, Brunner JL, Collins JP (2009) Spatial and temporal patterns of *Ambystoma tigrinum* virus (ATV) prevalence in tiger salamanders *Ambystoma tigrinum nebulosum*. Dis Aquat Org 85:1–6

Grizzle JM, Brunner CJ (2003) Review of largemouth bass virus. Fisheries 28:10–14

Grizzle JM, Altinok I, Fraser WA et al (2002) First isolation of largemouth bass virus. Dis Aquat Org 50:233–235

Groocock GH, Grimmett SG, Getchell RG et al (2008) A survey to determine the presence and distribution of largemouth bass virus in wild freshwater bass in New York State. J Aquat Anim Health 20:158–164

Haislip NA, Gray MJ, Hoverman JT et al (2011) Development and disease: how susceptibility to an emerging pathogen changes through anuran development. PLoS One 6:e22307

Hall EM, Goldberg CS, Brunner JL, Crespi EJ (2018) Seasonal dynamics and potential drivers of ranavirus epidemics in wood frog populations. Oecologia 188:1253–1262

Hall EM, Brunner JL, Hutzenbiler B, Crespi EJ (2020) Salinity stress increases the severity of ranavirus epidemics in amphibian populations. Proc R Soc B 287(1926):20200062

Hamed MK, Gray MJ, Miller DL (2013) First report of ranavirus in plethodontid salamanders from the Mount Rogers National Recreation Area, Virginia, USA. Herpetol Rev 44:455–457

Hanson LA, Petrie-Hansons L, Meals KO et al (2001) Persistence of large-mouth bass virus infection in a northern Mississippi reservoir after a die-off. J Aquat Anim Health 13:27–34

Hardman RH, Sutton WB, Irwin KJ et al (2020) Geographic and individual determinants of important amphibian pathogens in hellbenders (*Cryptobranchus alleganiensis*) in Tennessee and Arkansas, USA. J Wildl Dis 56(4):803–814

Harp EM, Petranka JW (2006) Ranavirus in wood frogs (*Rana sylvatica*): potential sources of transmission within and between ponds. J Wildl Dis 42:307–318

Hartmann AM, Maddox ML, Ossiboff RJ, Longo AV (2022) Sustained ranavirus outbreak causes mass mortality and morbidity of imperiled amphibians in Florida. EcoHealth 19(1):8–14. https://doi.org/10.1007/s10393-021-01572-6

Hausmann JC, Wack AN, Allender MC et al (2015) Experimental challenge study of FV3-like ranavirus infection in previously FV3-like ranavirus infected eastern box turtles (*Terrapene carolina carolina*) to assess infection and survival. J Zoo Wildl Med 46(4):732–746

Heard MJ, Smith KF, Ripp KJ et al (2013) The threat of disease increases as species move towards extinction. Conserv Biol 24:1378–1388

Hedrick RP, McDowell TS (1995) Properties of iridoviruses from ornamental fish. Vet Res 26:423–427

Heldstab A, Bestetti G (1982) Spontaneous viral hepatitis in a spur-tailed mediterranean land tortoise. J Zoo Anim Med 13:113–120

Herath J, Ellepola G, Meegaskumbura M (2021) Patterns of infection, origins, and transmission of ranaviruses among the ectothermic vertebrates of Asia. Ecol Evol 11(22):15498–15519

Hick PM, Subramaniam K, Thompson PM et al (2017) Molecular epidemiology of epizootic haematopoietic necrosis virus (EHNV). Virology 511:320–329

Holopainen R, Ohlemeyer S, Schuetze H et al (2009) Ranavirus phylogeny and differentiation based on major capsid protein, DNA polymerase and neurofilament triplet H1-like protein genes. Dis Aquat Org 85:81–91

Holopainen R, Subramaniam K, Steckler NK et al (2016) Genome sequence of a ranavirus isolated from pike-perch *Sander lucioperca*. Genome Announc 4(6):e01295–e01216. https://doi.org/10.1128/genomeA.01295-16

Homan RN, Bartling JR, Stenger RJ et al (2013) Detection of ranavirus in Ohio, USA. Herpetol Rev 44:615–618

Hoverman JT, Gray MJ, Miller DL (2011a) Anuran susceptibility to ranaviruses: role of species identity, exposure route and a novel virus isolate. Dis Aquat Org 98:97–107

Hoverman JT, Gray MJ, Haislip NA et al (2011b) Phylogeny, life history, and ecology contribute to differences in amphibian susceptibility to ranaviruses. EcoHealth 8:301–319

Hoverman JT, Gray MJ, Haislip NA et al (2012a) Widespread occurrence of ranavirus in pond breeding amphibian populations. EcoHealth 9:36–48

Hoverman JT, Mihaljevic JR, Richgels LD et al (2012b) Widespread co-occurrence of virulent pathogens within California amphibian communities. EcoHealth 9:288–292

Hsieh CY, Rairat T, Chou CC (2021) Detection of ranavirus by PCR and in situ hybridization in the American bullfrog (*Rana catesbeiana*) in Taiwan. Aquaculture 543:736955

Huang Y, Huang X, Liu H et al (2009) Complete sequence determination of a novel reptile iridovirus isolated from soft-shelled turtle and evolutionary analysis of *Iridoviridae*. BMC Genomics 10:224

Huang SM, Tu C, Tseng CH et al (2011) Genetic analysis of fish iridoviruses isolated in Taiwan during 2001–2009. Arch Virol 156:1505–1515

Hyatt AD, Gould AR, Zupanovic Z et al (2000) Comparative studies of piscine and amphibian iridoviruses. Arch Virol 145:301–331

Hyatt AD, Williamson M, Coupar BEH et al (2002) First identification of a ranavirus from green pythons (*Chrondropython viridis*). J Wildl Dis 38:239–252

Isidoro-Ayza M, Lorch JM, Grear DA et al (2017) Pathogenic lineage of Perkinsea associated with mass mortality of frogs across the United States. Sci Rep 7(1):1–10

Iwanowicz L, Densmore C, Hahn C et al (2013) Identification of largemouth bass virus in the introduced Northern snakehead inhabiting the Chesapeake Bay Watershed. J Aquat Anim Health 25:191–196

Jancovich JK, Davidson EW, Morado JF et al (1997) Isolation of a lethal virus from the endangered tiger salamander *Ambystoma tigrinum stebbinsi*. Dis Aquat Org 31:161–167

Jancovich JK, Davidson EW, Seiler A et al (2001) Transmission of the *Ambystoma tigrinum* virus to alternative hosts. Dis Aquat Org 46:159–163

Jancovich JK, Davidson EW, Parameswaran N et al (2005) Evidence for emergence of an amphibian iridoviral disease because of human-enhanced spread. Mol Ecol 14:213–224

Jancovich JK, Bremont M, Touchman JW et al (2010) Evidence for multiple recent host species shifts among the ranaviruses (family *Iridoviridae*). J Virol 84:2636–2647

Jancovich JK, Steckler NK, Waltzek TB (2015) In: Gray MJ, Chinchar VG (eds) Ranavirus taxonomy and phylogeny. Springer Online, Ranaviruses: lethal pathogens of ectothermic vertebrates. https://link.springer.com/book/10.1007/978-3-319-13755-1

Jensen NJ, Larsen JL (1982) The ulcus-syndrome in cod (*Gadus morhua*). IV. Transmission experiments with two viruses isolated from cod and *Vibrio anguillarum*. Nord Vet Med 34:136–142

Jensen NJ, Bloch B, Larsen JL (1979) The ulcus-syndrome in cod (*Gadus morhua*). III. A preliminary virological report. Nord Vet Med 31:436–442

Jensen BB, Ersboll AK, Ariel E (2009) Susceptibility of pike Esox lucius to a panel of Ranavirus isolates. Dis Aquat Org 83:169–179

Jensen BB, Holopainen R, Tapiovaara H et al (2011a) Susceptibility of pike-perch *Sander lucioperca* to a panel of ranavirus isolates. Aquaculture 313:24–30

Jensen BB, Reschova S, Cinkova K et al (2011b) Common carp (*Cyprinus carpio*) and goldfish (*Carassius auratus*) were not susceptible to challenge with ranavirus under certain challenge conditions. Bull Eur Assoc Fish Pathol 31:112–118

Jin Y, Bergmann SM, Mai Q et al (2022) Simultaneous isolation and identification of Largemouth bass virus and Rhabdovirus from moribund largemouth bass (*Micropterus salmoides*). Viruses 14(8):1643. https://www.mdpi.com/1999-4915/14/8/1643

John KR, George MR, Kar D et al (2016) Detection of Ranavirus infection in cultivated carps of Northeast India. Fish Pathol 51(Special-issue):S66–S74

Johnson AJ (2006) Iridovirus infections of captive and free-ranging chelonians in the United States. Ph.D. thesis, Veterinary Medicine. University of Florida, Gainesville, p 149

Johnson AJ, Pessier AP, Jacobson ER (2007) Experimental transmission and induction of rana-viral disease in western ornate box turtles (*Terrapene ornata ornata*) and red-eared sliders (*Trachemys scripta elegans*). Vet Path 44:285–297

Johnson AJ, Pessier AP, Wellehan JFX et al (2008) Ranavirus infection of free-ranging and captive box turtles and tortoises in the United States. J Wildl Dis 44:851–863

Johnson AJ, Wendland L, Norton TM et al (2010) Development and use of an indirect enzyme-linked immunosorbent assay for detection of iridovirus exposure in gopher tortoises (*Gopherus polyphemus*) and eastern box turtles (*Terrapene carolina carolina*). Vet Microbiol 142:160–167

Juhász T, Woynarovichne LM, Csaba G et al (2013) Isolation of ranavirus causing mass mortality in brown bullheads (*Ameiurus nebulosus*) in Hungary. Magyar Allatorvosok Lapja 135:763–768

Julian JT, Brooks RP, Glenney GW, Coll JA (2019) State-wide survey of amphibian pathogens in green frog (*Lithobates clamitans melanota*) reveals high chytrid infection intensities in con-structed wetlands. Herpetol Conserv Biol 14:199–211

Kanchanakhan S (1998) An ulcerative disease of the cultured tiger frog, *Rana tigrina*, in Thailand: virological examination. AAHRI News 7:1–2

Kayansamruaj P, Rangsichol A, Dong HT et al (2017) Outbreaks of ulcerative disease associated with ranavirus infection in barcoo grunter, Scortum barcoo (McCulloch & Waite). J Fish Dis 40(10):1341–1350. https://doi.org/10.1111/jfd.12606

Kik M, Martel A, Spitzen-van der Sluijs A et al (2011) Ranavirus associated mass mortality in wild amphibians in the Netherlands, 2010: a first report. Vet J 190:284–228

Kik M, Stege M, Boonyarittichaikij R et al (2012) Concurent ranavirus and *Batrachochytrium den-drobatidis* infection in captive frogs (*Phyllobates* and *Dendrobates* species), the Netherlands, 2012: a first report. Vet J 194:247–249

Kim S, Sim MY, Eom AH et al (2009) PCR detection of ranavirus in gold-spotted pond frogs (*Rana plancyi chosenica*) from Korea. Korean J Environ Biol 27(1):110–113

Kimble SJ, Karna AK, Johnson AJ et al (2014) Mosquitoes as a potential vector of ranavirus trans-mission in terrestrial turtles. EcoHealth. https://doi.org/10.1007/s10393-014-0974-3

Kimble SJA, Johnson AJ, Williams RN, Hoverman JT (2017) A severe ranavirus outbreak in cap-tive, wild-caught box turtles. EcoHealth 14(4):810–815

Kirschman L, Palis J, Fritz K et al (2017) Two ranavirus-associated mass-mortality events among larval amphibians in Illinois, USA. Herpetol Rev 48(4):779–782

Kolby JE, Smith KM, Berger L et al (2014) First evidence of amphibian chytrid fungus (*Batrachochytrium dendrobatids*) and ranavirus in Hong Kong amphibian trade. PLoS One 9:e90750

Kolby JE, Smith KM, Ramirez SD et al (2015) Rapid response to evaluate the presence of amphib-ian chytrid fungus (*Batrachochytrium dendrobatidis*) and ranavirus in wild amphibian popula-tions in Madagascar. PLoS One 10(6):e0125330

Kolesnik E, Obiegala A, Marschang RE (2017) Detection of *Mycoplasma* spp., herpesviruses, topiviruses, and ferlaviruses in samples from chelonians in Europe. J Vet Diagn Invest 29(6):820–832

Kolozsvary MB, Brunner JL (2016) Presence of ranavirus in a created temporary pool complex in southeastern New York, USA. Herpetol Rev 47(4):598–600

Kwon S, Park J, Choi WJ (2017) First case of ranavirus-associated mass mortality in a natural population of the Huanren frog (*Rana huanrenensis*) tadpoles in South Korea. Anim Cells Syst 21(5):358–364

Landsberg JH, Kiryu Y, Tabuchi M et al (2013) Co-infection by alveolate parasites and frog virus 3-like ranavirus during and amphibian larval mortality event in Florida, USA. Dis Aquat Org 105:89–99

Langdon JS (1989) Experimental transmission and pathogenitiy of epizootic hematopoietic necro-sis virus (EHNV) in redfin perch, *Perca fluviatilis* and 11 other teleosts. J Fish Dis 12:295–310

Langdon JS, Humphrey JD (1987) Epizootic haematopoietic necrosis a new viral disease in redfin perch *Perca fluviatilis* L. in Australia. J Fish Dis 10:289–298

Langdon JS, Humphrey JD, Copland J et al (1986a) The disease status of Australian salmonids: viruses and viral diseases. J Fish Dis 9:129–135

Langdon JS, Humphrey JD, Williams LM et al (1986b) First virus isolation from Australian fish: an iridovirus-like pathogen from redfin perch, *Perca fluviatilis* L. J Fish Dis 9:263–268

Langdon JS, Humphrey JD, Williams LM (1988) Outbreaks of an EHNV-like iridovirus in cultured rainbow trout, *Salmo gairdneri* Richardson, in Australia. J Fish Dis 11:93–96

Leceta J, Zapata A (1985) Seasonal changes in the thymus and spleen of the turtle, *Mauremys caspica*. Dev Compar Immunol 9(4):653–668

Leceta J, Zapata A (1986) Seasonal variations in the immune response of the tortoise *Mauremys caspica*. Immunology 57(3):483–487

Lei X-Y, Ou T, Zhu R-L et al (2012) Sequencing and analysis of the complete genome of *Rana grylio* virus (RGV). Arch Virol 157:1559–1564

Leimbach S, Schütze H, Bergmann SM (2014) Susceptibility of European sheatfish *Silurus glanis* to a panel of ranaviruses. J Appl Ichthyol 30:93–101

Lisachov AP, Lisachova LS, Simonov E (2022) First record of ranavirus (*Ranavirus* sp.) in Siberia, Russia. Herpetozoa 35:33–37

Liu F, Tian S, Feng Y et al (2023a) First report of a CMTV-like ranavirus in farmed *Percocypris pingi* in China. Aquaculture 574. https://doi.org/10.1016/j.aquaculture.2023.739701

Liu XD, Zhang YB, Zhang ZL et al (2023b) Isolation, identification and the pathogenicity characterization of a *Santee-Cooper ranavirus* and its activation on immune responses in juvenile largemouth bass (*Micropterus salmoides*). Fish Shellfish Immunol 135:108641

Ma J, Zeng L, Zhou Y et al (2014) Ultrastructural morphogenesis of an amphibian iridovirus isolated from Chinese giant salamander (*Andrias dacidianus*). J Comp Pathol 150:325–331

Ma HL, Peng C, Su YL et al (2016) Isolation of a Ranavirus-type grouper iridovirus in mainland China and comparison of its pathogenicity with that of a Megalocytivirus-type grouper iridovirus. Aquaculture 463:145–151. https://doi.org/10.1016/j.aquaculture.2016.05.032

Maclaine A (2019) Characterisation of ranaviral infection and its management in Australian lizards. PhD thesis, James Cook University. Available from: https://doi.org/10.25903/5eefdd7a25893

Maclaine A, Mashkour N, Scott J, Ariel E (2018) Susceptibility of eastern water dragons *Intellagama lesueurii lesueurii* to Bohle iridovirus. Dis Aquat Org 127:97–105

Maclaine A, Forzán MJ, Mashkour N et al (2019) Pathogenesis of Bohle iridovirus (Genus Ranavirus) in experimentally infected Juvenile eastern water dragons (*Intellagama lesueurii lesueurii*). Vet Pathol 56:465–475

Maclaine A, Wirth WT, McKnight DT et al (2020) Ranaviruses in captive and wild Australian lizards. Facets 5. https://doi.org/10.1139/FACETS-2020-0011

Majji S, LaPatra S, Long SM et al (2006) *Rana catesbeiana* virus Z (RCV-Z): a novel pathogenic ranavirus. Dis Aquat Org 73:1–11

Mao J, Hedrick RP, Chinchar VG (1997) Molecular characterization, sequence analysis, and taxonomic position of newly isolated fish iridoviruses. Virology 229:212–220

Mao JD, Green E, Fellers G et al (1999a) Molecular characterization of iridoviruses isolated from sympatric amphibians and fish. Virus Res 63:45–52

Mao JH, Wang J, Chinchar GD et al (1999b) Molecular characterization of a ranavirus isolated from largemouth bass *Micropterus salmoides*. Dis Aquat Org 37:107–114

Marschang RE, Hyndman TH (2022) Emerging infectious diseases of reptiles. In: Miller RE, Calle PP, Lamberski N (eds) Fowler's zoo and wild animal medicine, current therapy, vol 10. Elsevier, St. Louis, pp 441–446

Marschang RE, Becher P, Posthaus H et al (1999) Isolation and characterization of an iridovirus from Hermann's tortoises (*Testudo hermanni*). Arch Virol 144:1909–1922

Marschang RE, Braun S, Becher P (2005) Isolation of a ranavirus from a gecko (*Uroplatus fimbriatus*). J Zoo Wildl Med 36:295–300

Marschang RE, Stöhr AC, Blahak S et al (2013) Ranaviruses in snakes, lizards, and chelonians. In: Proceedings second international symposium on ranaviruses, Knoxville, TN, USA, 27–29 July

Marschang RE, Origgi FC, Stenglein MD et al (2021) Viruses and viral diseases of reptiles. In: Jacobson ER, Garner MM (eds) Infectious diseases and pathology of reptiles: color atlas and text, vol 1, 2nd edn. CRC Press, Boca Raton, pp 575–703

Martel A, Fard MS, van Rooij P et al (2012) Road-killed common toads (*Bufo bufo*) in Flanders (Belgium) reveal low prevalence of ranavirus and *Batrachochytrium dentrobatidis*. J Wildl Dis 48:835–839

Martinez-Silvestre A, Montori A, Oromi N et al (2017) Detection of a Ranavirus in introduced newts in Catalonia (NE Spain). Herpetol Notes 10:23–26

Mavian C, Lopez-Bueno A, Balseiro A et al (2012) The genome sequence of the emerging common midwife toad virus identifies an evolutionary intermediate within ranaviruses. J Virol 86:3617–3625

Mazzoni R, José de Mesquita A, Fleury LFF et al (2009) Mass mortality associated with frog virus 3-like ranavirus infection in farmed tadpoles, *Rana catesbeiana*, from Brazil. Dis Aquat Org 86:181–191

McKenzie CM, Piczak ML, Snymar HN et al (2019) First report of ranavirus mortality in a common snapping turtle *Chelydra serpentina*. Dis Aquat Org 132:221–227

Meddings JI (2011) Revelations in reptilian immunology: serology and sources of variation. Honours thesis. College of Public Health, Medical and Veterinary Science, James Cook University, Australia

Miaud C, Pozet F, Gaudin NCG et al (2016) Ranavirus causes mass die-offs of alpine amphibians in the Southwestern Alps, France. J Wildl Dis 52(2):242–252

Miller DL, Rajeev S, Gray MJ et al (2007) Frog virus 3 infection, cultured American bullfrogs. Emerg Infect Dis 13:342–343

Miller DL, Rajeev S, Brookins M et al (2008) Concurrent infection with *Ranavirus*, *Batrachochytrium dendrobatids*, and *Aeromonas* in a captive amphibian colony. J Zoo Wildl Med 39:445–449

Miller DL, Gray MJ, Rajeev S et al (2009) Pathological findings in larval and juvenile anurans inhabiting farm ponds in Tennessee, USA. J Wildl Dis 45:314–324

Miller D, Gray M, Storfer A (2011) Ecopathology of ranaviruses infecting amphibians. Viruses 3:2351–2373

Miller DL, Pessier AP, Hick P et al (2024) Pathology and diagnostics. In: Gray MJ, Chinchar VG (eds) Ranaviruses: lethal pathogens of ectothermic vertebrates, 2nd edn. Springer Online. In press

Millikin AR, Davis DR, Brown DJ et al (2023) Prevalence of ranavirus in spotted salamander (*Ambystoma maculatum*) larvae from created vernal pools in West Virginia, USA. J Wildl Dis 59(1):24–36

Milstein D (2011) Tadpoles' edema. First year research report to the Israel Nature and Parks Authority

Monson-Collar K, Hazard L, Dolcemascolo P (2013) A Ranavirus-related mortality recent and the first report of Ranavirus in New Jersey. Herpetol Rev 44:263–265

Moody NJG, Owens L (1994) Experimental demonstration of pathogenicity of a frog virus, Bohle iridovirus, for fish species, barramundi *Lates calcifer*. Dis Aquat Org 18:95–102

Mosher BA, Brand AB, Wiewel AN et al (2019) Estimating occurrence, prevalence, and detection of amphibian pathogens: insights from occupancy models. J Wildl Dis 55(3):563–575

Mu WH, Geng Y, Yu ZH et al (2018) FV3-like ranavirus infection outbreak in black-spotted pond frogs (*Rana nigromaculata*) in China. Microbial Pathog 123:111–114

Müller M, Zangger N, Denzler T (1988) Iridovirus-Epidemie bei der griechischen Landschildkröte (*Testudo hermanni hermanni*). Verhandl Ber 30. Int Symp Erkr Zoo- und Wildtiere, Sofia, 271–274

Murali S, Wu MF, Guo IC et al (2002) Molecular characterization and pathogenicity of a grouper iridovirus (GIV) isolated from yellow grouper, *Epinephelus awoara* (Temminck & Schlegel). J Fish Dis 25:91–100

Neal JW, Eggleton MA, Goodwin AE (2009) The effects of largemouth bass virus on a quality largemouth bass population in Arkansas. J Wildl Dis 45:766–771

Newman JC, Mota JL, Hardman RH et al (2019) Pathogen detection in Green Salamanders (*Aneides aeneus*) in South Carolina, USA. Herpetol Rev 50(3):503–505

O'Bryan CJ, Gray MJ, Brooks CS (2012) Further presence of ranavirus infection in amphibian populations of Tennessee, USA. Herpetol Rev 43:293–295

O'Connor KM, Rittenhouse TA, Brunner JL (2016) Ranavirus is common in wood frog (*Lithobates sylvaticus*) tadpoles throughout Connecticut, USA. Herpetol Rev 47:394–397

Olori JC, Netzband R, McKean N et al (2018) Multi-year dynamics of ranavirus, chytridiomycosis, and co-infections in a temperate host assemblage of amphibians. Dis Aquat Org 130(3):187–197

Paetow LJ, Pauli BD, McLaughlin J et al (2011) First detection of ranavirus in *Lithobates pipiens* in Quebec. Herpetol Rev 42:211–214

Park IK, Koo KS, Moon KY et al (2017) PCR detection of ranavirus from dead *Kaloula borealis* and sick *Hyla japonica* tadpoles in the wild. Korean J Herpetol 8:10–14

Park J, Grajal-Puche A, Roh NH et al (2021) First detection of ranavirus in a wild population of Dybowski's brown frog (*Rana dybowskii*) in South Korea. J Ecol Environ 45(1):1–7

Pasmans F, Blahak S, Martel A et al (2008) Ranavirus-associated mass mortality in imported red tailed knobby newts (*Tylototriton kweichowensis*): a case report. Vet J 175:257–259

Patla DA, St-Hilaire SO, Ray AN et al (2016) Amphibian mortality events and ranavirus outbreaks in the Greater Yellowstone Ecosystem. Herpetol Rev 47(1):50–54

Paull SH, Song SJ, McClure KM et al (2012) From superspreaders to disease hotspots: linking transmission across hosts and space. Front Ecol Environ 10:75–82

Peace A, O'Regan SM, Spatz JA et al (2019) A highly invasive chimeric ranavirus can decimate tadpole populations rapidly through multiple transmission pathways. Ecol Model 410. https://doi.org/10.1016/j.ecolmodel.2019.108777

Pearman PB, Garner TWJ (2005) Susceptibility of Italian agile frog populations to an emerging strain of Ranavirus parallels population genetic diversity. Ecol Lett 8:401–408

Pearman PB, Garner TWJ, Straub M et al (2004) Response of the Italian agile frog (*Rana latastei*) to a *Ranavirus*, frog virus 3: a model for viral emergence in naïve populations. J Wildl Dis 40:660–669

Peiffer LB, Sander S, Gabrielson K et al (2019) Fatal ranavirus infection in a group of zoo-housed Meller's chameleons (*Trioceros melleri*). J Zoo Wildl Med 50(3):696–705

Peñafiel-Ricaurte A, Price SJ, Leung WT et al (2023) Is *Xenopus laevis* introduction linked with Ranavirus incursion, persistence and spread in Chile? PeerJ 11:e14497

Penzes JJ, Pham HT, Benk M, Tijssen P (2015) Novel parvoviruses in reptiles and genome sequence of a lizard parvovirus shed light on *Dependoparvovirus* genus evolution. J Gen Virol 96:2769–2779

Perpiñán D, Blas-Machado U, Sánchez S, Miller DL (2016) Concurrent phaeohyphomycosis and *Ranavirus* infection in an eastern box turtle (*Terrapene carolina*) in Athens Georgia, USA. J Wildl Dis 52:742–745

Petranka JW, Murray SM, Kennedy CA (2003) Response of amphibians to restoration of a southern Appalachian wetland: perturbations confounded post-restoration assessment. Wetlands 23:278–290

Picco AM, Collins JP (2008) Amphibian commerce as a likely source of pathogen pollution. Conserv Biol 22:1582–1589

Picco AM, Brunner JL, Collins JP (2007) Susceptibility of the endangered California tiger salamander, *Ambystoma californiense*, to ranavirus infection. J Wildl Dis 43:286–290

Picco AM, Karam AP, Collins JP (2010) Pathogen host switching in commercial trade with management recommendations. EcoHealth 7:252–256

Plumb JA, Hanson LA (2011) Health maintenance and principal microbial diseases of cultured fishes. Wiley, Hoboken

Plumb JA, Zilberg D (1999a) Survival of largemouth bass iridovirus in frozen fish. J Aquat Anim Health 11:94–96

Plumb JA, Zilberg D (1999b) The lethal dose of largemouth bass virus in juvenile largemouth bass and the comparative susceptibility of striped bass. J Aquat Anim Health 11:246–252

Plumb JA, Grizzle JM, Young HE et al (1996) An iridovirus isolated from wild largemouth bass. J Aquat Anim Health 8:265–270

Plumb JA, Noyes AD, Graziono S et al (1999) Isolation and identification of viruses from adult largemouth bass during a 1997–1998 survey in the southeastern United States. J Aquat Anim Health 11:391–399

Pozet F, Morand M, Moussa A et al (1992) Isolation and preliminary characterization of a pathogenic icosahedral deoxyribovirus from the catfish *Ictalurus melas*. Dis Aquat Org 14:35–42

Prasankok P, Chutmongkonkul M, Kanchankhan S (2005) Characterization of iridovirus isolated from diseased marbled sleepy goby, Oxyeleotris marmoratus. In: Walker PJ, Lester RG, Bondad-Reantaso M (eds) Diseases in Asian aquaculture V. Manila, Asian Fisheries Society

Price SJ (2015) Comparative genomics of amphibian-like ranaviruses, nucleocytoplasmic large DNA viruses of poikilotherms. Evol Bioinforma 11(S2):71–82

Price SJ, Garner TWJ, Nichols RA et al (2014) Collapse of amphibian communities due to an introduced ranavirus. Curr Biol 24:2586–2591

Price SJ, Wadia A, Wright ON et al (2017) Screening of a long-term sample set reveals two *Ranavirus* lineages in British herpetofauna. PLoS One 12(9):e0184768

Puschendorf R, Wallace M, Chavarría MM et al (2019) Cryptic diversity and ranavirus infection of a critically endangered Neotropical frog before and after population collapse. Anim Conserv 22(5):515–524

Qin QW, Lam TJ, Sin YM et al (2001) Electron microscopic observations of a marine fish iridovirus isolated from brown-spotted grouper, *Epinephelus tauvina*. J Virol Methods 98(1):17–24. https://doi.org/10.1016/s0166-0934(01)00350-0

Qin QW, Chang SF, Ngoh-lim GH et al (2003) Characterization of a novel ranavirus isolated from grouper *Epinephelus tauvina*. Dis Aquat Org 53:1–9

Reshetnikov AN, Chestnut T, Brunner JL et al (2014) Detection of the emerging amphibian pathogens Batrachochytrium dendrobatidis and ranavirus in Russia. Dis Aquat Org 110(3):235–240

Richter SC, Drayer AN, Strong JR et al (2013) High prevalence of ranavirus infection in permanent constructed wetlands in eastern Kentucky, USA. Herpetol Rev 44:464–466

Ridenhour BJ, Storfer AT (2008) Geographically variable selection in *Ambystoma tigrinum* virus (*Iridoviridae*) throughout the western USA. J Evol Biol 21:1151–1159

Rivera B, Cook K, Andrews K et al (2019) Pathogen dynamics in an invasive frog compared to native species. EcoHealth 16:222–234

Roh N, Park J, Kim J (2022) Prevalence of Ranavirus infection in three anuran species across South Korea. Viruses 14(5):1073

Rosa GM, Sabino-Pinto J, Laurentino TG et al (2017) Impact of asynchronous emergence of two lethal pathogens on amphibian assemblages. Sci Rep 7(1):1–10

Rothermel BR, Travis ER, Miller DL et al (2013) High occupancy of stream salamanders despite high *Ranavirus* prevalence in a southern Appalachians watershed. EcoHeath 10:184–189

Ruder MG, Allison AB, Miller DL et al (2010) Pathology in practice. J Am Vet Med Assoc 237:783–785

Ruggeri J, Ribeiro LP, Pontes MR et al (2019) Discovery of wild amphibians infected with Ranavirus in Brazil. J Wildl Dis 55(4):897–902

Russell DM, Goldberg CS, Sprague L et al (2011) Ranavirus outbreaks in amphibian populations of northern Idaho. Herpetol Rev 42:223–225

Saucedo B, Hughes J, van Beurden SJ et al (2017) Complete genome sequence of frog virus 3, isolated from a strawberry poison frog (*Oophaga pumilio*) imported from Nicaragua into the Netherlands. Genome Announc 5(35):e00863–e00817

Saucedo B, Hughes J, Spitzen-Van Der Sluijs A et al (2018) Ranavirus genotypes in the Netherlands and their potential association with virulence in water frogs (*Pelophylax* spp.). Emerg Microb Infect 7(1):1–14

Saucedo B, Serrano JM, Jacinto-Maldonado M et al (2019) Pathogen risk analysis for wild amphibian populations following the first report of a ranavirus outbreak in farmed American Bullfrogs (*Lithobates catesbeianus*) from Northern Mexico. Viruses 11(1):26

Schloegel LM, Picco AM, Kilpatrick AM et al (2009) Magnitude of the US trade in amphibians and presence of *Batrachochytrium dendrobatids* and ranavirus infection in imported North American bullfrogs (*Rana catesbeiana*). Biol Conserv 142:1420–1426

Schloegel LM, Daszak P, Cunningham AA et al (2010) Two amphibian diseases, chytridiomycosis and ranaviral disease, are now globally notifiable to the World Health Organization for Animal Health (OIE): an assessment. Dis Aquat Org 92:101–108

Schock DM, Bollinger TK, Chinchar VG et al (2008) Experimental evidence that amphibian ranaviruses are multi-host pathogens. Copeia 1:133–143

Schock DM, Bollinger TK, Collins JP (2009) Mortality rates differ among amphibian populations exposed to three strains of a lethal ranavirus. Ecohealth 6(3):438–448

Schock DM, Ruthig GR, Collins JP et al (2010) Amphibian chytrid fungus and ranaviruses in the Northwest Territories, Canada. Dis Aquat Org 92:231–240

Scholz F, Vendramin N, Olesen NJ et al (2022) Experimental infection trials with European North Atlantic ranavirus (*Iridoviridae*) isolated from lumpfish (*Cyclopterus lumpus*, L.). J Fish Dis 45:1745–1756

Schönbächler K, Segner H, Amphimaque B, Friker B, Hofer A, Lange B, Stirn M, Pantchev N, Origgi FC, Hoby S (2022) Health assessment of captive and free-living European pond turtles (*Emys orbicularis*) in Switzerland. J Zoo Wildl Med 53(1):159–172

Schramm HL Jr, Davis JG (2006) Survival of largemouth bass from populations infected with largemouth bass virus and subjected to simulated tournament conditions. N Am J Fish Manag 26:826–832

Sharifian-Fard M, Pasmans F, Adriaensen C et al (2011) Ranavirosis in invasive bullfrogs, Belgium. Emerg Infect Dis 17:2371–2372

Sim RR, Allender MC, Crawford LK et al (2016) Ranavirus epizootic in captive eastern box turtles (*Terrapene carolina carolina*) with concurrent herpesvirus and mycoplasma infection: management and monitoring. J Zoo Wildl Med 47(1):256–270

Sivasankar P, John KR, George MR et al (2017) Characterization of a virulent ranavirus isolated from marine ornamental fish in India. Virus Dis 28(4):373–382. https://doi.org/10.1007/s13337-017-0408-2

Smalling KL, Bunnell JF, Cohl J et al (2018) An initial comparison of pesticides and amphibian pathogens between natural and created wetlands in the New Jersey Pinelands, 2014–16 (No. 2018-1077). US Geological Survey

Smith TC, Picco AM, Knapp R (2017) Ranaviruses infect mountain yellow-legged frogs (*Rana muscosa* and *Rana sierrae*) threatened by *Batrachochytrium dendrobatidis*. Herpetol Conserv Biol 12(1):149–159

Smith SN, Watters JL, Marhanka EC et al (2019) Investigating ranavirus prevalence in central Oklahoma, USA, amphibians. Herpetol Rev 50(3):508–512

Song WJ, Qin QW, Qiu J et al (2004) Functional genomics analysis of Singapore grouper iridovirus: complete sequence determination and proteomic analysis. J Virol 78(22):12576–12590

Southard GM, Fries LT, Terre DR (2009) Largemouth bass virus in Texas: distribution and management issues. J Aquat Anim Health 21:36–42

Souza MJ, Gray MJ, Colclough P et al (2012) Prevalence of infection by *Batrachochytrium dendrobatidis* and *Ranavirus* in eastern Hellbenders (*Cryptobranchus alleganeiensis alleganeiensis*). J Wildl Dis 48:560–566

Speare R, Smith JR (1992) An iridovirus-like agent isolated from the ornate burrowing frog, *Limnodynastes* in northern Australia. Dis Aquat Org 14:51–57

Speare R, Freeland WJ, Bolton SJ (1991) A possible iridovirus in erythrocytes of *Bufo marinus* in Costa Rica. J Wildl Dis 27:457–462

Spitzen-van der Sluijs A, Pasmans F, Struijk RP et al (2016) Course of an Isolated ranavirus outbreak in a *Pelobates fuscus* population in the Netherlands. J Herpetol Med Surg 26(3–4):117–121

Sriwanayos P, Subramaniam K, Stilwell NK et al (2020) Phylogenomic characterization of ranaviruses isolated from cultured fish and amphibians in Thailand. Facets 5(1):963–979

St Amour V, Wong WM, Garner TWJ et al (2008) Anthropogenic influence on prevalence of 2 amphibian pathogens. Emerg Infect Dis 12:1175–1176

Stagg HEB, Guðmundsdóttir S, Vendramin N et al (2020) Characterization of ranaviruses isolated from lumpfish *Cyclopterus lumpus* L. in the North Atlantic area: proposal for a new ranavirus species (European North Atlantic Ranavirus). J Gen Virol 101(2):198–207

Stark T, Laurijssens C, Weterings M et al (2014) Death in the clouds: ranavirus associated mortality in an assemblage of cloud forest amphibians in Nicaragua. Acta Herpetol 9:125–127

Stilwell N, Frasca S Jr, Farina L et al (2022) The effect of water temperature on Frog virus 3 disease in hatchery-reared pallid sturgeon *Scaphirhynchus albus*. Dis Aquat Org 148:73–86

Stöhr AC, Hoffmann A, Papp T et al (2013a) Long-term study of an infection with ranaviruses in a group of edible frogs (*Pelophylax* kl. *esculentus*) and partial characterization of two viruses based on four genomic regions. Vet J 197:238–244

Stöhr AC, Blahak S, Heckers KO et al (2013b) Ranavirus infections associated with skin lesions in lizards. Vet Res 44:84

Stöhr AC, Fleck J, Mutchmann F et al (2013c) Ranavirus infection in a group of wild caught Lake Urmia newts *Neurergus crocatus* importuned from Iraq to Germany. Dis Aquat Org 103:185–189

Stöhr AC, Heckers KO, Ball L et al (2013d) Coinfection with multiple viruses in European pond turtles (*Emys orbicularis*). In: Proceedings Association of Reptilian and Amphibian Veterinarians, Indianapolis, IN, USA

Stöhr AC, López-Bueno A, Blahak S et al (2015) Phyology and differentiation of reptilian and amphibian ranaviruses detected in Europe. PLoS One 10(2):e0118633

Storfer A, Alfaro ME, Ridenhour BJ et al (2007) Phylogenetic concordance analysis shows an emerging pathogen is novel and endemic. Ecol Lett 10:1075–1083

Subramaniam K, Toffan A, Cappellozza E et al (2016) Genomic sequence of a ranavirus isolated from short-finned eel (*Anguilla australis*). Genome Announc 4:e00843–e00816

Subramaniam K, Waltzek TB, Chinchar VG (2019) Genomic sequence of a Bohle iridovirus strain isolated from a diseased boreal toad (*Anaxyrus boreas boreas*) in a North American aquarium. Arch Virol 164:1923–1926

Sutton WB, Gray MJ, Hoverman JT et al (2015) Trends in ranavirus prevalence among plethodontid salamanders in the Great Smoky Mountains National Park. EcoHealth 12:320–329

Talbott K, Wolf TM, Sebastian P et al (2018) Factors influencing detection and co-detection of Ranavirus and *Batrachochytrium dendrobatidis* in Midwestern North American anuran populations. Dis Aquat Org 128(2):93–103

Tamukai K, Tokiwa T, Kobayashi H, Une Y (2016) Ranavirus in an outbreak of dermatophilosis in captive inland bearded dragons (*Pogona vitticeps*). Vet Dermatol 27:99–105e28

Tapiovaara H, Olesen NJ, Linden J et al (1998) Isolation of an iridovirus from pike-perch *Stizostedion lucioperca*. Dis Aquat Org 32:185–193

Teacher AGF, Cunningham AA, Garner TWJ (2010) Assessing the long-term impacts of Ranavirus infection on wild common frog populations. Anim Conserv 13:514–522

Thumsová B, Price SJ, González-Cascón V et al (2022) Climate warming triggers the emergence of native viruses in Iberian amphibians. Iscience 25(12):105541

Todd-Thompson M (2010) Seasonality, variation in species prevalence and localized disease for ranavirus in Cades Cove (Great Smoky Mountains National Park) amphibians. MSc thesis. University of Tennessee, Knoxville, TN, USA

Tornabene BJ, Blaustein AR, Briggs CJ et al (2018) The influence of landscape and environmental factors on ranavirus epidemiology in a California amphibian assemblage. Freshw Biol 63(7):639–651

Torrence SM, Green DE, Benson CJ et al (2010) A new ranavirus isolated from *Pseudacris clarkii* tadpoles in Playa wetlands in the southern High Plains, Texas. J Aquat Anim Health 22:65–72

Tsai CT, Ting JW, Wu MH et al (2005) Complete genome sequence of the grouper iridovirus and comparison of genomic organization with those of other iridoviruses. J Virol 79(4):2010–2023

Uhlenbrok C (2010) Detection of ranavirus infection in tortoises and characterization of virus isolates. Veterinary medical thesis. Justus-Liebig-University, Giessen, Germany

Une Y, Nakajinma K, Taharaguchi S et al (2009a) Ranavirus infection outbreak in the salamander (*Hynobius nebulosus*) in Japan. J Comp Pathol 141:310

Une Y, Sakuma A, Matsueda H et al (2009b) Ranavirus outbreak in North American bullfrogs (*Rana catesbeiana*), Japan. Emerg Infect Dis 15:1146–1147

Une Y, Nakajima K, Taharaguchi S et al (2009c) Ranavirus infection outbreak in the salamander (*Hynobius nebulosus*) in Japan. J Compar Pathol 4(141):310

Urgiles VL, Ramírez ER, Villalta CI et al (2021) Three pathogens impact terrestrial frogs from a high-elevation tropical hotspot. EcoHealth 18:451–464

USFWS (2011) National wild fish health survey database. USFWS, Washington, DC. www.fws.gov/wildfishsurvey/database/nwfhs. Accessed 20 May 2014

Uyehara IK, Gamble T, Cotner S (2010) The presence of *Ranavirus* in anuran populations at Itasca State Park, Minnesota, USA. Herpetol Rev 41:177–217

Vilaça ST, Bienentreu J-F, Brunetti CR et al (2019) Frog virus 3 genomes reveal prevalent recombination between ranavirus lineages and their origins in Canada. J Virol 93(20). https://doi.org/10.1128/jvi.00765-19

von Essen M, Leung WT, Bosch J et al (2020) High pathogen prevalence in an amphibian and reptile assemblage at a site with risk factors for dispersal in Galicia, Spain. PLoS One 15(7):e0236803

Vörös J, Herczeg D, Papp T et al (2020) First detection of ranavirus infection in amphibians in Hungary. Herpetol Notes 13:213–217

Waltzek TB, Miller DL, Gray MJ et al (2014) New disease records for hatchery-reared sturgeon: expansion of host range of frog virus 3 into pallid sturgeon, Scaphirhynchus albus. Dis Aquat Organ 111(3):219–227

Waltzek TB, Subramaniam K, Jancovich (2024) Ranavirus taxonomy and phylogeny. In: Gray MJ, Chinchar VG (eds) Ranaviruses: lethal pathogens of ectothermic vertebrates, 2nd edn. Springer Online. In press

Wang L, Bai B, Huang SQ et al (2017) QTL mapping for resistance to Iridovirus in Asian seabass using genotyping-by-sequencing. Marine Biotech 19(5):517–527

Warne RW, LaBumbard B, LaGrange S et al (2016) Co-infection by chytrid fungus and ranaviruses in wild and harvested frogs in the tropical Andes. PLoS One 11(1):e0145864

Watters JL, Davis DR, Yuri T, Siler CD (2018) Concurrent infection of *Batrachochytrium dendrobatidis* and ranavirus among native amphibians from northeastern Oklahoma, USA. J Aquat Anim Health 30(4):291–301

Wei J, Huang Y, Zhu W et al (2019) Isolation and identification of Singapore grouper iridovirus Hainan strain (SGIV-HN) in China. Arch Virol 164(7):1869–1872

Weir RP, Moody NJG, Hyatt AD (2012) Isolation and characterization of a novel Bohle-like virus from two frog species in the Darwin rural area, Australia. Dis Aquat Org 99:169–177

Weng SP, He JG, Wang XH et al (2002) Outbreaks of an iridovirus disease in cultured tiger frog, *Rana tigrina rugulosa*, in southern China. J Fish Dis 25:423–427

Westhouse RA, Jacobson ER, Harris RK et al (1996) Respiratory and pharyngo-esophageal iridovirus infection in a gopher tortoise (*Gopherus polyphemus*). J Wildl Dis 32:682–686

Whitfield SM, Geerdes E, Chacon I et al (2013) Infection and co-infection by the amphibian chytrid fungus and ranavirus in wild Costa Rican frogs. Dis Aquat Org 104:173–178

Whitfield SM, Alvarado-Barboza G, Abarca JG et al (2021) Ranavirus is widespread in Costa Rica and co-occurs with threatened amphibians. Dis Aquat Org 144:89–98

Whittington RJ, Reddacliff GL (1995) Influence of environmental temperature on experimental infection of redfin perch (*Perca fluviatilis*) and rainbow trout (*Oncorhynchus mykiss*) with epizootic haematopoietic nerosis virus, an Australian iridovirus. Aust Vet J 72:421–424

Whittington RJ, Philbey A, Reddacliff GL et al (1994) Epidemiology of epizootic haematopoietic necrosis virus (EHNV) infection in farmed rainbow trout, *Oncorhynchus mykiss* (Walbaum): findings based on virus isolation, antigen capture ELISA and serology. J Fish Dis 17:205–218

Whittington RJ, Kearns C, Hyatt AD et al (1996) Spread of epizootic haematopoietic necrosis virus (EHNV) in redfin perch (*Perca fluviatilis*) in southern Australia. Aust Vet J 73:112–114

Whittington RJ, Reddacliff LA, Marsh I et al (1999) Further observations on the epidemiology and spread of epizootic haematopoietic necrosis virus (EHNV) in farmed rainbow trout *Oncorhynchus mykiss* in southeastern Australia and a recommended sampling strategy for surveillance. Dis Aquat Org 35:125–130

Whittington RJ, Becker JA, Dennis MM (2010) Iridovirus infections in finfish – critical review with emphasis on ranaviruses. J Fish Dis 33 95–122

Winter JM, Mumm L, Adamovicz LA et al (2020) Characterizing the epidemiology of historic and novel pathogens in Blanding's turtles (*Emydoidea blandingii*). J Zoo Wildl Med 51(3):606–617

Winzeler ME, Hamilton MT, Tuberville TD, Lance SL (2015) First case of ranavirus and associated morbidity and mortality in an eastern mud turtle *Kinosternon subrubrum* in South Carolina. Dis Aquat Org 114:77–81

Winzeler ME, Williams RN, Kimble SJ (2016) Surveying for ranavirus in green frogs (*Lithobates clamitans*) at five locations in Indiana. J N Am Herpetol 5:23–26

Winzeler ME, Haskins DL, Lance SL, Tuberville TD (2018) Survey of aquatic turtles on the Savannah River Site, South Carolina, USA, for prevalence of ranavirus. J Wildl Dis 54(1):138–141

Wirth W (2020) Ranaviral infection in Australian freshwater turtles. Ph.D. thesis. College of Public Health, Medical and Veterinary Science, James Cook University, Australia

Wirth W, Schwarzkopf L, Schaffer J, Ariel E (2024) No Ranaviral DNA found in Australian freshwater turtles 2014–19 despite previous serologic evidence. J Wildlife Dis 60(3):683–690. https://doi.org/10.7589/JWD-D-23-00051

Wirth W, Schwarzkopf L, Skerratt LF, Ariel E (2018) Ranaviruses and reptiles. PeerJ 6:e6083

Wirth W, Schwarzkopf L, Skerratt LF et al (2019) Dose-dependent morbidity of freshwater turtle hatchlings, *Emydura macquarii krefftii*, inoculated with a ranavirus isolate (Bohle iridovirus, *Iridoviridae*). J Gen Virol 100(10):1431–1441

Wirth W, Lesbarrères D, Ariel E (2021) Ten years of ranavirus research (2010–2019): an analysis of global research trends. Facets 6(1):44–57

WOAH (World Organisation for Animal Health) (2021) Red sea bream iridoviral disease. In: Manual of diagnostic tests for aquatic animals. https://www.woah.org/fileadmin/Home/eng/Health_standards/aahm/current/2.3.07_RSIVD.pdf

WOAH (World Organisation for Animal Health) (2022) Infection with *Ranavirus* species. Aquatic Code Online Access. https://www.woah org/en/what-we-do/standards/codes-and-manuals/aquatic-code-online-access/?id=169&L=1&htmfile=chapitre_ranavirus.htm

Wolf K, Bullock GL, Dunbar CE et al (1969) Tadpole edema virus: a viscerotropic pathogen for anuran amphibians. J Infect Dis 118:253–262

Woodland JE, Brunner CJ, Noyes AD et al (2002a) Experimental oral transmission of largemouth bass virus. J Fish Dis 25:669–672

Woodland JE, Noyes AD, Grizzle JM (2002b) A survey to detect largemouth bass virus among fish from hatcheries in the southeastern USA. T Am Fish Soc 131:308–311

Wynne FJ (2019) Detection of ranavirus in endemic and threatened amphibian populations of the Australian Wet Tropics Region. Pac Conserv Biol 26:93–97

Xiao H, Liu M, Li S et al (2019) Isolation and characterization of a ranavirus associated with disease outbreaks in cultured hybrid grouper (♀ Tiger Grouper *Epinephelus fuscoguttatus* × ♂ Giant Grouper *E. lanceolatus*) in Guangxi, China. J Aquat Anim Health 31(4):364–370

Xu K, Zhu D, Wei Y et al (2010) Broad distribution of ranavirus in free-ranging *Rana dybrowskii* in Heilongjiang, China. EcoHealth 7:18–23

Youker-Smith TE, Whipps CM, Ryan SJ (2016) Detection of an FV3-like ranavirus in wood frogs (*Lithobates sylvaticus*) and green frogs (*Lithobates clamitans*) in a constructed vernal pool network in central New York State. Herpetol Rev 47:595–598

Youker-Smith TE, Boersch-Supan PH, Whipps CM, Ryan SJ (2018) Environmental drivers of ranavirus in free-living amphibians in constructed ponds. EcoHealth 15(3):608–618

Yu Z, Mou W, Geng Y et al (2020) Characterization and genomic analysis of a ranavirus associated with cultured black-spotted pond frogs (*Rana nigromaculata*) tadpoles mortalities in China. Transbound Emerg Dis 67(5):1954–1963

Zhang Q, Li Z, Jiang Y et al (1996) Preliminary studies on virus isolation and cell infection from diseased frog *Rana gyrlio*. Acta Hydrobiol Sinica 4:390–392

Zhang QY, Xiao F, Li ZQ et al (2001) Characterization of an iridovirus from the cultured pig frog, *Rana gyrlio*, with lethal syndrome. Dis Aquat Org 48:27–36

Zhang W, Duan C, Zhang H et al (2020) Widespread outbreaks of the emerging mandarinfish ranavirus (MRV) both in natural and ISKNV-FKC vaccinated mandarinfish *Siniperca chuatsi* in Guangdong, South China, 2017. Aquaculture 520:734989

Zhao R, Geng Y, Qin Z et al (2020) A new ranavirus of the Santee-Cooper group invades largemouth bass (*Micropterus salmoides*) culture in southwest China. Aquaculture 526:735363

Zhou ZY, Geng Y, Liu XX et al (2013) Characterization of a ranavirus isolated from the Chinese giant salamander (*Andrias davidiansu*, Blanchard, 1871) in China. Aquaculture 384–387:66–73

Zhu YQ, Wang XL (2016) Genetic diversity of ranaviruses in amphibians in China: 10 new isolates and their implications. Pak J Zool 48(1):107–114

Zilberg D, Grizzle JM, Plumb JA (2000) Preliminary description of lesions in juvenile largemouth bass injected with largemouth bass virus. Dis Aquat Org 39:143–146

Zupanovic Z, Lopez G, Hyatt AD et al (1998a) Giant toads, *Bufo marinus*, in Australia and Venezuela have antibodies against ranaviruses. Dis Aquat Org 32:1–8

Zupanovic Z, Musso C, Lopez G et al (1998b) Isolation and characterization of iridoviruses from the giant toad *Bufo marinus* in Venezuela. Dis Aquat Org 33:1–9

Ranavirus Ecology: From Individual Infections to Population Epidemiology to Community Impacts

Jesse L. Brunner, Andrew Storfer, Emily H. Le Sage, Trenton W. J. Garner, Matthew J. Gray, and Jason T. Hoverman

1 Introduction

Ranaviruses were thought to have little impact on populations of fish and amphibians for decades after their serendipitous discovery in primary kidney cell cultures of northern leopard frogs (*Rana pipiens*) in the 1960s (Chinchar et al. 2009; Granoff et al. 1966; Williams et al. 2005). This view dramatically changed in light of mounting examples of ranavirus-caused mass mortality events in fish and amphibians around the world (e.g., Cunningham et al. 1996; Hartmann et al. 2022; Langdon and Humphrey 1987; Price et al. 2014; von Essen et al. 2020). An increase in search effort, especially with more sensitive methods of detecting ranaviruses, has led not only to a clearer view of an expanding geographic and host range but also to a growing recognition that ranavirus infections frequently occur in populations with little or no apparent mortality (Brunner et al. 2021). The new picture that has emerged is that the outcomes of ranavirus infections are highly variable among individual hosts, among populations and species, and across landscapes. Understanding the causes of that variation remains a key goal in ranavirus research.

J. L. Brunner (✉) · A. Storfer
School of Biological Sciences, Washington State University, Pullman, WA, USA
e-mail: jesse.brunner@wsu.edu

E. H. Le Sage
Biology Department, Skidmore College, Saratoga Springs, NY, USA

T. W. J. Garner
Institute of Zoology, Zoological Society of London, London, UK

M. J. Gray
Center for Wildlife Health, School of Natural Resources, University of Tennessee, Knoxville, TN, USA

J. T. Hoverman
Department of Forestry and Natural Resources, Purdue University, West Lafayette, IN, USA

© The Author(s) 2025
M. J. Gray, V. G. Chinchar (eds.), *Ranaviruses*,
https://doi.org/10.1007/978-3-031-64973-8_7

Whereas our knowledge of the geographic distribution of ranaviruses and their host ranges have expanded dramatically in the last 20 years (Marschang et al. 2024, this volume), there remain critical gaps in our knowledge of ranavirus ecology. Most of what we know about ranavirus ecology stems from observations in the seasons, life stages, and taxa where hosts are easy to find and sample. For instance, much of what we know about the natural ecology of ranavirus in amphibians comes from epizootic events in pond-breeding amphibians (e.g., Harp and Petranka 2006; Ariel et al. 2009a; Price et al. 2014; Cunningham et al. 1996); in reptiles from serendipitous observations of mortality events (e.g., von Essen et al. 2020) or labor-intensive searches, such as with turtle-sniffing dogs (Allender et al. 2013a); and in fish from sporadic mass mortality events, especially in economically important species and settings (e.g., Chao and Pang 1997). What happens outside of these times and events we generally do not know; however, experiments have also been used to assess the importance of species, age, environmental variation, and ranavirus genotype, among other factors, on infection and disease dynamics. Beyond a few well-studied systems, we frequently lack the long-term data with which to determine how often and under what circumstances ranavirus-induced mortality events translate to declines in host populations (Gray et al. 2024, this volume). Thus, for students at the start of their careers, there is still plenty to be learned and a need for creative approaches to advance our understanding of ranavirus ecology.

In this chapter, we will build a picture of ranavirus ecology as we understand it, while highlighting important questions and opportunities for further research along the way. We will also emphasize the need to match data with process thinking, which requires both experiments and models to assign causation (Gray et al. 2024, this volume). Because ranavirus ecology and epidemiology are dynamic, we must think not just about some particular state of the system (i.e., our data as collected), but how the system got to that state (i.e., the processes at work). Moreover, models, implicitly or explicitly stated, are critical for improving our understanding of causation within the system. Finally, note that we organize this chapter in terms of scale, from individual hosts to populations to communities, with a focus on how processes and outcomes at one level might scale up to the next level.

2 Ranavirus Infections at the Individual Level

At the level of the individual host, there are three key attributes of infections that determine ecological or epidemiological dynamics and outcomes: (1) host susceptibility to infection, which we define as the probability of infection given an exposure, (2) the shedding or transmissibility of infections, and (3) the outcome of infections—from recovery and clearance of the virus to harboring persistent, sublethal infections, to morbidity and mortality. These are fundamental components of most parasite transmission models, so we discuss them collectively as we highlight factors that strongly influence individual infections.

2.1   The Influence of Exposure Dose

In general, the probability of infection increases with the dose of virus particles (virions) to which an individual is exposed, as has been demonstrated by numerous dose-response experiments (Table 1). This might be because each virion (or group of virions) has some small probability of causing an infection such that increasing the dose simply increases the number of opportunities (the single-hit model). Alternatively, if hosts can resist invading virus populations up to a certain threshold size, increasing doses will tend to increase the fraction of hosts whose threshold has been exceeded, leading to more infections or death (Box 1). In the wild, larger inocula might also contain greater genetic diversity of the virus, which could increase the chances of including a variant that replicates well in the host. We have little data on the genetic diversity of ranaviruses within host populations, although the occurrence of recombinant strains in the wild indicates different species and likely strains do co-occur (Candido et al. 2019; Epstein and Storfer 2015; Vilaça et al. 2019b).

The outcomes of infection—the probability and time to death—are also dose-dependent, even conditional on infections having been established (Brunner et al. 2005; Forzán Gómez 2015; Warne et al. 2011; Wirth et al. 2019). Moreover, the medial lethal dose (LD_{50}) is often, though not always, higher than the median infectious dose (ID_{50}; Table 1). Dose-dependent mortality must be due in part to larger doses leading to larger founding population sizes, giving the virus population a head start against the host's immune response (Box 1). However, simple models including host immune responses can produce complex host-virus dynamics (Mihaljevic et al. 2019) even before including the myriad of variables that influence viral replication (e.g., viral genotype, temperature) and host responses (e.g., species, stage, age, condition) that almost certainly influence these interactions. Novel hypotheses and rigorous empirical tests are needed to provide even a crude explanation for the dose dependence of infection outcomes.

> **Box 1 Ecological Models of Ranavirus in Hosts**
> Several simple models provide some helpful intuition about the dose-dependent nature of infections, as well as how other physical and biological factors might alter these relationships. Importantly, none are "right" in that they all ignore important biology and often predict patterns that are not observed in empirical data. But, they can be useful for clarifying our thinking and even serve as null models.
>
> **Models of Dose-Dependent Infection Probability**
>
> The *exponential or "single-hit" dose-response model* assumes that each infectious propagule (i.e., viable virion) has a small, independent probability of causing an infection, κ (Furumoto and Mickey 1967). We can then estimate the probability of becoming infected as the probability of *not* being infected by the n propagules one is exposed to:

(continued)

Box 1 (continued)

$$\Pr\left(\text{infection}|n\right) = 1 - \left(1 - \kappa\right)^{n}.$$

However, we usually do not know the actual number of propagules, but instead just the expected dose, D, that they are exposed to. If we assume this is Poisson distributed, we get this exponential model:

$$\Pr\left(\text{infection}|n\right) = 1 - e^{-\kappa D}.$$

This yields an asymmetric dose-response curve (Fig. 1). Changes in the host-virus interaction such as the type of infectious contact, host susceptibility, or virus strain either affect the exposure dose, D, or the per-virion chance of infection, κ, in essence shifting the lines left or right. However, this simple model can be modified to account for antagonism or synergism between propagules (Regoes et al. 2003), variability in susceptibility among individuals in a population (Ben-Ami et al. 2010), and clumping of propagules (Anand et al. 2022), all of which can change the shape of the curves.

If instead we focus on the *distribution of susceptibility*—defined as the threshold dose an individual can resist—in the host population, we again arrive at a dose-response relationship (Fig. 2). Typically, we might assume this distribution is normally distributed, at least on a scale of log-doses, such that some individuals are easily infected and others are very resistant, but most are somewhere in between. The proportion of individuals that become infected is thus the proportion of the (sample of the) population where the threshold has been exceed. Mathematically, this is the cumulative distribution

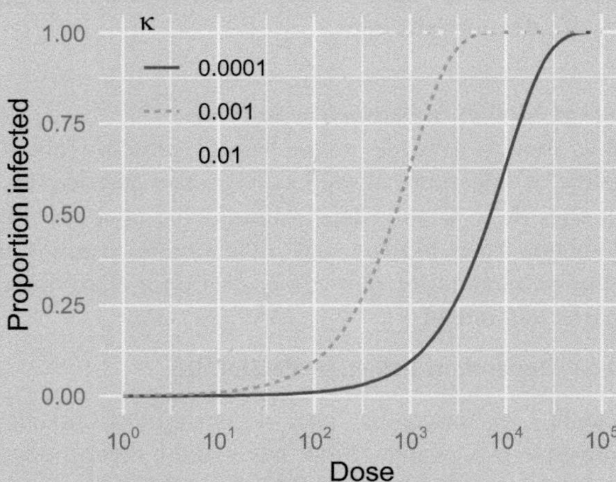

Fig. 1 Probability or proportion infected as a function of dose in the single-hit model. κ is the per-virion chance of infection

(continued)

Box 1 (continued)

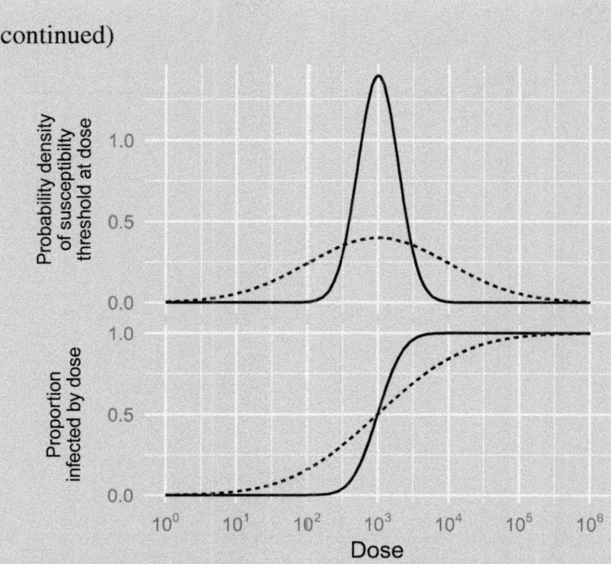

Fig. 2 Distribution of susceptibility, the exposure dose an individual can resist (top) and the resulting probability or portion infected (bottom) as a function of dose in a probit model. The dotted lines represent a population with a wider distribution of susceptibility than the population represented by the solid line

function (CDF) of the normal distribution (Fig. 2) or the probit (Bliss 1934). What is especially interesting and useful with this approach is that while the center of the distribution estimates the ID_{50} or LD_{50}, the slope of a probit regression (or, equivalently, a logistic regression) tells us about the variation in susceptibility of the hosts: flatter slopes imply a wider distribution of susceptibility and steeper slopes a more homogeneous population (e.g., Dwyer et al. 1997; Fig. 2). The reliance of this model on a threshold dose that can be resisted makes it more amenable to describing the probability of infections than the probability of mortality, which is the product of dynamic interactions.

Models of Exponentially Growing Virus Population

Brunner et al. (2005) assumed a simple model in which the virus population grows exponentially at rate r to some lethal size or density, V_{lethal}, from an initial starting population, V_0, corresponding to the amount of virus that initiates an infection (Fig. 3). Presumably, V_0 is some small fraction of the exposure dose. Anything that increases the dose, and thus the starting population size, or the intrinsic growth rate, r, is thus expected to reduce the time it takes the virus population to reach the lethal level (Fig. 3). While this model ignores host responses, we might assume that viral infections that attain a larger size or density sooner will be harder to clear, making them more likely to attain lethal sizes, sooner. But empirical data strongly contradict this null model:

(continued)

Box 1 (continued)

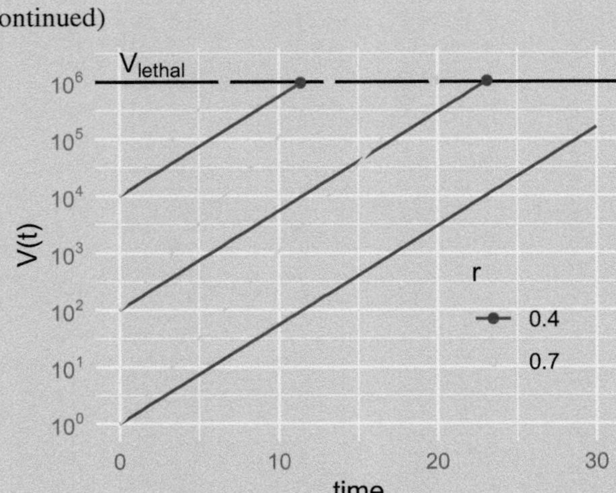

Fig. 3 The size of exponentially growing virus populations over time up to a presumed lethal size (V_{lethal}; points). The different y-intercepts represent different starting population sizes (V_0), and the different colored lines represent different intrinsic growth rates (r)

viral titers do not increase monotonically over time and vary by orders of magnitude among individual infections measured at the same time (e.g., Brunner et al. 2019; Ford et al. 2022), suggesting other, presumably host factors play a key role in constraining virus populations, and viral titers in dead tadpoles decline with time from exposure (e.g., Brunner and Collins 2009), contradicting the very existence of V_{lethal}.

Mihaljevic et al. (2019) constructed a more realistic suite model of an exponentially growing virus population actively reduced by (unspecified) immune responses that was able to capture the general features of the dynamics of viral titers in experimentally infected *L. catesbeianus* tadpoles (Brunner et al. 2019). However, these models were deterministic and so could not account for the large variation in titers among individuals, nor were they designed to explain mortality.

Finally, while we have less information on transmissibility—the capacity to infect others, often measured as the amount of virus shed onto host skin (collected by swabs or skin scrapes) or into the water (increasingly measured with environmental DNA)—the available data does suggest that more intense infections (i.e., larger virus populations in hosts) are also more infectious (Araujo et al. 2016; Brunner et al. 2007, 2019; Echaubard et al. 2010; Ford et al. 2022). In other words, ranaviruses, like many other pathogens, have load-dependent transmission (Wilber et al. 2017). Thus, hosts exposed to more virions are more likely to be infected, have more intense and presumably transmissible infections, and are more likely to die,

Table 1 Dose-response studies using at least two doses of a ranavirus and measuring infection or mortality

Class	Species	Stage	Virus	Route	Log$_{10}$ doses	Other variables tested	ID$_{50}$	LD$_{50}$	Study
Animals individually housed									
Amphibian	*Ambystoma maculatum*	L	FV3	WB	1.4, 2.4, 3.4, 4.4, 5.4	—	—	3.4	Brunner unpublished data
Amphibian	*A. mavortium*	L	ATV	WB	3, 4	Temperatures: 10, 18, 26 °C	0.8, <3, 1.9	2, <3, 6.8	Rojas et al. (2005)
Amphibian	*A. mavortium*	L	ATV	WB	2, 2.5, 3, 3.5, 4, 5	Host sources: FC, LC, SO	3.3, 2.7, 2.9	3.3, 2.8, 3.0	Brunner et al. (2005)
Amphibian	*Bufo bufo*	L	FV3	WB	2.5, 4.5	Virus isolates: BUK2, BUK3, RUK11, RUK13	4.6, 4.8, 4.6, 4.7	4.6, 8.1, 4.4, 5.5	Duffus et al. (2014)
Amphibian	*Lithobates catesbeianus*	L	FV3	WB	3, 5	—	2.4[a]	5.5[a]	Brunner et al. (2019)
Amphibian	*L. sylvaticus*	A	FV3	O	0.4, 1.4, 2.4, 3.4, 4.4, 5.4	—	2.9	2.9	Forzán Gómez (2015)
Amphibian	*L. sylvaticus*	L	FV3	WB	3.5, 4.5, 5.5	Salinity: 250, 1500, 3000 uS/cm	2.4, <3.5, 1.9	5, 4.7, 4.3	Hall et al. (2020)
Amphibian	*L. sylvaticus*	L	FV3	WB	1.8, 2.8, 3.8	—	2.4	—	Duffus et al. (2008)
Amphibian	*L. sylvaticus*	L	FV3	WB	1.4, 2.4, 3.4, 4.4, 5.4	—	< 1.4	2.6	Warne et al. (2011)
Amphibian	*Litoria caerulea*	J	FV3	I	3, 4.5	—	—	< 3	Cullen and Owens (2002)
Amphibian	*Osteopilus septentrionalis*	L	FV3	O	3, 4, 5, 6	Co-exposure: *Aplectana hamatospicula*, *B. dendrobatidis*	—	<0, 11.2	Ramsay and Rohr (2023)
Amphibian	*Rana latastei*	L	FV3	WB	2.6, 3.6, 4.6, 5.6, 6.6	—	—	< 2.6	Pearman et al. (2004)
Amphibian	*R. temporaria*	A	FV3	WB	4.5, 6, 7.5	—	4.6[a]	—	Ford et al. (2022)
Amphibian	*R. temporaria*	L	FV3	WB	2.5, 4.5	Virus isolates: BUK2, BUK3, RUK11, RUK13	3.1, 3.9, <0, <0	3.4, 3.8, 3.1, 3.2	Duffus et al. (2014)

(continued)

Table 1 (continued)

Class	Species	Stage	Virus	Route	$\log_{10}$ doses	Other variables tested	ID_{50}	LD_{50}	Study
Reptile	*Emydura macquarii*	H	FV3	I	1.3, 2.3, 3.3, 4.3, 5.3	—	2.5	4.4[b]	Wirth et al. (2019)
Animals housed in groups									
Amphibian	*L. catesbeianus*	L	FV3	WB	2.6, 4.6, 6.8	—	>6.8	>6.8	Alfaia et al. (2020)
Amphibian	*L. catesbeianus*	A	FV3	WB	2.6, 4.6, 6.8	—	>6.8	>6.8	Alfaia et al. (2020)
Amphibian	*L. sylvaticus*	L	FV3	WB	2, 3, 4	Host density: 6.6/L, 13.3/L		>4, <2	Echaubard et al. (2010)
Bony fish	*Micropterus dolomieu*	J	SCRV	I	1, 3, 5, 7	Virus isolates: 13–286, 13–295	< 1, <1	1, 1	Boonthai et al. (2018)
Bony fish	*M. dolomieu*	J	SCRV	WB	1, 2, 3, 4, 6, 7	Virus isolates: 13–286, 13–295	1.5, 1.5	2.8, 3.2	Boonthai et al. (2018)
Bony fish	*M. salmoides*	A	SCRV	I	4.1, 5.1, 6.1, 7.1, 8.1, 9.1, 10.1	Virus passage: 3rd, 5th	—	5.5, 5.9	Deng et al. (2011)
Bony fish	*M. salmoides*	J	SCRV	I	0.8, 1.8, 2.8, 3.8	—	NA	2.5	Plumb and Zilberg (1999)
Bony fish	*Percocypris pingi*	A	CMTV	I	3.3, 8.6	—	—	1.2	Liu et al. (2023)
Bony fish	*Silurus glanis*	J	EHNV	WB	5, 6	Host age: 8, 10 weeks	6.1, 6.5	6.1, 6.5	Abonyi et al. (2022)
Bony fish	*Triplophysa siluroides*	A	CMTV	I	3.9, 8.1	—	—	4.5	Deng et al. (2020)

Other variables may have been tested, as noted. Note that studies in which animals were group housed cannot exclude possible secondary transmission and re-exposure. Thus the doses listed represent minimal exposure doses. Stage: *A* adult, *L* larval, *J* juvenile, *H* hatchling. Route: *I* injection, *WB* water bath, *O* oral exposure. $\log_{10}$ doses are the total amount of virus given to animals by injection or orally or the concentration per mL in water baths, rounded to the nearest decimal place. ID_{50} and LD_{50} are the 50% infectious and lethal doses, respectively, as estimated from a logistic regression fit to the data presented in each source. Where the ID_{50} or LD_{50} is preceded by a less than or greater than sign, the data did not permit estimating a regression line (e.g., because all individuals were infected or uninfected or all survived or died), and so we present the lowest or highest exposure dose as a point of reference. See the original sources for sample sizes and other details

[a]Subsets of animals were destructively sampled at intervals throughout the study

[b]Animals were euthanized based on signs of infection, prior to death

and die sooner. Importantly, it can be useful to think of ranavirus infections not as distinct, binary states—uninfected or infected—and instead as occurring along a distribution of intensities that imply different consequences in terms of their transmissibility and outcomes.

2.2   The Influence of Temperature

Because ranaviruses and their hosts are ectotherms, temperature plays a key role in the dynamics of ranavirus infections (Table 2). All else being equal, growth rates of pathogens, including viruses, in culture increase with temperature up to some optimum where they rapidly decline to the virus' upper thermal limit (Altizer et al. 2013; Molnár et al. 2017). Different ranavirus species appear to have different thermal optima for growth (Ariel et al. 2009b; Grant et al. 2003; Rojas et al. 2005; Speare and Smith 1992), although the relationship between temperature and replication is usually characterized coarsely. Additionally, the kinetics and efficacy of ectotherm host immune responses to pathogens are also temperature-dependent (Herczeg et al. 2021; Raffel et al. 2006; Ribas et al. 2009). The outcome of ranavirus infections at different temperatures may thus depend on the relative performance of the virus and host at these temperatures (the "thermal mismatch" hypothesis, developed in the context of chytridiomycosis; Cohen et al. 2017, 2019). The majority of ranavirus literature suggest that warmer temperatures lead to greater infection and mortality in susceptibility host species (Ariel and Jensen 2009; Hall et al. 2018; Rayl and Allender 2020; Stilwell et al. 2022; Whittington and Reddacliff 1995). This has led several researchers to suggest that warming temperatures associated with climate change could result in greater ranavirus emergence (Price et al. 2019; Thumsová et al. 2022). However, some studies find that experimental infections were more lethal at *lower* temperatures (Allender et al. 2013b; Rojas et al. 2005). In reality, the influence of temperature on ranavirus-host interactions is likely nonlinear, where virus performance in hosts reaches an optimum temperature and then declines (Wirth and Ariel 2020). A consequence is that slight changes in temperature may have strong impacts on the outcome of infections. For example, Brand et al. (2016) showed that a 2 °C increase in temperature led to apparently persistent experimental FV3 infections in wood frogs (*Lithobates sylvaticus*) larvae becoming lethal. While experiments rarely use enough temperature levels over a wide-enough range to discern temperature optima for ranavirus growth in vitro or in vivo, let alone host performance (Molnár et al. 2017), we expect host and ranavirus optima, and their relative mismatch, likely vary with viral strain, host species, age class, and other factors (Table 2).

It is also important to note that the influence of temperature may be indirect, because temperature, or at least temperature extremes, or correlated environmental variables can be a physiologically stressful (Bayley et al. 2013; Ariel and Jensen 2009; Echaubard et al. 2014). Ranavirus-associated mortality events in fishes, for instance, often correspond with spikes in temperature and anoxia (Grizzle and

Table 2 Studies investigating the effect of various chemical, physical, and biological cofactors on ranavirus infections

Type of cofactor	Cofactor	Class	Species	Stage	Virus	Route	Immune response	Prevalence of infection	Rv titer	Mortality	Reference
Studies in the laboratory											
Physical	Temperature	Amphibian	*Ambystoma mavortium*	L	ATV	WB	—	—	↓	↓	Rojas et al. (2005)
Physical	Temperature	Amphibian	*Lithobates pipiens, L. sylvaticus*	L	FV3	WB	—	↑	—	↓	Echaubard et al. (2014)
Physical	Temperature	Amphibian	*L. sylvaticus, L. clamitans, Hyla chrysoscelis, A. maculatum*	L	FV3	WB	—	—	↑	↑	Brand et al. (2016)
Physical	Temperature	Amphibian	*Rana temporaria*	J	FV3	WB	—	—	—	↑	Price et al. (2019)
Physical	Temperature	Amphibian	*R. temporaria*	L	EHNV, CMTV, FV3	WB	—	~	~	↑	Bayley et al. (2013)
Physical	Temperature	Fish	*Esox lucius*	J	EHNV, CMTV, FV3	WB	—	—	—	↑	Bang Jensen et al. (2009)
Physical	Temperature	Fish	*Micropterus salmoides*	J	SCRV	I	—	—	~	↑	Grant et al. (2003)
Physical	Temperature	Fish	*Perca fluviatilis, Oncorhynchus mykiss*	A, J	EHNV	WB, I	—	—	—	↑	Whittington and Reddacliff (1995)
Physical	Temperature	Fish	*P. fluviatilis, O. mykiss*	J	EHNV	I,WB, Co	—	—	—	↷	Ariel and Jensen (2009)
Physical	Temperature	Fish	*Sander lucioperca*	J	EHNV, CMTV, FV3	I,WB, co	—	↓	—	↑	Bang Jensen et al. (2011b)
Physical	Temperature	Fish	*Silurus glanis*	J	EHNV, CMTV, FV3	WB	—	~	—	↑↓	Leimbach et al. (2014)
Physical	Temperature	Reptile	*Emydura macquarii krefftii*	H	FV3	I	—	↷	↷	~	Wirth and Ariel (2020)
Physical	Temperature	Reptile	*Trachemys scripta elegans*	A	FV3	I	—	~	↓	↓	Allender et al. (2013b)

Physical	Temperature	Reptile	*T. scripta elegans*	J	FV3	I	—	↑	~	↑	Rayl and Allender (2020)
Physical	Temperature	Reptile	*T. scripta elegans, Graptemys pseudogeographica, Pseudemys concinna*	J	FV3	I	—	~	~	↑	Allender et al. (2018)
Physical and biological	Temperature/ (coinfection with *Fc*/A*s*)	Fish	*M. dolomieu*	J	SCRV	WB	—	↑/~		↑/(↑/↓)	Boonthai et al. (2018)
Chemical	Atrazine/sodium nitrate	Amphibian	*A. mavortium*	L	ATV	WB	↓/↓	↑/↓	~/~	—	Forson and Storfer (2006)
Chemical	Atrazine/ chlorpyrifos	Amphibian	*A. mavortium*	L	ATV	WB	—	~/↑	—	↑/↑	Kerby and Storfer (2009)
Chemical	Carbaryl/ thiamethoxam	Amphibian	*L. sylvaticus*	L	FV3	WB	—	~/~	~/~	↑/~	Pochini and Hoverman (2017)
Chemical	Carbaryl	Amphibian	*L. sylvaticus*	L	FV3	WB	—	~	~	↓	Billet et al. (2021)
Chemical	Salinity	Amphibian	*L. sylvaticus*	L	FV3	WB	↓	~	↑	↑	Hall et al. (2020)
Chemical	Compounds from oil and gas drilling	Amphibian	*Xenopus laevis*	A	FV3	WB	↓	—	↑	—	Robert et al. (2019)
Chemical	Compounds from oil and gas drilling	Amphibian	*X. laevis*	L	FV3	WB	↓	—	↑	~	Robert et al. (2018)
Chemical	Atrazine	Amphibian	*X. laevis*	L, A	FV3	WB	↓↓	—	~ ~	↑ ~	Sifkarovski et al. (2014)
Chemical	Carbaryl	Amphibian	*X. laevis*	L, A	FV3	WB	↓↓	—	~ ~	—	De Jesús et al. (2017)
Chemical	Atrazine/ glyphosate	Reptile	*T. scripta elegans*	J	FV3	I	—	~/~		~/~	Goodman et al. (2021)

(continued)

Table 2 (continued)

Type of cofactor	Cofactor	Class	Species	Stage	Virus	Route	Immune response	Prevalence of infection	Rv titer	Mortality	Reference
Chemical and biological	Carbaryl/ predator	Amphibian	*A. mavortium*	L	ATV	WB	—	—	—	↑/↑	Kerby et al. (2011)
Chemical and biological	Salinity/low food	Amphibian	*L. sylvaticus*	L	FV3	WB	—	—	↑/↑	—	Hughey et al. (2023)
Chemical and biological	Glyphosate/ coinfection with *Bd*	Amphibian	*Cryptobranchus alleganiensis*	J	FV3	WB	—	~/~	—	~/~	Cusaac et al. (2021)
Biological	Coinfection with *Ech*	Amphibian	*H. versicolor*	L	FV3	WB	—	~	↓	↑	Wuerthner et al. (2017)
Biological	Predators	Amphibian	*L. clamitans, L. sylvaticus, Pseudacris feriarum, H. chrysoscelis*	L	FV3	WB	—	~	~	~	Haislip et al. (2012)
Biological	Predators/ crowding/low food	Amphibian	*L. sylvaticus*	L	FV3	WB	—	~/~/~	—	~/~/~	Reeve et al. (2013)
Biological	Coinfection with *Bd/Ah*	Amphibian	*Osteopilus septentrionalis*	L	FV3	O	—	—	↓/↓	~/~	Ramsay and Rohr (2023)
Studies in mesocosms											
Physical	Temperature	Fish	*Scaphirhynchus albus*	J	FV3	WB	—	↑	↑	↑	Stilwell et al. (2022)
Physical and biological	Temperature/ coinfection with *Bd*	Amphibian	*R. dalmatina, Bufo bufo*	L	FV3	WB	—	~↓/↓↓	~↓/~↓	↓~/~~	Herczeg et al. (2023)

Biological	Coinfection with Ech	Amphibian	H. versicolor, Anaxyrus americanus, L. pipiens, Ps. crucifer	L	FV3	Co	—	~	↓	~	Wuerthner et al. (2017)
Biological	Predators	Amphibian	H. versicolor, L. pipiens	L	FV3	Co	—	↓	—	↑	Gallagher et al. (2019)
Biological	Predators	Amphibian	Ps. crucifer, H. versicolor, An. americanus, L. pipiens, L. sylvaticus	L	FV3	Co	—	↓	↓	↑↓	DeBlieux and Hoverman (2022)
Biological	Predators/ crowding/low food	Amphibian	L. sylvaticus	L	FV3	Co	—	~/~/~	—	~ /↓/ ~	Reeve et al. (2013)

Stage: *A* adult, *L* larval, *J* juvenile, *H* hatchling. Route: *I* injection, *WB* water bath, *O*, oral exposure, *Co* co-housing with previously infected animals. Symbols represent an increase (↑), decrease (↓), increase and then decrease (↷), varied (↑↓), and small or no effect (~) of exposure to or an increase in the cofactor. When more than cofactor was studied, their effects are separated by a forward slash (/). Abbreviations for pathogens in coinfection are *Bd Batrachochytrium dendrobatidis, Fc Flavobacterium columnare, As Aeromonas salmonicida, Ech Echinoparyphium, Ah Aplectana hamatospicula*

Brunner 2003) or other changes to water quality (Inendino et al. 2005). Thus, while temperature clearly plays an important role in ranavirus infections, it remains an open question whether there are consistent patterns in how temperature influences the outcome or transmissibility of infections.

2.3 The Influence of Other Environmental Factors

In addition to temperature, a wide variety of environmental factors have been experimentally tested for their influence on ranavirus infections, from salinity, nutrient conditions, exposure to agricultural and other anthropogenically introduced chemicals, coinfections, and more (Table 2). A similarly diverse suite of environmental factors have been examined for their associations with ranavirus occurrence or mortality (e.g., St-Amour et al. 2008; Davis et al. 2020; North et al. 2015; Palomar et al. 2021; Peeler and Feist 2011; Richter et al. 2013; Smalling et al. 2022). There are many possible mechanisms by which environmental factors might influence the virus or host, but beyond temperature, most of the focus is at least implicitly on the host's capacity to effectively prevent or clear viral infections.

For instance, the stress-induced susceptibility hypothesis, which has been especially influential in the research on amphibian declines (Kiesecker 2011), posits that chronic exposure to stressful conditions is immunosuppressive (Carey 1993; Dhabhar 2014). For example, amphibian larvae raised in water containing pesticides are more likely to become infected and have more severe ranavirus infections than those in unpolluted water (Forson and Storfer 2006; Kerby and Storfer 2009; Pochini and Hoverman 2017). Similarly, tadpoles raised in high salinity environments were also more likely to die from infections (Hall et al. 2020). However, there are also a myriad of counterexamples of putative stressors (e.g., predators and even starvation levels of food restriction) that do not increase susceptibility to infections or worsen outcomes (Billet et al. 2021; Goodman et al. 2021; Haislip et al. 2012; Reeve et al. 2013). Given the multiplicity of factors considered, diversity of hosts, and range of settings and outcomes considered (Table 2), it is perhaps not surprising that the results of environmental factors on ranavirus infections are idiosyncratic. There is a real need for more research to decipher the environmental contexts that strongly influence ranavirus-host interactions.

Four measures would improve our understanding a great deal. First, researchers need to be much better at reporting the experimental conditions, sources of animals and virus isolates, and other details to facilitate comparison and meta-analyses. While the drivers of interstudy variation are unclear, seemingly small differences in experimental conditions or methodology might be the key to explaining differences in outcomes (Kumar et al. 2020). Second, to the extent possible given divergent objectives and systems, researchers might develop sets of standardized methods (e.g., modes of exposure, doses and durations, holding conditions) to facilitate clearer, stronger comparisons between studies. Third, we hope researchers will more accurately estimate and report the variability in key responses in addition to mean effect sizes. Heterogeneity in quantities like susceptibility and transmissibility can have important consequences for the probability and outcome of epidemics (Dwyer et al. 1997; Hickson and Roberts 2014; Lloyd-Smith et al. 2005). Lastly, we encourage researchers to make their raw data and method as widely available as possible (e.g., in data sharing sites) to facilitate future analyses and comparisons.

2.4 Differences Among Developmental Stages and Host Species

There is a tremendous amount of variation in susceptibility, transmissibility, and outcomes among host life history stages (e.g., larvae vs. metamorphosed stages; Haislip et al. 2011; Warne et al. 2011), populations (Pearman and Garner 2005), and species (Ariel et al. 2015; Brenes et al. 2014b; Cullen et al. 1995; Duffus et al. 2014; Hoverman et al. 2010, 2011), yet there are some patterns. For instance, amphibian larvae are more likely to be infected and die than metamorphosed or adult stages in North America (Haislip et al. 2011). However, this pattern is reversed in the UK (Cunningham et al. 1996; Duffus et al. 2013) and perhaps in Europe where most ranavirus-associated mortality events are in adult stages (Price et al. 2014; Rosa et al. 2019), though this may have more to do with different patterns of exposure than inherent differences in susceptibility.

Among North American amphibians, pond-breeding frogs, especially those with short larval periods and in the family Ranidae, tend to be more likely to become infected and die from experimental infections compared to other taxa (Hoverman et al. 2011). However, there are often challenges associated with generalizing. For instance, wood frogs (*L. sylvaticus*) are widely considered among the most susceptible species in North America based on controlled experiments (Hoverman et al. 2011) and frequently reported mass mortality events (Gahl and Calhoun 2008; Hall et al. 2016; Harp and Petranka 2006; Wheelwright et al. 2014). However, individuals do escape infections (Hall et al. 2020) or retain persistent infections (Crespi et al. 2015) in nature. Moreover, the wood frog's congener, the American bullfrog (*L. catesbeianus*), is among the most tolerant anurans to ranavirus infection (Brunner et al. 2019; Hoverman et al. 2011), suggesting there is only so much phylogenetic conservatism in susceptibility to and outcomes of ranavirus infection. Thus, we urge caution in making overly strong assumptions about the susceptibility, outcomes, or transmissibility of particular species or developmental stages.

Part of the problem is that except for rare studies using a suite of hosts and standardized methods (e.g., Hoverman et al. 2011), comparisons are often made among studies using different hosts and life stages, virus species or strains, housing conditions, and even methods of exposure, making it difficult to attribute differences in susceptibility or outcome to any particular factor (e.g., host species identity). Standardization would be very helpful. For example, a dose-response design is useful for establishing comparable measures of susceptibility across studies (Table 1).

It is also worth noting that we know little about the relative competence of host species or stages to transmit infections, though we would expect from first principles, and two empirical studies (Brunner et al. 2019; Ford et al. 2022), that it should track the intensity of infection in the host. One of the most important challenges for understanding ranavirus ecology is understanding which stages and species tend to amplify ranavirus infection within a community and which tend to act as carriers or maintain ranavirus infections within a community or spread it among habitats.

3 Ranavirus Transmission

Much of what we know of ranavirus epizootics still comes from mass mortality events (MMEs). Mortality can be quite rapid, such as the observation of an apparently healthy population of wood frog larvae all dying within 2 days (Wheelwright et al. 2014), to occurring over weeks and months in populations that are very large (e.g., in fish; Langdon et al. 1986) and those that are more diffuse in the environment (e.g., in turtles; Belzer and Seibert 2011). MMEs are often, but not always, observed in particular seasons (Brunner et al. 2015), perhaps reflecting the seasonal life cycle events of many of their hosts that caused increased host abundance or seasonal conditions (e.g., especially warm periods). It is worth noting, however, the biases inherent in observations of MMEs, including the fact that we rarely know the onset of a MME (Todd-Thompson 2010), many or even most MMEs probably go unobserved, and most observations come from well-studied or easily observed systems and places. As emphasized in Gray et al. (2024, this volume), longitudinal studies with frequent sampling are essential to understand ranavirus dynamics in amphibian communities (Todd-Thompson 2010). Moreover, ranavirus epizootics can occur with little or no apparent mortality (e.g., Greer et al. 2009; O'Connor et al. 2016). It can therefore be difficult to infer the dynamics of ranavirus epizootics from observations of MMEs.

Compartment epidemic models—so-called SIR models—are useful guides to thinking about the dynamics of epidemics. They can use averages of individual-level quantities—susceptibility, outcomes, and transmissibility—combined with assumptions about how hosts move and interact and thus transmit infections to project population-level epidemic dynamics (Gray et al. 2024, this volume). These models are necessarily vast simplifications and are usually restricted to single species occurring in closed populations with a single pathogen of interest and assume that all individuals within a compartment are identical (Wilber et al. 2016). However, SIR models are very useful for focusing our thinking on the processes governing the flow of individuals into the "infected" compartment due to transmission and the flow out as animals die or recover by clearing infections (Gray et al. 2024, this volume).

Transmission is at the core of infectious disease dynamics. It can be thought of in two parts: (1) the rate at which susceptible hosts make contact with potentially infectious materials (e.g., infected hosts or virions in water) and (2) the probability of becoming infected given such a contact, which is a function of susceptibility, by our definition, and the competence of the infected host. Since the probability of infection is dose-dependent (see above), we can instead think of this second part as at least scaling with the number of virions that are transferred to the susceptible host by a potentially infectious contact or from the environment.

Before we proceed, we must note that much of what we know about transmission is largely built on first principles and mathematical models. Empirical data on transmission usually comes from laboratory settings and mesocosm experiments where artificial conditions and experimental artifacts (e.g., animals housed in small habitats with little or no habitat structure and extremely simplified biotic communities) likely dominate the results. It is important to acknowledge that even the simple question of which routes of transmission are most important in nature (Peace et al. 2019), under which conditions or settings, is still largely uncertain empirically. Studies of transmission under natural conditions are difficult, but necessary for a more accurate picture of ranavirus transmission.

3.1 Transmission from the Environment

Ranaviruses are transmitted to aquatic hosts via water containing infectious particles, as has been demonstrated by innumerable water-bath exposures (e.g., Bang Jensen et al. 2011a; Hoverman et al. 2010; Whittington and Reddacliff 1995) and experiments with co-housed, but physically separated or sequentially housed pairs of infected and susceptible hosts (e.g., Jancovich et al. 1997; Brenes et al. 2014a). The importance of water-borne transmission in nature, however, is less clear. Virions that are shed into water by infectious hosts are diluted into what may be quite large volumes of water, even in smaller pools and wetlands, at least if we assume that the water is well-mixed. Thus, we might think of the rate of transmission through the water as a constant rate of contact (i.e., being *in* the water) with low concentrations of virions in the "virus pool." This would imply low rates of water-borne transmission early in the epizootic, with rates increasing as infections become more common and the virus pool grows (Brunner and Yarber 2018; Peace et al. 2019). Note that in captive settings, this same amount of virus would be shed into a much smaller volume of water, reaching concentrations that might be very likely to cause infections.

If we relax the assumption that the waterbody is well-mixed and instead assume virions accumulate in local "hot spots" around infected hosts, then transmission would depend on contact with these areas *near* hosts. As "near" becomes smaller and smaller distances, this converges on transmission by direct contact *with* infected individual hosts. The same would seem to be true of virions accumulating in micro-habitats in the environment (e.g., pond substrates, burrows), such as those shed from carcasses (see below). Lastly, while it seems obvious that water-borne transmission is less important for terrestrial species and stages because they contact water less often, they may still become infected when visiting water bodies to breed, oviposit, or rehydrate. Transmission from water, soil, or pond substrates might prove important for spillover to terrestrial species—indeed, transmission between classes of hosts is possible through the water in experimental settings (Brenes et al. 2014a)—but we expect that it may be less important than direct contact of hosts in the terrestrial environment. For instance, amphibians often cluster during hibernation and estivation, and some species engage in courtship, breeding, and territorial defense in the terrestrial environment (Wells 2007).

Transmission from the water or substrates depends on ranavirus accumulating and persisting in the environment. However, ranavirus in the environment is subject to physical, chemical, and biotic inactivation or degradation (reviewed in Brunner and Yarber 2018). Persistence times, often expressed as the time required for 90% of the virus to be inactivated (T90s), are quite varied: days to months under semi-natural conditions in water or substrates. These T90s largely depend on temperature—T90s only exceed a few days when temperatures are low, around 4 °C—and the presence of microbes and even zooplankton, both of which inactivate and remove ranavirus particles (Johnson and Brunner 2014; Munro et al. 2016; Nazir et al. 2012). As most ranavirus epizootics seem to occur in spring, summer, or fall (Brunner et al. 2015), temperatures are likely in the range where viruses are degraded fairly rapidly (T90s in the range of 2–5 days under naturalistic conditions), meaning ranavirus shed into the environment probably does not remain active for long. This, in addition to the dilution of shed virus, would seem to imply that the rate of transmission through the water (or environment more broadly) is low, at least until there are many individuals—live or dead—shedding moderately high amounts of virus into the environment (Brunner and Yarber 2018; Peace et al. 2019).

In captive settings, we might expect water-borne transmission to be either very unimportant or very important, depending on the details of husbandry. Water treatments that inactivate ranavirus (e.g., ultraviolet filtering) and thus reduce persistence times would minimize the potential for transmission through water. However if animals are housed in stagnant, untreated water, virions might persist and accumulate to concentrations that virtually ensure water-borne transmission.

3.2 Transmission by Direct Contact

Mathematical models suggest ranavirus transmission via direct or close contact seems to be more important to epizootics in nature (Brunner and Yarber 2018; Peace et al. 2019). First, direct or close contacts likely transfer more virus to the susceptible host than virus diluted in larger bodies of water. Even short, approximately 1 s contacts with infected individuals or handling susceptible individuals with ranavirus-contaminated gloves can have a moderately high chance of transmission (Brunner et al. 2007; Gray et al. 2018). Second, animals often aggregate to some degree, at least in aquatic species and stages, ensuring many repeated contacts over time. In one mesocosm experiment, wood frog larvae made ≥ 50 contacts per larva per day under moderately low densities (Brunner et al. 2017). Such estimates are likely inflated due to necessarily artificial conditions (i.e., small size of mesocosms) as a similar study with eastern newts (*Notophthalmus viridescens*) found that habitat complexity (presence of artificial plants) reduced contact rates (Malagon et al. 2020). Nonetheless, it is apparent that contacts, and thus opportunities for transmission, still accumulate rapidly over time in aquatic amphibians. We know little of ranaviruses in fish populations, but the same logic should hold.

It is an open question how often terrestrial hosts of ranavirus (e.g., adult or direct developing amphibians, most reptiles), which range over much larger spaces, make casual contacts with one another and thus whether low contact rates potentially limit transmission through this route. From first principles, we would expect breeding aggregations would be an ideal setting for transmission as hosts come into close contact. There may thus be annual pulses of transmission among adults (Duffus et al. 2019). Note that it is also possible that susceptible individuals might avoid infected individuals, as was demonstrated in juvenile wood frogs (Le Sage et al. 2022), which would reduce the occurrence of potentially infectious contacts. In any case, compartment models fit to available data suggest that transmission by close contact contributes much more to new infections than water-borne transmission for at least the early and middle stages of epizootics (Brunner and Yarber 2018; Peace et al. 2019).

Ranaviruses are also easily transmitted by consuming infected tissues, which may occur during aggressive interactions between conspecifics up to and including cannibalism or necrophagy (Le Sage et al. 2019; Harp and Petranka 2006; Brunner et al. 2007; Pearman et al. 2004). On first principles, consumption seems an especially efficient method of transferring large numbers of virions to the consumer (Hoverman et al. 2010). Indeed, the probability of infection when an animal consumes even small amounts of infectious tissues is quite high (Hoverman et al. 2010). Moreover, cannibalistic tiger salamander larvae (*A. mavortium nebulosum*) do not reduce their cannibalism in the presence of ranavirus-infected conspecifics (Parris et al. 2005), so cannibals and scavenging animals both seem to be at a high risk of infection. However, whether transmission by consumption plays a small or large role in epizootics, at least during the initial stages of epizootics, is not clear. There are a few key factors that will determine its importance.

First, for cannibalism to contribute substantially to transmission, the victims must be shared among cannibals (Rudolf and Antonovics 2007), or one carcass would lead to only a single infection. Similarly, aggressive interactions and cannibalism also tend to be *directed* such that larger individuals are aggressive to or cannibalize (much) smaller individuals, and not vice versa. This, too, would accumulate infections in larger size classes, but not among the rest of the population. Carcass sharing during scavenging is probably more common, and so this does not seem to present a barrier to transmission by necrophagy.

Second, while scavenging and necrophagy can be common, even among putatively herbivorous tadpoles (Altig et al. 2007; Whiles et al. 2006), the density of infectious carcasses on which they can feed is initially low. If individuals randomly encounter these rare carcasses, then the overall rate of transmission by necrophagy would be low. If instead individuals seek out carcasses to scavenge, transmission from infectious carcasses would be much higher. Thus, scavenging behavior is quite important to transmission rates and needs further investigation.

Finally, we must consider the time that carcasses persist in the environment. Reports of carcass removal or degradation rates are intermittent in the literature but often suggest fairly rapid removal. For instance, in one study half of snake carcasses along a road were removed in 8 h and all by the next day (Degregorio et al. 2011). Carcasses of larval amphibians might persist for only 1–2 days, especially when scavenged by invertebrates (Le Sage et al. 2019; Regester and Whiles 2006). Fish carcasses have been shown to persist for days to months (Parmenter and Lamarra 1991). Only when persistence times are long and consumption rates are high (assuming random encounters) does transmission by necrophagy become important in epidemic models of ranavirus (Brunner and Yarber 2018; Peace et al. 2019). Still, transmission through necrophagy most certainly does occur and may be especially important in certain systems and time periods. For instance, in northern latitudes with cooler water temperatures and scavenger depauperate systems, carcasses could play an important role in extending the duration of ranavirus outbreaks (Peace et al. 2019).

Necrophagy and predation of infected individuals may also play a role in interclass transmission (Belzer and Seibert 2011; Price et al. 2014). Price et al. (2014), for instance, isolated a ranavirus from a snake (*Natrix maura*) that had consumed an infected newt during a MME. While such contacts may be rare, because they are intimate contacts with large doses of virus, they would tend to maximize the chance of transmission when they do occur.

Finally, the detection of ranavirus in *Aedes* sp. mosquitoes during a ranavirus outbreak in eastern box turtles (*T. c. carolina*) and one *Ochlerotatus* sp. found engorging on a turtle led to the hypothesis that mosquitoes might serve as mechanical or biological vectors of ranavirus (Kimble et al. 2015). However, the lack of detection in mosquito species known to feed on reptiles and amphibians (*A. canadensis*, *Culex erraticus*, *C. territans*, and *Uranotaenia sapphirina*) during an active epizootic in box turtles and amphibians suggests infections in mosquitos are rare (Lopez Vargas et al. 2020). Moreover, it has yet to be established that infected mosquitos can infect naive hosts. If, however, this route of transmission is demonstrated, it might provide another explanation for interclass transmission.

4 Ranavirus Epizootics in and Among Host Populations

4.1 Transmission Dynamics and Mortality

Ranaviruses often cause epizootic mortality, but how well this mortality tracks the dynamics of infection is, surprisingly, unclear. In the simplest sense, our compartment models assume a constant flow of individuals from the susceptible compartment into the infected one (i.e., transmission) and then out due to host death or recovery via viral clearance. Laboratory experiments suggest that ranavirus infections usually lead to mortality or, more rarely, recovery (depending on the dose of exposure, host condition, species identity, etc.) within a period of days to weeks. Thus, we presume that ranavirus-related mortality during an epizootic trails ranavirus infections by a period of days to weeks; the accumulated mortality can be viewed as a lagging indicator of the accumulation of infections (e.g., Todd-Thompson 2010). Parameterized epizootic models of ranavirus tend to predict very rapid epizootics followed by rapid die-offs that involve nearly the entire population (e.g., Brunner and Yarber 2018; Earl and Gray 2014; Peace et al. 2019 but see). Moreover, results from mesocosm experiments are largely consistent with these predictions of rapid transmission followed by rapid mortality (Harp and Petranka 2006; Reeve et al. 2013; Brunner et al. 2017). However, the few longitudinal studies of ranavirus epizootics in the wild suggest slower, longer, and more variable epizootics with mortality not clearly tracking prevalence (Greer et al. 2009; Hall et al. 2018). While models fit to relevant data predict rapidly increasing prevalence and mortality over a period of weeks, Hall et al. (2018) observed low-intensity infections in wood frog larvae in vernal pools persisted at moderate prevalence for several weeks before both prevalence and intensity of infections rapidly increased up to and during mortality events (Hall et al. 2018). Rather than the classic epizootic dynamics of compartment models, their results were consistent with mortality occurring when pond temperatures rose or late pre-metamorphic developmental stages were attained (Hall et al. 2018; Todd-Thompson 2010). Similarly, ranavirus-related mortality events in turtles have been associated with cold snaps (Agha et al. 2017), and those in fishes are often associated with elevated temperature and hypoxia (Grizzle and Brunner 2003; Goldberg 2002), though we generally lack data on prevalence or intensities of infection over time in wild reptiles and fishes to infer when transmission occurred. Collectively, these results suggest that ranavirus transmission and mortality are uncoupled, at least in some settings (e.g., species, stages, or conditions). Perhaps, then, environmental conditions or their physiological consequences, including rates of development, explain the pattern noted at the start of this chapter that ranavirus epizootics sometimes occur with little or no obvious mortality. More fine-grained, longitudinal, well-replicated studies are needed to determine with any confidence that this pattern is common and repeatable.

4.2 Varying Outcomes Among Host Populations

There have been numerous studies comparing ranavirus prevalence or mortality among populations with the goal of understanding how environmental conditions or settings affect the outcome of ranavirus epizootics, especially in pond-breeding amphibians. The chemical environment (Gahl and Calhoun 2010; Hall et al. 2020; Smalling et al. 2022), biotic environment (Youker-Smith et al. 2018), physical structure (Greer and Collins 2008; Hoverman et al. 2012a), position in the landscape (Gahl and Calhoun 2008; Tornabene et al. 2018; Youker-Smith et al. 2018), community composition (Bienentreu et al. 2022; Bielby et al. 2021; North et al. 2015; Snyder et al. 2023), and distance to or impact of human activities (St-Amour et al. 2008; North et al. 2015) have all been investigated as playing key roles in the occurrence or outcome of ranavirus among populations. Given the diversity of study design, factors considered, and potential causal pathways for these environmental factors to influence the outcome of ranavirus epizootics, it is difficult to draw out generalities. Instead, we highlight one common and influential hypothesis for why environmental conditions impact ranavirus epizootics: the "stress-induced suscepti-bility" hypothesis.

The stress-induced susceptibility hypothesis posits that because chronic stress, especially from anthropogenic alterations to the environment, can be immunosup-pressive (Dhabhar 2002; Hofer and East 2012), populations experiencing these stressors should experience more and more severe ranavirus epizootics (Carey 1993). There are many examples that are at least consistent with this idea. For instance, several ranavirus-associated mortality events in turtles have been observed when the turtles were moved or held in captivity as part of translocation programs (Cozad et al. 2020; Farnsworth and Seigel 2013; Kimble et al. 2017), and crowding and perhaps stress seem to increase the impact of LMBV infections on largemouth bass (Inendino et al. 2005; Schramm et al. 2006). But overall, the results of studies of putative stressors are idiosyncratic, even at the individual level (Table 2) let alone population-level outcomes, and there are not yet clear, overarching conclusions such as "anthropogenic stressors cause increased ranavirus prevalence or cause more severe epizootics."

There are several reasons why we might expect weak or varied relationships between putative stressors and ranavirus epizootics among populations. First, it is important to note that laboratory experiments are designed to detect differences among treatments, keeping animals under ideal conditions with all other factors held constant. In nature, where little is constant and conditions are rarely ideal, the influence of some putative stressor of interest may recede into the background noise or be swamped by animals that are experiencing the effect of multiple stressors. Second, it is important to recognize that populations exist in communities of inter-acting species embedded in complex environments and biotic interactions that may have a stronger influence on the dynamics of ranavirus epizootics than stress-induced differences in susceptibility (e.g., behavioral responses to predators chang-ing transmission, or the most susceptible hosts dying of starvation). For instance, the presence of invertebrate (non-host) predators in an assemblage of tadpole spe-cies has been shown to reduce infection prevalence under mesocosm conditions (Gallagher et al. 2019; DeBlieux and Hoverman 2022). This "healthy herds" effect appears to be mediated by virus-induced vulnerability to predation such that infected individuals are more easily captured by predators (DeBlieux and Hoverman 2019). Finally, there can be important feedbacks between the processes influenced by stressors. For instance, chronic exposure to stressors might mean that individuals are more susceptible to infection, which would presumably make epizootics larger and more severe, but if those more easily infected individuals die more quickly or with less intense infections, this would presumably reduce rates of transmission and thus make epizootics smaller and less severe (see Lafferty and Holt 2003 for a theoretical discussion). Careful thinking about how or whether individual effects of stress scale to population-level dynamics and outcomes is essential, and empirical studies that help us understand how these effects manifest are sorely needed.

Disentangling the possible drivers of ranavirus epizootics and, perhaps separately, mortality events requires longitudinal data to understand processes (i.e., fates and rates) rather than simply the state of the system at a particular time point (Gray et al. 2024, this volume). For instance, prevalence estimated at one or two time points might be consistent with any number of possible dynamics before, during, and after, but more frequent estimates can help us see how prevalence changes relative to our expectations of infection dynamics and mortality (Hall et al. 2018; Todd-Thompson 2010). Even better would be tracking the status of marked animals to infer the rate of gain or loss of infections in the marked group, or, with individual marks, even the likely fate of infections. If ranavirus infections are quickly lethal, marked infected individuals should rapidly disappear from future surveys, and if they persist for long periods, then our expectations must be revised. While estimating epidemiological rates can be difficult under natural settings, we are confident that creative researchers can identify solutions. For instance, while repeated sampling from amphibian larvae is difficult if not impossible, it may be feasible in larger bodied adult amphibians as well as fishes and reptiles. Enclosed sentinel animals can help estimate the rate at which susceptible individuals become infected, or force of infection, in a location (Halliday et al. 2007; Heisey et al. 2006). Additionally, pairing prevalence data with infection intensities can help researchers infer whether the individuals are recently infected or recovering and whether they might be slightly or highly infectious to others. Moreover, it may be feasible and ethical in certain places and settings (e.g., mesocosms) to experimentally manipulate the variable(s) thought to drive susceptibility or infection outcomes and observe epizootic dynamics and mortality. We also suspect that aquaculture and ranaculture facilities might be a rich environment in which to observe "natural experiments."

Finally, it is worth noting that the impacts of ranavirus infections extend beyond direct disease-induced death; ranavirus-infected individuals may be smaller and less likely to survive later challenges (e.g., overwintering, predation; DeBlieux and Hoverman 2019) or less likely to successfully reproduce. For instance, populations of common frogs (*R. temporaria*) in the UK with a history of ranavirus appear to have a truncated age structure, presumably due to increased adult mortality from ranavirus infections (Campbell et al. 2018). A population-projection model suggests these effects lead to smaller populations that are more vulnerable to stochastic losses (Campbell et al. 2018). Similarly, differences in the breeding phenology lead to greater impacts on female Bosca's newts (*Lissotriton boscai*) than males, which elevates the chances of extinctions (Rosa et al. 2019). While difficult to accomplish, studies tracking animals that survive infections over longer periods would help establish the comprehensive costs of ranavirus infection.

4.3   Population-Level Impacts of Ranavirus Outbreaks

Given the uncertainty about the causes and dynamics of ranavirus epizootics, it will not be surprising to learn that we still know little about the long-term impacts of ranavirus outbreaks on populations beyond observations of mass mortality events. Longitudinal studies of common frogs (*R. temporaria*) in the UK and amphibian communities in the Iberian Peninsula find that ponds with ranavirus mortality events decline over years and decades (Price et al. 2014; Teacher et al. 2010). However, we should point to the contrary case of wood frogs in the well-studied Yale Meyer forest in the Northeast USA. Here, ranavirus has been present and apparently causing mortality events in wood frog larvae in vernal wetlands for decades (Hall et al. 2016), with no pattern of decline in pond occupancy (Rowland et al. 2022). What explains these contrasting trends? At least in the UK, ranaviruses tend to kill adults rather than larvae (Duffus et al. 2013), which are more important demographically (Biek et al. 2002). Alternatively, it may be that rising temperatures are leading to more or more severe epizootics in some places (Price et al. 2019; Thumsová et al. 2022) or perhaps more lethal viruses are causing the declines in the UK and western Europe (Price et al. 2014, 2016). Or, as we discuss below, the trajectory may have more to do with connectivity among host populations allowing for demographic rescue.

One pattern that does seem consistent among longitudinal studies of host populations is that declines are often associated with repeated die-offs. In Northern Spain where whole communities of amphibians have declined in a national park, declines occurred in wetlands with recurrent ranavirus-associated mortality (Price et al. 2014). Single die-off events, while catastrophic for the fraction of the population in the affected site, might be akin to many other sources of mass mortality that occur commonly in the ecology of hosts (e.g., pond drying, hypoxia events), from which they are demographically capable of population recovery via sufficient recruitment (Petranka et al. 2007). Earl and Gray (2014) estimated that it would take 5 consecutive years of catastrophic mortality in wood frog tadpole populations for a robust isolated population to go extinct. Even with eastern box turtles in the eastern US and common frogs (*Rana temporaria*) and common toads (*Bufo bufo*) in the UK, where the adults rather than larvae or juveniles are impacted by ranavirus outbreaks, the greatest demographic impacts appear to be associated with repeated ranavirus epizootics (Belzer and Seibert 2011; Teacher et al. 2010; Campbell et al. 2018; Duffus et al. 2019). This recurrence then places a spotlight on how ranaviruses persist between epizootics, especially when they kill most of their host population (see Box 2). It should also encourage us to think about the ecology of ranaviruses at the landscape level.

Box 2 The Spark That Starts an Epizootic
Epizootics are presumably initiated when infection is (re)introduced into a susceptible population. Circumstantial evidence suggests that sublethally infected adult amphibians bring ranavirus back to ponds where the infection spreads among naive larvae, some of which then retain sublethal infections, restarting the cycle (Brunner et al. 2004; Crespi et al. 2015). However, we cannot rule out environmental persistence, at least over winters in colder environments (Brunner and Yarber 2018). Ranavirus persists in lesions of the swim bladder of largemouth bass for over a year after a mortality event (Hanson et al. 2001), which might be the source of recurrent epizootics.

When we broaden our view to multi-host communities, there are many more potential routes of introduction. Interspecific predation or necrophagy might play an important role in the introduction of ranavirus into terrestrial species and reptiles. For instance, CMTV was isolated from one of several water snakes in the genus *Natrix* found consuming dead newts (*Lissotriton boscai*) during a mass mortality event in Spain (Ayres 2012; Price et al. 2014). Circumstantial evidence suggests a series of ranavirus-related mortality events in eastern box turtles (*Terrapene carolina carolina*) were a result of some turtles consuming ranavirus-infected amphibian carcasses (Belzer and Seibert 2011). How ranavirus is introduced into fishes is, at least at this point, purely speculative.

5   Ranavirus Across the Landscape

Ranaviruses are presumably moved between habitat patches across a landscape with their hosts, at least in the case of amphibians (Brunner et al. 2004; Crespi et al. 2015). For instance, ranavirus was detected in 21 of 34 constructed vernal pools in Virginia, USA, within several years of their construction (Millikin et al. 2023) and as soon as a year after construction of an experimental array of pools in New York, USA (Youker-Smith et al. 2016), which suggests the virus was introduced by colonizing hosts (Box 2). It is also possible that ranaviruses could be translocated among sites by humans on recreational gear, footwear, or research equipment if not disinfected (Gray et al. 2018; Casais et al. 2019). At a larger scale, ranavirus appears to be invading Chile via the introduction of *Xenopus laevis*, although it has apparently not yet spilled over into native amphibians (Soto-Azat et al. 2016; Peñafiel-Ricaurte et al. 2023). The role of invasive species introducing or maintaining ranavirus in a landscape was highlighted in a recent study of the impacts of a landscape-scale effort to eradicate invasive American bullfrogs (*R. catesbeiana*) in the US Southwest (Hossack et al. 2023). It revealed that ranavirus and the fungal pathogen, *Batrachochytrium dendrobatidis*, disappeared from wetlands in which bullfrogs were eradicated, although the native amphibian populations did not increase in response (Hossack et al. 2023). The repeated occurrence of ranavirus outbreaks in

American bullfrog farms and naturalized populations around the world suggest a widespread and ongoing risk of introduction and spread into native taxa (Flechas et al. 2023; Mazzoni et al. 2009; Miller et al. 2007; Roh et al. 2022; Saucedo et al. 2019; Sriwanayos et al. 2020), although other farmed and invasive species may also play a similar role (e.g., Rivera et al. 2019). However, it is important to note that ranavirus can also be common across landscapes of native amphibians (Crespi et al. 2015; Gahl and Calhoun 2008; Hoverman et al. 2012b; Mosher et al. 2019; Patla et al. 2016; Tornabene et al. 2018; Vilaça et al. 2019a), even in low diversity communities (Bienentreu et al. 2022). Lastly, there are many studies finding no ranavirus among amphibian populations or communities (e.g., Hanlon et al. 2016; Martel et al. 2013), suggesting low prevalence or perhaps the absence of the virus. Carter (2018) found that ranavirus prevalence in adult amphibians declined to <1% following consecutive drought years at an ephemeral pond-breeding site in the southern Appalachian Mountains that previously experienced reoccurring die-offs of wood frog and marbled salamander (*Ambystoma opacum*) larvae due to ranavirus (Green et al. 2002). Thus, hydroperiod might play a role in ranavirus persistence at sites, where amplification by larvae facilitates transmission to adults and reservoir species. In summary, ranavirus prevalence among habitat patches for amphibians, or at least research sites, can be very common to apparently absent, and we simply do not know why.

There is a dearth of information about ranaviruses in reptiles across landscapes. Most reports of ranavirus in wild reptiles come from incidental observations (e.g., Alves de Matos et al. 2011; Farnsworth and Seigel 2013; Johnson et al. 2008; McKenzie et al. 2019; Winzeler et al. 2018) or moribund animals brought into veterinary facilities (e.g., Allender et al. 2006, 2011). In those few studies where reptile populations were surveyed for ranaviruses, estimates of prevalence vary quite a lot. For instance, ranavirus were very common in agamid lizards in several locations northern Queensland, Australia (Maclaine et al. 2020), and in squamates in Spain (von Essen et al. 2020), though the sample sizes in Spain were quite low (e.g., five of seven viperine snakes [*Natrix maura*] and a single Bocage's wall lizard [*Podarcis bocagei*]). Serological surveys of Australian crocodiles, snakes, lizards, and tortoises suggest that exposure to ranaviruses in northern Queensland is common (Ariel 1997). Most studies in wild chelonians in North America, however, suggest prevalence is very low outside of mortality events (Allender et al. 2009, 2011; Butterfield et al. 2019; Carstairs 2019; Currylow et al. 2014; Winzeler et al. 2018), although in at least one area in Virginia, USA, ranavirus is relatively common in eastern box turtles (*T. c. carolina*)—persisting over several years in the absence of apparent mortality—as well as in the eastern fence lizard (*Sceloporus undulatus*; Goodman et al. 2013; Goodman et al. 2018). As with amphibians, why ranavirus is common in some reptile populations and rare in others is unclear.

In fishes, most reports of ranavirus come from aquaculture and most originate with investigations of mortality events (e.g., Deng et al. 2011, 2020; Dong et al. 2011; George et al. 2015; Whittington et al. 1994). However, there are two systems in which we have broad spatial and temporal sampling: the Santee-Cooper ranavirus in the USA and epizootic hematopoietic necrosis virus (EHNV) in Australia.

The Santee-Cooper ranavirus causes infrequent die-offs in wild fish in the USA, particularly largemouth bass (*Micropterus salmoides*), especially impacting larger adults (Grizzle and Brunner 2003; Maceina and Grizzle 2006). However, mortality events may not have large impacts on the host populations, perhaps because impacts are focused on the larger adults (Grizzle and Brunner 2003; Neal et al. 2009). Moreover, LMBV can occur at moderate to high prevalence without apparent mortality (Grizzle and Brunner 2003; Leis et al. 2018). Indeed, it is widespread among the water bodies and rivers surveyed even with no history of mortality events (Goldberg et al. 2003; Groocock et al. 2008; Maceina and Grizzle 2006; Southard et al. 2009). Prevalence of LMBV and mortality are elevated among fish released after being caught due to increased transmission in live wells and associated stress (Leis et al. 2018; Schramm et al. 2006), but otherwise there are no obvious patterns to its distribution or the occurrence of LMBV-related mortality.

EHNV was isolated during a mortality event in redfin perch (*Perca fluviatilis*) in artificial impoundments in Victoria, Australia (Langdon et al. 1986). Over the following decades, the virus spread upstream in the river system in which it was first observed, but also into other river systems and lakes, including into farmed rainbow trout (*Oncorhynchus mykiss*; Langdon et al. 1988), suggesting anthropogenic assistance in its spread (Becker et al. 2019; Whittington et al. 1996, 2010). It was generally found at low prevalence outside of mortality events. In one well-studied lake, EHNV was apparently absent from adult redfin perch for 2 years in September and October, but appeared in the young of the year in December, although it is unknown whether the virus was simply at very low prevalence in the population, was in some natural reservoir, or was introduced (Becker et al. 2019). Surprisingly, the frequency of epizootics declined over time, and the virus, while notifiable, has not been detected in wild fish in over a decade (Becker et al. 2019). So, as with ranaviruses in amphibians, we are left with more questions than answers about the landscape epidemiology of ranaviruses in fish.

6 Conclusion

It has been 8 years since the publication of the first edition of this book and chapter and several decades since we collectively began researching the ecology of ranaviruses. Given this timespan and the burgeoning literature around ranaviruses, it might appear that most if not all the important questions have been answered. However, we hope this chapter has called attention to many important open questions. For instance, there remains a great deal to understand about linking environmental conditions (e.g., temperature, stressors, etc.) to processes (e.g., transmission, virulence), as opposed to outcomes (e.g., proportion dead).

Such studies will help more clearly elucidate epidemiological dynamics of ranaviruses under different environmental conditions and their significance to conservation.

There has also been a clear disconnect between our expectations of how ranavirus epizootics *should* work from a bottom-up perspective based on individual outcomes in laboratory studies and a top-down perspective based on observed population and landscape-level outcomes. The bottom-up perspective tends to predict epizootics that progress rapidly and cause nearly complete mortality, while the observations at larger scales often, though not always, suggest slower and more variability dynamics and infections without apparent mortality. Though we lack data to make strong conclusions in most cases, we are clearly missing important elements of how ranavirus ecology actually works.

Finally, and perhaps most importantly, we must also recognize that there are missing links in our understanding of the ranavirus life cycle. For instance, almost all our knowledge of ranaviruses in amphibians in nature come from aggregations of hosts in breeding ponds, but what happens outside of these easily observed times and places? What happens after an epizootic in pond-breeding amphibians when young of the year disperse? What happens in the interval between mass mortality events? Is there continued transmission at some low rate as hosts occasionally contact each other in the terrestrial environment? Are sublethal, inapparent infections sufficiently long-lasting that they do, as we suspect, maintain ranavirus in a population when transmission is rare or impossible? If so, what triggers the switch to an epizootic? How do ranaviruses move between populations? We have similar questions about how ranavirus epidemiology works in reptiles and fishes outside of sporadic mortality events. It is also well established that ranaviruses do occur in captive populations and the live animal trade (e.g., Cheng et al. 2014; Maclaine et al. 2020; Picco et al. 2010; Robert et al. 2007; Schloegel et al. 2009; Stöhr et al. 2013). But if ranaviruses are so easily transmitted and so frequently cause morbidity and mortality, why are they not more common?

Thus, there is a great deal of important work to do to understand ranavirus ecology, which we hope this chapter inspires researchers to undertake. Answering these and related questions will put us in a much stronger position not just to understand ranavirus ecology but to manage or even mitigate the impacts of ranavirus on their hosts. This is especially important in high-value economic settings and for conservation of threatened hosts.

Acknowledgments We extend our thanks to Jacob Kerby and Thomas Raffel for their careful reading and helpful suggestions and comments that improved this chapter. We thank the following organizations for providing funds to support Open Access publishing of the second edition: University of Tennessee's (UT) Open Publishing Support Fund, UT School of Natural Resources, UT Center for Wildlife Health, Washington State University Libraries, University of Mississippi Medical Center, Gordon State College, Association of Reptile and Amphibian Veterinarians, and the Global Ranavirus Consortium.

References

Abonyi F, Varga Á, Sellyei B, Eszterbauer E, Doszpoly A (2022) Juvenile Wels catfish (Silurus glanis) display age-related mortality to European Catfish Virus (ECV) under experimental conditions. Viruses 14:1832

Agha M, Price SJ, Nowakowski AJ, Augustine B, Todd BD (2017) Mass mortality of eastern box turtles with upper respiratory disease following atypical cold weather. Dis Aquat Org 124:91–100

Alfaia SR, Cândido M, Sousa RLMD, Harakava R, Cassiano LL, Martins AMCRPDF, Ferreira CM (2020) Experimental Frog Virus 3 infection using Brazilian strain: amphibians susceptibility. Braz J Vet Res Anim Sci 57:e169134

Allender MC, Fry MM, Irizarry AR, Craig L, Johnson AJ, Jones M (2006) Intracytoplasmic inclusions in circulating leukocytes from an eastern box turtle (*Terrapene carolina carolina*) with iridoviral infection. J Wildl Dis 42:677–684

Allender MC, Abd-Eldaim M, Kuhns A, Kennedy M (2009) Absence of Ranavirus and herpesvirus in a survey of two aquatic turtle species in Illinois. J Herpetol Med Surg 19:16–20

Allender MC, Abd-Eldaim M, Schumacher J, McRuer D, Christian LS, Kennedy M (2011) PCR prevalence of ranavirus in free-ranging eastern box turtles (*Terrapene carolina carolina*) at rehabilitation centers in three southeastern US States. J Wildl Dis 47:759–764

Allender MC, Mitchell MA, McRuer D, Christian S, Byrd J (2013a) Prevalence, clinical signs, and natural history characteristics of frog virus 3-like infections in eastern box turtles (*Terrapene carolina carolina*). Herpetol Conserv Biol 8:308–320

Allender MC, Mitchell MA, Torres T, Sekowska J, Driskell EA (2013b) Pathogenicity of frog virus 3-like virus in red-eared slider turtles (*Trachemys scripta elegans*) at two environmental temperatures. J Comp Pathol 149:356–367

Allender MC, Barthel AC, Rayl JM, Terio KA (2018) Experimental transmission of frog virus 3-like ranavirus in juvenile chelonians at two temperatures. J Wildl Dis 54:716–725

Altig R, Whiles MR, Taylor CL (2007) What do tadpoles really eat? Assessing the trophic status of an understudied and imperiled group of consumers in freshwater habitats. Freshw Biol 52:386–395

Altizer S, Ostfeld RS, Johnson PT, Kutz S, Harvell CD (2013) Climate change and infectious diseases: from evidence to a predictive framework. Science 341:514–519

Alves de Matos AP, Caeiro MF, Papp T, Matos BA, Correia AC, Marschang RE (2011) New viruses from Lacerta monticola (Serra da Estrela, Portugal): further evidence for a new group of nucleo-cytoplasmic large deoxyriboviruses. Microsc Microanal 17:101–108

Anand S, Krishan J, Sreekanth B, Mayya YS (2022) A comprehensive modelling approach to estimate the transmissibility of coronavirus and its variants from infected subjects in indoor environments. Sci Rep 12:14164

Araujo A, Kirschman L, Warne RW (2016) Behavioural phenotypes predict disease susceptibility and infectiousness. Biol Lett 12:20160480

Ariel E (1997) Pathology and serological aspects of Bohle iridovirus infections in six selected water-associated reptiles in North Queensland. Dissertation, James Cook University

Ariel E, Jensen BB (2009) Challenge studies of European stocks of redfin perch, *Perca fluviatilis* L., and rainbow trout, *Oncorhynchus mykiss* (Walbaum), with epizootic haematopoietic necrosis virus. J Fish Dis 32:1017–1025

Ariel E, Kielgast J, DVart HE, Larsen K, Tapiovaara H, Bang Jensen B, Holopainen R (2009a) Ranavirus in wild edible frogs *Pelophylax* kl. *esculentus* in Denmark. Dis Aquat Org 85:7–14

Ariel E, Nicolajsen N, Christophersen MB, Holopainen R, Tapiovaara H, Jensen BB (2009b) Propagation and isolation of ranaviruses in cell culture. Aquaculture 294:159–164

Ariel E, Wirth W, Burgess G, Scott J, Owens L (2015) Pathogenicity in six Australian reptile species following experimental inoculation with Bohle iridovirus. Dis Aquat Org 115:203–212

Ayres C (2012) Scavenging in the genus *Natrix*. Acta Herpetol 7:2012

Bang Jensen B, Ersboll AK, Ariel E (2009) Susceptibility of pike *Esox lucius* to a panel of *Ranavirus* isolates. Dis Aquat Org 83:169–179

Bang Jensen B, Reschova S, Cinkova K, Ariel E, Vesely T (2011a) Common carp (*Cyprinus carpio*) and goldfish (*Carassius auratus*) were not susceptible to challenge with ranavirus under certain challenge conditions. Bull Eur Assoc Fish Pathol 31:112–118

Bang Jensen B, Holopainen R, Tapiovaara H, Ariel E (2011b) Susceptibility of pike-perch *Sander lucioperca* to a panel of ranavirus isolates. Aquaculture 313:24–30

Bayley AE, Hill BJ, Feist SW (2013) Susceptibility of the European common frog *Rana temporaria* to a panel of ranavirus isolates from fish and amphibian hosts. Dis Aquat Org 103:171–183

Becker JA, Gilligan D, Asmus M, Tweedie A, Whittington RJ (2019) Geographic distribution of epizootic haematopoietic necrosis virus (EHNV) in freshwater fish in south eastern Australia: lost opportunity for a notifiable pathogen to expand its geographic range. Viruses 11:315

Belzer W, Seibert S (2011) A natural history of *Ranavirus* in an eastern box turtle population. Turtle Tortoise Newsl 15:18–25

Ben-Ami F, Ebert D, Regoes RR (2010) Pathogen dose infectivity curves as a method to analyze the distribution of host susceptibility: a quantitative assessment of maternal effects after food stress and pathogen exposure. Am Nat 175:106–115

Biek R, Funk WC, Maxell BA, Mills LS (2002) What is missing in amphibian decline research: insights from ecological sensitivity analysis. Conserv Biol 16:728–734

Bielby J, Price SJ, Monsalve-CarcaÑo C, Bosch J (2021) Host contribution to parasite persistence is consistent between parasites and over time, but varies spatially. Ecol Appl 31:e02256

Bienentreu J-F, Schock D, Greer AL, Lesbarreres D (2022) Ranavirus amplification in low-diversity amphibian communities. Front Vet Sci 9:755426

Billet LS, Wuerthner VP, Hua J, Relyea RA, Hoverman JT (2021) Population-level variation in infection outcomes not influenced by pesticide exposure in larval wood frogs (*Rana sylvatica*). Freshw Biol 66:1169–1181

Bliss CI (1934) The method of probits. Science 79:38–39

Boonthai T, Loch TP, Yamashita CJ, Smith GD, Winters AD, Kiupel M, Brenden TO, Faisal M (2018) Laboratory investigation into the role of largemouth bass virus (Ranavirus, Iridoviridae) in smallmouth bass mortality events in Pennsylvania rivers. BMC Vet Res 14:62

Brand MD, Hill RD, Brenes R, Chaney JC, Wilkes RP, Grayfer L, Miller DL, Gray MJ (2016) Water temperature affects susceptibility to ranavirus. EcoHealth 13:350–359

Brenes R, Gray MJ, Waltzek TB, Wilkes RP, Miller DL (2014a) Transmission of ranavirus between ectothermic vertebrate hosts. PLoS One 9:e92476

Brenes R, Miller DL, Waltzek TB, Wilkes RP, Tucker JL, Chaney JC, Hardman RH, Brand MD, Huether RR, Gray MJ (2014b) Susceptibility of fish and turtles to three ranaviruses isolated from different ectothermic vertebrate classes. J Aquat Anim Health 26:118–126

Brunner JL, Collins JP (2009) Testing assumptions of the trade-off theory of the evolution of parasite virulence. Evol Ecol Res 11:1169–1188

Brunner JL, Yarber CM (2018) Evaluating the importance of environmental persistence for *Ranavirus* transmission and epidemiology. In: Malmstrom C (ed) Environmental virology and virus ecology. Elsevier

Brunner JL, Schock DM, Collins JP, Davidson EW (2004) The role of an intraspecific reservoir in the persistence of a lethal ranavirus. Ecology 85:560–566

Brunner JL, Richards K, Collins JP (2005) Dose and host characteristics influence virulence of ranavirus infections. Oecologia 144:399–406

Brunner JL, Schock DM, Collins JP (2007) Transmission dynamics of the amphibian ranavirus *Ambystoma tigrinum* virus. Dis Aquat Org 77:87–95

Brunner JL, Storfer A, Gray MJ, Hoverman JT (2015) Ranavirus ecology and evolution: from epidemiology to extinction. In: Gray MJ, Chinchar VG (eds) Ranaviruses: lethal pathogens of ecothermic vertebrates. Springer International Publishing, Cham

Brunner JL, Beaty L, Guitard A, Russel D (2017) Heterogeneities in the infection process drive ranavirus transmission. Ecology 98:576–582

Brunner JL, Olson AD, Rice JG, Meiners SE, Le Sage MJ, Cundiff JA, Goldberg CS, Pessier A (2019) Ranavirus infection dynamics and shedding in American bullfrogs: consequences for spread and detection in trade. Dis Aquat Org 135:135–150

Brunner JL, Olson DH, Gray MJ, Miller DL, Duffus ALJ (2021) Global patterns of ranavirus detections. Facets 6:912–924

Butterfield MM, Davis DR, Madison JD, Kerby JL (2019) Surveillance of ranavirus in false map turtles (*Graptemys pseudogeographica*) along the lower Missouri River, USA. Herpetol Rev 50:76–78

Campbell LJ, Garner TWJ, Tessa G, Scheele BC, Griffiths AGF, Wilfert L, Harrison XA (2018) An emerging viral pathogen truncates population age structure in a European amphibian and may reduce population viability. PeerJ 6:e5949

Candido M, Tavares LS, Alencar ALF, Ferreira CM, Queiroz SRDA, Fernandes AM, Sousa RLMD (2019) Genome analysis of Ranavirus frog virus 3 isolated from American Bullfrog (Lithobates catesbeianus) in South America. Sci Rep 9:17135

Carey C (1993) Hypothesis concerning the causes of the disappearance of boreal toads from the mountains of Colorado. Conserv Biol 7:355–362

Carstairs SJ (2019) Evidence for low prevalence of ranaviruses in Ontario, Canada's freshwater turtle population. PeerJ 7:e6987

Carter ED (2018) Implications of drought and ranavirus on an amphibian community in the Great Smoky Mountains National Park. MS thesis, University of Tennessee, Knoxville

Casais R, Larrinaga AR, Dalton KP, Dominguez Lapido P, Marquez I, Becares E, Carter ED, Gray MJ, Miller DL, Balseiro A (2019) Water sports could contribute to the translocation of ranaviruses. Sci Rep 9:2340

Chao C-B, Pang VF (1997) An outbreak of an iridovirus-like infection in cultured grouper (*Epinephelus* spp.) in Taiwan. J Chin Soc Vet Sci 23:411–422

Cheng K, Jones ME, Jancovich JK, Burchell J, Schrenzel MD, Reavill DR, Imai DM, Urban A, Kirkendall M, Woods LW, Chinchar VG, Pessier AP (2014) Isolation of a Bohle-like iridovirus from boreal toads housed within a cosmopolitan aquarium collection. Dis Aquat Org 111:139–152

Chinchar VG, Hyatt AD, Miyazaki T, Williams T (2009) Family *Iridoviridae*: poor viral relations no longer. In: Van Etten JL (ed) Current topics in microbiology and immunology, vol 328: lesser known large dsDNA viruses. Springer, Berlin

Cohen JM, Venesky MD, Sauer EL, Civitello DJ, McMahon TA, Roznik EA, Rohr JR (2017) The thermal mismatch hypothesis explains host susceptibility to an emerging infectious disease. Ecol Lett 20:184–193

Cohen JM, McMahon TA, Ramsay C, Roznik EA, Sauer EL, Bessler S, Civitello DJ, Delius BK, Halstead N, Knutie SA, Nguyen KH, Ortega N, Sears B, Venesky MD, Young S, Rohr JR (2019) Impacts of thermal mismatches on chytrid fungus Batrachochytrium dendrobatidis prevalence are moderated by life stage, body size, elevation and latitude. Ecol Lett 22:817

Cozad RA, Norton TM, Aresco MJ, Allender MC, Hernandez SM (2020) Pathogen surveillance and detection of ranavirus (frog virus 3) in translocated gopher tortoises (Gopherus polyphemus). J Wildl Dis 56:679–683

Crespi EJ, Rissler LJ, Mattheus NM, Engbrecht K, Duncan SI, Seaborn T, Hall EM, Peterson JD, Brunner JL (2015) Geophysiology of wood frogs: landscape patterns of prevalence of disease and circulating hormone concentrations across the eastern range. Integr Comp Biol 55:602–617

Cullen BR, Owens L (2002) Experimental challenge and clinical cases of Bohle iridovirus (BIV) in native Australian anurans. Dis Aquat Org 49:83–92

Cullen CR, Owens L, Whittington RJ (1995) Experimental infection of Australian anurans (*Limnodynastes terraereginae* and *Litoria latopalmata*) with Bohle iridovirus. Dis Aquat Org 23:83–92

Cunningham AA, Langton TES, Bennet PM, Lewin JF, Drury SEN, Gough RE, MacGregor SK (1996) Pathological and microbiological findings from incidents of unusual mortality of the common frog (*Rana temporaria*). Philos Trans R Soc Lond Ser B Biol Sci 351:1539–1557

Currylow AF, Johnson AJ, Williams RN (2014) Evidence of ranavirus infections among sympatric larval amphibians and box turtles. J Herpetol 48:117–121

Cusaac JPW, Carter ED, Woodhams DC, Robert J, Spatz JA, Howard JL, Lillard C, Graham AW, Hill RD, Reinsch S, McGinnity D, Reeves B, Bemis D, Wilkes RP, Sutton WB, Waltzek TB, Hardman RH, Miller DL, Gray MJ (2021) Emerging pathogens and a current-use pesticide: potential impacts on eastern hellbenders. J Aquat Anim Health 33:24–32

Davis DR, Ferguson KJ, Schwarz MS, Kerby JL (2020) Effects of agricultural pollutants on stress hormones and viral infection in larval salamanders. Wetlands 40:577–586

De Jesús AF, Lawrence BP, Robert J (2017) Long term effects of carbaryl exposure on antiviral immune responses in *Xenopus laevis*. Chemosphere 170:169–175

DeBlieux TS, Hoverman JT (2019) Parasite-induced vulnerability to predation in larval anurans. Dis Aquat Org 135:241–250

DeBlieux TS, Hoverman JT (2022) Pathogens and predators: examining the separate and combined effects of natural enemies on assemblage structure. Oecologia 200:307

Degregorio BA, Hancock TE, Kurz DJ, Yue S (2011) How quickly are road-killed snakes scavenged? Implications for underestimates of road mortality. J N C Acad Sci 127:184–188

Deng GC, Li SJ, Xie J, Bai JJ, Chen KC, Ma DM, Jiang XY, Lao HH, Yu LY (2011) Characterization of a ranavirus isolated from cultured largemouth bass (*Micropterus salmoides*) in China. Aquaculture 312:198–204

Deng L, Geng Y, Zhao R, Gray MJ, Wang K, Ouyang P, Chen D, Huang X, Chen Z, Huang C, Zhong Z, Guo H, Fang J (2020) CMTV-like ranavirus infection associated with high mortality in captive catfish-like loach, Triplophysa siluroides, in China. Transbound Emerg Dis 67:1330–1335

Dhabhar FS (2002) Stress-induced augmentation of immune function: the role of stress hormones, leukocyte trafficking, and cytokines. Brain Behav Immun 16:785–798

Dhabhar FS (2014) Effects of stress on immune function: the good, the bad, and the beautiful. Immunol Res 58:193–210

Dong W, Zhang X, Yang C, An J, Qin J, Song F, Zeng W (2011) Iridovirus infection in Chinese giant salamanders, China, 2010. Emerg Infect Dis 17:2388–2389

Duffus ALJ, Pauli BD, Wozney K, Brunetti CR, Berrill M (2008) Frog virus 3-like infections in aquatic amphibian communities. J Wildl Dis 44:109–120

Duffus ALJ, Nichols RA, Garner TWJ (2013) Investigations into the life history stages of the common frog (*Rana temporaria*) affected by an amphibian ranavirus in the United Kingdom. Herpetol Rev 44:260–263

Duffus ALJ, Nichols RA, Garner TWJ (2014) Experimental evidence in support of single host maintenance of a multihost pathogen. Ecosphere 5:142

Duffus ALJ, Garner TWJ, Nichols RA, Standridge JP, Earl JE (2019) Modelling ranavirus transmission in populations of common frogs (Rana temporaria) in the United Kingdom. Viruses 11:556

Dwyer G, Elkinton JS, Buonaccorsi JP (1997) Host heterogeneity in susceptibility and disease dynamics: tests of a mathematical model. Am Nat 150:685–707

Earl JE, Gray MJ (2014) Introduction of ranavirus to isolated wood frog populations could cause local extinction. EcoHealth 11:581. https://doi.org/10.1007/s10393-014

Echaubard P, Little K, Pauli B, Lesbarrères D (2010) Context-dependent effects of ranaviral infection on northern leopard frog life history traits. PLoS One 5:e13723

Echaubard P, Leduc J, Pauli B, Chinchar VG, Robert J, Lesbarrères D (2014) Environmental dependency of amphibian-ranavirus genotypic interactions: evolutionary perspectives on infectious diseases. Evol Appl 7:723–733

Epstein B, Storfer A (2015) Comparative genomics of an emerging amphibian virus. Genes Genomes Genet 6:15–27

Farnsworth S, Seigel R (2013) Responses, movements, and survival of relocated box turtles during construction of the intercounty connector highway in Maryland. Transp Res Rec 2362:1–8

Flechas SV, Urbina J, Crawford AJ, Gutiérrez K, Corrales K, Castellanos LA, González MA, Cuervo AM, Catenazzi A (2023) First evidence of ranavirus in native and invasive amphibians in Colombia. Dis Aquat Org 153:51–58

Ford CE, Brookes LM, Skelly E, Sergeant C, Jordine T, Balloux F, Nichols RA, Garner TWJ (2022) Non-lethal detection of frog virus 3-like (RUK13) and common midwife toad virus-like (PDE18) ranaviruses in two UK-native amphibian species. Viruses 14:2635

Forson DD, Storfer A (2006) Atrazine increases ranavirus susceptibility in the tiger salamander, *Ambystoma tigrinum*. Ecol Appl 16:2325–2332

Forzán Gómez MDJ (2015) Pathogenesis of frog virus 3 (*Ranavirus* sp, *Iridoviridae*) in the wood frog, *Rana sylvatica* (*Lithobates sylvaticus*). Dissertation, University of Prince Edward Island

Furumoto WA, Mickey R (1967) A mathematical model for the infectivity-dilution curve of tobacco mosaic virus: theoretical considerations. Virology 32:216–223

Gahl MK, Calhoun AJK (2008) Landscape setting and risk of ranavirus mortality events. Biol Conserv 141:2679–2689

Gahl MK, Calhoun AJK (2010) The role of multiple stressors in ranavirus-caused amphibian mortalities in Acadia National Park wetlands. Can J Zool 88:108–121

Gallagher SJ, Tornabene BJ, DeBlieux TS, Pcchini KM, Chislock MF, Compton ZA, Eiler LK, Verble KM, Hoverman JT (2019) Healthy but smaller herds: predators reduce pathogen transmission in an amphibian assemblage. J Anim Ecol 88:1613–1624

George MR, John KR, Mansoor MM, Saravanakumar R, Sundar P, Pradeep V (2015) Isolation and characterization of a ranavirus from koi, *Cyprinus carpio* L., experiencing mass mortalities in India. J Fish Dis 38:389–403

Goldberg TL (2002) Largemouth bass virus: an emerging problem for warmwater fisheries? In: Philipp DP, Ridgway MS (eds) American Fisheries Society symposium 31, Bethesda, Maryland

Goldberg TL, Coleman DA, Grant EC, Inendino KR, Philipp DP (2003) Strain variation in an emerging iridovirus of warm-water fishes. J Virol 77:8812–8818

Goodman RM, Miller DL, Ararso YT (2013) Prevalence of ranavirus in Virginia turtles as detected by tail-clip sampling versus oral-cloacal swabbing. Northeast Nat 20:325–332

Goodman RM, Hargadon KM, Carter ED (2018) Detection of ranavirus in eastern fence lizards and eastern box turtles in Central Virginia. Northeast Nat 25:391–398

Goodman RM, Carter ED, Miller DL (2021) Influence of herbicide exposure and ranavirus infection on growth and survival of juvenile red-eared slider turtles (Trachemys scripta elegans). Viruses 13:1440

Granoff A, Came PE, Breeze DC (1966) Viruses and renal carcinoma of *Rana pipiens* I. The isolation and properties of virus from normal and tumor tissue. Virology 29:133–148

Grant EC, Philipp DP, Inendino KR, Goldberg TL (2003) Effects of temperature on the susceptibility of largemouth bass to largemouth bass virus. J Aquat Anim Health 15:215–220

Gray MJ, Spatz JA, Carter ED, Yarber CM, Wilkes RP, Miller DL (2018) Poor biosecurity could lead to disease outbreaks in animal populations. PLoS One 13:e0193243

Gray MJ et al (2024) Design and analysis of ranavirus studies: insights into planning surveillance, modeling host-pathogen dynamics, and performing risk analyses. In: Ranaviruses: emerging pathogens of ectothermic vertebrates. Springer Nature Switzerland, Cham

Green ED, Converse KA, Schrader AK (2002) Epizootiology of sixty-four amphibian morbidity and mortality events in the USA, 1996–2001. Ann N Y Acad Sci 969:323. https://doi.org/10.1111/j.1749-6632.2002.tb04400.x

Greer AL, Collins JP (2008) Habitat fragmentation as a result of biotic and abiotic factors controls pathogen transmission throughout a host population. J Anim Ecol 77:364–369

Greer AL, Brunner JL, Collins JP (2009) Spatial and temporal patterns of *Ambystoma tigrinum* virus (ATV) prevalence in tiger salamanders (*Ambystoma tigrinum nebulosum*). Dis Aquat Org 85:1–6

Grizzle JM, Brunner CJ (2003) Review of largemouth bass virus. Fisheries 28:10–14

Groocock GH, Grimmett SG, Getchell RG, Wooster GA, Bowser PR (2008) A survey to determine the presence and distribution of largemouth bass virus in wild freshwater bass in New York State. J Aquat Anim Health 20:158–164

Haislip NA, Gray MJ, Hoverman JT, Miller DL (2011) Development and disease: how susceptibility to an emerging pathogen changes through anuran development. PLoS One 6:e22307

Haislip NA, Hoverman JT, Miller DL, Gray MJ (2012) Natural stressors and disease risk: does the threat of predation increase amphibian susceptibility to ranavirus? Can J Zool 90:893–902

Hall EM, Crespi EJ, Goldberg CS, Brunner JL (2016) Evaluating environmental DNA-based quantification of ranavirus infection in wood frog populations. Mol Ecol Resour 16:423–433

Hall EM, Goldberg CS, Brunner JL, Crespi EJ (2018) Seasonal dynamics and potential drivers of ranavirus epidemics in wood frog populations. Oecologia 188:1253–1262

Hall EM, Brunner JL, Hutzenbiler B, Crespi EJ (2020) Salinity stress increases the severity of ranavirus epidemics in amphibian populations. Proc R Soc B Biol Sci 287:20200062

Halliday JE, Meredith AL, Knobel DL, Shaw DJ, Bronsvoort BM, Cleaveland S (2007) A framework for evaluating animals as sentinels for infectious disease surveillance. J R Soc Interface 4:973–984

Hanlon SM, Henson JR, Patillio B, Weeks D, Kerkby JL, Moore JE (2016) No occurrence of ranaviruses in reptiles from Wapanocca National Wildlife Refuge in Arkansas, USA. Herpetol Rev 47:606–607

Hanson LA, Petrie-Hanson L, Meals KO, Chinchar VG, Rudis M (2001) Persistence of largemouth bass virus infection in a northern Mississippi reservoir after a die-off. J Aquat Anim Health 13:27–34

Harp EM, Petranka JW (2006) Ranavirus in wood frogs (*Rana sylvatica*): potential sources of transmission within and between ponds. J Wildl Dis 42:307–318

Hartmann AM, Maddox ML, Ossiboff RJ, Longo AV (2022) Sustained ranavirus outbreak causes mass mortality and morbidity of imperiled amphibians in Florida. EcoHealth 19:8–14

Heisey DM, Joly DO, Messier F (2006) The fitting of general force-of-infection models to wildlife disease prevalence data. Ecology 87:2356–2365

Herczeg D, Ujszegi J, Kásler A, Holly D, Hettyey A (2021) Host-multiparasite interactions in amphibians: a review. Parasit Vectors 14:296

Herczeg D, Holly D, Kásler A, Bókony V, Papp T, Takács-Vágó H, Ujszegi J, Hettyey A (2023) Amphibian larvae benefit from a warm environment under simultaneous threat from chytridiomycosis and ranavirosis. Oikos 2023:e09953

Hickson RI, Roberts MG (2014) How population heterogeneity in susceptibility and infectivity influences epidemic dynamics. J Theor Biol 350:70–80

Hofer H, East ML (2012) Stress and immunosuppression as factors in the decline and extinction of wildlife populations: concepts, evidence and challenges. In: New directions in conservation medicine: applied cases of ecological health. Oxford University Press, New York, pp 109–133

Hossack BR, Oja EB, Owens AK, Hall D, Cobos C, Crawford CL, Goldberg CS, Hedwall S, Howell PE, Lemos-Espinal JA, MacVean SK, McCaffery M, Mosley C, Muths E, Sigafus BH, Sredl MJ, Rorabaugh JC (2023) Empirical evidence for effects of invasive American Bullfrogs on occurrence of native amphibians and emerging pathogens. Ecol Appl 33:e2785

Hoverman JT, Gray MJ, Miller DL (2010) Anuran susceptibilities to ranaviruses: role of species identity, exposure route, and a novel virus isolate. Dis Aquat Org 89:97–107

Hoverman JT, Gray MJ, Haislip NA, Miller DL (2011) Phylogeny, life history, and ecology contribute to differences in amphibian susceptibility to ranaviruses. EcoHealth 8:301–319

Hoverman JT, Gray MJ, Miller DL, Haislip NA (2012a) Widespread occurrence of ranavirus in pond-breeding amphibian populations. EcoHealth 9:36–48

Hoverman JT, Mihaljevic JR, Richgels KL, Kerby JL, Johnson PT (2012b) Widespread co-occurrence of virulent pathogens within California amphibian communities. EcoHealth 9:288–292

Hughey MC, Warne R, Dulmage A, Reeve RE, Curtis GH, Whitfield K, Schock DM, Crespi E (2023) Diet- and salinity-induced modifications of the gut microbiota are associated with differential physiological responses to ranavirus infection in *Rana sylvatica*. Philos Trans R Soc Lond B Biol Sci 378:20220121

Inendino KR, Grant EC, Philipp DP, Goldberg TL (2005) Effects of factors related to water quality and population density on the sensitivity of juvenile largemouth bass to mortality induced by viral infection. J Aquat Anim Health 17:304–314

Jancovich JK, Davidson EW, Morado JF, Jacobs BL, Collins JP (1997) Isolation of a lethal virus from the endangered tiger salamander *Ambystoma tigrinum stebbinsi*. Dis Aquat Org 31:161–167

Johnson AF, Brunner JL (2014) Persistence of an amphibian ranavirus in aquatic communities. Dis Aquat Org 111:129–138

Johnson AJ, Pessier AP, Wellehan JFX, Childress A, Norton TM, Stedman NL, Bloom DC, Belzer W, Titus VR, Wagner R, Brooks JW, Spratt J, Jacobson ER (2008) Ranavirus infection of free-ranging and captive box turtles and tortoises in the United States. J Wildl Dis 44:851–863

Kerby JL, Storfer A (2009) Combined effects of atrazine and chlorpyrifos on susceptibility of the tiger salamander to *Ambystoma tigrinum* virus. EcoHealth 6:91–98

Kerby JL, Hart AJ, Storfer A (2011) Combined effects of virus, pesticide, and predator cue on the larval tiger salamander (*Ambystoma tigrinum*). EcoHealth 8:46

Kiesecker JM (2011) Global stressors and the global decline of amphibians: tipping the stress immunocompetency axis. Ecol Res 26:897–908

Kimble SJ, Karna AK, Johnson AJ, Hoverman JT, Williams RN (2015) Mosquitoes as a potential vector of ranavirus transmission in terrestrial turtles. EcoHealth 12:334–338

Kimble SJA, Johnson AJ, Williams RN, Hoverman JT (2017) A severe ranavirus outbreak in captive, wild-caught box turtles. EcoHealth 14:810–815

Kumar R, Malagon DA, Carter ED, Miller DL, Bohanon ML, Cusaac JPW, Peterson AC, Gray MJ (2020) Experimental methodologies can affect pathogenicity of Batrachochytrium salamandrivorans infections. PLoS One 15(9):e0235370

Lafferty KD, Holt RD (2003) How should environmental stress affect the population dynamics of disease? Ecol Lett 6:654–664

Langdon JS, Humphrey JD (1987) Epizootic haematopoietic necrosis, a new viral disease in redfin perch, *Perca fluviatilis* L., in Australia. J Fish Dis 10:289–297

Langdon JS, Humphrey JD, Williams LM, Hyatt AD, Westbury HA (1986) First virus isolation from Australian fish: An iridovirus-like pathogen from redfin perch, *Perca fluviatilis* L. J Fish Dis 9:263–268

Langdon JS, Humphrey JD, Williams LM (1988) Outbreaks of an EHNV-like iridovirus in cultured rainbow trout, *Salmo gairdneri* Richardson, in Australia. J Fish Dis 11:93–96

Le Sage MJ, Towey BD, Brunner JL (2019) Do scavengers prevent or promote disease transmission? The effect of invertebrate scavenging on Ranavirus transmission. Funct Ecol 33:1342–1350

Le Sage EH, Diamond M, Crespi EJ (2022) Ranavirus infection-induced avoidance behaviour in wood frog juveniles: do amphibians socially distance. Biol Lett 18:20220359

Leimbach S, Schütze H, Bergmann SM (2014) Susceptibility of European sheatfish to a panel of ranaviruses. J Appl Ichthyol 30:93–101

Leis E, McCann R, Standish I, Bestul A, Odom T, Finnerty C, Bennie B (2018) Comparison of lethal and nonlethal sampling methods for the detection of largemouth bass virus (LMBV) from largemouth bass in the upper Mississippi River. J Aquat Anim Health 30:217

Liu F, Tian S, Feng Y, Qin Z, Geng Y, Ou Y, Chen D, Huang X, Guo H, Zuo Z, Deng H, Lai W (2023) First report of a CMTV-like ranavirus in farmed Percocypris pingi in China. Aquaculture 574:739701

Lloyd-Smith JO, Schreiber SJ, Kopp PE, Getz WM (2005) Superspreading and the effect of individual variation on disease emergence. Nature 438:355–359

Lopez Vargas NA, Adamovicz L, Willeford B, Allan BF, Allender MC (2020) Lack of molecular detection of frog virus 3-like ranavirus (FV3) in mosquitoes during natural outbreak and non-outbreak conditions. Facets 5:812–820

Maceina MJ, Grizzle JM (2006) The relation of largemouth bass virus to largemouth bass population metrics in five Alabama reservoirs. Trans Am Fish Soc 135:545–555

Maclaine A, Wirth WT, McKnight DT, Burgess GW, Ariel E (2020) Ranaviruses in captive and wild Australian lizards. Facets 5:758–768

Malagon DA, Melara LA, Prosper OF, Lenhart S, Carter ED, Fordyce JA, Peterson AC, Miller DL, Gray MJ (2020) Host density and habitat structure influence host contact rates and *Batrachochytrium salamandrivorans* transmission. Sci Rep 10:5584

Marschang RE et al (2024) Ranavirus distribution and host range. In: Ranaviruses: emerging pathogens of ectothermic vertebrates. Springer Nature Switzerland, Cham

Martel A, Adriaensen C, Sharifian-Fard M, Spitzen-van der Sluijs A, Louette G, Baert K, Crombaghs B, Dewulf J, Pasmans F (2013) The absence of zoonotic agents in invasive bullfrogs (*Lithobates catesbeianus*) in Belgium and the Netherlands. EcoHealth 10:344–347

Mazzoni R, de Mesquita AJ, Fleury LFF, de Brito WMED, Nunes IA, Robert J, Morales H, Coelho ASG, Barthasson DL, Galli L (2009) Mass mortality associated with a frog virus 3-like ranavirus infection in farmed tadpoles *Rana catesbeiana* from Brazil. Dis Aquat Org 86:181–191

McKenzie CM, Piczak ML, Snyman HN, Joseph T, Theijin C, Chow-Fraser P, Jardine CM (2019) First report of ranavirus mortality in a common snapping turtle Chelydra serpentina. Dis Aquat Org 132:221–227

Mihaljevic JR, Greer AL, Brunner JL (2019) Evaluating the within-host dynamics of *Ranavirus* infection with mechanistic disease models and experimental data. Viruses 11:396

Miller DL, Rajeev S, Gray MJ, Baldwin CA (2007) Frog virus 3 infection, cultured American bullfrogs. Emerg Infect Dis 13:342–343

Millikin AR, Davis DR, Brown DJ, Woodley SK, Coster S, Welsh A, Kerby JL, Anderson JT (2023) Prevalence of ranavirus in spotted salamander (*Ambystoma maculatum*) larvae from created vernal pools in West Virginia, USA. J Wildl Dis 59:24

Molnár PK, Sckrabulis JP, Altman KA, Raffel TR (2017) Thermal performance curves and the metabolic theory of ecology—a practical guide to models and experiments for parasitologists. J Parasitol 103:423–439

Mosher BA, Brand AB, Wiewel ANM, Miller DAW, Gray MJ, Miller DL, Grant EHC (2019) Estimating occurrence, prevalence, and detection of amphibian pathogens: insights from occupancy models. J Wildl Dis 53:563–575

Munro J, Bayley AE, McPherson NJ, Feist SW (2016) Survival of frog virus 3 in freshwater and sediment from an English lake. J Wildl Dis 52:138–142

Nazir J, Spengler M, Marschang RE (2012) Environmental persistence of amphibian and reptilian ranaviruses. Dis Aquat Org 98:177–184

Neal JW, Eggleton MA, Goodwin AE (2009) The effects of largemouth bass virus on a quality largemouth bass population in Arkansas. J Wildl Dis 45:766–771

North AC, Hodgson DJ, Price SJ, Griffiths AG (2015) Anthropogenic and ecological drivers of amphibian disease (ranavirosis). PLoS One 10:e0127037

O'Connor KM, Rittenhouse TAG, Brunner JL (2016) Ranavirus is common in wood frog (*Lithobates sylvaticus*) tadpoles throughout Connecticut. Herpetol Rev 47:394–397

Palomar G, Jakóbik J, Bosch J, Kolenda K, Kaczmarski M, Jośko P, Roces-Díaz JV, Stachyra P, Thumsová B, Zieliński P, Pabijan M (2021) Emerging infectious diseases of amphibians in Poland: distribution and environmental drivers. Dis Aquat Org 147:1–12

Parmenter RR, Lamarra VA (1991) Nutrient cycling in a freshwater marsh: the decomposition of fish and waterfowl carrion. Limnol Oceanogr 36:976–987

Parris MJ, Storfer A, Collins JP, Davidson EW (2005) Life-history responses to pathogens in tiger salamander (*Ambystoma tigrinum*) larvae. J Herpetol 39:366–372

Patla D, St-Hilaire S, Ray A, Hossack BR, Peterson CR (2016) Amphibian mortality events and ranavirus outbreaks in the Greater Yellowstone Ecosystem. Herpetol Rev 47:50–54

Peace A, O'Regan SM, Spatz JA, Reilly PN, Hill RD, Carter ED, Wilkes RP, Waltzek TB, Miller DL, Gray MJ (2019) A highly invasive chimeric ranavirus can decimate tadpole populations rapidly through multiple transmission pathways. Ecol Model 410:108777

Pearman PB, Garner TWJ (2005) Susceptibility of Italian agile frog populations to an emerging strain of *Ranavirus* parallels population genetic diversity. Ecol Lett 8:401–408

Pearman PB, Garner TWJ, Straub M, Greber UF (2004) Response of the Italian agile frog (Rana latastei) to a ranavirus, frog virus 3: a model for viral emergence in naive populations. J Wildl Dis 40:660–669

Peeler EJ, Feist SW (2011) Human intervention in freshwater ecosystems drives disease emergence. Freshw Biol 56:705–716

Peñafiel-Ricaurte A, Price SJ, Leung WTM, Alvarado-Rybak M, Espinoza-Zambrano A, Valdivia C, Cunningham AA, Azat C (2023) Is *Xenopus laevis* introduction linked with Ranavirus incursion, persistence and spread in Chile. PeerJ 11:e14497

Petranka JW, Harp EM, Holbrook CT, Hamel JA (2007) Long-term persistence of amphibian populations in a restored wetland complex. Biol Conserv 138:371–380

Picco AM, Karam AP, Collins JP (2010) Pathogen host switching in commercial trade with management recommendations. EcoHealth 7:252–256

Plumb JA, Zilberg D (1999) The lethal dose of largemouth bass virus in juvenile largemouth bass and the comparative susceptibility of striped bass. J Aquat Anim Health 11:246–252

Pochini KM, Hoverman JT (2017) Reciprocal effects of pesticides and pathogens on amphibian hosts: the importance of exposure order and timing. Environ Pollut 221:359–366

Price S, Garner T, Nichols R, Balloux F, Ayres C, Mora-Cabello de Alba A, Bosch J (2014) Collapse of amphibian communities due to an introduced *Ranavirus*. Curr Biol 24:2586–2591

Price SJ, Garner TWJ, Cunningham AA, Langton TES, Nichols RA (2016) Reconstructing the emergence of a lethal infectious disease of wildlife supports a key role for spread through translocations by humans. Proc R Soc B 283:20160952

Price SJ, Leung WTM, Owen CJ, Puschendorf R, Sergeant C, Cunningham AA, Balloux F, Garner TWJ, Nichols RA (2019) Effects of historic and projected climate change on the range and impacts of an emerging wildlife disease. Glob Chang Biol 25:2648–2660

Raffel TR, Rohr JR, Kiesecker JM, Hudson PJ (2006) Negative effects of changing temperature on amphibian immunity under field conditions. Funct Ecol 20:819–828

Ramsay C, Rohr JR (2023) Identity and density of parasite exposures alter the outcome of coinfections: implications for management. J Appl Ecol 60:205–214

Rayl JM, Allender MC (2020) Temperature affects the host hematological and cytokine response following experimental ranavirus infection in red-eared sliders (Trachemys scripta elegans). PLoS One 15:e0241414

Reeve BC, Crespi EJ, Whipps CM, Brunner JL (2013) Natural stressors and ranavirus susceptibility in larval wood frogs (*Rana sylvatica*). EcoHealth 10:190–200

Regester KJ, Whiles MR (2006) Decomposition rates of salamander (*Ambystoma maculatum*) life stages and associated energy and nutrient fluxes in ponds and adjacent forest in southern Illinois. Copeia 2006:640–649

Regoes RR, Hottinger JW, Sygnarski L, Ebert D (2003) The infection rate of *Daphnia magna* by *Pasteuria ramosa* conforms with the mass-action principle. Epidemiol Infect 131:957–966

Ribas L, Li MS, Doddington BJ, Robert J, Seidel JA, Kroll JS, Zimmerman LB, Grassly NC, Garner TW, Fisher MC (2009) Expression profiling the temperature-dependent amphibian response to infection by *Batrachochytrium dendrobatidis*. PLoS One 4:e8408

Richter SC, Drayer AN, Strong JR, Kross CS, Miller DL, Gray MJ (2013) High prevalence of ranavirus infection in permanent constructed wetlands in eastern Kentucky, USA. Herpetol Rev 44:464–466

Rivera B, Cook K, Andrews K, Atkinson MS, Savage AE (2019) Pathogen dynamics in an invasive frog compared to native species. EcoHealth 16:222–234

Robert J, Abramowitz L, Gantress J, Morales HD (2007) *Xenopus laevis*: a possible vector of ranavirus infection? J Wildl Dis 43:645–652

Robert J, McGuire CC, Kim F, Nagel SC, Price SJ, Lawrence BP, De Jesús Andino F (2018) Water contaminants associated with unconventional oil and gas extraction cause immunotoxicity to amphibian tadpoles. Toxicol Sci 166:39

Robert J, McGuire CC, Nagel S, Lawrence BP, Andino FJ (2019) Developmental exposure to chemicals associated with unconventional oil and gas extraction alters immune homeostasis and viral immunity of the amphibian Xenopus. Sci Total Environ 671:644–654

Roh N, Park J, Kim J, Kwon H, Park D (2022) Prevalence of ranavirus infection in three anuran species across South Korea. Viruses 14:1073

Rojas S, Richards K, Jancovich JK, Davidson EW (2005) Influence of temperature on ranavirus infection in larval salamanders, *Ambystoma tigrinum*. Dis Aquat Org 63:95–100

Rosa GM, Bosch J, Martel A, Pasmans F, Rebelo R, Griffiths RA, Garner TWJ (2019) Sex-biased disease dynamics increase extinction risk by impairing population recovery. Anim Conserv 22:579

Rowland FE, Schyling ES, Freidenburg LK, Urban MC, Richardson JL, Arietta AZA, Rodrigues SB, Rubinstein AD, Benard MF, Skelly DK (2022) Asynchrony, density dependence, and persistence in an amphibian. Ecology 103:e3696

Rudolf VH, Antonovics J (2007) Disease transmission by cannibalism: rare event or common occurrence. Proc R Soc B 274:1205–1210

Saucedo B, Serrano JM, Jacinto-Maldonado M, Leuven RSEW, Rocha García AA, Méndez Bernal A, Gröne A, van Beurden SJ, Escobedo-Bonilla CM (2019) Pathogen risk analysis for wild amphibian populations following the first report of a ranavirus outbreak in farmed American bullfrogs (Lithobates catesbeianus) from northern Mexico. Viruses 11:26

Schloegel LM, Picco AM, Kilpatrick AM, Davies AJ, Hyatt AD, Daszak P (2009) Magnitude of the U.S. trade in amphibians and presence of *Batrachochytrium dendrobatidis* and ranavirus infection in imported North American bullfrogs (*Rana catesbeiana*). Biol Conserv 142:1420–1426

Schramm HL, Walters AR, Grizzle JM, Beck BH, Hanson LA, Rees SB (2006) Effects of live-well conditions on mortality and largemouth bass virus prevalence in largemouth bass caught during summer tournaments. N Am J Fish Manag 26:812–825

Sifkarovski J, Grayfer L, De Jesus Andino F, Lawrence BP, Robert J (2014) Negative effects of low dose atrazine exposure on the development of effective immunity to FV3 in Xenopus laevis. Dev Comp Immunol 47:52–58

Smalling KL, Mosher BA, Iwanowicz LR, Loftin KA, Boehlke A, Hladik ML, Muletz-Wolz CR, Córtes-Rodríguez N, Femmer R, Campbell Grant EH (2022) Site- and individual-level contaminations affect infection prevalence of an emerging infectious disease of amphibians. Environ Toxicol Chem 41:781–791

Snyder PW, Ramsay CT, Harjoe CC, Khazan ES, Briggs CJ, Hoverman JT, Johnson PTJ, Preston D, Rohr JR, Blaustein AR (2023) Experimental evidence that host species composition alters host-pathogen dynamics in a ranavirus-amphibian assemblage. Ecology 104:e3885

Soto-Azat C, Peñafiel-Ricaurte A, Price SJ, Sallaberry-Pincheira N, García MP, Alvarado-Rybak M, Cunningham AA (2016) *Xenopus laevis* and emerging amphibian pathogens in Chile. EcoHealth 13:775–783

Southard GM, Fries LT, Terre DR (2009) Largemouth bass virus in Texas: distribution and management issues. J Aquat Anim Health 21:36–42

Speare R, Smith JR (1992) An iridovirus-like agent isolated from the ornate burrowing frog *Limnodynastes ornatus* in northern Australia. Dis Aquat Org 14:51–57

Sriwanayos P, Subramaniam K, Stilwell NK, Imnoi K, Popov VL, Kanchanakhan S, Polchana J, Waltzek TB (2020) Phylogenomic characterization of ranaviruses isolated from cultured fish and amphibians in Thailand. Facets 5:963–979

St-Amour V, Wong WM, Garner TW, Lesbarrères D (2008) Anthropogenic influence on prevalence of 2 amphibian pathogens. Emerg Infect Dis 14:1175–1176

Stilwell NK, Frasca S, Farina LL, Subramaniam K, Imnoi K, Viadanna PH, Hopper L, Powell J, Colee J, Waltzek TB (2022) Effect of water temperature on frog virus 3 disease in hatchery-reared pallid sturgeon Scaphirhynchus albus. Dis Aquat Org 148:73–86

Stöhr AC, Blahak S, Heckers KO, Wiechert J, Behncke H, Mathes K, Günther P, Zwart P, Ball I, Rüschoff B (2013) Ranavirus infections associated with skin lesions in lizards. Vet Res 44:1–10

Teacher AGF, Cunningham AA, Garner TWJ (2010) Assessing the long-term impact of ranavirus infection in wild common frog populations. Anim Conserv 13:514–522

Thumsová B, Price SJ, González-Cascón V, Vörös J, Martínez-Silvestre A, Rosa GM, Machordom A, Bosch J (2022) Climate warming triggers the emergence of native viruses in Iberian amphibians. Iscience 25:105541

Todd-Thompson M (2010) Seasonality, variation in species prevalence, and localized disease for *Ranavirus* in Cades Cove (Great Smoky Mountains National Park) amphibians. MS thesis, University of Tennessee, Knoxville

Tornabene BJ, Blaustein AR, Briggs CJ, Calhoun DM, Johnson PTJ, McDevitt-Galles T, Rohr JR, Hoverman JT (2018) The influence of landscape and environmental factors on ranavirus epidemiology in a California amphibian assemblage. Freshw Biol 63:639–651

Vilaça ST, Grant SA, Beaty L, Brunetti CR, Congram M, Murray DL, Wilson CC, Kyle CJ (2019a) Detection of spatiotemporal variation in ranavirus distribution using eDNA. Environ DNA, vol 2, p 210

Vilaça ST, Bienentreu JF, Brunetti CR, Lesbarrères D, Murray DL, Kyle CJ (2019b) Frog virus 3 genomes reveal prevalent recombination between ranavirus lineages and their origins in Canada. J Virol 93:10–128

von Essen M, Leung WTM, Bosch J, Pooley S, Ayres C, Price SJ (2020) High pathogen prevalence in an amphibian and reptile assemblage at a site with risk factors for dispersal in Galicia, Spain. PLoS One 15:e0236803

Warne RW, Crespi EJ, Brunner JL (2011) Escape from the pond: stress and developmental responses to ranavirus infection in wood frog tadpoles. Funct Ecol 25:139–146

Wells KD (2007) The ecology and behavior of amphibians. University of Chicago Press

Wheelwright NT, Gray MJ, Hill RD, Miller DL (2014) Sudden mass die-off of a large population of wood frog (*Lithobates sylvaticus*) tadpoles in Maine, USA, likely due to ranavirus. Herpetol Rev 45:240–242

Whiles MR, Lips KR, Pringle CM, Kilham SS, Bixby RJ, Brenes R, Connelly S, Colon-Gaud JC, Hunte-Brown M, Huryn AD, Montgomery C, Peterson S (2006) The effects of amphibian population declines on the structure and function of neotropical stream ecosystems. Front Ecol Environ 4:27–34

Whittington RJ, Reddacliff GL (1995) Influence of environmental temperature on experimental infection of redfin perch (*Perca fluviatilis*) and rainbow trout (*Oncorhynchus mykiss*) with epizootic haematopoietic necrosis virus, an Australian iridovirus. Aust Vet J 72:421–424

Whittington RJ, Philbey A, Reddacliff GL, Macgown AR (1994) Epidemiology of epizootic hematopoietic necrosis virus (EHNV) infection in farmed rainbow trout, *Oncorhynchus mykiss* (Walbaum): findings based on virus isolation, antigen capture ELISA and serology. J Fish Dis 17:205–218

Whittington RJ, Kearns C, Hyatt AD, Hengstberger S, Rutzou T (1996) Spread of epizootic haematopoietic necrosis virus (EHNV) in redfin perch (*Perca fluviatilis*) in southern Australia. Aust Vet J 73:112–114

Whittington RJ, Becker JA, Dennis MM (2010) Iridovirus infections in finfish – critical review with emphasis on ranaviruses. J Fish Dis 33:95–122

Wilber MQ, Langwig KE, Kilpatrick AM, McCallum HI, Briggs CJ (2016) Integral Projection Models for host-parasite systems with an application to amphibian chytrid fungus. Methods Ecol Evol 7:1182. https://doi.org/10.1111/2041-210X.12561

Wilber MQ, Knapp RA, Toothman M, Briggs CJ (2017) Resistance, tolerance and environmental transmission dynamics determine host extinction risk in a load-dependent amphibian disease. Ecol Lett 20:1169–1181

Williams T, Barbosa-Solomieu V, Chinchar VG (2005) A decade of advances in iridovirus research. Adv Virus Res 65:173–248

Winzeler ME, Haskins DL, Lance SL, Tuberville TD (2018) Survey of aquatic turtles on the Savannah River Site, South Carolina, for prevalence of ranavirus. J Wildl Dis 54:138–141

Wirth W, Ariel E (2020) Temperature-dependent infection of freshwater turtle hatchlings, Emydura macquarii krefftii, inoculated with a ranavirus isolate (Bohle iridovirus, Iridoviridae). Facets 5:821–830

Wirth W, Schwarzkopf L, Skerratt LF, Tzamouzaki A, Ariel E (2019) Dose-dependent morbidity of freshwater turtle hatchlings, Emydura macquarii krefftii, inoculated with Ranavirus isolate (Bohle iridovirus, Iridoviridae). J Gen Virol:jgv001324

Wuerthner VP, Hua J, Hoverman JT (2017) The benefits of coinfection: trematodes alter disease outcomes associated with virus infection. J Anim Ecol 86:921–931

Youker-Smith TE, Whipps CM, Ryan SJ (2016) Detection of an FV3-like ranavirus in wood frogs (*Lithobates sylvaticus*) and green frogs (*Lithobates clamitans*) in a constructed vernal pool network in central New York State. Herpetol Rev 47:595–598

Youker-Smith TE, Boersch-Supan PH, Whipps CM, Ryan SJ (2018) Environmental drivers of ranavirus in free-living amphibians in constructed ponds. EcoHealth 15:608–618

Pathology and Diagnostics

Debra L. Miller, Allan P. Pessier, Paul Hick, Richard J. Whittington, and María J. Forzán

1 Ranaviral Disease

1.1 Introduction

Ranaviruses were detected and diagnosed as disease agents in amphibians in the 1960s, fish in the 1970s, and reptiles in the 1980s (Marschang et al. 2024). Since these initial cases, ranaviruses have been linked to numerous epizootic mortality events in these three classes of vertebrates (Marschang et al. 2024). Although long-term population data are generally lacking, there is empirical evidence that ranaviruses can cause population declines in amphibians (Teacher et al. 2010; Beebee 2012; Price et al. 2014), the nearly complete loss of entire age classes of fish and

D. L. Miller (✉)
One Health Initiative, Center for Wildlife Health and Department of Biomedical and Diagnostic Sciences, University of Tennessee, Knoxville, TN, USA
e-mail: dmille42@utk.edu

A. P. Pessier
College of Veterinary Medicine, Washington State University, Pullman, WA, USA
e-mail: apessier@wsu.edu

P. Hick
Elizabeth Macarthur Agricultural Institute, Department of Primary Industries, Menangle, NSW, Australia
e-mail: paul.hick@dpi.nsw.gov.au

R. J. Whittington
Sydney School of Veterinary Science, Faculty of Science, University of Sydney, Sydney, NSW, Australia
e-mail: richard.whittington@sydney.edu.au

M. J. Forzán
Department of Veterinary Sciences, University of Wyoming, Laramie, WY, USA
e-mail: mforzan@uwyo.edu

© The Author(s) 2025
M. J. Gray, V. G. Chinchar (eds.), *Ranaviruses*,
https://doi.org/10.1007/978-3-031-64973-8_8

amphibians due to ranavirus outbreaks have been reported (Petranka et al. 2003; Todd-Thompson 2010; Waltzek et al. 2014; Wheelwright et al. 2014), and mathematical models suggest that the introduction of ranaviruses to an isolated population of a highly susceptible species could result in extirpation within 5 years (Earl and Gray 2014). We also know that some species of global conservation concern are highly susceptible (Geng et al. 2010; Sutton et al. 2014a; Waltzek et al. 2014).

Additionally, there is a growing impact of ranaviruses in farmed fish and amphibians (Mazzoni et al. 2009; Waltzek et al. 2014; Wei et al. 2019; Zhao et al. 2020) and on recreational fisheries (Grizzle and Brunner 2003). Given these concerns, all species of ranaviruses are listed as notifiable agents by the World Organization for Animal Health (WOAH, formerly OIE, WOAH 2023) when they cause morbidity or mortality in amphibians.

Despite the global awareness of the existence of ranaviruses, their distribution and their effects on host populations and international commerce remain poorly understood. There is limited information on the mechanisms that affect host–ranavirus interactions and factors that lead to mortality events (Gray et al. 2009). The epidemiology of ranaviruses in different species and geographic contexts requires further assessment because there is evidence that some ranaviruses (e.g., epizootic hematopoietic necrosis virus, EHNV) may establish less virulent relationships with host populations over time and may not necessarily spread in a catastrophic manner (Becker et al. 2019). Various field studies and controlled experiments are expanding our knowledge of ranaviruses (Marschang et al. 2024; Brunner et al. 2024; Jancovich et al. 2024); however, more research is needed. Properly designed studies (Gray et al. 2024) are a first step to investigating hypotheses associated with ranavirus emergence. Use of appropriate diagnostic techniques is key to identifying ranavirus infections and determining the effects of infection on host species, including the detection of subclinical infections which can be a source of pathogen spread. Understanding the differential diagnoses for ranaviral disease is important, and combining infection data with pathological and environmental information is essential to confirm that ranavirus is the etiologic disease agent.

In this updated chapter, we begin with an overview of the gross and microscopic lesions associated with ranaviral disease followed by a discussion of the current diagnostic tests used by research and veterinary diagnostic laboratories worldwide. We point out the limitations of certain diagnostic techniques and identify needed areas of improvement. Finally, we briefly discuss research into treatment and vaccine development for ranaviruses.

1.2 *Ranaviral Disease Pathology*

1.2.1 **Field and Clinical Findings**

In amphibians, outbreaks of ranaviral disease are most often observed in larvae and recently metamorphosed animals (Green et al. 2002; Docherty et al. 2003; Balseiro et al. 2009, 2010; Forzán et al. 2019; Hartmann et al. 2022); however, outbreaks that include adult animals are increasingly recognized (Cunningham et al. 2007; Cheng et al. 2014; Hartmann et al. 2022). Earl and Gray (2014) demonstrated that ranavirus-associated mortality of larvae or metamorphs was sufficient to cause population declines in highly susceptible species. Researchers have demonstrated a strong correlation between infection prevalence and mortality in laboratory experiments with frog virus 3 (FV3) (see Haislip et al. 2011; Hoverman et al. 2011; Brenes et al. 2014a); thus, high infection prevalence during field surveillance may be an indicator of an impending die-off (Gray et al. 2024).

Mortality events often present as sudden and massive deaths across multiple species (Todd-Thompson 2010; Wheelwright et al. 2014). Deaths may continue for weeks, with later deaths due to individuals succumbing to secondary bacterial or fungal infections (Jancovich et al. 1997; Cunningham et al. 2007; Miller et al. 2008; Cheng et al. 2014). Field signs vary but lethargy is commonly reported in all classes (Table 1). Chelonians can have additional signs of respiratory distress (Ruder et al. 2010; Farnsworth and Seigel 2013; Wirth et al. 2019). Experimental infection with Bohle iridovirus (BIV) in juvenile eastern water dragons *(Intellagama lesueurii lesueurii)* described anorexia, loss of balance, hemorrhages, and epidermal ulcerations (Maclaine et al. 2018). Prior to death, diseased larval and adult amphibians and fish may exhibit erratic swimming, loss of buoyancy, and loss of righting reflex (Mao et al. 1999; Bollinger et al. 1999; Zilberg et al. 2000; Geng et al. 2010; Miller et al. 2011). Fatally infected adult amphibians often develop oral hemorrhage and increased skin shedding with decreased dermatophagy (Forzán et al. 2015). Recently, a vestibular syndrome has been observed in cultured adult bullfrogs (*Lithobates catesbeianus*) in Brazil (R. Mazzoni, CRMV-GO, Brazil, personal communication) and may explain these changes in coordination, although histologic examination of infected animals will be needed if causality is to be proposed. In an aquaculture setting, the disease often presents as an acute mass mortality event. Affected grouper (*Epinephelus* spp.) infected with Singapore grouper iridovirus (SGIV) have reduced appetite and activity (Xiao et al. 2019) typical of "sleepy grouper disease" (Chua et al. 1994), and skin lesions sometimes precede death in farmed largemouth bass (*Micropterus salmoides*) (Deng et al. 2011).

1.2.2   Gross Pathology

The appearance of disease in individual animals reflects the systemic distribution of the virus and associated host response. Clinical disease is typically acute and can affect a high proportion of the population. In wild populations, the acute course of disease and rapid mortality might prevent detection of the disease event (Gray et al. 2024), or outbreaks might present as a large number of dead individuals. When sampling during an outbreak of mortalities, clinically affected individuals are preferentially collected because they are most likely to yield a diagnosis. Affected individuals present with skin and oral hemorrhages, edema, and necrosis as the most common gross lesions; however, the presentation of these changes can vary depending on the species affected and whether exposure to environmental stressors or other pathogens also occurs (Table 1).

Table 1 Examples of clinical signs and gross changes that can be observed in individuals with natural or experimentally induced ranaviral disease

Class	Lesion
Amphibian larvae	Loss of buoyancy; erratic swimming; anorexia; swelling (edema) of the body, head, legs, and internal soft tissues; external hemorrhages (especially around the vent, periocular, gular region, legs); occasional internal hemorrhages (especially pronephros, liver, spleen)
Anuran adults	Lethargy; anorexia with decreased dermatophagy; loss of buoyancy and erratic swimming (aquatic species); swelling (edema) of legs, feet, body, and internal soft tissues; skin ulcers; dermal, oral, and internal hemorrhages; friable (necrotic) organs
Caudate adults	Lethargy; anorexia; loss of buoyancy and erratic swimming (aquatic species); hemorrhages (especially on tail and plantar surfaces of feet); swelling (edema); skin ulcers; internal hemorrhages; friable (necrotic) organs; necrosis of extremities (Chinese giant salamanders)
Fish	Loss of buoyancy; erratic swimming; anorexia; red swollen gills; hemorrhages (especially periocular, fat bodies, swim bladder); overinflated swim bladder; friable (necrotic) organs; multiple pale foci in liver; ulcerative skin lesions; inappetence
Chelonians	Respiratory difficulty; anorexia; oral and esophageal hemorrhage; oral and esophageal necrosis; oral and esophageal necrotic pseudomembrane formation; swelling (edema or rarely necrosis) of head, neck, legs, internal soft tissues, periocular; skin ulcers; friable (necrotic) organs; hemorrhages (especially internal)
Lizards	Lethargy; anorexia; oral necrotic plaques; skin ulcers; friable (necrotic) organs; occasional internal hemorrhages; and edema
Snakes	Lethargy; anorexia; oral and nasal ulcers

Amphibians

Hemorrhages and erythema are common in the skin of anurans and caudates (Fig. 1a, b). Hemorrhages are most often present on the ventral surfaces near the vent, rear legs, and gular regions, but can also be observed around the eyes, ear drum, tongue, tail, and feet (Balseiro et al. 2009; Cheng et al. 2014; Cunningham et al. 2007; Docherty et al. 2003; Geng et al. 2010; Jerrett et al. 2015; Kik et al. 2011; Meng et al. 2014; Sutton et al. 2014a; Forzán et al. 2017). Raised skin plaques or polyps have been described in tiger salamanders (*Ambystoma tigrinum*) and Chinese giant salamanders (Jancovich et al. 1997; Bollinger et al. 1999; Geng et al. 2010). Other findings in the skin of anurans and caudates can include ulceration or rough discolored gray areas (e.g., Bollinger et al. 1999; Cunningham et al. 2007; Kik et al. 2011). Swelling of the legs, body, and head from accumulation of fluid (i.e., edema) within the tissues, lymph sacs, and body cavity is commonly seen in amphibians and is especially evident in larvae (e.g., see Wolf et al. 1968; Miller et al. 2011; Meng et al. 2014).

Internally, hemorrhage and necrosis are common findings, especially in the gastrointestinal wall, spleen, pronephros and mesonephros (kidney), and liver (Fig. 1c, d). The cause of cell death may be associated with apoptosis or virus replication (Grayfer et al. 2024). Necrosis may present as generalized friable organs or as discrete pale foci scattered throughout an organ. Splenomegaly and hepatomegaly have

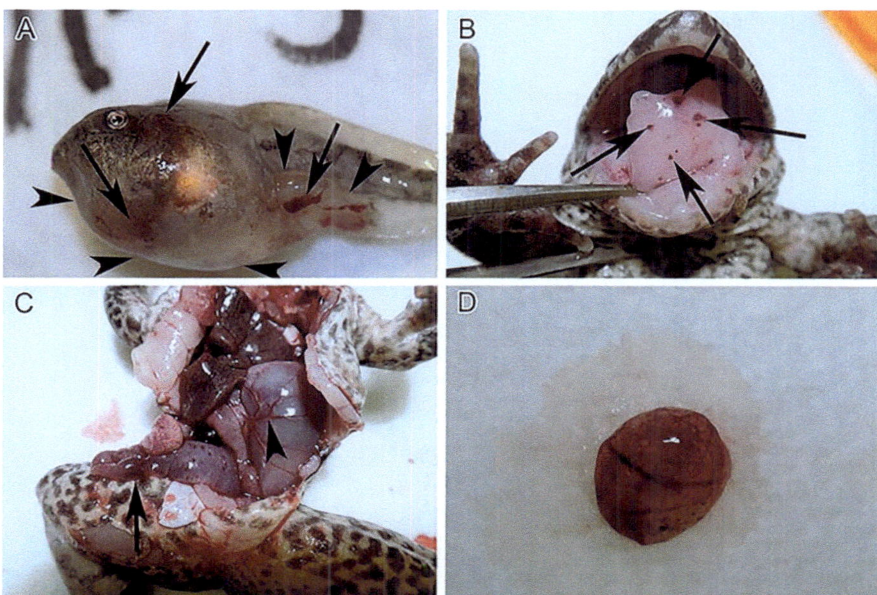

Fig. 1 Gross lesions seen in amphibians with ranaviral disease. (**a**) Hemorrhages (*arrows*) and edema (*arrowheads*) in a wood frog (*Lithobates sylvaticus*) tadpole experimentally challenged with an FV3 ranavirus. (**b**) Tongue hemorrhages (*arrows*) in a dusky gopher frog (*Lithobates sevosus*) experimentally challenged with an FV3 ranavirus. (**c**) Intestinal hemorrhage (*arrow*) and congested blood vessels (*arrowhead*) in a dusky gopher frog experimentally challenged with an FV3 ranavirus. (**d**) Friable and hemorrhagic spleen of a dusky gopher frog experimentally challenged with an FV3 ranavirus

also been reported (Kik et al. 2011) and may be related to congestion and hemorrhage or multifocal necrosis of lymphoid tissue (Forzán et al. 2017). Intestinal hemorrhage has been seen in mortality events and in experimentally challenged anurans and caudates (Bollinger et al. 1999; Geng et al. 2010; Cheng et al. 2014; Meng et al. 2014; Forzán et al. 2017). In Brazil, hemorrhage and necrosis have been seen in the inner ear (vestibular region) of ranavirus-positive bullfrogs displaying vestibular syndrome (R. Mazzoni, CRMV-GO, personal communication).

Fish

In fish, multifocal cutaneous hemorrhages (Fig. 2a; Waltzek et al. 2014; Deng et al. 2020; Stillwell et al. 2022; Liu et al. 2023) and red swollen gills (Mao et al. 1999) are seen on external examination. Internally, hemorrhages may occur in any organ including fat and swim bladder, and organs may be friable (Fig. 2b, Waltzek et al. 2014; Zilberg et al. 2000; Liu et al. 2023). Reddacliff and Whittington (1996) provided detailed descriptions of lesions due to EHNV in redfin perch (*Perca fluviatilis*) and rainbow trout (*Oncorhynchus mykiss*). Sick fish were dark, stopped eating, and sometimes were ataxic. Gross lesions included a swollen abdomen with swelling of the spleen and kidney and multiple pale foci that were sometimes present in

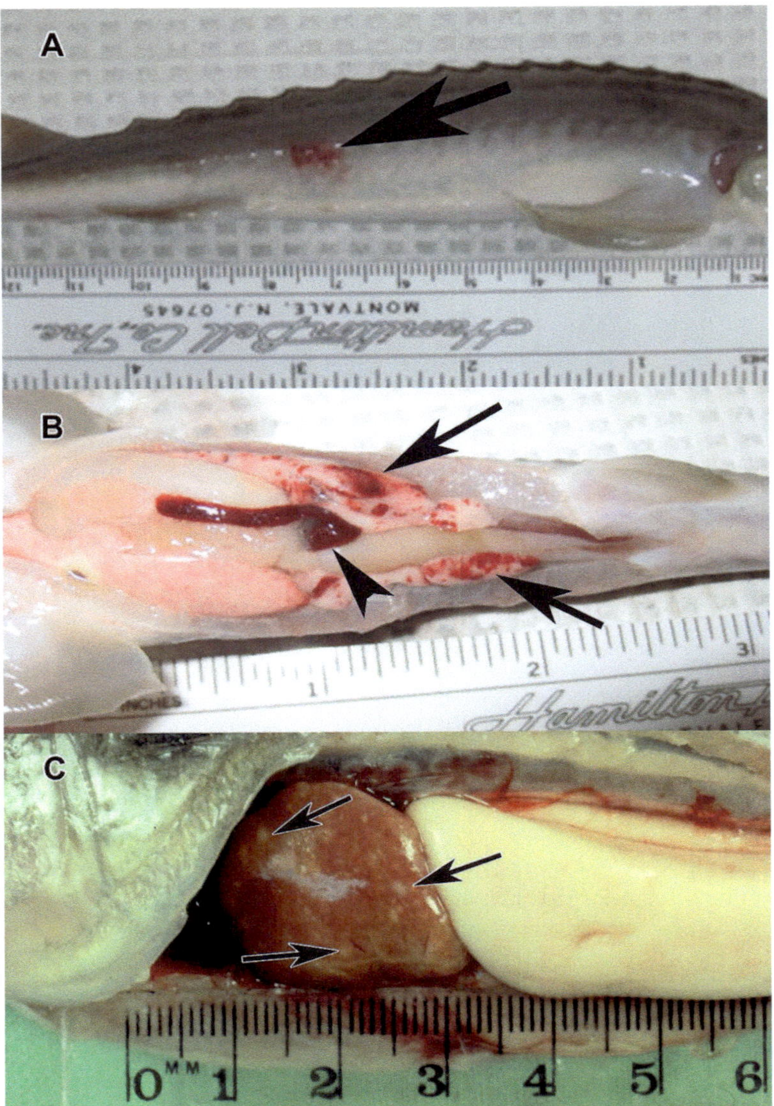

Fig. 2 Gross lesions seen in fish with ranaviral disease. (**a**) Pallid sturgeon (*Scaphirhynchus albus*) with cutaneous hemorrhage (*arrow*) due to an FV3 ranavirus. Photo by Thomas B Waltzek, University of Florida. (**b**) Hemorrhage in the fat (*arrows*) and spleen (*arrowhead*) of a pallid sturgeon with FV3 ranavirus. Photo by Thomas B Waltzek, Washington State University. (**c**) Multifocal hepatic necrosis evidenced by areas of pale discoloration in the liver (*arrows*) and ecchymotic hemorrhage in the retroperitoneum of an adult redfin perch (*Perca fluviatilis*) infected with EHNV

the liver. Over-inflation, hyperemia, or yellow exudate affecting the swim bladder are described in largemouth bass virus (LMBV, species *Santee Cooper ranavirus*) infections (Plumb et al. 1999; Hanson et al. 2001; Grizzle and Brunner 2003). Liver pallor, hyperemia of skin, and peritonitis were reported in largemouth bass (*Micropterus salmoides*) experimentally challenged (IP route) with Santee-Cooper ranavirus (SCRV, Zilberg et al. 2000; Grant et al. 2003). Within this same species, skin ulceration and splenomegaly were reported in China (Deng et al. 2011), and in experimental infections there was necrosis of spleen and liver (Xu et al. 2023). Hybrid mandarin fish (*Siniperca scherzeri* x *Siniperca chuatsi*) in China had marked ascites and exophthalmia (Dong et al. 2017), while Barcoo grunter (*Scortum barcoo*) in Thailand had ulceration and necrosis of the kidney, spleen, and gill (Kayansamruaj et al. 2017) as a result of SCRV infection. A common midwife toad virus (CMTV)-like ranavirus caused skin ulcerations and hemorrhages as well as swollen spleens and kidneys in farmed *Percocypris pingi* (Liu et al. 2023). In contrast, no lesions were described in the first reports of ornamental fish with doctor fish virus and guppy virus 6 infections (Hedrick and McDowell 1995).

Reptiles
Chelonians can develop periocular swelling, ulceration, and necrosis of the oral cavity, swollen head and extremities, ocular and nasal discharges, and occasional skin ulcerations and hepatic necrosis (Fig. 3, e.g., Johnson et al. 2008; Ruder et al. 2010; Wirth et al. 2023). Multifocal to confluent tan friable areas (necrosis) may be seen internally (especially in the GI and respiratory tracts). Occasionally, hemorrhages of the GI tract can be seen and may be the only change observed in aquatic turtles (Fig. 3d, D. Miller, personal observation); however, it often is unclear if this is due to secondary infections and can be seen with other pathogens (e.g., herpesvirus infection). In lizards, skin lesions are common and include gray discoloration, ulcerative and necrotizing dermatitis, and hyperkeratosis (Stöhr et al. 2013; Maclain et al. 2019), but splenic necrosis and hepatocellular inclusions have also been reported (Maclain et al. 2019). Ulceration of the oral mucosa has been reported in green pythons (*Chondropython viridis*) (Hyatt et al. 2002).

1.2.3   Histopathology

Necrosis of the hematopoietic tissues, vascular endothelium, and epithelial cells, hemorrhage, and intracytoplasmic basophilic inclusion bodies are microscopic lesions that can be seen in all hosts, with inclusion bodies seen less consistently in some species (Table 2 and Fig. 4; Reddacliff and Whittington 1996; Cunningham et al. 2007; Allender et al. 2013b; Bayley et al. 2013; Cheng et al. 2014; Jerrett et al. 2015; Waltzek et al. 2014; Forzán et al. 2017; Maclain et al. 2019; Stillwell et al. 2022; Wirth et al. 2023). The liver, spleen, and kidney (including pronephros and mesonephros) are most commonly affected in fatal cases and can involve both the hematopoietic and non-hematopoietic components of these tissues (Fig. 4a–c). In amphibians, a wide tissue tropism has been observed, and additional changes that

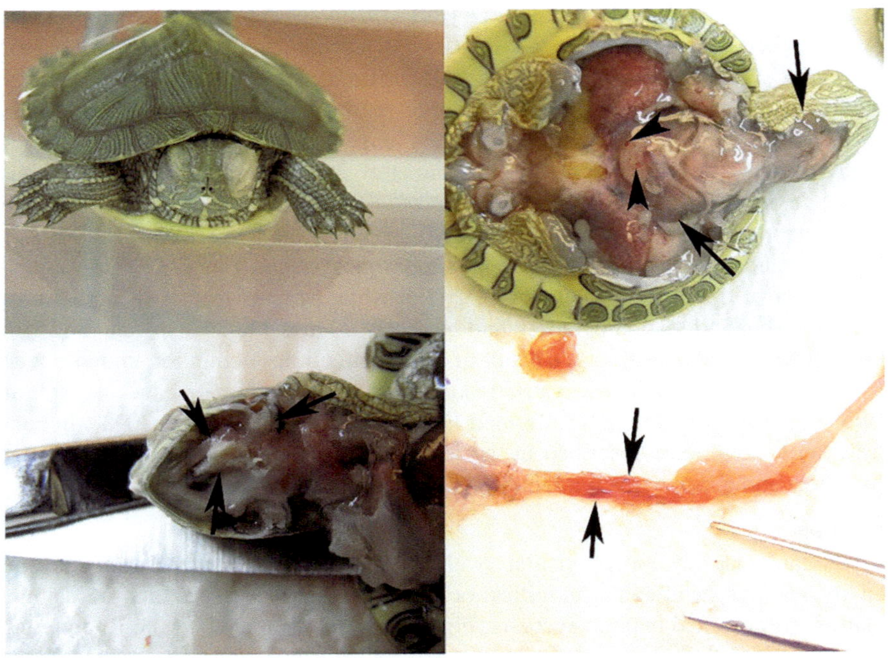

Fig. 3 Gross lesions in a red-eared slider (*Trachemys scripta elegans*) experimentally challenged with an FV3. (**a**) Periocular swelling. The swelling is bilateral but more prominent around the left eye of this turtle. (**b**) Internal soft tissue edema (*arrows*) and hemorrhages (*arrowheads*). (**c**) Necrotic plaques (*arrows*) on the oral mucosa. (**d**) Hemorrhage (*arrows*) of the intestinal mucosa

have been reported include degeneration and ulceration of the epidermis (Cunningham et al. 2007; Geng et al. 2010; Cheng et al. 2014; Meng et al. 2014; Forzán et al. 2017), necrosis of the GI mucosa (Bollinger et al. 1999; Forzán et al. 2017), necrosis of lymphoid tissue (Bollinger et al. 1999; Balseiro et al. 2009; Meng et al. 2014; Forzán et al. 2017), necrosis of hematopoietic tissue in the bone marrow (Forzán et al. 2017), necrosis of neuroepithelial tissue (Docherty et al. 2003), non-suppurative encephalitis and necrosis (M. Forzán, unpublished data), skeletal muscle degeneration (Miller et al. 2008), necrosis of the pancreas (Balseiro et al. 2010; Jerrett et al. 2015; Kik et al. 2011), multicentric hemorrhage, vestibular hemorrhage and necrosis (R. Mazzoni, CRMV-GO, personal communication), perineural hemorrhage and necrosis (Jerrett et al. 2015), and an ocular malformation (Burton et al. 2008).

In chelonians, common findings include fibrinoid vasculitis, myositis, and necrotizing pharyngitis, esophagitis, stomatitis, bone marrow necrosis, and, in one species of turtle, gonadal necrosis and inflammation (Fig. 4d; Johnson et al. 2007, 2008; Ruder et al. 2010; Allender et al. 2013b; Wirth et al. 2023). The necrotizing stomatitis and pharyngitis often found in chelonians overlap significantly with those seen with herpesvirus and adenovirus infections (Johnson et al. 2005; Rivera et al. 2009). In lizards, necroulcerative glossitis, hepatic necrosis, ulcerative dermatitis, bone marrow necrosis, and secondary infections have been reported (Marschang

Table 2 Examples of histopathological changes that can be observed in individuals with natural or experimentally induced ranaviral disease

Organ	Histopathological change
Kidney (including pronephros of larvae and mesonephros of adult amphibians and fish)	Degeneration or necrosis (tubular epithelial cells and glomeruli), intracytoplasmic inclusions (primarily in tubular epithelial cells), necrosis of hematopoietic tissue
Liver	Degeneration or necrosis (sinusoids, melanomacrophage centers, hepatocytes), intracytoplasmic inclusions (hepatocytes), necrosis of hematopoietic tissue
Spleen	Necrosis, intracytoplasmic inclusions (leukocytes)
Pancreas	Necrosis, intracytoplasmic inclusions within exocrine cells
Muscle	Degeneration of muscle fibers, hemorrhage
Skin (especially lizards, fish, and adult amphibians)	Erosion, ulceration, hemorrhage, intracytoplasmic inclusions (epithelial cells)
Thymus, lymphoid tissue	Depletion, apoptosis, necrosis
Gastrointestinal tissue	Apoptosis, necrosis of epithelial cells, intracytoplasmic inclusions
Vessels	Necrosis (endothelial cells)
Upper respiratory tract (especially chelonians and snakes)	Necrosis of epithelial cells, intracytoplasmic inclusion (epithelial cells)
Bone marrow (chelonians, lizards and amphibians)	Necrosis, intracytoplasmic inclusions (hematopoietic cells)
Gonads (chelonians)	Necrosis and inflammation

et al. 2005; Behncke et al. 2013; Stöhr et al. 2013; Maclain et al. 2019, Fig. 5). Necrosis of the pharyngeal submucosa, ulceration of the nasal mucosa, and hepatic degeneration and necrosis have been reported in snakes (green pythons, *Morelia viridis*) (Hyatt et al. 2002).

In fish, additional findings have also been reported. In pallid sturgeon (*Scaphirhynchus albus*) experimentally challenged with FV3, Stilwell et al. (2022) reported decreased hematopoietic cell populations in the spleen and pericardial lymphomyeloid tissue, disruption of blood vessels within gill lamellae and disruption of cutaneous epithelial cells. Borzym and Maj-Paluch (2015) reported similar depletion in hematopoietic tissue of the kidney and destruction of the gill lamellae, as well as hypertrophy of epithelial cells within the gills of rainbow trout (*Oncorhynchus mykiss Walbaum*) and European perch (*Perca fluviatilis L.*) experimentally challenged with EHNV. Another example is that focal necrosis of gastrointestinal (GI) mucosal epithelial cells was reported in largemouth bass (*Micropterus salmoides*) experimentally challenged (IP route) with SCRV (Zilberg et al. 2000; Grant et al. 2003).

1.2.4   Subclinical Infection

Subclinical infections may play an important role in the epidemiology of ranaviruses (Brunner et al. 2024). It is instructive to note that the original amphibian ranavirus isolates were from animals that were, with the exception of one with a renal tumor, ostensibly healthy (Granoff et al. 1966). Subclinical infections have been

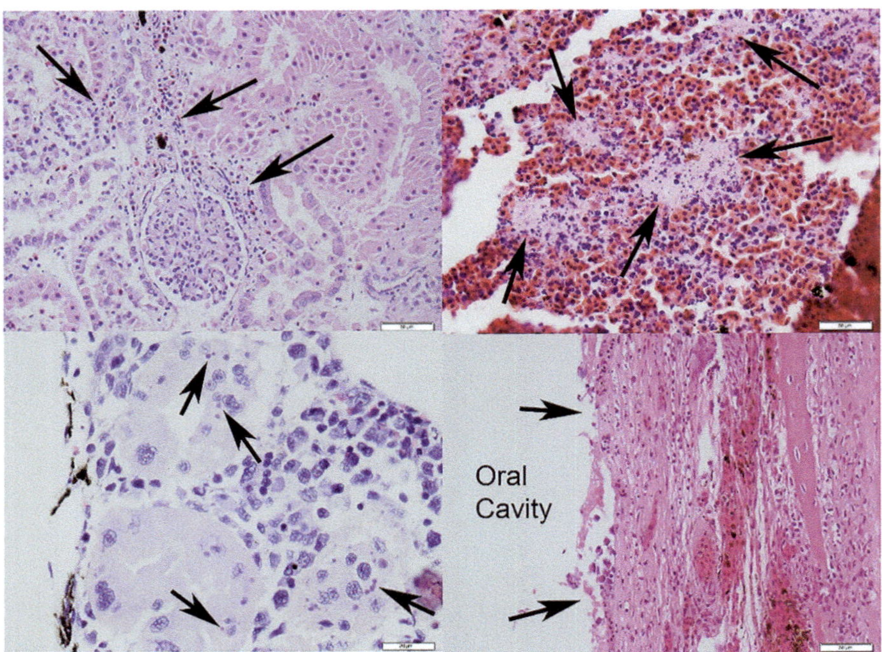

Fig. 4 Histopathological changes seen in individuals with ranaviral disease from experimental challenges with FV3. (**a**) Necrosis (*arrows*) of the hematopoietic tissue within the kidney of a pallid sturgeon (*Scaphirhynchus albus*). (**b**) Necrosis (*arrows*) of the spleen of a dusky gopher frog (*Lithobates sevosus*). (**c**) Intracytoplasmic inclusion bodies (*arrows*) within the renal tubular epithelial cells of a southern leopard frog (*Lithobates sphenocephalus*). (**d**) Necrosis (*arrows*) of the mucosa of the oral cavity of a red-eared slider (*Trachemys scripta elegans*)

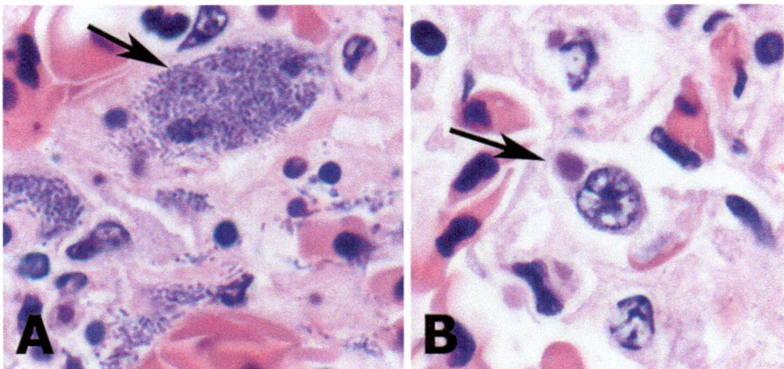

Fig. 5 Histopathology of ranaviral disease with superimposed secondary bacterial infection in an American bullfrog (*Lithobates catesbeianus*). (**a**) Bacterial colonies and necrosis in the spleen (*arrows*) can obscure subtle evidence of ranavirus infection. (**b**) Careful examination of the histologic sections from this individual revealed rare intracytoplasmic inclusion bodies highly suggestive of ranavirus infection

detected in wild amphibians (Gray et al. 2007, 2009; Rothermel et al. 2013), chelonians (Allender et al. 2013b; Goodman et al. 2013; Ariel et al. 2017), lizards, snakes (Ariel et al. 2017), and fish (Goldberg 2002; Whittington et al. 2010; Waltzek et al. 2014). Additionally, subclinical infections have been experimentally produced in amphibians (Brunner et al. 2004; Harp and Petranka 2006; Robert et al. 2007, 2011), chelonians (Johnson et al. 2007; Brenes et al. 2014a, b), and fish (Bang Jensen et al. 2011; Becker et al. 2013; Brenes et al. 2014a, b). Ana Balseiro (SERIDA, personal communication) has observed positive immunohistochemical staining in renal glomeruli of adult common midwife toads (*Alytes obstetricans*) with no clinical signs of disease. In most of the studies above, it is unknown whether the subclinical infections would have developed into clinical disease, because individuals were euthanized or released into the environment. It is likely that subclinical infections represent early stages of ranaviral disease in some cases, while in other cases, infections are resolved or persistent infections are maintained.

2 Diagnostic Testing

2.1 *World Organization for Animal Health (WOAH) Standards*

The WOAH (formerly OIE) provides recommendations and protocols on diagnostic testing for various animal pathogens (https://www.woah.org). One goal of the WOAH is to provide procedures to declare an international shipment of animals or a site, region, or nation as pathogen free for a particular agent. Agents of concern to the WOAH are called notifiable and deemed to be a risk to international commerce or human health. Ranaviruses that infect amphibians and EHNV are listed as notifiable; hence sampling and diagnostic procedures are provided in the Manual of Diagnostic Tests for Aquatic Animals published by the WOAH (WOAH 2023). Gray et al. (2024) discuss how to perform a risk analysis for introduction of ranavirus into an area following WOAH procedures.

Considering that the scope of WOAH is to detect ranaviruses for the purpose of international trade, the sampling procedures and diagnostic techniques they outline may not be applicable for all investigations. In general, we recommend that the procedures in Gray et al. (2024) be followed for determining required sample size and designing studies for ranaviruses. The diagnostic procedures recommended by WOAH also may not be applicable for all regions of the world or may be cost prohibitive. Below, we review the majority of diagnostic techniques used to detect ranavirus infection and determine if individuals are in a diseased state. Table 3 provides guidance as to what techniques can be used given general study directions. In general, we recommend that investigators consult with experts that routinely perform ranavirus diagnostics so that appropriate techniques are chosen to address the particular goals of a study. Importantly, the appropriate sample collecting and preservation procedures can depend on the diagnostic technique that is used. The Global

Table 3 Examples of diagnostic tests that might be used based upon the study goal

Tests[a]	Study goal					
	Detection/ surveillance Prevalence studies	Investigating mortality events	Phylogenetic studies	Ecological studies	Viral morphometric studies	Host immune studies
Conventional PCR	X	X		X	X	X
qPCR	X	X		X	X	X
DNA sequencing (e.g., next generation, Sanger)	X	X	X	X	X	X
Ag-capture ELISA	X	X		X	X	X
Virus isolation		X	X		X	X
Serology	X	X				X
Bioassay		X			X	X
Histology, cytology		X		X	X	X
IHC, ISH, EM		X			X	X

[a]*PCR* polymerase chain reaction, *qPCR* quantitative real-time PCR, *IHC* immunohistochemistry, *Ag-capture ELISA* antigen capture enzyme linked immuno-sorbent assay, *ISH* in situ hybridization, *EM* electron microscopy (which might include SEM, TEM, or immunoEM)

Ranavirus Consortium (GRC) maintains a list of laboratories on their website that routinely perform ranavirus diagnostics (http://www.ranavirus.org/).

2.2   Matching Diagnostic Tests with Study Goal

Investigations into ranaviruses have various purposes, and, depending on the goal, some diagnostic techniques may be more appropriate than others. Moreover, the diagnostic needs for field vs. controlled studies can be different. In general, most investigations can be classified as detection of ranavirus for facilitating trade according to WOAH guidelines (Gray et al. 2024), detection for mapping distribution and estimating prevalence or incidence (Gray et al. 2024), isolation for phylogenetic classification (Jancovich et al. 2015), viral morphometrics and host immune response (Jancovich et al. 2024; Grayfer et al. 2024), ecological factors related to emergence and risk factors for disease outbreaks (Brunner et al. 2024), as well as diagnosis of mortality events (Marschang et al. 2024). Although techniques are constantly being developed and improved, common diagnostic tools used for ranavirus investigations include those that can detect ranavirus and those that can detect the host response to infection. For example, polymerase chain reaction (PCR) simply detects the presence of nucleic acid for a virus-specific sequence, but does not provide evidence as to whether the virus is active (i.e., able to replicate and cause disease). Virus isolation, on the other hand, demonstrates the presence of infectious (active) virus, but it requires highly specialized laboratories to be performed. Other tests (e.g., histology, cytology, gene expression, antibody production) detect the host's response to the infection. Techniques that demonstrate the presence of virus within a lesion such as electron microscopy (EM), immunohistochemistry (IHC), and in situ hybridization (ISH) are particularly useful for demonstrating an association between the presence of the virus and the development of lesions or responses to disease. In general, investigations with the primary goal of ranavirus detection (e.g., surveillance studies) will use PCR or antigen detection techniques, phylogenetic studies will use virus isolation and genomic sequencing, ecological studies will use molecular modalities and histology, and viral morphometric studies, host immune studies, and mortality investigations may use all techniques (Table 3). Additionally, confirmation of ranavirus as the etiologic agent of a die-off requires use of multiple techniques and identification of associated histopathological changes within tissues. Importantly, diagnosis of ranaviral disease cannot be inferred solely with infection data or the observation of gross signs.

Along with study goals, the type of sample may dictate the type of test that can be performed (Table 4). Nonlethal samples generally include swabs, tail or toe clips, and blood because these are generally easily collected (Greer and Collins 2007; Gray et al. 2012). Positive detection of ranaviruses in swabs and toe/tail clips is usually best in the later stages of infection, so keep in mind that when used for surveillance, subclinical individuals can yield false-negative results (Gray et al. 2012; Sutton et al. 2014b; Forzán et al. 2017; Cozad et al. 2020). Certification for freedom

Table 4 Various specimens used for *Ranavirus* testing, the type of test that can be performed, and the limitations of the test result

Specimen	Test	Limitations
Swab	PCR, virus isolation	False positives (environmental contamination); total DNA may be minimal; no histology
Tail or toe clip	PCR, virus isolation	False positives (environmental contamination); no histology
Whole body or internal organs	PCR, virus isolation, histology, IHC, ISH, EM	Dead/humanely killed animals
Fixed tissue (e.g., 10% neutral buffered formalin; 70–95% ethanol)	PCR, histology, IHC, ISH, EM	No virus isolation; electron microscopy is possible but suboptimal if formalin or ethanol fixed or if using formalin-fixed paraffin-embedded tissue
Blood	ELISA if serum separated, differential cell count if blood smear is prepared	Best obtained from live animals; can be difficult to obtain; often cannot obtain large enough quantity from small individuals

from infection in fish species requires a representative sample processed for tests of visceral tissues. Regardless, sampling materials should correspond to the requirements of the laboratory processing the samples, for example, some laboratories may prefer a certain type of swab (e.g., wood vs. polypropylene) or storage or transport method (e.g., dry vs. frozen vs. fixed in ethanol). Thus, it is always best to first contact the laboratory for guidelines on submission. Swabbing should be done by firmly but not forcefully swiping (one to multiple times) the swab along the surface to be tested. Surfaces that are swabbed for ranaviruses typically include the oral cavity, cloaca, or skin lesions (Pessier and Mendelson 2010). Swabbing the vent might provide evidence of intestinal shedding. Tail or toe clips also can be effective at detecting ranavirus infection and might result in fewer false-negative test results compared to swabs (Gray et al. 2012). For many salamander species, there are natural breakpoints near the tip of the tail where light pressure can be applied to autotomize the tail and collect tissue without cutting (Sutton et al. 2014b). Swabs or tissues can be stored frozen or in ethanol.

Disposable gloves should be worn and changed between animals to minimize cross- contamination of samples and to prevent unintentionally transmitting the virus among animals. Additionally, animals should not be co-housed, and processing in the field should occur on a clean surface (Fig. 6). For tail or toe clips, sterile instruments must be used to avoid sample contamination. We recommend using a different sterile (e.g., autoclaved) set of instruments for each animal. Although

Fig. 6 (continued) salamander species, there are natural break points in the tail that can easily yield a tail sample by simply applying gentle pressure near a break point and can be done without removing the animal from the bag (**e**). When testing for concurrent *Batrachochytrium dendrobatidis* and *Batrachochytrium salamandrivorans* infections, swabbing can be accomplished within the containers as well (**e** *inset*). All equipment, waders, and boots should be disinfected (e.g., with 1% Nolvasan®) before leaving a field site (**f**)

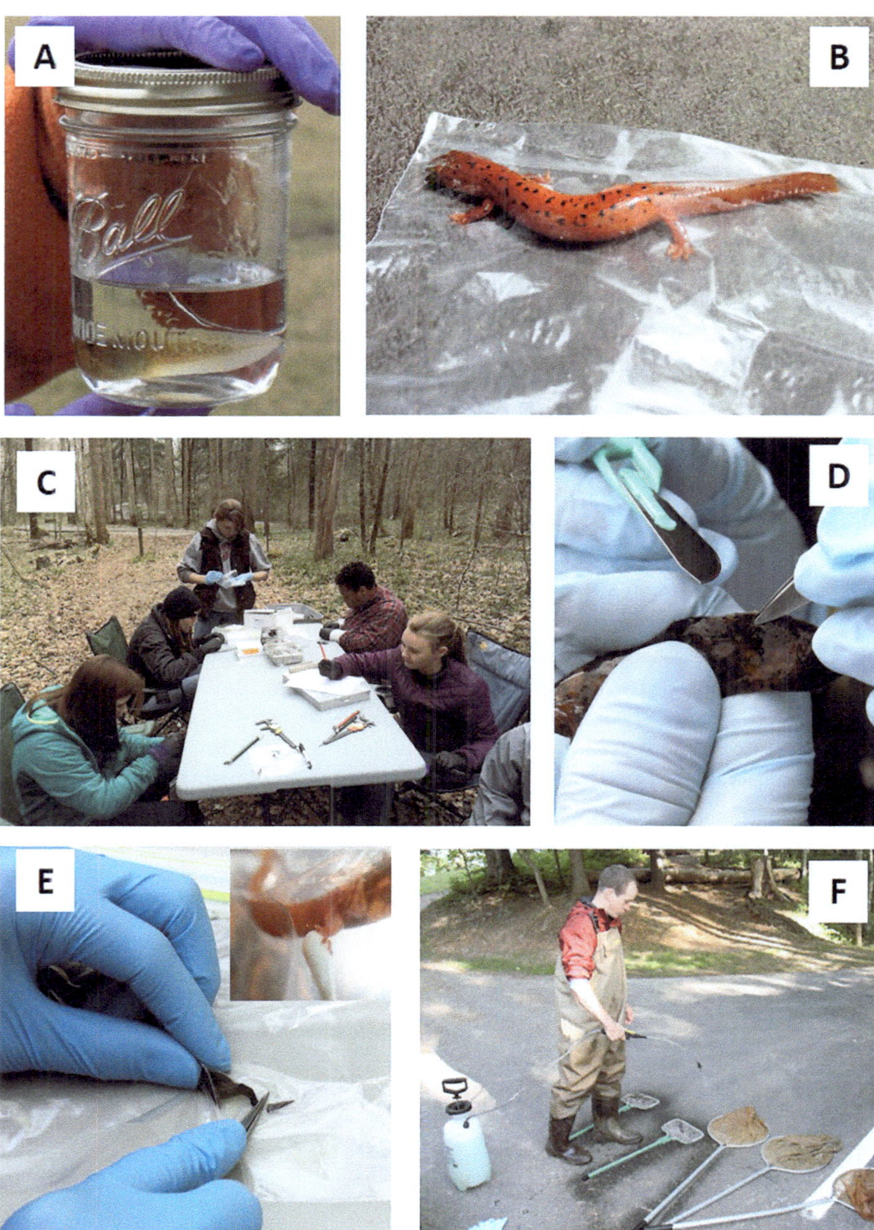

Fig. 6 Field sampling station demonstrating sterile collection techniques. Animals should be placed in separate containers, such as glass jars containing sterile water for aquatic stages (**a**) or plastic sealable bags for terrestrial stages (**b**). A portable table can serve as a processing station that can be easily disinfected after each field site (**c**). A sterile set of instruments should be used for each animal. For some species (e.g., Hellbenders, *Cryptobranchus alleganiensis*), a small slice can be collected from the dorsal tail using a sterile disposable scalpel and forceps (**d**). For many

ranavirus can be quickly inactivated by various disinfectants (Byran et al. 2009), cross-contamination (and thus false-positive results with molecular testing) can occur if ranavirus DNA is not degraded. While degradation studies have not been performed for ranaviruses, autoclaving, flaming, and long duration soaking (>12 h) in full strength bleach (6% NaOCl) have been used to degrade the DNA of other pathogens (Cashin et al. 2008). Some researchers are investigating the use of liver aspirates (Forzan and Wood 2013); however, the feasibility in field collection and the need for expertise in collection may limit their use. Blood can be collected from multiple locations, including the caudal vein (fish, caudates, snakes), subcarapacial sinus or occipital venus sinus (turtles, Martínez-Silvestre et al. 2002; Allender et al. 2011), abdominal and facial veins (anurans, Forzan and Wood 2013), heart (fish, anurans, caudates, snakes), and ducts of cuvier (fish).

Detailed reviews of sample collection protocols for amphibian necropsy and mortality events are available (Green et al. 2009; Pessier and Mendelson 2010). In brief, submission of whole moribund animals sent to a diagnostic laboratory in less than 24 hours is preferred. However, if this is not possible, tissue samples may be collected and submitted. It is important to realize that if a sample is submitted for only one test, the result will only be positive or negative for that one pathogen by that one test. If positive, the pathogen may or may not have played a role in morbidity and mortality. If negative, the cause of the morbidity and mortality will remain undetermined. Thus, ideally multiple samples should be submitted, and multiple diagnostic tests performed. At minimum, tissue collection should include the major organs (liver, kidney, spleen, lungs/gills, heart, skin, digestive tract) and any lesions noted. A set of these tissues should be submitted fresh or frozen for pathogen testing, and a set should be submitted fixed (e.g., 10% buffered formalin) for histological evaluation. Samples should be triple packaged in leak-proof containers and shipped following the guidelines of the carrier (e.g., http://www.fedex.com/ downloads/hk_english/packagingtips/pointers.pdf). Autolyzed samples are of little diagnostic value, but, if they are all that remain, it may be possible to glean some information. It is best to contact the diagnostic laboratory before submission of any samples, but especially before submitting autolyzed samples. As mentioned above, a list of diagnostic laboratories can be found on the GRC website. Importantly, researchers should disinfect footwear and sampling gear that comes in contact with potentially infected animals or water that contains ranavirus. Nolvasan® (1%, Fort Dodge Animal Health, Fort Dodge, Iowa, USA) and 0.2% sodium hypochlorite are effective at inactivating ranavirus, as well as other pathogens such as the amphibian chytrid fungus (Bryan et al. 2009; Gold et al. 2013).

2.3   Diagnostic Tests

Confirmation of the presence of ranavirus in host tissues can be achieved using various methods, including virus isolation, EM, antigen-capture enzyme-linked immunosorbent assay (Ag-capture ELISA), IHC, ISH, and PCR. Of these tests, all may be performed on specimens collected from dead organisms with minimal postmortem change, and most can be performed on specimens collected from live organisms (Table 4). Some tests (e.g., PCR) may yield results even with advanced autolysis, but this is not ideal. For all tests, tissues need to be treated in a test-specific manner to expose the virus and thus allow for its detection. Several validated methods that incorporate automated, semi-automated, and manual homogenization of tissues have been described and compare the extraction efficiencies of ranavirus from tissues (Whittington and Steiner 1993; Rimmer et al. 2012). Extraction efficiency is especially important if viral loads are low in tissues, such as in subclinical infections. Additionally, all tests should include positive and negative controls. For example, positive PCR controls typically include extracted DNA from a virus isolate and known infected animal, whereas negative controls typically include extracted DNA from a known negative animal and DNA grade water. In this scenario, the controls verify that the test was effective and had low contamination likelihood. Reagents for diagnostic testing for ranaviruses (control DNA, anti-ranavirus antibodies) can be obtained from one of the authors (RW); DNA plasmid sets to establish a standard curve for quantitative PCR testing can also be obtained from commercial laboratories such as Pisces Molecular (Boulder, Colorado, USA).

2.3.1 Polymerase Chain Reaction

Both conventional and quantitative real-time PCR (qPCR) are used to detect ranavirus DNA. The results of molecular assays need to be interpreted based on the validation data and with consideration for which ranaviruses are in scope, for example, the fish ranaviruses SCRV/LMBV and SGIV are not detected by tests for amphibian ranaviruses, although some tests will detect amphibian and some fish (EHNV, short-finned eel ranavirus [SERV]) ranaviruses (Stilwell et al. 2018). The PCR assays can be performed on a variety of specimens, including fresh tissue, formalin-fixed paraffin-embedded tissues, and swabs. In the most commonly used assays, the major capsid protein (MCP) gene is targeted (Mao et al. 1997; Hyatt et al. 2000; Kattenbelt et al. 2000; Marsh et al. 2002; Pallister et al. 2007; Rimmer et al. 2012; Stillwell et al. 2018). The MCP gene is highly conserved, which makes it a desirable region to target for identifying the presence of ranavirus DNA; however, it has limitations if the goal is to explore subtle genomic or phylogenic differences among isolates within a particular ranavirus species (Jancovich et al. 2015). Other targets for PCR have included the neurofilament triplet H1 protein (Holopainen et al. 2009), DNA polymerase (Holopainen et al. 2009), and an intergenic variable region (Jancovish et al. 2005).

The sensitivity of PCR may vary depending on the type of specimen tested. In general, it is considered that tissue samples from internal organs represent infection status better than nonlethal sampling techniques such as tail clips, toe clips, and swabs (Greer and Collins 2007; Gray et al. 2012). Given that ranavirus tropism differs among tissue types and host postexposure duration (Robert et al. 2011; Ma et al. 2014), it is expected that PCR test results will depend on the tissue used. In frogs infected with FV3, ranavirus is first detected in the kidney and intestines (Robert et al. 2011) and then the liver, spleen, and other major organs. Thus, testing different tissues can provide evidence of infection severity. Commonly, researchers test for infection using a homogenate of different tissues (e.g., Hoverman et al. 2011) to increase detection probability. Although it may seem obvious, during a mortality event, it is best to obtain samples from dead or dying individuals as animals not exhibiting clinical signs or gross lesions may yield negative results, even if they are sharing a small habitat with infected individuals (e.g., Forzán and Wood 2013).

As mentioned before, testing for ranavirus in nonlethal samples can lead to false-negative and false-positive results when compared to whole animal or liver samples (Greer and Collins 2007; Gray et al. 2012; Forzán et al. 2017). False-negative results may be caused by insufficient virus in swabs or tail and toe clips, whereas false-positive results could be caused by virus on the outside of the animal being detected (Gray et al. 2012). Regardless, nonlethal sampling techniques can be useful for ranavirus surveillance if animal collections are not allowed or population abundance is low.

It is important to note that PCR assays are not perfect, implying that some rate of false-positive and false-negative results are expected even when a laboratory procedure with excellent analytical characteristics is performed without error and according to ISO17025 accreditation. Accurate interpretation of PCR results requires estimation of the diagnostic sensitivity and diagnostic specificity of the assay so that the positive and negative predictive value of the result can be calculated (Greiner and Gardner 2000). These characteristics are measured when a test protocol is validated for a specific sample type and host species. Occasionally, when the quantity of virus present is low, different results from the same sample can occur. At minimum, we recommend that all samples tested by qPCR are run in duplicate (although triplicate is preferred) on the same qPCR machine, and only samples with consecutive positive results are declared positive. If one sample appears positive and the other negative, a third sample can be run, and the declaration of infection made based on the majority of the results. Ongoing research (E. Grant, US Geological Survey, and D. Miller) is estimating detection probabilities of qPCR using the protocol of Picco et al. (2007) and double sampling procedures. Alternative molecular detection techniques such as loop-mediated isothermal amplification (LAMP) are under evaluation for use in aquaculture and are considered desirable for potential point of care detection of SGIV (Yu et al. 2022) and LMBV (Zhu et al. 2020). Lateral flow biosensors using aptamers have also been developed for point-of-care detection of SGIV (Liu et al. 2021).

Conventional PCR

Conventional PCR has been used to detect ranaviral infection in fresh or fixed tissue specimens from surveillance studies and mortality events, as well as the identification of cultured virus (Miller et al. 2007; Gray et al. 2009; Meng et al. 2014). Phylogenetic mapping of banked DNA sequences (GenBank) representing isolates from throughout the world reveals significant sequence identity within the amphibian ranavirus and EHNV MCP gene, which has been exploited in designing primers for detection of ranaviruses by PCR. In addition, sequence polymorphisms within the MCP can be used to distinguish some isolates to the species level (e.g., FV3, EHNV) by restriction digestion of amplicons from conventional PCR (Marsh et al. 2002; Holopainen et al. 2009). Conventional PCRs for detection of ranaviruses and restriction endonuclease analysis (REA) for subtyping are described in the WOAH Manual of Diagnostic Tests for Aquatic Animals (WOAH 2023) although sequencing of amplicons is a now much more accessible and informative technique.

Amplification of non-ranaviral DNA can occur. Thus, while non-sequencing in endemic regions may be elected, it is generally advised to at least sequence a subset of the conventional PCR products to confirm positive results. This is especially important when new hosts are affected or ranavirus is detected in regions previously thought to be free of these viruses. Sequencing of PCR products can also be informative for preliminary genus-level virus identification (e.g., FV3 or *Ambystoma tigrinum* virus [ATV]).

Quantitative Real-Time PCR

The advent of qPCR has provided significant advances in the study of various pathogens and their virulence. Studies have found that qPCR is more sensitive than conventional PCR as it can detect lower viral loads and be more sensitive than virus isolation (e.g., Pallister et al. 2007; Jaramillo et al. 2012). Pallister et al. (2007) found that qPCR could differentiate between a number of ranaviruses that infect fish, amphibians, and reptiles and was especially effective at distinguishing European from Australian ranaviruses based on the application of different hydrolysis probes. A useful protocol for amphibians was reported by Picco et al. (2007), which amplifies a 70-bp region of the ranavirus MCP gene; Allender et al. (2013a) developed a similar assay for chelonians; and Stilwell et al. (2018) developed and initiated the validation pathway for a protocol that is able to detect 33 different isolates from various species of fish, amphibians, and reptiles. Duplex and multiplex protocols have also been developed to simultaneously detect multiple pathogens (Standish et al. 2018). Viral load can be estimated when the genomic DNA in samples is quantified, and equal amounts of DNA used from each specimen tested (Leung et al. 2017). Viral load is predicted by enter ing the cycle threshold (Ct) value for a sample into a regression equation (called the standard curve) that relates Ct values and known virus quantities for the PCR, system that is used for testing (Yuan et al. 2006). Importantly, Ct values on different PCR systems are not equivalent; thus, conversion of Ct values to a standard unit of virus quantity using a standard curve is necessary for interpretation among studies. Typically, the standard unit of measurement is a $\log_{10}$-transformed value of virus concentration per unit of genomic DNA or tissue. For example, Brenes et al. (2014a) reported virus levels as plaque-forming units (PFU) per 0.25 µg of genomic DNA. Results of PCR systems can vary among runs; thus, it is ideal to estimate a standard curve for each run (i.e., plate). Guidelines are available for reporting results for quantitative PCR assays (Bustin et al. 2010; WOAH 2023). Slope and intercept parameters for the standard curve can be averaged among multiple independent runs for a more robust estimate of viral load, similar to model averaging. Standard curves should be published with qPCR results.

Given that pathogen load often is positively related to morbidity in animals, qPCR might provide insight into the disease state (clinical vs. subclinical), but this remains to be shown for ranaviruses. Caution must be exercised when interpreting high Ct values (e.g., at or near threshold), because they can represent amplification or fluorescence artifacts or cross-contamination (Caraguel et al. 2011). Conventional PCR, followed by DNA sequencing or virus isolation, can be used to verify viral DNA presence within samples having high Ct values. Additionally, a standard curve for the PCR system can be used to identify a Ct threshold where the sample is declared PCR-positive. For example, Brenes et al. (2014a, b) conservatively declared a positive result if the Ct value was less than the lower bound of a 95% confidence interval at a predicted virus quantity of zero. In general, data on the diagnostic sensitivity and specificity of published qPCR tests for ranavirus are lacking. However, the test described by Jaramillo et al. (2012) has been validated for EHNV and is included in the 2023 edition of the WOAH Manual. Also listed are Palister et al. (2007 and the protocol which is the basis for the one used at the WOAH Reference Laboratory) and Stilwell et al. (2018, which is partially validated for *Ranavirus* isolates across classes of hosts). Similarly, Allender et al. (2013a) validated the qPCR for detection of FV3 virus in eastern box turtles (*Terrapene carolina carolina*).

Differentiation of Ranavirus Species and Strains

In most routine diagnostic and research situations, existing PCR assays based on the MCP gene work well for determining the presence or absence of a ranavirus or to categorize a virus to a species (e.g., FV3, Jancovich et al. 2015) or even a subtype such as differentiation of European and Australian variants related to EHNV (Marsh et al. 2002). However, in other instances such as epidemiologic investigations that need to determine the distribution of different ranavirus strains or translocation programs for wildlife species where it may be important to determine if the same ranavirus is present in both source and destination populations, these assays are not as useful because the MCP gene is so highly conserved (Jancovich et al. 2015). As an example, it is now well documented that different strains of FV3 as determined by genomic restriction endonuclease analysis (REA) frequently have identical DNA sequences within regions of the MCP gene commonly used in diagnostic conventional PCR (Schock et al. 2008; Duffus and Andrews 2013).

The laboratory techniques used to differentiate specific ranavirus strains such as virus isolation and purification followed by genomic REA are not available in all laboratories and are not practical for some studies; therefore, there is a need to find and validate methods for rapid strain identification. One possible approach is genotyping using conventional PCR and DNA sequencing for genes such as the intergenic variable region (Jancovich et al. 2005; Weir et al. 2012) or neurofilament triplet H1-like protein (Holopainen et al. 2009; Cheng et al. 2014) that contain variable repeating regions. By using consistent approaches for selection of sequencing targets, researchers ensure that advances in the understanding of ranavirus phylogeny and epidemiology can be more rapidly obtained. Complete genome analysis is increasingly possible and provides a tool for molecular epidemiology (Hick et al. 2017).

2.3.2 Antigen-Capture ELISA

The analytical sensitivity of Ag-capture ELISA applied to fish tissue homogenates for detection of EHNV was 10^3–10^4 TCID$_{50}$/ml. Antigen-capture ELISA is useful for diagnosis and surveillance because it can be applied quickly (compared to virus isolation) and inexpensively (compared to molecular assays) to test large numbers of samples. Relative to virus isolation, the diagnostic specificity and sensitivity of Ag-capture ELISA was 100% and 60%, respectively (Whittington and Steiner 1993).

2.3.3 Virus Isolation

Virus isolation using well-characterized cell lines that are commercially available from cell repositories is another approach for determining the presence of ranavirus. Additional advantages of cell culture techniques include demonstration of viable virus and amplification of it for further characterization. Culture specimens also may yield better PCR results and ultimately better products for sequencing than tissue specimens. Cell lines including bluegill fry (BF-2), fathead minnow (FHM), and epithelioma papulosum cyprini (EPC) cells are suitable for virus isolation (McClenahan et al. 2005). EHNV replicates in many fish cell lines including fathead minnow (FHM), rainbow trout gonad (RTG), bluegill fry (BF-2), and Chinook salmon embryo (CHSE-214) at a range of temperatures from 15 to 22 °C (Langdon et al. 1986; Crane et al. 2005; Ariel et al. 2009; WOAH 2023). European catfish virus can be isolated using EPC cells, FHM, and channel catfish ovary (CCO) cells at 15–25 °C, but BF-2 cells were tenfold more sensitive than EPC and CCO cells (Ahne et al. 1989; Pozet et al. 1992). SCRV was originally isolated from largemouth bass in FHM cells (Plumb et al. 1996) and subsequently in BF-2, EPC, and CCO cells at 25–32 °C (McClenahan et al. 2005). SGIV can be propagated on a range of cells including FHM and BF-2, but more rapid cytopathic effect (CPE) and higher titers were observed in grouper (GP) embryo cells (Qin et al. 2003).

Ranaviruses commonly found in amphibians can be isolated on some fish cells as well as amphibian cells. Incubation temperature can vary and is a critical consideration for optimum results. For example, ATV reproduces on FHM, RTG, and bullfrog tongue cells at 25 °C (Docherty et al. 2003), FV3 on FHM cells or frog embryo fibroblasts at 27 °C (Cunningham et al. 1996, 2007), and EPC at 24 °C (Ariel et al. 2009), and CMTV can be propagated on EPC at 15 °C (Balseiro et al. 2009). Ma et al. (2014) and Geng et al. (2010) found that Chinese giant salamander virus (CGSV; also designated as Chinese giant salamander iridovirus, GSIV) could be propagated on EPC at 25 °C and 20 °C, respectively.

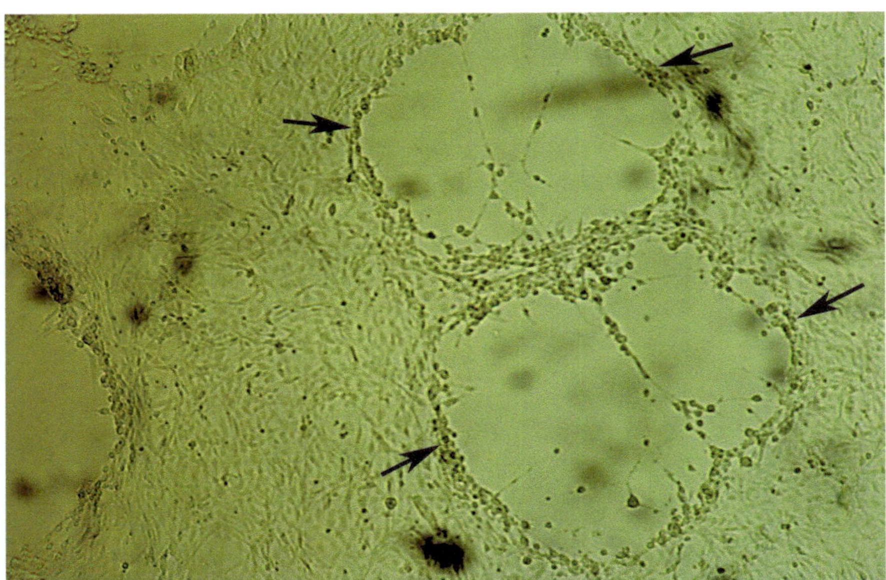

Fig. 7 A typical cytopathic effect (CPE) of EHNV in bluefill fry (BF-2) cells showing multiple foci of lysis of the cell monolayer. Note the rounding of cells on the margins of the lytic areas (*arrows*). After a few days, destruction of the entire monolayer can be expected

All ranaviruses produce a similar CPE with focal lysis of cells, which typically is followed within a few days by destruction of the entire monolayer (Fig. 7). Ranavirus CPE presents as a net-like appearance on monolayers of FHM cells (G. Chinchar, personal communication); however, the appearance of CPE is not sufficient to indicate that a ranavirus is the cause. It is also noteworthy that a blind passage is sometimes required before CPE is detected. When CPE is observed in cell monolayers after passage, several techniques can be used to confirm the presence of a ranavirus (Jaramillo et al. 2012; Stilwell et al. 2018). Suitable techniques include indirect fluorescent antibody stain, Ag-capture ELISA, and PCR (WOAH 2023b). Alternatively, viral nucleic acid sequencing or restriction enzyme digestion can demonstrate the presence of a ranavirus and allow for differentiation between species and strains (Hyatt et al. 2000; Marsh et al. 2002). The virus isolation technique can be adapted to determine the quantity of infectious virus (i.e., the 50% tissue culture infective dose, $TCID_{50}$; Rojas et al. 2005).

Virus isolation requires that viability of the virus is maintained from the field to the laboratory. Thus, this technique requires an appropriate cold chain throughout transport and transfer, or false-negative results could occur (WOAH 2023). Autolyzed tissues also should be avoided. Virus isolation using cell culture has a very high analytical sensitivity, but is not as sensitive as real-time PCR. For example, detection of EHNV using BF-2 cells was 100-fold less sensitive than qPCR (Jaramillo et al. 2012). When used for the purpose of surveillance and to detect subclinical infection, the sensitivity of virus isolation requires preparation techniques that ensure maximum release of cell-associated virus from the tissues (Whittington and Steiner 1993). Recently, modification of these methods using high-throughput laboratory technologies such as bead-beating was shown to be compatible with virus isolation and increased the sensitivity of qPCR for detection of EHNV (Rimmer et al. 2012). Ranaviruses are abundant in the liver, kidney, and spleen of clinically affected fish, amphibians, and reptiles. For SCRV, it was suggested that gills, swim bladder, and posterior kidney should be used as a minimum sample (Beck et al. 2006), while a pool of kidney, liver, and spleen is preferred for EHNV.

Virus isolation provides an important tool for basic studies of the replication and morphogenesis of ranaviruses. For example, after initial descriptions of GCSV based on diagnostic tests including isolation on EPC cells (Geng et al. 2010), additional important information about the virus was acquired using EM and proteomic and RNA knockdown studies in EPC cells (Li et al. 2014b; Ma et al. 2014). These types of studies can provide valuable information on virus replication, which may in turn prove useful for developing methods of treatment and control of ranavirus infection.

2.3.4 Detecting Antibodies: Serology

An ELISA for detection of antibodies was developed using a combination of anti-ranavirus antibodies and species-specific anti-immunoglobulin reagents and has been used to detect a specific adaptive immune responses to ranavirus infection in the serum of redfin perch (*Perca fluviatilis*), rainbow trout (*Oncorhynchus mykiss*), and the cane toad (*Bufo marinus*; Whittington et al. 1994, 1997; Whittington and Reddacliff 1995; Whittington and Speare 1996). Seroconversion occurred when redfin perch and rainbow trout were injected with inactivated EHNV (Whittington et al. 1994; Whittington and Reddacliff 1995). Serological approaches that target a long-term adaptive immune response to ranavirus infection potentially offer a sensitive method for differentiating host populations with endemic ranavirus infections from areas that are ranavirus free. For example, Whittington et al. (1999) reported a small proportion of seropositive adult rainbow trout in EHNV-infected farmed fish populations. Similarly, anti-ranavirus antibodies have been detected in populations of free-living anurans (Whittington et al. 1997; Zupanovic et al. 1998); and humoral immune responses have been measured in comprehensive investigations of the pathogenesis of ranavirus infection in *Xenopus laevis* (Gantress et al. 2003).

Given the success in fish, ELISA testing for an antibody response is being applied to reptiles. Ariel (1997) utilized methods developed by Hengstberger et al. (1993) to detect anti-BIV antibodies in wild reptiles in Australia. More recently, Johnson et al. (2007, 2010) and Allender (2012) used ELISA for surveillance and laboratory investigations to test for the presence of anti-ranavirus antibodies in the plasma of various chelonians in the USA.

Serological testing is an easy and cost-effective technique, although it also presents limits (Jaramillo et al. 2017). The application of serological tests is limited by the cross-reactivity of antibodies to all ranaviruses and the absence of a secondary test method to evaluate sensitivity and specificity. Virus neutralization tests cannot be used because antibodies with neutralizing activity to EHNV were not evoked following immunization of rabbits (Hedrick et al. 1992) or from mouse monoclonal antibodies (Monini and Ruggeri 2002). Current knowledge of the amphibian, reptile, and piscine immune systems indicates limitations to serological techniques that are imposed by a lack of affinity maturation, poor immune memory, and temperature dependence of the response (McLoughlin and Graham 2007). For this reason, the WOAH does not currently support serology as a useful strategy, although both antigen and antibody ELISAs are listed for infection with EHNV (WOAH 2021).

2.3.5   Testing in Animals: Bioassay

A bioassay or experimental transmission trial is an important part of the diagnostic process that is used in cases where a novel ranavirus is detected or disease is detected in a new host or novel ecological setting. An example of this can be found in Waltzek et al. (2014), which describes a mortality event in pallid sturgeon caused by an FV3-like ranavirus. In this case, experimental transmission of the virus isolated from the die-off event was used to infect pallid sturgeon in the laboratory, resulting in a mortality event mimicking the one from which the virus was originally isolated. Subsequent sequencing of the virus revealed it was an FV3 strain.

2.3.6   Examining the Tissues

Histopathology and Cytology
Histopathology is an essential tool for investigating cases of disease with unknown etiology. Characteristic lesions can provide preliminary evidence of certain pathogens and guidance in selecting subsequent diagnostic tests. Histopathology is also important for determining the relevance of ranavirus infection detected during a surveillance study or disease event. Once death of the organism occurs, tissues breakdown (autolyze) quickly, especially in warm conditions, which can obscure microscopic detail. Thus, histology is performed best when using morbid specimens that were collected alive, humanely euthanized (AVMA 2020), and preserved in a fixative immediately after death. A 9:1 ratio of neutral buffered formalin (10%) to tissue is generally used for fixation (Pessier and Mendelson 2010). Ethanol (70%) or other fixatives (e.g., Davidson's or Bouin's) may be preferred for fish tissues. Small animals (e.g., 10 cm or less) may be placed whole in a fixative if shipping to a diagnostic laboratory. A small (1–2 cm) incision into the coelomic cavity will aid in preserving internal tissues. In small fish, this is accomplished by removing the tail and operculum. Larger animals may be opened (necropsied), and representative samples collected from all tissues (see Sect. 2.2), so lesions can be observed grossly and differential diagnoses considered. At the time of necropsy, it is important to collect a second set of samples to be stored fresh (for immediate testing) or frozen (for

future testing) if histopathology results indicate that lesions consistent with ranavirus infection are present and confirmatory testing is necessary or to rule out the presence of concurrent pathogens. To prevent cross-contamination, a different set of sterile instruments should be used for each individual.

Often cases of ranaviral disease are not straightforward and require the use of additional diagnostic methods such as IHC, ISH, EM, or molecular methods for confirmation. The observation of characteristic intracytoplasmic inclusion bodies is an inconsistent finding, especially in chelonians (DeVoe et al. 2004; Johnson et al. 2007, 2008). In other cases, subtle lesions of ranaviral infection can be obscured by secondary bacterial or fungal infection (Fig. 5).

Although not a definitive test of disease, cytology can be used to document changes in the innate immune system through blood cell counts and for detection of viral inclusions. Blood may be collected as described above (see Sect. 2.2), and air-dried blood smears stained and subjected to cytological examination with oil-immersion light microscopy. Cytoplasmic inclusion bodies may be observed within the cytoplasm of leukocytes in reptiles and amphibians (Allender et al. 2006; Forzán et al. 2016). Although not definitive, inclusion bodies are suggestive of ranavirus infection, when corresponding clinical, gross, and histopathological changes are present. When possible, immunohistochemistry can be applied to unstained blood smears to confirm the ranaviral identity of the inclusions (Forzán et al. 2016). It should be noted that viral inclusions in erythrocytes may represent erythrocytic iridoviruses, which may be a different genus within the family *Iridoviridae* (Wellehan et al. 2008; Grosset et al. 2014).

Tests for Visualizing the Virus

There are several persistent questions regarding the pathogenicity of ranaviruses: Is ranavirus causing the lesion, where is the virus in subclinical infections, and what cells are infected by the virus, and how does this vary by host species and developmental stage? A key factor in pathogenesis is identifying the cell types targeted by the virus. This can be done with histology; however, cellular changes are not always visible, such as with subclinical infections. Additionally, attributing cellular changes due to ranavirus can be challenging with histology, especially in cases of concurrent infection with other pathogens. Visualizing the virus (or its nucleic acid) within cells can be useful for attributing cellular changes to ranavirus (Miller and Gray 2010) and can be accomplished using IHC, in situ hybridization (ISH), and EM.

Immunohistochemistry

For IHC, enzyme-conjugated, virus-specific antisera are used to demonstrate the intracellular location of viral antigens via an enzyme (e.g., horseradish peroxidase or alkaline phosphatase) catalyzed reaction (Fig. 8). Additional approaches have been developed to detect EHNV and European catfish virus (ECV) antigen using polyclonal antisera in Ag-capture ELISA and immunoelectronmicroscopy (Steiner et al. 1991; Hengstberger et al. 1993; Hyatt et al. 1991; Whittington and Steiner 1993; Ahne et al. 1998). Secondary antibodies suitable for immunofluorescence or enzymatic detection enable the detection of ranavirus antigens in tissue sections and cell cultures (Reddacliff and Whittington 1996). Some antisera are directed against

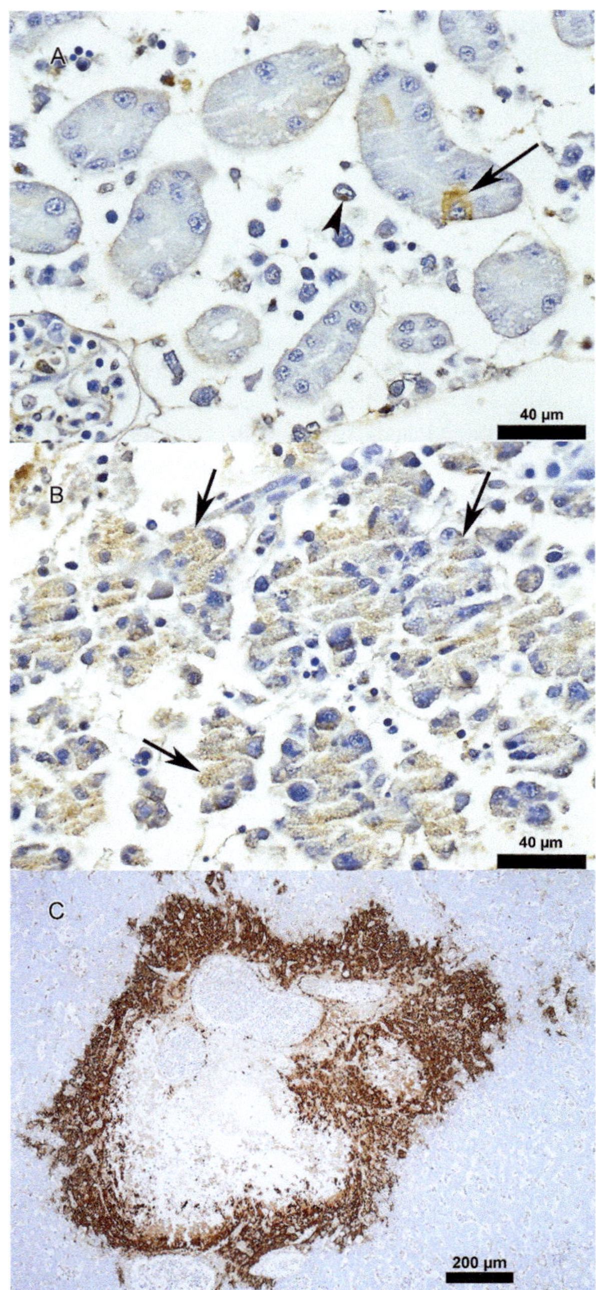

Fig. 8 (**a**) Immunohistochemical staining for ranavirus in renal tubular epithelial cells (*arrow*) of a Southern leopard frog (*Lithobates sphenocephala*). An intracytoplasmic inclusion body within a hematopoietic cell (*arrowhead*) also stains positive. (**b**) Positive staining (*arrows*) is present throughout the spleen of the Southern leopard frog. (**c**) Positive staining in the liver of a redfin perch (*Perca fluviatilis*)

the MCP and are cross-reactive; thus, all ranaviruses can be detected, but it is not possible to distinguish between ranaviruses (Hedrick et al. 1992; Ahne et al. 1998). Other antisera have been developed against purified virions (Balseiro et al. 2009), but again, it might not be possible to distinguish between ranaviruses. Effective use of immunohistochemical staining has been reported for amphibians (see Hyatt et al. 2002; Cunningham et al. 2008; Balseiro et al. 2009, 2010; Bayley et al. 2013). Balseiro et al. (2009, 2010) used IHC to detect the presence of ranavirus in various cells and found that glomeruli appear to have the most positive staining in larval toads but ganglia (particularly in the skeletal muscle) displayed the most positive staining in juvenile newts.

Since the publication of the first edition of this monograph, the use of IHC has increased both in research studies and diagnostic investigations (e.g., Forzán et al. 2017; Maclain et al. 2019; Wirth et al. 2023), partly due to the availability of high-quality antibodies provided by WOAH's reference laboratory in Sydney, Australia (Whittington and Deece 2004).

IHC is particularly useful when only fixed tissues are available, such as in diagnostic investigation of mass die-off events where shipping fresh or frozen tissues was not possible. Although PCR can be run on preserved samples and provide information on the presence or absence of ranavirus, the technique cannot verify the virus' involvement in a particular lesion (and thus possibly the cause of death). IHC provides supportive evidence of lesion etiology by microscopically visualizing the association of a pathogen with a lesion.

In Situ Hybridization

Another approach to visualizing the virus is by ISH, which uses molecular probes to localize specific nucleic acid sequences within fixed tissue sections. For example, specific staining was observed in the kidney and spleen of Malabar grouper, (*Epinephelus malabaricus*) infected with Singapore grouper iridovirus (SGIV), a Ranavirus which affects marine fish and has approximately 73% MCP gene sequence homology with FV3 (Huang et al. 2004). As with IHC, the last few years have seen an increased in the use of ISH in experimental infections and diagnostic investigations of ranavirus disease. One of the most used ISH protocols, described in Maclaine et al. (2018), uses ACDBio probes (https://acdbio.com) and has been used to detect ranaviral nucleic acid in frog, chelonian and lizard species, and pallid sturgeon (Stilwell 2017; Maclaine et al. 2018; Forzán et al. 2019; Pfeiffer et al. 2019; Stilwell et al. 2022; Wirth et al. 2023) (Fig. 9).

Electron Microscopy

Electron microscopy (EM) is used for visual confirmation of the identity of cultured virus and for visualizing the virus within tissue sections. Verification of the cultured product involves negative staining and assessment via scanning EM (SEM), whereas preserved tissues are used to visualize the virus within the tissues by transmission EM (TEM). For example, Burton et al. (2008) used TEM to examine the ultrastructure of tissues from a malformed eye of a preserved specimen. Electron microscopy revealed that the sample contained viral particles consistent with members of the family *Iridoviridae*, which allowed further characterization of ranavirus using PCR

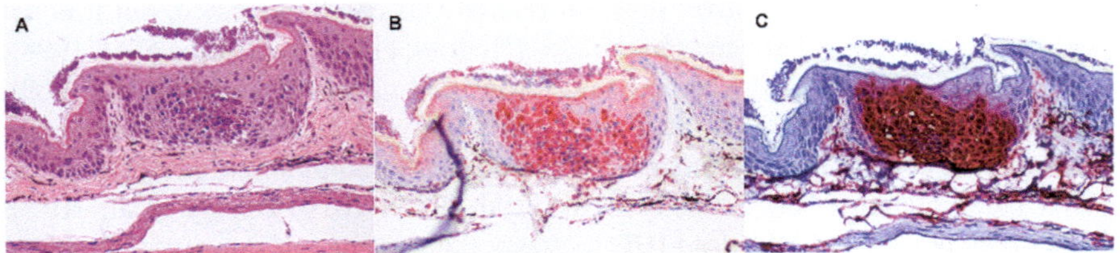

Fig. 9 Ranavirus infection in the epidermis of a water dragon: (**a**) hematoxylin and eosin (H&E) staining highlights the necrosis, (**b**) immunohistochemistry (IHC), and (**c**) in situ hybridization (ISH) demonstrate the presence of the virus in the lesion. (Modified from Maclain et al. (2019). Copyright © The Authors 2019)

and DNA sequencing. Similarly, histological examination may reveal structures suggestive of intracellular viral inclusions, and EM can be used to verify the particles are viral origin (Cunningham et al. 1996, 2007; Balseiro et al. 2010; Cheng et al. 2014; Meng et al. 2014). In general, identification is only reliably made at the family (*Iridoviridae*) level; however, Hyatt et al. (2000, 2002) reported variations in size among members of the family, suggesting that more specific identification might be possible.

As early as the 1980s, cryoelectron microscopy (cryo-EM) has been used to examine the ultrastructure of the surface of viruses (see Adrian et al. 1984). This technique uses rapid freezing of purified virus, thus avoiding alterations from chemical preservatives. During the procedure, multiple photos are taken and reconstructed to yield a 3D image of the virus. Researchers have been using cryo-EM to evaluate the role of various surface proteins in viral assembly and replication (see Yan et al. 2009; Whitley et al. 2010; Tran et al. 2011). It is possible that these findings may provide insight into treatment protocols or vaccine development.

Other methods include negative contrast EM and immunoEM (see WOAH 2023). Negative contrast EM can be used as a screening process in diagnostic laboratories equipped with an electron microscope. However, as with TEM, it does not identify the virus to the genus. When gold-labeling is combined with labeling by ranavirus-specific antibodies (i.e., immunoEM), it allows for confirming the virus genus as *Ranavirus*.

2.4 Test Validation and Efficiency

2.4.1 Gold Standards and Limitations of Diagnostic Tests

There is no single gold standard test for ranavirus; rather the tests vary based upon the question asked. In the case of disease diagnosis, laboratory results should be interpreted in conjunction with knowledge of the clinical status of the animal, pathological findings, and observed population effects. For example, isolation of ranavirus from tissues along with corresponding necropsy findings (gross and histopathological findings) and ISH/IHC provides strong evidence supportive of ranavirus as an etiology of death. Similarly, the use of diagnostic tests for surveillance or certification of infection-free individuals or zones requires a statistically valid sampling protocol (Gray et al. 2024). Calculation of the sample size required to assess freedom from infection requires knowledge of the minimum expected prevalence and a diagnostic test with known sensitivity and specificity. For example, the prevalence of EHNV infection in rainbow trout can be as low as 4% (Whittington et al. 1994, 1999). Gray et al. (2024) provide recommendations on sample sizes required to detect ranavirus given a range of assumed infection prevalence, and useful online calculators have been developed by epidemiologists for a wide range of applications (https://epitools.ausvet.com.au).

Given that diagnostic tests (e.g., qPCR) are imperfect, estimating the diagnostic sensitivity and specificity of tests is important to accurately interpret results. The manner in which different tests are applied can influence the outcomes of prevalence studies (Wynne et al. 2020). Methods used by laboratories to demonstrate the validity of laboratory test results are outlined in an introductory chapter in the Manual of Diagnostic Tests for Aquatic Animals (WOAH 2023). The WOAH guidelines outline the principles of using a laboratory quality management system to ensure that the results of tests are reliable by documenting the complete suite of measures required to minimize and identify false-positive and false-negative results. In general, several laboratories working in collaboration can estimate false-positive and false-negative results by testing the same samples, which can be used to estimate test sensitivity and specificity. This is accomplished by distributing control samples (i.e., known positive and negative) and standard reagents among laboratories in a blind experimental design and estimating error rates among laboratories. An additional way to evaluate laboratory proficiency is to participate in round-robin testing of sets of blind samples and comparing results to the known composition of the samples and with results from other laboratories running similar sets.

3   Treatment and Vaccine Development

Treatment and vaccination for ranavirus infection are probably most applicable in captive populations; thus, these may be most useful in zoological collections or for conservation programs targeting rare species. The impact of ranaviruses on several fish species important to aquaculture is driving vaccine research for this purpose (Zhang et al. 2020). Although at present there are no vaccines available for ranaviruses in fish, there has been research to develop vaccines for LMBV in largemouth bass (Zhang et al. 2023; Yi et al. 2020; Yao et al. 2022). Vaccine technologies including DNA are also been evaluated for SGIV in grouper (Yu et al. 2019). Vaccines for amphibians are in the early stages of research (Chen et al. 2018). Current treatments for ranavirus infections are limited. Allender et al. (2012) reported the possibility of using guanine analogue antiviral drugs (acyclovir and valacyclovir) to treat chelonians for infection with iridoviruses and herpesviruses. Recently, Li et al. (2014a) reported antiviral activity of DNA aptamers when treating SGIV. Heat treatment is effective at inactivating many pathogens of ectothermic vertebrates such as *Batrachochytrium dendrobatidis* (Woodhams et al. 2003). However, the effectiveness of heat as a treatment for ranavirus likely varies among host species and viral strains. Rojas et al. (2005) reported that salamanders housed at elevated temperatures (26 °C) were more likely to survive exposure to ATV than those at lower temperatures. Similarly, Allender et al. (2013b) found greater pathogenicity of an FV3 at 22 °C compared to 28 °C in red-eared sliders (*Trachemys scripta elegans*). However, several other studies report faster replication and greater pathogenicity of ranaviruses at warmer temperatures (Whittington and Reddacliff 1995; Grant et al. 2003; Ariel and Jensen 2009; Ariel et al. 2009; Bayley et al. 2013; Stilwell et al. 2022). Given that most ranaviruses do not replicate above 32 °C (Chinchar 2002; Ariel et al. 2009), elevating body temperature above this threshold may be useful for some host species. Research is needed to determine the effectiveness of heat treatment, duration required for inac tivation, and if there is variability among host species and viral strains.

To date, the development of vaccines against iridoviruses has primarily been focused on fish species within the aquaculture industry. Some vaccines are commercially available, yet not applicable to all species or cross-reactive to other iridoviruses (Oh et al. 2014). Although live vaccines are often used, DNA vaccines are showing promise. For example, Zhang et al. (2012) found that turbot (*Scophthalmus maximus*) vaccinated with DNA vaccines were more likely to survive infection with rock bream iridovirus than unvaccinated fish. Similarly, Caipang et al. (2006) reported evidence of immunity against infection with red sea bream iridovirus (megalocytivirus) following administration of a DNA vaccine in red sea bream (*Pagrus major*). One aspect of vaccination that often limits its use is that it is delivered by intramuscular injection, which is labor intensive. Other researchers have been exploring more feasible delivery methods, such as oral formulations (Tamaru et al. 2006).

4 Summary Section with Final Recommendations

Ranaviral disease is devastating to susceptible hosts and causes hemorrhage, ulceration, edema, and organ necrosis. Although lesions vary among classes of ectothermic vertebrates, endothelial cell necrosis with subsequent hemorrhage is one of the changes that occurs across all classes. New technologies have enabled application of diagnostic techniques, such as IHC and ISH, that allow us to visualize viral antigens within tissues. Advances in techniques might include laser dissection of lesions followed by IHC and PCR, as well as 3D sequential tomography.

There are limitations to all diagnostic tests; thus, it is important that investigators and researchers use multiple tests for accurate diagnosis and make the distinction between subclinical infection and disease. The most appropriate test to use is dependent upon the question that needs to be answered. For example, if one wants to determine whether an animal harbors a ranavirus, then PCR, qPCR, ISH/IHC, and virus isolation all might provide an answer. Importantly, there is a difference between pathogen detection and determining that a pathogen is associated with the observed clinical disease. There also are different tests recommended for determining prevalence of current or past infection (e.g., qPCR and antibody ELISA, respectively). A typical approach to determine infection status is to apply tests with known sensitivity and specificity to a population using a statistically valid sampling strategy to demonstrate freedom from infection or disease at a minimum expected prevalence (Cameron and Baldock 1998). High-throughput qPCR laboratory methodologies are well suited to handle the large sample sizes required for these surveys. However, we still do not have a test that will verify that an individual animal is free from ranavirus infection, especially using nonlethal sampling. For example, if an animal tests positive by PCR but then negative when retested 2 weeks later, it is unknown if the animal has cleared the virus or if the virus remains hidden in the animal's system. Latent infections by ranavirus may be possible.

We also need a cost-efficient, validated method for detecting and identifying different strains within a given ranavirus species. Current methods for doing so are not widely available and can be cost prohibited; however, future success of reintroduction or translocation programs may hinge on our ability to differentiate between strains within source and destination populations. Likewise, strain identification will be important to include in future ranavirus reporting and mapping programs, as these data can be used in epidemiological studies. Point-of-care qPCR methods, which may be used in situ and yield result in minutes, have been proposed as an alternative to the more traditional in-lab tests. Validations of these methods are crucial to ensure their reliability, however, before they can be applied to management or diagnostic settings.

There are many different protocols available for various tests. The WOAH has guidelines for standardizing tests among laboratories. With the wide range of host species and sample types encountered during ranavirus studies, it is important to develop and share validation data for diagnostic techniques. We recommend that laboratories studying ranaviruses work together to standardize detection protocols. An alternative to standardizing protocols around the world is the implementation of a ring trial (also known as a round-robin), where blinded samples are tested at many different laboratories internationally. With the support of Environment and Climate Change Canada, a round-robin validation test was conducted in 2020–2021. Over 40 countries around the world volunteered to participate and received a report highlighting how their results compared to those of other labs; a summary of the outcomes is available online (www.diagnostics.salamanderfungus.org/2020-round-robin-summary). A round-robin performed every 2 or 3 years would help ensure the reliability and comparability of results produced by laboratories working with different protocols and materials.

There remains much to learn regarding the pathology of ranaviral disease and diagnostic testing for ranavirus infection and disease. Given our current understanding of pathogenesis and recent advances in the genetics of ranavirus and immune response of the host, we should be able to develop effective management and treatment modalities for use in conservation programs, commercial and zoological facilities, and aquaria. Further, our understanding of the disease process coupled with our growing knowledge of the ecology and epidemiology of ranaviruses provides a basis for development of management plans for aquatic ecosystems.

Acknowledgements We thank Robert Ossiboff and Thomas Waltzek for their careful review of this chapter. We thank the following organizations for providing funds to support Open Access publishing of the second edition: University of Tennessee's (UT) Open Publishing Support Fund, UT School of Natural Resources, UT Center for Wildlife Health, Washington State University Libraries, University of Mississippi Medical Center, Gordon State College, Association of Reptile and Amphibian Veterinarians, and the Global Ranavirus Consortium.

References

Adrian M, Dubochet J, Lepault J, McDowall AW (1984) Cryo-electron microscopy of viruses. Nature 308:32–36

Ahne W, Schlotfeldt HJ, Thomsen I (1989) Fish viruses: isolation of an icosahedral cytoplasmic deoxyribovirus from sheatfish *Silurus glanis*. J Veterinary Med Ser B 36:333–336

Ahne W, Bearzotti M, Bremont M, Essbauer S (1998) Comparison of European systemic piscine and amphibian iridoviruses with epizootic haematopoietic necrosis virus and frog virus 3. J Vet Med 45:373–383

Allender MC (2012) Characterizing the epidemiology of ranavirus in North American chelonians: diagnosis, surveillance, pathogenesis, and treatment. Dissertation, University of Illinois at Urbana-Champaign, Urbana-Champaign, p 219

Allender MC, Fry MM, Irizarry AR, Craig L, Johnson AJ, Jones M (2006) Intracytoplasmic inclusions in circulating leukocytes from an eastern box turtle (*Terrapene carolina carolina*). J Wildl Dis 42:677–684

Allender MC, Abd-Eldaim M, Schumacher J, McRuer D, Christian LS, Kennedy M (2011) PCR prevalence of Ranavirus in free-ranging eastern box turtles (*Terrapene carolina carolina*) at rehabilitation centers in three southeastern US states. J Wildl Dis 47:759–764

Allender MC, Mitchell MA, Yarborough J, Cox S (2012) Pharmacokinetics of a single oral dose of acyclovir and valacyclovir in North American box turtles (*Terrapene* sp). J Vet Pharmacol Ther 36:205–208

Allender MC, Bunick D, Mitchell MA (2013a) Development and validation of Taqman quantitative PCR for detection of frog virus 3-like virus in eastern box turtles (*Terrapene carolina caro lina*). J Virol Methods 188:121–125

Allender MC, Mitchell MA, Torres T, Sekowska J, Riskell EA (2013b) Pathogenicity of frog virus 3-like virus in red-eared slider turtles at two environmental temperatures. J Comp Pathol 149:356–367

Ariel E (1997) Pathology and serological aspects of Bohle iridovirus infections in six selected water-associated reptiles in North Queensland. Ph.D. Dissertation, James Cook University, Townsville City, p 214

Ariel E, Jensen BB (2009) Challenge studies of European stocks of redfin perch, *Perca fluviatilis* L., and rainbow trout, *Oncorhynchus mykiss* (Walbaum), with epizootic haematopoietic necrosis virus. J Fish Dis 32(12):1017–1025

Ariel E, Nicolajsen N, Christophersen MB, Holopainen R, Tapiovaara H, Jensen BB (2009) Propagation and isolation of ranaviruses in cell culture. Aquaculture 294:159–164

Ariel E, Elliott E, Meddings JI, Miller J, Santos MB, Owens L (2017) Serological survey of Australian native reptiles for exposure to ranavirus. Dis Aquat Org 126:173–183

AVMA (2020) AVMA guidelines for the euthanasia of animals, 2020th edn. American Veterinary Medical Association, p 121

Balseiro A, Dalton KP, del Cerro A, Marquez I, Cunningham AA, Parra F, Prieto JM, Casais R (2009) Pathology, isolation and molecular characterization of a ranavirus from the common midwife toad *Alytes obstetricans* on the Iberian Peninsula. Dis Aquat Org 84:95–104

Balseiro A, Dalton KP, del Cerro A, Marquez I, Parra F, Prieto JM, Casais R (2010) Outbreak of common midwife toad virus in alpine newts (*Mesotriton alpestris cyreni*) and common midwife toads (*Alytes obstetricans*) in Northern Spain: a comparative pathological study of an emerging ranavirus. Vet J 186:256–258

Bang Jensen B, Holopainen R, Tapiovaara H, Ariel E (2011) Susceptibility of pike-perch *Sander lucioperca* to a panel of ranavirus isolates. Aquaculture 313:24–30

Bayley AE, Hill BJ, Feist SW (2013) Susceptibility of the European common frog *Rana temporaria* to a panel of ranavirus isolates from fish and amphibian hosts. Dis Aquat Org 103:171–183. https://doi.org/10.3354/dao02574

Beck BH, Bakal RS, Brunner CJ, Grizzle JM (2006) Virus distribution and signs of disease after immersion exposure to largemouth bass virus. J Aquat Anim Health 18:176–183

Becker JA, Tweedie A, Gilligan D, Asmus M, Whittington RJ (2013) Experimental infection of Australian freshwater fish with epizootic haematopoietic necrosis virus (EHNV). J Aquat Anim Health 25:66–76. https://doi.org/10.1080/08997659.2012.747451

Becker JA, Gilligan D, Asmus M, Tweedie A, Whittington RJ (2019) Geographic distribution of Epizootic haematopoietic necrosis virus (EHNV) in freshwater fish in South Eastern Australia: lost opportunity for a notifiable pathogen to expand its geographic range. Viruses-Basel 11(4):315

Beebee T (2012) Impact of Ranavirus on garden amphibian populations. Herpetol Bull 120:1–3

Behncke H, Stöhr AC, Heckers K, Ball I, Marschang RE (2013) Mass-mortality in green striped tree dragons (*Japalura splendida*) associated with multiple viral infections. Vet Rec 173:248. https://doi.org/10.1136/vr.101545

Bollinger TK, Mao J, Schock D, Brigham RM, Chinchar VG (1999) Pathology, isolation, and preliminary molecular characterization of a novel iridovirus from tiger salamanders in Saskatchewan. J Wildl Dis 35:413–429

Borzym E, Maj-Paluch J (2015) Experimental infection with epizootic haematopoietic necrosis virus (*EHNV*) of rainbow trout (*Oncorhynchus mykiss Walbaum*) and European perch (*Perca fluviatilis L.*). Bull Vet Inst Pulawy 59:473–477. https://doi.org/10.1515/bvip-2015-0070

Brenes R, Gray MJ, Waltzek TB, Wilkes RP, Miller DL (2014a) Transmission of ranavirus between ectothermic vertebrate hosts. PLoS One 9(3):e92476. https://doi.org/10.1371/journal.pone.0092476

Brenes R, Miller DL, Waltzek TB, Wilkes RP, Tucker JL, Chaney JC, Hardman RH, Brand MD, Huether RR, Gray MJ (2014b) Susceptibility of fish and turtles to three ranaviruses isolated from different ectothermic vertebrate classes. J Aquat Anim Health 26(2):118–126. https://doi.org/10.1080/08997659.2014.886637

Brunner JL, Schock DM, Davidson EW, Collins JP (2004) Intraspecific reservoirs: complex life cycles and the persistence of a lethal ranavirus. Ecology 85:560–566

Brunner JL, Storfer A, LeSage EH, Garner TWJ, Gray MJ, Hoverman JT (2024) Ranavirus ecology: from individual infections to population epidemiology to community impacts. In: Gray MJ, Chinchar VG (eds) Ranaviruses: lethal pathogens of ectothermic vertebrates. Springer, New York

Bryan LK, Baldwin CA, Gray MJ, Miller DL (2009) Efficacy of select disinfectants at inactivating Ranavirus. Dis Aquat Org 84(2):89–94

Burton EC, Miller DL, Styer EL, Gray MJ (2008) Amphibian ocular malformation associated with frog virus 3. Vet J 177:442–444

Bustin SA, Beaulieu JF, Huggett J, Jaggi R, Kibenge FSB, Olsvik PA, Penning LC, Toegel S (2010) MIQE precis: practical implementation of minimum standard guidelines for fluorescence-based quantitative real-time PCR experiments. BMC Mol Biol 11:74. https://doi.org/10.1186/1471-2199-11-74

Caipang CMA, Takano T, Hirono I, Aoki T (2006) Genetic vaccines protect red seabream, *Pagrus major*, upon challenge with red seabream iridovirus (RSIV). Fish Shellfish Immunol 21:130–138

Cameron AR, Baldock FC (1998) A new probability formula for surveys to substantiate freedom from disease. Prev Vet Med 34:1–17

Caraguel CGB, Stryhn H, Gagne N, Dohoo IR, Hammell KL (2011) Selection of a cutoff value for real-time polymerase chain reaction results to fit a diagnostic purpose: analytical and epidemiologic approaches. J Vet Diagn Invest 23:2–15

Cashins S, Skerratt FL, Alford RA (2008) Sodium hypochlorite denatures the DNA of the amphibian chytrid fungus *Batrachochytrium dendrobatidis*. Dis Aquat Org 80:63–67

Chen ZY, Li T, Gao ZC, Wang CF, Zhang QY (2018) Protective immunity induced by DNA vaccination against Ranavirus infection in Chinese Giant Salamander *Andrias davidianus*. Viruses 10:52. https://doi.org/10.3390/v10020052

Cheng K, Jones MEB, Jancovich JK, Burchell J, Schrenzel MD, Reavill DR, Imai DI, Urban A, Kirkendall M, Woods LW, Chinchar VG, Pessier AP (2014) Isolation of a Bohle-like iridovirus from boreal toads housed within a cosmopolitan aquarium collection. Dis Aquat Org 111(2):139–152

Chinchar VG (2002) Ranaviruses (family *Iridoviridae*): emerging cold-blooded killers. Arch Virol 147:447–470

Chua FH, Ng ML, Ng KL, Loo JJ, Wee JY (1994) Investigation of outbreaks of a novel disease, 'Sleepy Grouper Disease', affecting the brown-spotted grouper, *Epinephelus tauvina* *Forskal*. J Fish Dis 17(4):417–427

Cozad RA, Norton TM, Aresco MJ, Allender MC, Hernandez SM (2020) Pathogen surveillance and detection of Ranavirus (*Frog virus 3*) in translocated gopher tortoises (*Gopherus polyphemus*). J Wildl Dis 56(3):679–683. https://doi.org/10.7589/2019-02-053

Crane MSJ, Young J, Williams L (2005) Epizootic haematopoietic necrosis virus (EHNV): growth in fish cell lines at different temperatures. Bull Eur Assoc Fish Pathol 25:228–231

Cunningham AA, Langton TES, Bennett PM, Lewin JF, Drury SEN, Gough RE, Macgregor SK (1996) Pathological and microbiological findings from incidents of unusual mortality of the common frog (*Rana temporaria*). Philos Trans R Soc Lond Ser B Biol Sci 351:1539–1557

Cunningham AA, Hyatt AD, Russell P, Bennett PM (2007) Emerging epidemic diseases of frogs in Britain are dependent on the source of ranavirus agent and the route of exposure. Epidemiol Infect 135:1200–1212. https://doi.org/10.1017/S0950268806007679

Cunningham AA, Tems CA, Russell PH (2008) Immunohistochemical demonstration of Ranavirus antigen in the tissues of infected frogs (*Rana temporaria*) with systemic haemorrhagic or cutaneous ulcerative disease. J Comp Pathol 138:3–11

Deng G, Li S, Xie J, Bai J, Chen K, Ma D, Jiang X, Lao H, Yu L (2011) Characterization of a ranavirus isolated from cultured largemouth bass (Micropterus salmoides) in China. Aquaculture 312:198–204

Deng L, Geng Y, Zhao R, Gray MJ, Wang K, Ouyang P, Chen D, Huang X, Chen Z, Huang C, Zhong Z, Guo H (2020) Fang J (2020) CMTV-like ranavirus infection associated with high mortality in captive catfish-like loach, *Triplophysa siluroides*, in China. Transbound Emerg Dis 67:1330–1335. https://doi.org/10.1111/tbed.13473

DeVoe R, Geissler K, Elmore S, Rotstein D, Lewbart G, Guy J (2004) Ranavirus-associated morbidity and mortality in a group of captive eastern box turtles (*Terrapene carolina carolina*). J Zoo Wildl Med 35:534–543

Docherty DE, Meteyer CU, Wang J, Mao J, Case ST, Chinchar VG (2003) Diagnostic and molecular evaluation of three iridovirus-associated salamander mortality events. J Wildl Dis 39(3):556–566

Dong CF, Wang ZM, Weng SP, He JG (2017) Occurrence of a lethal ranavirus in hybrid mandarin (Siniperca scherzeri x Siniperca chuatsi) in Guangdong, South China. Vet Microbiol 203:28–33

Duffus AL, Andrews AM (2013) Phylogenetic analysis of a frog virus 3-like ranavirus found at a site with recurrent mortality and morbidity events in southeastern Ontario, Canada: partial major capsid protein sequence alone is not sufficient for fine-scale differentiation. J Wildl Dis 49:464–467

Earl JE, Gray MJ (2014) Introduction of ranavirus to isolated wood frog populations could cause local extinction. EcoHealth. https://doi.org/10.1007/s10393-014-0950-y

Farnsworth SD, Seigel FA (2013) Responses, movements, and survival of relocate box turtles during construction of the Intercounty Connector Highway in Maryland. J Transp Res 2362:1–8

Forzan MJ, Wood J (2013) Low detection of ranavirus DNA in wild postmetamorphic green frogs, *Rana (Lithobates) clamitans*, despite previous or concurrent tadpole mortality. J Wildl Dis 49:879–886

Forzán MJ, Jones KM, Vanderstichel RV, Wood J, Kibenge FS, Kuiken T, Wirth W, Ariel E, Daoust PY (2015) Clinical signs, pathology and dose-dependent survival of adult wood frogs, Rana sylvatica, inoculated orally with frog virus 3 Ranavirus sp. Iridoviridae J Gener Virol 96:1138–1149

Forzán MJ, Smith TG, Vanderstichel RV, Hogan NS, Gilroy CV (2016) Hematologic reference intervals for Rana sylvatica (Lithobates sylvaticus) and effect of infection with Frog Virus 3 (Ranavirus sp., Iridoviridae). Vet Clin Pathol 45(3):430–443

Forzán MJ, Jones KM, Ariel E, Whittington RJ, Wood J, Markham RF, Daoust PY (2017) Pathogenesis of frog virus 3 (Ranavirus, Iridoviridae) infection in wood frogs (Rana sylvatica). Vet Pathol 54:531–548

Forzán MJ, Bienentreu J, Schock DM, Lesbarrères D (2019) Multi-tool diagnosis of an outbreak of ranavirosis in amphibian tadpoles in the Canadian boreal forest. Dis Aquat Org 135:33–41

Gantress J, Maniero GD, Cohen N, Robert J (2003) Development and characterization of a model system to study amphibian immune responses to iridoviruses. Virology 311:254–262

Geng Y, Wang KY, Zhou ZY, Li CW, Wang J, He M, Yin ZQ, Lai WM (2010) First report of a ranavirus associated with morbidity and mortality in farmed Chinese giant salamanders (*Andrias davidianus*). J Comp Pathol. https://doi.org/10.1016/j.jcpa.2010.11.012

Gold KK, Reed PD, Bemis DA, Miller DL, Gray MJ, Souza MJ (2013) Efficacy of common disinfectants and terbinafine HCl at inactivating the growth of *Batrachochytrium dendrobatidis* in culture. Dis Aquat Org 107:77–81

Goldberg TL (2002) Largemouth bass virus: an emerging problem for warmwater fisheries? Am Fish Soc Symp 31:411–416

Goodman RM, Miller DL, Ararso YT (2013) Prevalence of ranavirus in Virginia turtles as detected by tail-clip sampling versus oral-cloacal swabbing. Northeast Nat 20:325–332

Granoff A, Came PE, Breeze DC (1966) Viruses and renal carcinoma of Rana Pipiens .I. Isolation and properties of virus from normal and tumor tissue. Virology 29:133–148

Grant EC, Philipp DP, Inendino KR (2003) Effects of temperature on the susceptibility of largemouth bass to largemouth bass virus. J Aquat Anim Health 15:215–220

Gray MJ, Miller DL, Schmutzer AC, Baldwin CA (2007) Frog virus 3 prevalence in tadpole populations inhabiting cattle-access and non-access wetlands in Tennessee, U.S.A. Dis Aquat Org 77:97–103

Gray MJ, Miller DL, Hoverman JT (2009) First report of ranavirus infecting lungless salamanders. Herpetol Rev 40(3):316–319

Gray MJ, Miller DL, Hoverman JT (2012) Reliability of non-lethal surveillance methods for detecting ranavirus infection. Dis Aquat Org 99:1–6. https://doi.org/10.3354/dao02436

Gray MJ, Brunner JL, Earl JE, Wirth W, Peace A, Ariel E (2024) Design and analysis of ranavirus studies: insights into planning surveillance, modeling host-pathogen dynamics, and performing risk analyses. In: Gray MJ, Chinchar VG (eds) Ranaviruses: lethal pathogens of ectothermic vertebrates. Springer, New York

Grayfer L, Edholm E-S, Chinchar VG, Sang Y, Robert J (2024) Immune defenses against ranavirus infections. In: Gray MJ, Chinchar VG (eds) Ranaviruses: lethal pathogens of ectothermic vertebrates. Springer, New York

Green DE, Converse KA, Schrader AK (2002) Epizootiology of sixty-four amphibian morbidity and mortality events in the USA, 1996-2001. Ann N Y Acad Sci 969:323–339

Green DE, Gray MJ, Miller DL (2009) Disease monitoring and biosecurity. In: Dodd CK (ed) Amphibian ecology and conservation: a handbook of techniques. Oxford University Press, Oxford

Greer AL, Collins JP (2007) Sensitivity of a diagnostic test for amphibian ranavirus varies with sampling protocol. J Wildl Dis 43:525–532

Greiner M, Gardner IA (2000) Epidemiologic issues in the validation of veterinary diagnostic tests. Prev Vet Med 45:3–22

Grizzle JM, Brunner CJ (2003) Review of largemouth bass virus. Fisheries 28(11):10–14. https://doi.org/10.1577/1548-8446(2003)28[10:ROLBV]2.0.CO;2

Grosset C, Wellehan JF, Owens SD, McGraw S, Gaffney PM, Foley J, Childress AL, Yun S, Malm K, Groff JM, Paul-Murphy J, Weber ES (2014) Intraerythrocytic iridovirus in central bearded dragons (*Pogona vitticeps*). J Vet Diagn Invest 26:354–364

Haislip NA, Gray MJ, Hoverman JT, Miller DL (2011) Development and disease: how susceptibility to an emerging pathogen changes through anuran development. PLoS One 6(7):e22307. https://doi.org/10.1371/journal.pone

Hanson LA, Petrie-Hanson L, Meals KO, Chinchar VG, Rudis M (2001) Persistence of largemouth bass virus infection in a northern Mississippi reservoir after a die-off. J Aquat Anim Health 13:27–34

Harp EM, Petranka JW (2006) Ranavirus in wood frogs (*Rana sylvatica*): potential sources of transmission within and between ponds. J Wildl Dis 42:307–318

Hartmann AM, Maddox ML, Ossiboff RJ, Longo AV (2022) Sustained Ranavirus outbreak causes mass mortality and morbidity of imperiled amphibians in Florida. EcoHealth 19:8–14

Hedrick RP, McDowell TS (1995) Properties of iridoviruses from ornamental fish. Vet Res 26:423–427

Hedrick RP, McDowell TS, Ahne W, Torhy C, De Kinkelin P (1992) Properties of three iridovirus-like agents associated with systemic infections of fish. Dis Aquat Org 13:203–209

Hengstberger SG, Hyatt AD, Speare R, Coupar BEH (1993) Comparison of epizootic haematopoietic necrosis and Bohle iridoviruses, recently isolated Australian iridoviruses. Dis Aquat Org 15:93–107

Hick PM, Subramaniam K, Thompson PM, Waltzek TB, Becker JA, Whittington RJ (2017) Molecular epidemiology of Epizootic haematopoietic necrosis virus (EHNV). Virology 511:320–329

Holopainen R, Ohlemeyer S, Schutze H, Bergmann SM, Tapiovaara H (2009) Ranavirus phylogeny and differentiation based on major capsid protein, DNA polymerase and neurofilament triplet H1-like protein genes. Dis Aquat Org 85:81–91

Hoverman JT, Gray MJ, Haislip NA et al (2011) Phylogeny, life history, and ecology contribute to differences in amphibian susceptibility to ranaviruses. EcoHealth 8:301–319

Huang C, Zhang X, Gin KYH, Qin QW (2004) *In situ* hybridization of a marine fish virus, Singapore grouper iridovirus with a nucleic acid probe of major capsid protein. J Virol Methods 117:123–128

Hyatt AD, Eaton BT, Hengstberger S, Russel G (1991) Epizootic haematopoietic necrosis virus: detection by ELISA, immunohistochemistry and immunoelectronmicroscopy. J Fish Dis 14:605–617

Hyatt AD, Gould AR, Zupanovic Z, Cunningham AA, Hengstberger S, Whittington RJ, Kattenbelt J, Coupar BEH (2000) Comparative studies of piscine and amphibian iridoviruses. Arch Virol 145:301–331

Hyatt AD, Williamson M, Coupar BEH, Middleton D, Hengstberger SG, Gould AR, Selleck P, Wise TG, Kattenbelt J, Cunningham AA, Lee J (2002) First identification of a ranavirus from green pythons (*Chondropython viridis*). J Wildl Dis 38(2):239–252

Jancovich JK, Davidson EW, Morado JF, Jacobs BL, Collins JP (1997) Isolation of a lethal virus from the endangered tiger salamander *Ambystoma tigrinum stebbinsi*. Dis Aquat Org 31:161–167

Jancovich JK, Davidson EW, Parameswaran N, Mao J, Chinchar VG, Collins JP, Jacobs BC, Storfer A (2005) Evidence for emergence of an amphibian iridoviral disease because of human-enhanced spread. Mol Ecol 14:213–224

Jancovich JK, Steckler N, Waltzek TB (2015) Ranavirus taxonomy and phylogeny. In: Gray MJ, Chinchar VG (eds) Ranaviruses: lethal pathogens of ectothermic vertebrates. Springer, New York

Jancovich JK, Qin Q, Zhang Q-Y, Chinchar VG (2024) Ranavirus replication: new studies provide answers to old questions. In: Gray MJ, Chinchar VG (eds) Ranaviruses: lethal pathogens of ectothermic vertebrates. Springer, New York

Jaramillo D, Tweedie A, Becker JA, Hyatt A, Crameri S, Whittington RJ (2012) A validated quantitative polymerase chain reaction assay for the detection of ranaviruses (Family *Iridoviridae*) in fish tissue and cell cultures, using EHNV as a model. Aquaculture 356:186–192

Jaramillo D, Peeler EJ, Laurin E, Gardner IA, Whittington RJ (2017) Serology in finfish for diagnosis, surveillance, and research: a systematic review. J Aquat Anim Health 29(1):1–14

Jerrett RJ IV, Whittington, Weir RP (2015) Pathology of a Bohle-like virus infection in two Australian frog species (Litoria splendida and Litoria caerulea). J Comp Pathol 152:248e259. https://doi.org/10.1016/j.jcpa.2014.12.007

Johnson AJ, Pessier AP, Wellehan JF, Brown R, Jacobson ER (2005) Identification of a novel herpesvirus from a California desert tortoise (*Gopherus agassizii*). Vet Microbiol 111:107–116

Johnson AJ, Pessier AP, Jacobson ER (2007) Experimental transmission and induction of ranaviral disease in western ornate box turtles (*Terrapene ornata ornata*) and red-eared sliders (*Trachemys scripta elegans*). Vet Pathol 44:285–297

Johnson AJ, Pessier AP, Wellehan JFX, Childress A, Norton TM, Stedman NL, Bloom DC, Belzer W, Titus VR, Wagner R, Brooks JW, Spratt J, Jacobson ER (2008) Ranavirus infection of free- ranging and captive box turtles and tortoises in the United States. J Wildl Dis 44:851–863

Johnson AJ, Wendland L, Norton TM, Belzer B, Jacobson ER (2010) Development and use of an indirect enzyme-linked immunosorbent assay for detection of iridovirus exposure in gopher tortoises (*Gopherus polyphemus*) and eastern box turtles (*Terrapene carolina carolina*). Vet Microbiol 142:160–167

Kattenbelt JA, Hyatt AD, Gould AR (2000) Recovery of ranavirus dsDNA from formalin-fixed archival material. Dis Aquat Org 39:151–154

Kayansamruaj P, Rangsichol A, Dong HT, Rodkhum C, Maita M, Katagiri T, Pirarat N (2017) Outbreaks of ulcerative disease associated with ranavirus infection in barcoo grunter, Scortum barcoo (McCulloch & Waite). J Fish Dis 40:1341–1350

Kik M, Martel A, Spitzen-vander Sluijs A, Pasmans F, Wohlsein P, Grone A, Rijks JM (2011) Ranavirus-associated mass mortality in wild amphibians, The Netherlands, 2010: a first report. Vet J 190:284–286

Langdon JS, Humphrey JD, Williams LM, Hyatt AD, Westbury HA (1986) First virus isolation from Australian fish: an iridovirus-like pathogen from redfin perch, *Perca fluviatilis* L. J Fish Dis 9:263–268

Leung WTM, Thomas-Walters L, Garner TWJ, Balloux F, Durrant C, Price SJ (2017) A quantitative-PCR based method to estimate ranavirus viral load following normalisation by reference to an ultraconserved vertebrate target. J Virol Methods 249:147–155. https://doi.org/10.1016/j.jviromet.2017.08.016

Li P, Yan Y, Wei S, Weiv J, Gao R, Huang X, Huang Y, Jiang G, Qin Q (2014a) Isolation and characterization of a new class of DNA aptamers specific binding to Singapore grouper iridovirus (SGIV) with antiviral activities. Virus Res 188:146–154

Li W, Zhang X, Weng S, Zhao G, He J, Dong C (2014b) Virion-associated viral proteins of a Chinese giant salamander (*Andreas davidianus*) iridovirus (genus *Ranavirus*) and functional study of the major capsid protein (MCP). Vet Microbiol 172:129–139

Liu J, Zhang X, Zheng J, Yu Y, Huang X, Wei J, Mukama O, Wang S, Quin Q (2021) A lateral flow biosensor for rapid detection of Singapore grouper iridovirus (SGIV). Aquaculture 541. https://doi.org/10.1016/j.aquaculture.2021.736756

Liu F, Tian S, Feng Y, Qin Z, Geng Y, Ou Y, Chen D, Huan X, Guo H, Zuo Z, Deng H, Lai W (2023) First report of a CMTV-like Ranavirus in farmed *Percocypris pingi* in China. Aquaculture 574. https://doi.org/10.1016/j.aquaculture.2023.739701

Ma J, Zeng L, Zhou Y, Jiang N, Zhang H, Fan Y, Meng Y, Xu J (2014) Ultrastructural morphogenesis of an amphibian iridovirus isolated from Chinese giant salamander (*Andrias davidianus*). J Comp Pathol 150:325–331

Maclaine A, Mashkour N, Scott J, Ariel E (2018) Susceptibility of eastern water dragons Intellagama lesueurii lesueurii to Bohle iridovirus. Dis Aquat Org 127:97–105

Maclaine A, Forzan MJ, Mashkour N, Scott J, Ariel E (2019) Pathogenesis of Bohle Iridovirus (Genus Ranavirus) in experimentally infected Juvenile Eastern Water Dragons (Intellagama lesueurii lesueurii). Vet Pathol 56:465–475. https://doi.org/10.1177/0300985818823666

Mao J, Hedrick RP, Chinchar VG (1997) Molecular characterization, sequence analysis, and taxonomic position of newly isolated fish iridoviruses. Virology 229:212–220

Mao JD, Green E, Fellers G, Chinchar VG (1999) Molecular characterization of iridoviruses isolated from sympatric amphibians and fish. Virus Res 63:45–52

Marschang RE, Braun S, Becher P (2005) Isolation of a ranavirus from a gecko (*Uroplatus fimbriatus*). J Zoo Wildl Med 36:295–300

Marschang RE, Meddings J, Waltzek TB, Hick P, Allender MC, Wirth W, Duffus ALJ (2024) Distribution and host range of ranaviruses. In: Gray MJ, Chinchar VG (eds) Ranaviruses: lethal pathogens of ectothermic vertebrates. Springer, New York

Marsh IB, Whittington RJ, O'Rourke B, Hyatt AD, Chisholm O (2002) Rapid differentiation of Australian, European and American ranaviruses based on variation in major capsid protein gene sequence. Mol Cell Probes 16:137–151

Martínez-Silvestre A, Perpiñán D, Marco I, Lavín S (2002) Venipuncture technique of the occipital venous sinus in freshwater aquatic turtles. J Herpetol Med Surg 12:31–32

Mazzoni R, de Mesquita AJ, Fleury LFF, deBrito WMED, Nunes IA, Robert J, Morales H, Coelho ASG, Barthasson DL, Galli L, Catroxo MHB (2009) Mass mortality associated with a frog virus 3-like Ranavirus infection in farmed tadpoles Rana catesbeiana from Brazil. Dis Aquat Org 86:181–191

McClenahan SD, Beck BH, Grizzle JM (2005) Evaluation of cell culture methods for detection of largemouth bass virus. J Aquat Anim Health 17:365–372

McLoughlin MF, Graham DA (2007) Alphavirus infections in salmonids–a review. J Fish Dis 30:511–531

Meng Y, Ma J, Jiang N, Zeng LB, Xiao HB (2014) Pathological and microbiological findings from mortality of the Chinese giant salamander (*Andrias davidianus*). Arch Virol 159:1403–1412. https://doi.org/10.1007/s00705-013-1962-6

Miller DL, Gray MJ (2010) Amphibian decline and mass mortality: the value of visualizing ranavirus in tissue sections. Vet J 186:133–134

Miller DL, Rajeev S, Gray MJ, Baldwin C (2007) Frog virus 3 infection, cultured American bullfrogs. Emerg Infect Dis 13:342–343

Miller DL, Rajeev S, Brookins M, Cook J, Whittington L, Baldwin CA (2008) Concurrent infection with *Ranavirus, Batrachochytrium dendrobatidis*, and *Aeromonas* in a captive anuran colony. J Zoo Wildl Med 39:445–449

Miller DL, Gray MJ, Strofer A (2011) Ecopathology of ranaviruses infecting amphibians. Viruses 3:2351–2373. https://doi.org/10.3390/v3112351

Monini M, Ruggeri FM (2002) Antigenic peptides of the epizootic hematopoietic necrosis virus. Virology 297:8–18

Oh SY, Oh MJ, Nishizawa T (2014) Potential for a live red seabream iridovirus (RSIV) vaccine in rock bream *Oplegnathus fasciatus* at a low rearing temperature. Vaccine 32:363–368

Pallister J, Gould A, Harrison D, Hyatt A, Jancovich J, Heine H (2007) Development of real-time PCR assays for the detection and differentiation of Australian and European ranaviruses. J Fish Dis 30:427–438

Pessier AP, Mendelson JR (2010) A manual for control of infectious diseases in amphibian survival assurance colonies and reintroduction programs. IUCN/SSC Captive Breeding Specialist Group, Apple Valley

Petranka JW, Murray SM, Apple Valley MN, Kennedy CA (2003) Response of amphibians to restoration of a southern Appalachian wetland: perturbations confound post-restoration assessment. Wetlands 23:278–290

Pfeiffer LB, Sander S, Gabrielson K, Pessier AP, Allender MC, Waltzek T, Subramaniam K, Stilwell N, Adamovicz L, Bronson E, Mangus LM (2019) Fatal *Ranavirus* infection in a group of zoo-housed Meller's chameleons (*Trioceros melleri*). J Zoo Wildl Med 50(3):696–705. https://doi.org/10.1638/2018-0044

Picco AM, Brunner JL, Collins JP (2007) Susceptibility of the endangered California tiger salamander, *Ambystoma californiense*, to ranavirus infection. J Wildl Dis 43:286–290

Plumb JA, Grizzle JM, Young HE, Noyes AD, Lamprecht S (1996) An iridovirus isolated from wild largemouth bass. J Aquat Anim Health 8:265–270

Plumb JA, Noyes AD, Graziano S, Wang J, Mao JH, Chinchar VG (1999) Isolation and identification of viruses from adult largemouth bass during a 1997-1998 survey in the southeastern United States. J Aquat Anim Health 11:391–399

Pozet F, Morand M, Moussa A, Torhy C, de Kinkelin P (1992) Isolation and preliminary characterization of a pathogenic icosahedral deoxyribovirus from the catfish *Ictalurus melas*. Dis Aquat Org 14:35–42

Price SJ, Garner TWJ, Nichols RA et al (2014) Collapse of amphibian communities due to an introduced Ranavirus. Curr Biol 24:2586–2591. http://www.cell.com/currentbiology/pdfEx- tended/S0960-9822(14)01149-X

Qin QW, Chang SF, Ngoh-Lim GH, Gibson-Kueh S, She C, Lam TJ (2003) Characterization of a novel ranavirus isolated from grouper *Epinephelus tauvina*. Dis Aquat Org 53:1–9

Reddacliff LA, Whittington RJ (1996) Pathology of epizootic haematopoietic necrosis virus (EHNV) infection in Rainbow Trout (*Oncorhynchus mykiss* Walbaum) and Redfin Perch (*Perca fluviatilis* L.). J Comp Pathol 115:103–115

Rimmer AE, Becker JA, Tweedie A, Whittington RJ (2012) Validation of high throughput methods for tissue disruption and nucleic acid extraction for ranaviruses (family *Iridoviridae*). Aquaculture 338:23–28

Rivera S, Wellehan JF, McManamon R, Innis CJ, Garner MM, Raphael BL, Gregory CR, Latimer KS, Rodriguez CE, Diaz-Figueroa O, Marlar AB, Nyaoke A, Gates AE, Gilbert K, Childress AL, Risatti GR, Frasca S (2009) Systemic adenovirus infection in Sulawesi tortoises (*Indotestudo forstenii*) caused by a novel siadenovirus. J Vet Diagn Invest 21:415–426

Robert J, Abramowitz L, Gantress J, Morales HD (2007) *Xenopus laevis*: a possible vector of ranavirus infection? J Wildl Dis 43:645–652

Robert J, George E, De Jesús AF, Chen G (2011) Waterborne infectivity of the ranavirus frog virus 3 in *Xenopus laevis*. Virology 417(2):410–417

Rojas S, Richards K, Jancovich JK, Davidson E (2005) Influence of temperature on Ranavirus infection in larval salamanders *Ambystoma tigrinum*. Dis Aquat Org 63:95–100

Rothermel BB, Travis ER, Miller DL, Hill RL, McGuire JL, Yabsley MJ (2013) High occupancy of stream salamanders despite high Ranavirus prevalence in a Southern Appalachians Watershed. EcoHealth 10:184–189

Ruder MG, Allison AB, Miller DL, Keel MK (2010) Pathology in practice: ranaviral disease in a box turtle. J Am Vet Med Assoc 237:783–785

Schock DM, Bollinger TK, Chinchar VG, Jancovich JK, Collins JP (2008) Experimental evidence that amphibian ranaviruses are multi-host pathogens. Copeia 2008:133–143

Standish I, Leis E, Schmitz N, Credico J, Erickson S, Bailey J, Kerby J, Phillips K, Lewis T (2018) Optimizing, validating, and field testing a multiplex qPCR for the detection of amphibian pathogens. Dis Aquat Org 129:1–13. https://doi.org/10.3354/dao03230

Steiner KA, Whittington RJ, Petersen RK, Hornitzky C, Garnet H (1991) Purification of epizootic hematopoietic necrosis virus and its detection using ELISA. J Virol Methods 33:199–210

Stilwell NK (2017) Ranaviruses in aquaculture: genetic diversity, improved molecular tools, and experimental analysis of husbandry factors influencing morbidity. University of Florida. Dissertation p 146

Stilwell NK, Whittington RJ, Hick PM, Becker JA, Ariel E, Van Beurden S, Vendramin N, Olesen NJ, Waltzek TB (2018) Partial validation of a TaqMan real-time quantitative PCR for the detection of ranaviruses. Dis Aquat Org 128:105–116

Stilwell NK, Frasca S Jr, Farina LL, Subramaniam K, Imnoi K, Viadanna PH, Hopper L, Powell J, Colee J, Waltzek TB (2022) Effect of water temperature on frog virus 3 disease in hatchery-reared pallid sturgeon *Scaphirhynchus albus*. Dis Aquat Org 148:73–86. https://doi.org/10.3354/dao03645

Stöhr AC, Blahak S, Heckers KO, Wiechert J, Behncke H, Mathes K, Gunther P, Zwart P, Ball I, Marschang RERB (2013) Ranavirus infections associated with skin lesions in lizards. Vet Res 44:84. http://www.veterinaryresearch.org/content/44/1/84

Sutton WB, Gray MJ, Hardman RH, Wilkes RP, Kouba A, Miller DL (2014a) High susceptibility of the endangered dusky gopher frog to ranavirus. Dis Aquat Org 112(1):9–16

Sutton WB, Gray MJ, Hoverman JT, Secrist RG, Super P, Hardman RH, Tucker JL, Miller DL (2014b) Trends in ranavirus prevalence among plethodontid salamanders in the Great Smoky Mountains National Park. EcoHealth. https://doi.org/10.1007/s10393-014-0994-z

Tamaru Y, Ohtsuka M, Kato K, Manabe S, Kuroda K, Sanada M, Ueda M (2006) Application of the arming system for the expression of the 380R antigen from red sea bream iridovirus (RSIV) on the surface of yeast cells: a first step for the development of an oral vaccine. Biotechnol Prog 22:949–953

Teacher AGF, Cunningham AA, Garner WJ (2010) Assessing the long-term impact of ranavirus infection in wild common frog populations. Anim Conserv 13:514–522

Todd-Thompson M (2010) Seasonality, variation in species prevalence, and localized disease for ranavirus in Cades Cove (Great Smoky Mountains National Park) amphibians. Master Thesis, University of Tennessee, Knoxville. http://trace.tennessee.edu/utk_gradthes/665. Accessed 2 Sep 2014

Tran BN, Chen L, Liu Y, Wu J, Velazquez-Campoy A, Sivaraman J, Hew CL (2011) Novel histone H3 binding protein ORF158L from the Singapore grouper iridovirus. J Virol 85:9159–9166. https://doi.org/10.1128/JVI.02219-10

Waltzek TB, Miller DL, Gray MJ, Drecktrah B, Briggler JT, MacConnell B, Hudson C, Hopper L, Friary J, Yun SC, Malm KV, Weber ES, Hedrick RP (2014) Expansion of the host range of frog virus 3 into hatchery-reared pallid sturgeon *Scaphirhynchus albus*. Dis Aquat Org 111:219–227. https://doi.org/10.3354/dao02761

Wei J, Huang Y, Zhu W, Li C, Huang X, Qin Q (2019) Isolation and identification of Singapore grouper iridovirus Hainan strain (SGIV-HN) in China. Arch Virol 164(7):1869–1872. https://doi.org/10.1007/s00705-019-04268-z

Weir RP, Moody NJ, Hyatt AD, Crameri S, Voysey R, Pallister J, Jerrett IV (2012) Isolation and characterization of a novel Bohle-like virus from two frog species in the Darwin rural area, Australia. Dis Aquat Org 99:169–177

Wellehan JFX, Strik NI, Stacy BA, Childress AL, Jacobson ER, Telford SR (2008) Characterization of an erythrocytic virus in the family *Iridoviridae* from a peninsula ribbon snake (*Thamnophis sauritus sackenii*). Vet Microbiol 131:115–122

Wheelwright NT, Gray MJ, Hill RD, Miller DL (2014) Sudden mass die-off of a large population of wood frog (*Lithobates sylvaticus*) tadpoles in Maine, USA, likely due to ranavirus. Herpetol Rev 45:240–242

Whitley DS, Yu K, Sample RC, Sinning A, Henegar J, Norcross E, Chinchar VG (2010) Frog virus 3 ORF 53R, a putative myristoylated membrane protein, is essential for virus replication *in vitro*. Virology 405:448–456

Whittington R, Deece K (2004) Aquatic animal health subprogram: development of diagnostic and reference reagents for epizootic haematopoietic necrosis virus of finfish. The University of Sydney and Fisheries Research and Development Corporation, Sydney/Canberra

Whittington RJ, Reddacliff GL (1995) Influence of environmental temperature on experimental infection of redfin perch (*Perca fluviatilis*) and rainbow trout (*Oncorhynchus mykiss*) with epizootic haematopoietic necrosis virus, an Australian iridovirus. Aust Vet J 72:421–424

Whittington RJ, Speare R (1996) Sensitive detection of serum antibodies in the cane toad *Bufo marinus*. Dis Aquat Org 26:59–65

Whittington RJ, Steiner KA (1993) Epizootic haematopoietic necrosis virus (EHNV): improved ELISA for detection in fish tissues and cell cultures and an efficient method for release of antigen from tissues. J Virol Methods 43:205–220

Whittington RJ, Philbey A, Reddacliff GL, Macgown AR (1994) Epidemiology of epizootic haematopoietic necrosis virus (EHNV) infection in farmed rainbow trout, Oncorhynchus mykiss (Walbaum): findings based on virus isolation, antigen capture ELISA and serology. J Fish Dis 17:205–218

Whittington RJ, Kearns C, Speare R (1997) Detection of antibodies against iridoviruses in the serum of the amphibian *Bufo marinus*. J Virol Methods 68:105–108

Whittington RJ, Reddacliff LA, Marsh I, Kearns C, Zupanovic Z, Callinan RB (1999) Further observations on the epidemiology and spread of epizootic haematopoietic necrosis virus (EHNV) in farmed rainbow trout (*Oncorhynchus mykiss*) in southeastern Australia and a recommended sampling strategy for surveillance. Dis Aquat Org 35:125–130

Whittington RJ, Becker JA, Dennis MM (2010) Iridovirus infections in finfish—critical review with emphasis on ranaviruses. J Fish Dis 33:95–122. https://doi.org/10.1111/j.1365-2761.2009.01110.x

Wirth W, Schwarzkopf L, Skerratt LF, Tzamouzaki A, Ariel E (2019) Dose-dependent morbidity of freshwater turtle hatchlings, Emydura macquarii krefftii, inoculated with Ranavirus isolate (Bohle iridovirus, Iridoviridae). J Gen Virol 100:1431–1441

Wirth W, Forzan MJ, Schwarzkopf L, Ariel E (2023) Pathogenesis of Bohle iridovirus infection in Krefft's freshwater turtle hatchlings (Emydura macquarii krefftii). Vet Pathol 60:139–150. https://doi.org/10.1177/03009858221122591

Wolf K, Bullock GL, Dunbar CE, Quimby MC (1968) Tadpole edema virus: a viscerotropic pathogen for anuran amphibians. J Infect Dis 118:253–262

Woodhams DC, Alford RA, Marantelli G (2003) Emerging disease of amphibians cured by elevated body temperature. Dis Aquat Org 55:65–67

World Organization for Animal Health (WOAH) (2021) Chapter 2.1.3. Infection with Ranavirus. In Manual of diagnostic tests for aquatic animals. Accessed online (July 2024): https://www.woah.org/fileadmin/Home/eng/Health_standards/aahm/current/2.1.03_RANAVIRUS.pdf

World Organization for Animal Health (WOAH) (2023) Aquatic animal health code. Accessed online at https://www.woah.org/en/what-we-do/standards/codes-and-manuals/aquatic-code-online-access/

World Organization for Animal Health (WOAH) (2023b) Chapter 2.3.2. Infection with Epizootic haematopoietic necrosis virus. In Manual of diagnostic tests for aquatic animals. Accessed online (July 2024): https://www.woah.org/fileadmin/Home/eng/Health_standards/aahm/current/2.3.02.EHNV.pdf

Wynne F, Puschendorf R, Knight ME, Price SJ (2020) Choice of molecular assay determines ranavirus detection probability and inferences about prevalence and occurrence. Dis Aquat Org 141:139–147. https://doi.org/10.3354/dao03518

Xiao H, Liu M, Li S, Shi D, Zhu D, Ke K, Xu Y, Dong D, Zhu L, Yu Q, Li P (2019) Isolation and characterization of a ranavirus associated with disease outbreaks in cultured hybrid grouper (female Tiger Grouper Epinephelus fuscoguttatus × male Giant Grouper E. lanceolatus) in Guangxi, China. J Aquat An Health 31:364–370. https://doi.org/10.1002/aah.10090

Xu WH, Zhang ZM, Lai FX, Yang JH, Qin QW, Huang YH, Huang XH (2023) Transcriptome analysis reveals the host immune response upon LMBV infection in largemouth bass (Micropterus salmoides). Fish Shellfish Immunol 137:108753

Yan X, Yu Z, Zhang P, Battisti AJ, Holdaway HA, Chipman PR, Bajaj C, Bergoin M, Rossmann MG, Baker TS (2009) The capsid proteins of a large, icosahedral dsDNA virus. J Mol Biol 385:1287–1299. https://doi.org/10.1016/j.jmb.2008.11.002

Yao JY, Zhang CS, Yuan ZM, Huang L, Hu DY, Yu Z, Yin WL, Lin LY, Pan XY, Yang GL, Wang CF, Shen JY, Zhang HQ (2022) Oral vaccination with recombinant *Pichia pastoris* expressing Iridovirus major capsid protein elicits protective immunity in largemouth bass (*Micropterus salmoides*). Front Immunol, Sec Microbial Immunol 13. https://doi.org/10.3389/fimmu.2022.852300

Yi W, Zhang X, Zeng K, Xie DF, Song C, Tam K, Liu ZJ, Zhou T, Li W (2020) Construction of a DNA vaccine and its protective effect on largemouth bass (*Micropterus salmoides*) challenged with largemouth bass virus (LMBV). Fish Shellfish Immunol 106:103–109

Yu NT, Zheng XB, Liu ZX (2019) Protective immunity induced by DNA vaccine encoding viral membrane protein against SGIV infection in grouper. Fish Shellfish Immunol 92:649–654

Yu Y, Yang Z, Wang L, Sun F, Lee M, Wen Y, Qin Q, Yue GH (2022) LAMP for the rapid diagnosis of iridovirus in aquaculture. Aquacult Fish 7(2):158–165. https://doi.org/10.1016/j.aaf.2021.08.002

Yuan JS, Reed A, Chen F, Stewart CN (2006) Statistical analysis of real-time PCR data. BMC Bioinform 7:85. https://doi.org/10.1186/1471-2105-7-85

Zhang M, Hu YH, Ziao ZZ, Sun Y, Sun L (2012) Construction and analysis of experimental DNA vaccines against megalocytivirus. Fish Shellfish Immunol 33:1192–1198

Zhang W, Dua C, Zhang H, Weng S, He J, Dong C (2020) Widespread outbreaks of the emerging mandarin fish ranavirus (MRV) both in natural and ISKNV-FKC vaccinated mandarin fish *Siniperca chuatsi* in Guangdong, South China. Aquaculture 520. https://doi.org/10.1016/j.aquaculture.2020.734989

Zhang M, Chen X, Xue M, Jiang N, Li Y, Fan Y, Zhang P, Liu N, Xiao Z, Zhang Q, Zhou Y (2023) Oral vaccination of largemouth bass (*Micropterus salmoides*) against Largemouth Bass Ranavirus (LMBV) using yeast surface display technology. Animals. https://doi.org/10.3390/ani13071183

Zhao R, Geng Y, Qin Z, Wang K, Ouyang P, Chen D, Huang X, Zuo Z, He C, Guo H (2020) A new ranavirus of the Santee-Cooper group invades largemouth bass (*Micropterus salmoides*) culture in Southwest China. Aquaculture 526:735363. https://doi.org/10.1016/j.aquaculture.2020.735363

Zhu Q, Wang Y, Feng J (2020) Rapid diagnosis of largemouth bass ranavirus in fish samples using the loop-mediated isothermal amplification method. Mol Cell Probes 52:101569. https://doi.org/10.1016/j.mcp.2020.101569

Zilberg D, Grizzle JM, Plumb JA (2000) Preliminary description of lesions in juvenile largemouth bass injected with largemouth bass virus. Dis Aquat Org 39:143–146

Zupanovic Z, Lopez G, Hyatt AD, Green B, Bartran G, Parkes H, Whittington RJ, Speare R (1998) Giant toads *Bufo marinus* in Australia and Venezuela have antibodies against 'ranaviruses'. Dis Aquat Org 32:1–8

Design and Analysis of Ranavirus Studies: Insights into Planning Surveillance, Modeling Host-Pathogen Dynamics, and Performing Risk Analyses

Matthew J. Gray, Jesse L. Brunner, Julia E. Earl, Wytamma Wirth, Angela Peace, and Ellen Ariel

1 Surveillance

Designing surveillance studies and analyzing data, whether epidemiological or any other sort, requires care and thought. While we might wish for a simple decision tree telling us what tests to use, these paths are often much murkier than we hope. Sometimes, perhaps often, methods that may "work" with our data produce answers that are not what we seek or expect. We should view experimental design and statistics as tools to help us understand rather than oracles that tell us what is or is not true or real. Thus, rather than introducing a blizzard of statistical tests and procedures,

Supplementary Information The online version contains supplementary material available at https://doi.org/10.1007/978-3-031-64973-8_9.

M. J. Gray (✉)
Center for Wildlife Health, School of Natural Resources, University of Tennessee, Knoxville, TN, USA
e-mail: mgray11@utk.edu

J. L. Brunner
School of Biological Sciences, Washington State University, Pullman, WA, USA

J. E. Earl
School of Biological Sciences, Louisiana Tech University, Ruston, LA, USA

W. Wirth
Peter Doherty Institute for Infection and Immunity, University of Melbourne, Parkville, VIC, Australia

A. Peace
Department of Mathematics and Statistics, Texas Tech University, Lubbock, TX, USA

E. Ariel
College of Public Health, Medical and Veterinary Sciences, James Cook University, Townsville, QLD, Australia

M. J. Gray, V. G. Chinchar (eds.), *Ranaviruses*,
https://doi.org/10.1007/978-3-031-64973-8_9

"

along with a litany of cases where one should or should not use them, we will present a few basic principles and tools and focus on interpreting the results. For this reason, we present a few equations and restrict ourselves to R functions that more closely correspond to those equations. There are many statistical packages that will do a lot of the hard work for you. In particular, we recommend the very complete epiR package, which has functions for almost every analysis. However, before proceeding with any analysis, you should be clear on what it is you are trying to learn from it and, we hope, have a little bit of intuition about what you expect to find and why.

1.1   What Are We Screening for During Surveillance?

Choose a handful of scientific studies that test samples for *Ranavirus*, and you are likely to find that they are actually screening for very different things. For some purposes, simply finding *Ranavirus* DNA is sufficient. Scientists commonly use DNA-based methods—often conventional or real-time qPCR, but increasingly iso-thermal methods like LAMP are being developed and used—to detect *Ranavirus* DNA in host tissues or swabs, but also from the environment itself (environmental DNA or eDNA). It is important to note that the mere presence of pathogen DNA does not by itself indicate that viable virus is present, nor that hosts are infected with the virus. To demonstrate infection, scientists rely on virus isolation to establish that viable virus is present, and histopathology and microscopy, sometimes with immu-nohistochemical or related methods, to establish that the virus is present and repli-cating in host tissues. To demonstrate disease, or ranavirosis, requires physical examination often paired with histopathology to demonstrate pathogenesis associ-ated with infection (see Miller et al. 2024).

The statistics we present here can apply to any of these binary data—presence or absence of the pathogen's genome, the pathogen itself, viable pathogen, infections, or disease—with one exception: they do not apply when we are trying to detect events, such as epidemics or die-offs. In these cases, the samples (i.e., time points) are not interchangeable, and our probability of detecting an event changes relative to the timing of the unobserved or partially observed event. There are not yet any clear, universal guidance in these cases without making strong assumptions about how the events unfold, when they occur, etc.

1.2   Detecting Ranavirus or Establishing Freedom
from Ranavirus

Frequently, the goal of surveillance is to establish whether ranavirus or ranavirosis is present or absent. Or, relatedly, we may wish to establish with a certain degree of confidence that a population is *free of* ranavirus. Of course, if the pathogen or disease is found, we have a clear answer (assuming we trust our diagnostic test)—it is here!—but what can we make of their apparent absence? Your intuition is likely that the more samples we collect that are negative, the more confident we can be that the virus or disease is absent. Stated another way, we may not be able to say with complete certainty that the pathogen we are looking for is not present, but we can place upper limits on how common it could be given how many negative samples we have collected.

Let us formalize this intuition in two settings: (1) large populations where we can assume that our removing individuals to sample does not affect the prevalence in the rest of the population (e.g., most settings in the wild) and (2) small populations with known size where this assumption does not hold (e.g., most captive populations).

1.2.1   Large Populations

- *Approach*

We can describe the upper 95th percentile of the prevalence, p, in a large population from which we have tested n samples, all of which are negative, as:

$$p = 1 - 0.05^{1/n} \tag{1}$$

For instance, if we had a sample size of $n = 30$ and none tested positive, we could be 95% confident that the prevalence was at or below $p = 1 - 0.05^{(1/30)} = 0.095$ or 9.5%. Similarly, in R, you can use the *dbinom()* function to find the probability of observing $x = 0$ positives out of a sample of size $= 30$ with different prevalences or values of probability. For example, *dbinom(x = 0, size = 30, prob = 0.095) = 0.05.*

There is a shortcut, an approximation, that works surprisingly well when n gets larger than about 30:

$$p \approx \frac{3}{n} \tag{2}$$

See that $3/30 = 0.1$, which is a pretty good estimate of the value we just calculated.

We can also re-arrange the formula, above, to arrive at the sample size required to have a 95% chance of detecting the pathogen or disease, assuming the prevalence is p:

$$n = \frac{\log(0.05)}{\log(1-p)}, \tag{3}$$

which can be helpful in designing surveillance programs. So, if we want to have a 95% chance of detecting $p \leq 0.05$, we could calculate $\log(0.05)/\log(0.95) = 58.4$, which would round up to 59. Or, again, we could use the shortcut approximation of $3/p = 3/0.05 = 60$, which is quite close.

- *Explanation*

Equations 1, 2, and 3 can be used on their own—simply plug in the right numbers—but understanding their derivation and assumptions can help us not just understand when they are most (or least) appropriate but provide us with some clearer intuition about what they mean. Let us begin by assuming that we are collecting samples from a population large enough that our sampling does not affect prevalence in it. That is, if we removed 30 individuals, none of which were positive, it does not appreciably enrich the fraction of the rest of the population that is positive. If this assumption is reasonable, then we can use the binomial distribution to describe the probability of observing k positives out of n samples, with the probability, p, any given sample is positive (i.e., assuming random sampling p = prevalence) as:

$$\Pr(k|p,n) = \binom{n}{k} p^k (1-p)^{(n-k)},\qquad(4)$$

where $\binom{n}{k}$ is the binomial coefficient:

$$\binom{n}{k} = \frac{n!}{k!(n-k)!}.$$

If $k = 0$, then this binomial term simplifies to 1 and the whole equation simplifies to:

$$\Pr(0|p,n) = (1-p)^n\qquad(5)$$

As we can see, as n increases the probability of observing $k = 0$ decreases for any given value of $p > 0$. And, equivalently, as n increases, the observation of $k = 0$ is consistent only with smaller and smaller values of p.

If we set the probability of observing $k = 0$ to some predetermined value, often $\Pr(0|p,n) = 0.05$, or 5%, we can find the value of p that is consistent with our data (i.e., $k = 0$ of n samples positive).

$$\Pr(0|p,n) = 0.05 = (1-p)^n$$
$$0.05^{1/n} = 1-p$$
$$p = 1 - 0.05^{1/n}$$

This gives us the formula for prevalence, p, that would give us a 5% chance of getting zero positives out of a sample size of n (Eq. 1). Or, equivalently, this means we can be 95% confident that the prevalence in the population is equal to or less than p.

We can also rearrange this equation to determine the sample size required to detect the pathogen or disease with, for instance, 95% confidence, given an expected level of prevalence, p (Eq. 3).

$$0.05 = \left(1-p\right)^{n}$$
$$\ln\left(0.05\right) = n \times \ln\left(1-p\right)$$
$$n = \frac{\ln\left(0.05\right)}{\ln\left(1-p\right)}$$

For instance, if we wanted to have 95% confidence in detecting a pathogen that was at most 5% prevalent ($p = 0.05$), we would calculate:

$$n = \frac{\ln\left(0.05\right)}{\ln\left(1-0.05\right)} = \frac{\ln\left(0.05\right)}{\ln\left(0.95\right)} = \frac{-2.9957}{-0.0513} = 58.404$$

implying that we need a sample size of 59 or larger.

This derivation, taking the log of both sides of our equation, relates to a rule of thumb that provides a reasonable approximation of the upper bound of p, assuming we desired 95% confidence and $n \gtrsim 30$. We start with the logarithm of both sides, as above:

$$\ln\left(0.05\right) = n \times \ln\left(1-p\right)$$

but then we insert two approximations. First since $\ln(0.05) = -2.9957 \approx -3$, we substitute -3 into the left-hand size of the equation. Second, because $\ln(1-p) \approx -p$ when p is small, we substitute $-p$ into the right-hand size of the equation. This yields:

$$-3 \approx -np$$
$$p \approx \frac{3}{n}$$

or the "rule of three" (Eq. 2). This approximation suggests that with $n = 30$ and no positives, we can be 95% confident that the prevalence is less than or equal to $p \approx 3/30 = 0.1$, which is quite close to $p = 1 - 0.05^{(1/30)} = 0.09503$. And, with $0.05 \approx 3/n \rightarrow n \approx 3/0.05 = 60$, we approximate the sample size needed to be 95% certain of detecting infection or disease if the prevalence were 5%.

1.2.2 Small Populations

- *Approach*

In small populations, it is more appropriate to consider the *number* of positive individuals rather than prevalence because prevalence is not continuous (i.e., 1 of 10 $\rightarrow$ 10% prevalence, 2 of 10 $\rightarrow$ 20% prevalence, etc.). In addition, prevalence is affected by our sampling in small populations; removing individuals to test changes the prevalence in the remaining population. We thus need a different distribution, the hypergeometric, where we consider the probability that there are K positive individuals in a population of size N, rather than the prevalence.

We can describe the probability of getting zero positives from the n samples we collected as:

$$\Pr(0) = \frac{\binom{N-K}{n}}{\binom{N}{n}}, \tag{6}$$

where the parentheses represent the binomial expansion (this is a special case of the hypergeometric distribution; see below). There is unfortunately no simple expression to find the upper 95th percentile on the number of positive individuals in the population, K, but we can let the computer do the hard work for us. For instance, with a population of $N = 15$, where we sample $n = 6$ individuals, we can find the probability of observing zero positives if $K = 1, 2, 3, \ldots, 10$ individuals in the population were positive using the *dhyper()* function in R and finding the value of K for which this probability is ≤ 0.05. In this example, the code is:

```
N <- 15
n <- 6
K <- 1:10
```

```
cbind(K, Pr = dhyper(x=0, m=K, n=N-K, k=n))
```

```
         K         Pr
 [1,]    1 0.6000000000
 [2,]    2 0.3428571429
 [3,]    3 0.1846153846
 [4,]    4 0.0923076923
 [5,]    5 0.0419580420
 [6,]    6 0.0167832168
 [7,]    7 0.0055944056
 [8,]    8 0.0013986014
 [9,]    9 0.0001998002
[10,]   10 0.0000000000
```

Since the fifth element, 0.0419580420 is just below our cutoff of 5%, we can have $\geq 95\%$ confidence that $K = 5$ or less.

We can similarly find the sample size, n, required to have a certain confidence of detecting at least one positive (i.e., not getting all negatives) if we set K to some value. For instance, with $K = 3$ positives in a population of size $N = 15$, we can use the following code:

```
N <- 15
K <-3
n <- 1:10

cbird(n, Pr = dhyper(x=0, m=K, n=N-K, k=n) )
```

```
          n          Pr
 [1,]   1 0.80000000
 [2,]   2 0.62857143
 [3,]   3 0.48351648
 [4,]   4 0.36263736
 [5,]   5 0.26373626
 [6,]   6 0.18461538
 [7,]   7 0.12307692
 [8,]   8 0.07692308
 [9,]   9 0.04395604
[10,]  10 0.02197802
```

and choose the row where Pr is $\leq 5\%$ (i.e., when $n = 9$).

- *Explanation*

The large sample, binomial distribution approach assumes that every sample has the same probability of being positive as every other sample, which is reasonable if we were sampling from a very large population or if, for some reason, we were sampling with replacement, such as collecting non-lethal samples for ranavirus testing. Researchers are increasingly working with finite populations of known size, such as in captive collections or in trade, and lethal samples for ranavirus testing (from the liver or kidneys) are common. In smaller populations, sampling *with* removal, the probabilities of being positive are no longer interchangeable. Imagine sampling a population of $N = 10$ with $K = 1$ positive individual. The first sample would have a one in ten chance of being positive, but after we removed five negative individuals, the next sample would have a one in five chance of being positive; the probability has doubled. Or instead, if we had already sampled the one positive individual, every subsequent sample would have zero probability of being positive. In such finite situations where populations are sampled without replacement, the binomial assumption is no longer a reasonable simplification, and we instead use a hypergeometric distribution.

In a population of size N with K positive individuals, the probability of obtaining k positives from a sample of size n is:

$$\Pr(k|K,N,n) = \frac{\binom{K}{k}\binom{N-K}{n-k}}{\binom{N}{n}}, \tag{7}$$

where we employ the binomial coefficient three times. When $k = 0$, this simplifies somewhat to:

$$\Pr\left(0 \mid K, N, n\right) = \frac{\dbinom{N-K}{n}}{\dbinom{N}{n}},$$

which we saw in Eq. 6. While there is no simple formula for K given a sample size of n out of a population of size N and an assumed $\Pr(0 \mid K, N, n)$, the logic is the same as above, and R functions can do the calculations for us. For instance, we can calculate the probability of observing $k = 0$ positive samples out of $n = 10$ samples tested from a population of size $N = 20$, as:

```
K <- 1:10
N <- 20
n <- 10
# Note that the R function is parameterized slightly differently
data.frame(K=K, Pr = round( dhyper(x=0, m=K, n=N-K, k=n), 3) )
```

```
      K    Pr
1     1 0.500
2     2 0.237
3     3 0.105
4     4 0.043
5     5 0.016
6     6 0.005
7     7 0.002
8     8 0.000
9     9 0.000
10   10 0.000
```

We could be approximately 95% confident that $K \leq 4$ out of the 20 individuals in the population.

1.3 Estimating Precision on Infection Prevalence Estimates

The inferences in the preceding sections about how common the pathogen or disease might be in a population given a certain number of negative samples are just a special case of understanding the precision with which we estimated prevalence. A larger sample size increases the precision of our estimate and should thus yield narrower bounds on the values of prevalence that are consistent with our data.

There is a myriad of methods for approximating the confidence interval of a proportion, especially in the case of large (or infinite) populations using a binomial distribution. They range from fairly imprecise (i.e., the normal approximation that is common in introductory statistics textbooks, which often produces cutoffs <0 or >1) to quite good (e.g., Wilson's (1927) method). Rather than enter into this sometimes bewildering fray, we instead suggest a simple, coherent approach that produces intuitive results. In any case, most methods produce similar intervals, and we would caution against placing too much weight on any particular value or cutoff.

1.3.1 Large Populations

- *Approach*

While there is no simple closed-form formula for it, we can use an *R* function for the beta distribution (see below for more details), particularly the quantile function of the beta, to find an interval on the prevalence within a large population. For instance, if we collected 30 samples, 1 of which was positive and 29 that were negative, we could find the 95% confidence interval as:

```
round(
  qbeta(p=c(0.025, 0.975), # the quantiles of the 95th percentile
        shape1=1+1, # the number of positives + 1
        shape2=1+29), # the number of negatives + 1
  3)
```

```
[1] 0.008 0.167
```

Few researchers are familiar with the beta distribution, so it is worth some introduction. It is a continuous distribution bounded between 0 and 1 that is useful for describing the distribution of prevalence. It has two parameters, often called α and β, but sometimes referred to as shape1 and shape2. If we had no information about what prevalence might be, meaning that every possible value is equally likely, we could describe this as a beta distribution with $\alpha = \beta = 1$ (see the flat line gray line in the Fig. 1).

After collecting 30 samples, 6 of which are positive, our estimate of prevalence would be 6/30 = 0.2. To find the consistency of different values of prevalence with these data (see the curve in Fig. 1), we can simply add the positives to α and the negatives to β in the beta distribution.

Notice that our simple estimate of prevalence, 6/30 = 0.2, is most consistent with our data, but then the probability density tapers off rapidly with much lower or higher values of prevalence.

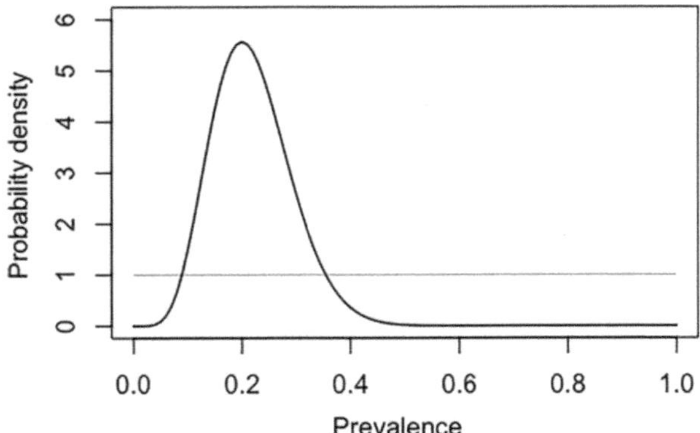

Fig. 1 Beta distributions describing the probability (really probability *density*, *y*-axis) that prevalence takes on any particular value (*x*-axis). The flat gray line is if we had no information on prevalence and thought any value was equally probable. This corresponds to the parameters $\alpha = \beta = 1$. The black curve represents the situation if we had collected 30 samples and found that 6 were positive and 24 were negative, corresponding to $\alpha = 1 + 6 = 7$ and $\beta = 1 + 24 = 25$

We can find a specific interval, such as a 95% interval, around our estimate of prevalence using the cumulative distribution function for the beta, which is qbeta() in *R*:

```
#95% CI on prevalence given 6 positive and 24 negative tests
qbeta(p=c(0.025, 0.975), shape1=1+6, shape2=1+24)
```

```
[1] 0.09594217 0.37473217
```

Note that this interval stems from a Bayesian perspective on our estimate of prevalence. It turns out that the beta distribution is the conjugate prior for a binomial likelihood, which means that the posterior is the same form as the prior, just with the parameters modified by the data. If that does not mean anything to you, do not fret; you can use the method without understanding priors! Moreover, do not get too worried about Bayesian versus frequentist approaches; all the intervals you calculate will be quite similar, even if they are conceptually different. However, one potential feature of this Bayesian approach is that if you had prior information about the prevalence (e.g., from an earlier survey), or were willing to assume that some values were much less likely than others (e.g., you feel confident that prevalence is less than 0.5), you could include that information in the prior values of α and β. This would produce more informative estimates of the prevalence and the interval that are consistent with your data and prior understanding of the system; however, using prior information is not necessary.

1.3.2   Small Populations

The approach to estimating the distribution of prevalence values consistent with our data in small, closed populations (e.g., in captive settings or trade) is similar to that for large populations; only we use a hypergeometric distribution. This will require a slightly more overtly Bayesian approach, but the code is still simple. We will take this in steps.

First, let us again imagine that we collected a sample of 30 individuals from a population of size 50. Prior to examining our results, we think that each possible value of K, the actual number of infected individuals in the population, is equally likely.

```
n <- 30 # sample size
N <- 50 # population size
Ks <- 0:N # possible numbers of positives in population

prior <- 1/length(Ks) # N+1 equally probable values of K
```

A plot of the prior against the possible values of K would yield a flat line or, actually, because K can only take on discrete values, a bar plot with bars of equal heights (Fig. 2).

In a Bayesian analysis, the *posterior* probability of any given value of K given our prior expectation and data (i.e., $z = 6$ positives out of n samples from a population of size N) is the *likelihood* of observing the data, given a particular value of K, times the prior probability that K took on this particular value:

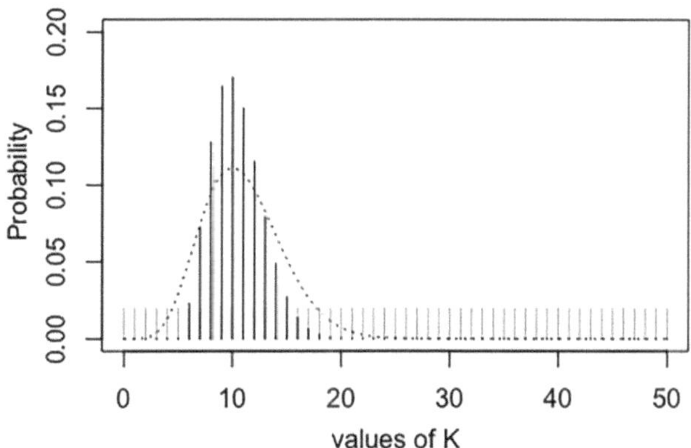

Fig. 2   The probability of different numbers of infected or diseased individuals, K, in a population of size $N = 50$. Note that because we are considering individuals rather than proportions of the population, the distribution is discrete. The light gray bars represent the case where we have no prior information or expectation about the commonness of infection in the population. The darker bars represent the case where we have collected 30 samples from this population without replacement and found that 6 were positive. The dotted line represents the equivalent beta distribution for the same data if we had instead been sampling from a very large population

$$\Pr(K|x,n,N) = \frac{\Pr(x|K,n,N) \times \Pr(K)}{\text{normalizing term}}$$

The likelihood is the term on the left of the numerator, and the prior is the term on the right. Note that we need to divide by some normalizing term so that this is a proper probability distribution. In discrete cases this normalizing term is simply the summation of the numerator for $K = 0, 1, 2, \ldots, N$.

$$\Pr(K|x,n,N) = \frac{\Pr(x|K,n,N) \times \Pr(K)}{\sum_{K=0}^{N} \Pr(x|K,n,N) \times \Pr(K)}$$

The likelihood for each possible value of K, in terms of code, is just the probability of observing six positives with a sample size of n from a hypergeometric distribution.

```
likelihood <- dhyper(x=6, # The number of sample positive
                     m=Ks, # possible numbers of positives
                     n=N-Ks, # number of negatives, given Ks
                     k=n) # sample size

cbind(Ks, likelihood)[1:21,] # likelihoods for first 21
values of K
```

```
         Ks    likelihood
 [1,]     0 0.0000000000
 [2,]     1 0.0000000000
 [3,]     2 0.0000000000
 [4,]     3 0.0000000000
 [5,]     4 0.0000000000
 [6,]     5 0.0000000000
 [7,]     6 0.0373661953
 [8,]     7 0.1188924397
 [9,]     8 0.2101354747
[10,]     9 0.2701741818
[11,]    10 0.2800586031
[12,]    11 0.2464515707
[13,]    12 0.1895781313
[14,]    13 0.1297113530
[15,]    14 0.0797549535
[16,]    15 0.0443083075
[17,]    16 0.0222807489
[18,]    17 0.0101276131
[19,]    18 0.0041431145
[20,]    19 0.0015138303
[21,]    20 0.0004883324
```

Notice that the likelihood of observing six positives is zero if K is less six, which we hope makes some intuitive sense. Likewise, there is a low or zero probability of observing six positives if K is very large.

Next, we can calculate the posterior, the results of which are plotted in Fig. 2.

```
posterior <- likelihood*prior/sum(likelihood*prior)

cbind(Ks, posterior)[1:21,] # posterior prob for 1st 21
values of K
```

```
        Ks      posterior
 [1,]    0 0.0000000000
 [2,]    1 0.0000000000
 [3,]    2 0.0000000000
 [4,]    3 0.0000000000
 [5,]    4 0.0000000000
 [6,]    5 0.0000000000
 [7,]    6 0.0227127854
 [8,]    7 0.0722679535
 [9,]    8 0.1277294062
[10,]    9 0.1642235223
[11,]   10 0.1702316999
[12,]   11 0.1498038959
[13,]   12 0.1152337661
[14,]   13 0.0788441558
[15,]   14 0.0484785012
[16,]   15 0.0269325007
[17,]   16 0.0135432003
[18,]   17 0.0061560001
[19,]   18 0.0025183637
[20,]   19 0.0009201714
[21,]   20 0.0002968295
```

Notice that the most probable value of K is 10, which corresponds to a prevalence of $10/50 = 20\%$, which is the same prevalence as in the sample ($6/30 = 20\%$).

1.4 Assumptions About Detecting Ranavirus and Ranavirosis

All our estimates of prevalence or confidence that ranavirus is absent are built on the critical assumptions that we sample randomly—in essence that positive and negative samples are equally likely to be sampled—and that our diagnostic test is essentially perfect. Let us delve into these separately.

1.4.1 Sampling Biases

All the statistical methods presented assume that we are sampling randomly (or at least haphazardly), without biases in who is sampled. If we were more likely to sample infected individuals than uninfected individuals, we would, of course, have a bias in our sample. If we were to ignore this bias, then our inferences about infection could be quite off. For example, imagine we were sampling during a die-off and captured primarily or solely dead or moribund individuals. Assuming that ranavirus were the cause of this event, we would expect that essentially all our samples would test positive for ranavirus. The problem is extrapolating this biased die-off sample to the population, and inferring that prevalence was near 100%. Note that such a bias could be in the opposite direction, where uninfected individuals were more likely to be sampled, perhaps because infected individuals are more cryptic or dead individuals decompose or are scavenged rapidly.

It is sometimes possible to correct for such biases if we have some independent estimates of them (e.g., from prior data, capture-mark-reencounter studies; see Cooch et al. 2012 for a discussion), but more often we simply must acknowledge their presence and interpret our estimates with caution. More often the best solution is to identify and correct for such biases *before* collecting samples.

1.4.2 Detection Biases

Our diagnostic tests are rarely perfect. They may have false negatives, meaning less than perfect sensitivity (Se), and false positives, meaning less than perfect specificity (Sp). Depending on the goals of your research, false negatives or false positives may be more concerning. While there are no universal solutions or optima, it can be helpful to consider the source of these spurious results. False-positive qPCR results might come from sample contamination in the field or laboratory with *Ranavirus* DNA. For example, it is possible to contaminate samples if gloves are not changed between infected and uninfected individuals (Gray et al. 2018). Contamination of an assay also could occur with qPCR standards without proper well labeling or pipetting technique. Field blanks, extraction controls, and no-template PCR controls are meant to help a researcher detect when contamination occurs. In these cases, the false positives come from outside of the diagnostic test itself; the primer and probe specificity for carefully designed qPCR assays usually precludes amplification of non-target DNA. Immunological methods as well as clinical diagnosis of disease also suffer from false positives. Here, the background signal (e.g., from cross-reactive antibodies or nonspecific signs of infection) from uninfected individuals can look very much like, and quantitatively overlap with, positive individuals. Training and experience can reduce these errors, but never to zero.

False negatives can also stem from error—hence the need for positive controls in most assays—but also because the probability of detection is inherently quantitative; it takes a certain amount of pathogen DNA, viable pathogen, or pathology to ensure detection (see Brunner 2020 for discussion). Beyond developing good protocols, researchers may also be able to design their studies to ensure better sensitivity. For instance, researchers often hold ranavirus-exposed animals for several days or weeks to let infections develop to detectable levels. Moreover, lethal samples (i.e., organ tissue) will likely result in greater detection of ranavirus compared to nonlethal samples (i.e., swabs, tail-clips; Gray et al. 2012).

It may be useful as well to revisit the question of what it is you are trying to detect. If you are primarily interested in ranavirosis, then a test that has a non-trivial chance of missing a low-level infection may not be much of an issue. However, if instead you are working to detect sublethal, inapparent infections, you will need very sensitive methods of detection and might consider using a two-step process: an initial test that is very sensitive, but prone to false positives, followed by an independent and more specific test to verify the initial results (Miller et al. 2024).

In principle, it is possible to establish the diagnostic sensitivity and specificity of a test or assay and then correct results with these values, but in practice there is no good reason to think that these values are constant among species, between life history stages, or even over time. We thus caution against using estimates of sensitivity and specificity, or the corrected estimates of prevalence or other quantities, as if they are well established or even constant from setting to setting or study to study. Rather, it is worth bearing in mind that poor data on the presence or absence of ranavirus or disease (or any other response) are never really improved by clever statistics.

All that said, there are several useful frameworks for adjusting estimates or probabilities of observing k positives with outside estimates of Se and Sp. We point the reader to Cameron and Baldock (1998) for the formulae (or the Electronic Supplement 1, which includes R code) and to the epiR package in R for the calculations.

1.5   *Interpreting Prevalence*

While prevalence is more commonly estimated than incidence during surveillance studies, it is simply a "snap shot" of the proportion of a population that are infected or diseased at a given time. It is difficult to interpret these snap shots in the absence of biological context. Indeed, interpreting *any* statistical result necessarily invokes some implicit or explicit model of how the system is thought to work. In the case of prevalence, we can think of it as the ratio of the positive individuals relative to the population as a whole. The number (or density or fraction) of positive individuals is a product of both the gain and loss of those positives. The population presumably gains positives through the process of transmission and infection, which is governed by the amount of direct or indirect exposure to the ranavirus as well as by the probability that individuals are infected given an exposure (one meaning for the term "susceptibility"). Various factors influence the probability of successful ranavirus transmission, and those factors are rarely static (Brunner et al. 2024). Additionally, a population loses positive individuals as they recover or die. We expect prevalence to change throughout a year and during the course of an epidemic, thus the timing of when we collect samples can be very relevant to interpreting prevalence estimates.

Prior data and context can be very helpful in interpreting prevalence. For example, if experimental exposures show that a species dies rapidly following ranavirus exposure, then high prevalence would be most consistent with sampling during (or soon after) a period of high transmission or during a ranavirus-induced die-off. Several studies have reported species-level susceptibility under controlled conditions (e.g., Hoverman et al. 2011; Brenes et al. 2014b; Brunner et al. 2024). Biological context also can be gleaned from the density of the population and the timing of the survey relative to the phenology of the organism. For example, observing low prevalence and a dense population of amphibian larvae early in spring would be more consistent with the virus recently being introduced rather than an outbreak already occurring.

Lastly, it is important to recognize that infected individuals may be more or less likely to be detected or captured than uninfected individuals, which can bias prevalence estimates (Cooch et al. 2012). For instance, moribund fish and tadpoles are often found near the surface, making them more easily detected, which would inflate prevalence estimates, whereas sick turtles may move less and have lower detection probabilities resulting in underestimates of prevalence. Variation in detection probabilities through time (e.g., developmental stages) and among locations also can lead to apparent differences in prevalence that do not reflect actual differences in the proportion infected. See Mosher et al. (2019) for an example of how accounting for detection probabilities can substantially improve estimates of prevalence.

1.6 Comparing Prevalence Estimates and Other Logistic Regressions

There are many ways of comparing counts or contingencies such as prevalence data. Many classic approaches, such as χ^2-tests, make strong assumptions about the numbers in each cell or whether sums across rows are fixed or not. Logistic regression is an equivalent, but more flexible approach, when we know the number of "successes" (i.e., number positive) out of the number of "trials" (i.e., number tested). These analyses model the probability of success as a linear function of the variables of interest, whether they are discrete (e.g., two samples in different times or places) or continuous (e.g., $\log_{10}$ (dose) of ranavirus exposure). The one potentially tricky bit is that they use a logit "link" function to map the linear model into probability space (see next section).

1.6.1 Approach

The basic form of the logistic regression is:

$$\text{logit}(p) = \beta_0 + \beta_1 x_1 + \beta_2 x_2 ..., \tag{8}$$

where p is the probability of being infected and the β s are the effects of the predictors of interest, designated x_1, x_2, etc. For any linear model that you might construct in a typical regression, you can also use in a logistic regression.

Let us use two examples to illustrate how this can work. First, imagine our data from before, where we found 6 of 30 samples were positive from wetland A, and then imagine we collected 25 samples from a neighboring wetland B where 12 samples were positive. Our question is whether there is strong evidence that these two sites have different prevalence; after all, sometimes, by chance, you might get more or fewer positives from a sample.

```
( df <- data.frame(wetland = c("A","B"),
                   Pos = c(6, 12),
                   Neg = c(24, 13)) )
```

```
    wetland Pos  Neg
1        A   6   24
2        B  12   13
```

```
m1 <- glm(cbind(Pos, Neg) ~ wetland,
          data=df,
          family = "binomial")
```

```
summary(m1)
```

```
Call:
glm(formula = cbind(Pos, Neg) ~ wetland, family = "binomial",
    data = df)

Deviance Residuals:
[1]  0  0

Coefficients:
            Estimate Std. Error z value Pr(>|z|)
(Intercept)  -1.3863     0.4564  -3.037  0.00239 **
wetlandB      1.3063     0.6071   2.152  0.03143 *
---
Signif. codes:  0 '***' 0.001 '**' 0.01 '*' 0.05 '.' 0.1 ' ' 1

(Dispersion parameter for binomial family taken to be 1)

    Null deviance: 4.9038e+00  on 1  degrees of freedom
Residual deviance: 1.0658e-14  on 0  degrees of freedom
AIC: 11.125

Number of Fisher Scoring iterations: 3
```

Running the model is relatively straightforward. We can give the generalized linear model function, glm(), two columns (the cbind() function binds columns together), the first with the number of positive samples and the second the number of negatives corresponding with the wetlands (rows).

We can interpret the coefficients more easily if we exponentiate them, yielding estimates of the *odds* of being positive. In this case the odds of being positive in wetland A are exp(−1.3863) = 0.25, which is essentially 6:24 or the ratio of the number of positives to the number of negatives. The expected odds of being positive in wetland B is exp(−1.3863 + 1.3063) = 0.923, which is 12:13.

We can also interpret the exponentiated coefficient for wetland B as being the increased odds of being positive if the sample comes from wetland B *relative* to wetland A. Thus exp(1.3063) = 3.692 is roughly (12/13)/(6/24). That is, in this example wetland B has 3.69-fold higher odds of a sample being positive compared to wetland A. If we had included a continuous variable in our model, for wetland salinity, the exponentiated parameter would be interpreted as the change in odds for a one-unit change in the predictor.

As a second example, let us consider our laboratory dose-response experiment with a ranavirus. We individually exposed larval wood frogs (*Lithobates sylvaticus*) to one of five doses of the virus, from ~25 to 250,000 plaque-forming units (pfu) per mL, and then recorded whether each individual ($n = 11$–12 per dose) died (Warne et al. 2011). We expected that the probability of death would tend to increase with increasing exposure to ranavirus, but our specific questions were as follows: (a) how rapidly (i.e., what is the slope of this relationship?) and (b) at what dose would we predict a 50% chance of death (the LD_{50})? These are the data and plotted in Fig. 3.

```
( df2 <- data.frame( Dose = c(1.40, 2.40, 3.40, 4.40, 5.40),
                     Died = c(   2,    3,   11,   11, 12),
                     N =    c(  12,   11,   12,   12, 12)) )
```

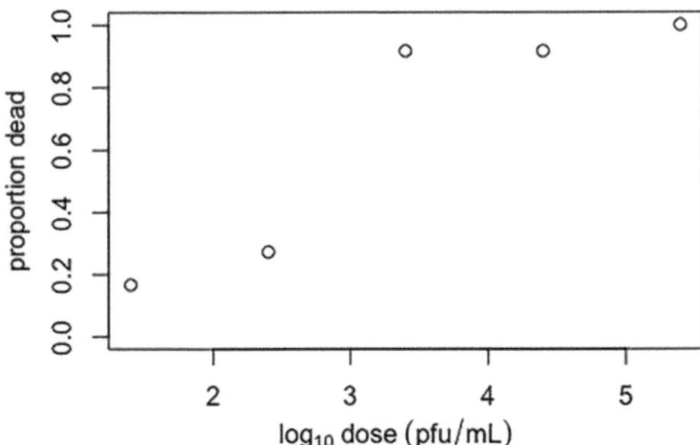

Fig. 3 The proportion of wood frog larvae that died when water bath exposed to one of five doses of a ranavirus. (Data from Warne et al. 2011)

```
   Dose Died  N
1  1.4    2 12
2  2.4    3 11
3  3.4   11 12
4  4.4   11 12
5  5.4   12 12
```

```
plot(Died/N ~ Dose, data=df2,
    xlab=expression( log[10]~dose~(pfu/mL) ),
    ylab="proportion dead",
    ylim=c(0,1))
```

Constructing a logistic regression is virtually the same as above. Note how we provide the function two columns, representing number of successes (=Died) and failures (=survived = N-Died).

```
m2 <- glm(cbind(Died, N-Died) ~ Dose,
        data=df2,
        family = "binomial")
summary(m2)
```

```
Call:
glm(formula = cbind(Died, N - Died) ~ Dose, family =
"binomial",
    data = df2)

Deviance Residuals:
      1        2        3        4        5
 0.5533  -1.0053   1.0944  -0.6384   0.4291

Coefficients:
            Estimate Std. Error z value Pr(>|z|)
(Intercept)  -4.4848     1.2446  -3.604 0.000314 ***
Dose          1.7317     0.4363   3.969 7.2e-05 ***
---
Signif. codes:  0 '***' 0.001 '**' 0.01 '*' 0.05 '.' 0.1 ' ' 1

(Dispersion parameter for binomial family taken to be 1)


    Null deviance: 38.0898  on 4  degrees of freedom
Residual deviance:  3.1061  on 3  degrees of freedom
AIC: 16.048

Number of Fisher Scoring iterations: 5
```

In this case the exponentiated coefficient, exp(1.7317) = 5.7, is interpreted as the increased odds of death with every one-unit increase in the predictor, which is an order of magnitude in dose. This also implies that with an increase in exposure dose of *two* orders of magnitude, the odds of death increase exp(1.7317 × 2) = 31.9-fold.

This seems like an enormous increase in the odds of death, and it is, but note that this is *relative* to where we started before increasing the dose by two orders of magnitude. If, for instance, we started at a dose of 1 (=10^0 = 1 pfu/mL) and then added two orders of magnitude (=10^{0+2} = 100 pfu/mL), the odds of death would be:

```
exp(-4.4848 + 1.7317*(0+2))
```

```
[1] 0.3600905
```

which is still less than equal odds of dying. The expected *probability* of death at this dose (10^2 pfu/mL) is as follows:

```
exp(-4.4848 + 1.7317*(0+2))/( 1 + exp(-4.4848 + 1.7317*(0+2)) )
```

```
[1] 0.2647548
```

Equivalently, we can use the predict() function for this estimate.

```
predict(m2,
        newdata=data.frame(Dose = 2), # <- predict at a dose of 2
        type = "response") # default is "link"
```

```
        1
0.2647569
```

The predict() function is also useful to draw a graph showing the predicted line (Fig. 4).

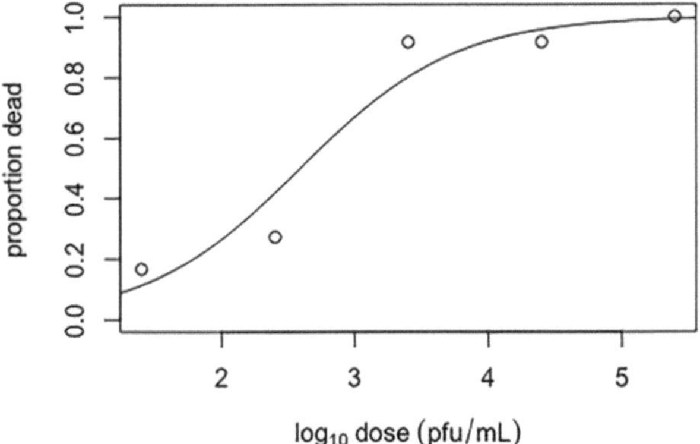

Fig. 4 The proportion of wood frog larvae that died when water bath exposed to one of five doses of a ranavirus (circles) as well as the best-fit line from a logistic regression fit to these data. (From Warne et al. 2011)

```
# our raw data
plot(Died/N ~ Dose, data=df2,
     xlab=expression( log[10]~dose~(pfu/mL) ),
     ylab="proportion dead",
     ylim=c(0,1))
# generate predictions
preds <- data.frame(Dose = seq(from=1, to=6, length.out=50))
preds$Prob <- predict(m2,
                      newdata=preds,
                      type="response")
# plot predicted line
lines(x=preds$Dose, y=preds$Prob, type = "l")
```

At what dose do we get an expected 50% chance of death? This ends up being estimated as:

$$\mathrm{LD}_{50} = -\frac{\beta_0}{\beta_1} \tag{9}$$

or in our case $-(-4.4848)/1.7317 = 2.59$. Recall that this is on the $\log_{10}$-scale, so this, the estimated LD_{50}, is $10^{2.59}$ or 3.89×10^2 pfu/mL.

Note that the *slope* of the dose-response relationship also tells us something about the underlying distribution of susceptibility (here, susceptibility to death from exposure). In general, a steeper slope implies less variability in the population (Ben-Ami et al. 2010). See Electronic Supplement 2 for a visual explanation of the relationship between the slope of a dose-response study and the underlying distribution of susceptibility in the population.

1.6.2 Explanation

Logistic regression connects binomial data (i.e., the number of "successes" out of a number of "trials") to a linear model using the logit link. This is important because a typical linear regression is unbounded, meaning it can produce values well below zero and well above one, but the binomial likelihood part of the model requires a bounded probability. The logit link transforms these unbounded values from the linear model into bounded probability estimates.

The logit transform is simply the log-odds of success. So, if p is the probability of success, then the odds of success are $p/(1 - p)$, and the log-odds are $\log[p/(1 - p)]$. In a logistic regression, we are saying that the log-odds are a linear function of our predictor(s):

$$\log\left(\frac{p}{1-p}\right) = \beta_0 + \beta_1 x_1 + \dots \tag{10}$$

From this equation, we can see that exponentiating the linear model (or a part of it) gives the predicted odds (or change in odds), which is useful for interpreting the coefficients of a logistic regression.

To convert the log-odds (or logit) of p or the results of our linear model back to probability space, we can use the inverse of the logit transform, which is also known as the logistic transform:

$$\log\left(\frac{p}{1-p}\right) = \beta_0 + \beta_1 x_1 + \cdots$$

$$\frac{p}{1-p} = \exp\left(\beta_0 + \beta_1 x_1 + \cdots\right) \tag{11}$$

$$p = \frac{\exp\left(\beta_0 + \beta_1 x_1 + \cdots\right)}{1 - \exp\left(\beta_0 + \beta_1 x_1 + \cdots\right)} = \frac{1}{1 + \exp\left(-\left[\beta_0 + \beta_1 x_1 + \cdots\right]\right)}$$

If the linear model produces some very large value, then this equates to a probability very close to one, and a linear model that produces very small values equates to probabilities very close to zero.

We can find the value of the predictor(s), such as the dose, expected to produce a 50% chance of death (or other outcomes) by inserting $p = 0.5$ (or any other value of interest such as $p = 0.9$ if we were interested in the LD_{90}) in the left-hand side of the logit equation and re-arranging we get Eq. 9:

$$\log\left(\frac{0.5}{1-0.5}\right) = \beta_0 + \beta_1 x_1$$

$$\log(1) = 0 = \beta_0 + \beta_1 x_1$$

$$x = \frac{\beta_0}{\beta_1}$$

So, with our parameters, we get $LD_{50} = -(-4.4848/1.7317) = 2.59$ pfu/mL.

1.7 Incidence

Infection prevalence is useful when describing the distribution of ranaviruses among regions and host species, but it does not convey information about risk or rates of infection. Infection incidence is the rate at which individuals become infected with a pathogen (i.e., the number of new cases that occur in a specified time period; Wobeser 2006). In small captive populations, it may be possible to determine how many individual animals become infected over short intervals. Consider, for example, a study with two surveys. The initial survey found that 2 of 50 animals were infected, and a second survey at the end of the week found that 10 individuals were infected. Often incidence is presented as the number of new cases (here, 8) per time period *out of the whole population*, which in our example would be 8 out of 50 or 16% per week (assuming infected individuals did not die or recover in between). However, some sources (e.g., veterinary epidemiologists; Dohoo et al. 2003; Wobeser 2006) will present incidence as the number of new cases per unit time *out of those at risk*, in which case incidence would be 16.7% (=8 new cases/48 at risk). This second version is synonymous with the force of infection, or per capita risk of infection, which is commonly considered (and distinguished from "incidence") in epidemic models (Hens et al. 2012). In any case, be sure to clearly present how you derive incidence as the definition is not universal. Also, if populations are not closed (i.e., immigration and birth or emigration and death occur), calculations of incidence rates need to be adjusted for the time at risk (see Dohoo et al. 2003 for details). Estimating incidence in wild populations can be difficult because we generally cannot track the fate or infection status of individuals.

Two approaches can help us estimate incidence in wild populations. First, one might use sentinels, uninfected individuals that are introduced into the environment (e.g., tadpoles in cages placed in a wetland) and regularly screened for infection to estimate force of infection. Sentinel species should be highly susceptible to ranavirus, so as to have a high chance of exhibiting an infection if given a reasonable exposure. One must also consider the assumption that the risk of exposure is the same for the sentinels as the free-living individuals (McCallum 2008).

A second approach is capture-mark-reencounter (CMR) studies where individuals are given unique marks and released. During subsequent encounter periods (e.g., trapping, netting), the researcher records the new and recaptured individuals and determines their infection status. Given that individuals are released, infection status must be determined using nonlethal methods (St-Amour and Lesbarrères 2007; Gray et al. 2012). Ultimately, CMR models estimate the probability of individuals changing infection status while accounting for imperfect detection of the pathogen and imperfect recapture probability of the host. We are not aware of any studies with ranavirus infections that have taken either approach.

1.8 Infection Intensity (Ranavirus Load)

1.8.1 Approach

Infections occur along a gradation of intensities, which may be relevant in our understanding of the dynamics or outcome of infections or epidemics. For instance, viral titers (from plaque-assays or $TCID_{50}$ assays) or viral gene quantities (from quantitative real-time PCR) are likely to be low soon after infection or carrier species that maintain persistent sublethal infections. So how can we analyze such quantitative information?

The first step is to recognize that viral quantities, Q, are usually lognormally distributed. We can think of the amount of virus at a given time in a growing virus

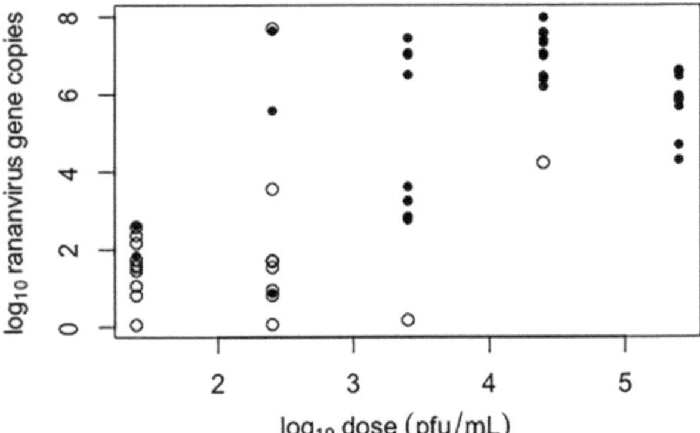

Fig. 5 The intensity of infections, described as the number of ranavirus gene copies per sample, detected with a qPCR reaction in animals that survived (open circles) or died (filled circles) in the dose-response study by Warne et al. (2011)

infection as the realization of a number of independent, *multiplicative* steps. Thus, the log of viral quantities, $\log(Q)$, can be thought of as the sum of the log of those multiplicative effects and so should be approximately normally distributed. Note that the base of the logarithm is not important, except for interpretation. Most researchers use $\log_{10}$ so that a change in one unit corresponds to a change in an order of magnitude. Thus, most statistical models that assume normally distributed responses may be appropriate for analyzing viral quantities.

Consider the real-time quantitative PCR (qPCR) data from the dose-response study with *L. sylvaticus* larvae introduced above (Fig. 5).

```
df3 <- data.frame(
  Dose=c(1.4, 1.4, 1.4, 1.4, 1.4, 1.4, 1.4, 1.4, 1.4, 1.4, 1.4, 1.4,
         2.4, 2.4, 2.4, 2.4, 2.4, 2.4, 2.4, 2.4, 2.4, 2.4, 2.4,
         3.4, 3.4, 3.4, 3.4, 3.4, 3.4, 3.4, 3.4, 3.4, 3.4, 3.4, 3.4,
         4.4, 4.4, 4.4, 4.4, 4.4, 4.4, 4.4, 4.4, 4.4, 4.4, 4.4, 4.4,
         5.4, 5.4, 5.4, 5.4, 5.4, 5.4, 5.4, 5.4, 5.4, 5.4, 5.4, 5.4),
  Q=c(0.06, 0.82, 1.06, 1.46, 1.55, 1.61, 1.74, 1.85, 2.17, 2.37,
      2.59, 2.62,
      0.07, 0.82, 0.87, 0.95, 1.54, 1.7, 1.72, 3.56, 5.58, 7.63, 7.69,
      0.18, 2.76, 2.79, 2.84, 2.85, 3.24, 3.26, 3.62, 6.5, 7.01,
7.06, 7.44,
      4.23, 6.2, 6.38, 6.46, 6.46, 6.99, 7.05, 7.32, 7.4, 7.57,
7.59, 7.98,
      4.3, 4.69, 5.68, 5.83, 5.84, 5.88, 5.91, 5.96, 6.46, 6.55,
6.6, 6.62),
  Died=c(0, 0, 0, 0, 0, 0, 0, 1, 0, 0, 0, 1,
         0, 0, 1, 0, 0, 0, 0, 0, 1, 1, 0,
         0, 1, 1, 1, 1, 1, 1, 1, 1, 1, 1, 1,
         0, 1, 1, 1, 1, 1, 1, 1, 1, 1, 1, 1,
         1, 1, 1, 1, 1, 1, 1, 1, 1, 1, 1, 1)
)
```

We are interested in whether the $\log_{10}(Q)$ in the larvae that died (filled circles in Fig. 5) varies with the dose to which the larvae were exposed. If there were a quantity of ranavirus in the host that was lethal and hosts die when that quantity is exceeded—a simple null model of virus-induced death we have considered (Brunner et al. 2005)—then there should be no clear relationship between the dose of exposure and the quantity in the larvae upon death. We might use a simple linear regression of the $\log_{10}(Q)$ against the $\log_{10}(Dose)$:

```
m3 <- lm(Q ~ Dose,
         data=df3[df3$Died==1,]) # <- just those that died
summary(m3)
```

```
Call:
lm(formula = Q ~ Dose, data = df3[df3$Died == 1, ])

Residuals:
    Min      1Q  Median      3Q     Max
-3.2751 -1.4101 -0.0122  1.4921  3.4849

Coefficients:
            Estimate Std. Error t value Pr(>|t|)
(Intercept)   2.2115     1.0140   2.181  0.03561 *
Dose          0.8057     0.2374   3.393  0.00166 **
---
Signif. Codes:  0 '***' 0.001 '**' 0.01 '*' 0.05 '.'
0.1 ' ' 1

Residual standard error: 1.677 on 37 degrees of freedom
Multiple R-squared:  0.2373,    Adjusted R-squared:  0.2167
F-statistic: 11.51 on 1 and 37 DF,  p-value: 0.001659
```

From these results it seems clear that there is an effect of dose on at least the $\log_{10}$ quantity of virus found in those larvae that died. It is a good idea to try to visualize what your model is telling you, especially relative to the original data.

```
# plot raw data
plot(Q ~ Dose,
     data=df3[df3$Died == 1,], # <- just those that died
     xlab=expression(log[10]~dose~(pfu/mL)),
     ylab=expression(log[10]~rananvirus~gene~copies)
)
```

```
# generate predicted values
preds <- data.frame(Dose = seq(from=1, to=6, length.out=50))
preds$Q <- predict(m3,
                    newdata=preds,
                    type="response")
# plot the predicted line
lines(x=preds$Dose, y=preds$Q, type = "l")
```

So, it appears that we are capturing the general pattern of increasing viral quantities with increasing exposure dose (Fig. 6), though there is a good deal of deviation from the model's predictions. That is, there is a pattern, but we are certainly not capturing the general features of the data.

1.8.2 Complications and Interpretation

This is a perfectly reasonable model, but we need to be aware of two important considerations. The first issue is how to interpret our model. That is, what do the coefficients for the (Intercept) and ($\log_{10}$) dose mean, biologically? It can be useful to convert from the log-scale to the typical linear scale. Our model says that the expectation of the $\log_{10}$(Dose) is:

$$\mathbb{E}\left(\log_{10}\left(Q_i\right)\right) = \beta_0 + \beta_{\text{dose}} \log_{10}\left(Dose_i\right), \tag{12}$$

where (Intercept) in the R output corresponds to β_0 and Dose to β_{dose}. If we exponentiate both sides, we get:

$$\mathbb{E}\left(Q_i\right) = 10^{\beta_0 + \beta_{\text{dose}} \log_{10}\left(Dose_i\right)} = 10^{\beta_0} Dose_i^{\beta_{\text{dose}}}.$$

On this scale, our model does not seem so simple and linear. First, note that 10^{β_0} is the expected pathogen load at death in larvae exposed to a dose of zero (we will come back to this in a moment). We will call this the baseline viral load, which in our example is $10^{2.2115} = 162.7$ copies. Second, as we increase the dose of inoculum,

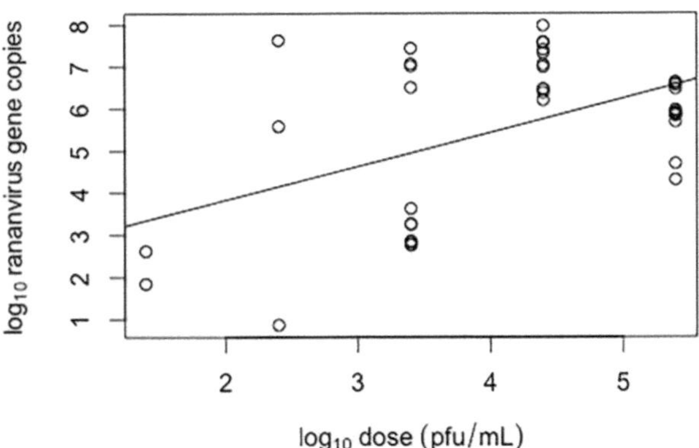

Fig. 6 The intensity of infections in animals that died in the dose-response study by Warne et al. (2011), as well as the best-fit linear regression line fit to these data

we raise it to the *power* of β_{dose}. If β_{dose} were zero, then this second term would always be equal to one, meaning that the expected ranavirus load in a larva is unchanged by the dose of inoculum—a scenario that is clearly inconsistent with these data! If $\beta_{dose} = 1$, then the expected load increases virion-per-virion with the exposure dose. So, we have two values of β_{dose} with clear biological meaning, and the expected relationship between dose and ranavirus load in dead larvae is only linear under these two cases; elsewhere it is a nonlinear power relationship.

In our example, a dose of $\log_{10}(100)$, or 2, would increase the viral load by $2^{\beta_{dose}}$, or $2^{0.8057} = 1.75$ times the baseline, up to $162.7 \times 1.75 = 284.7$ copies. A dose of $\log(1000)$, or 3, would then increase the viral load $3^{0.8057} = 2.42$ times the baseline, up to $162.7 \times 2.42 = 393.7$ copies. The effect of dose increases with the order of magnitude of the dose, but less than linearly. In other words, the effect of exposure dose seems to get relatively weaker and weaker as the dose increases in our system. We may be satisfied with this formulation, but it is important to recognize it does *not* produce a coefficient that can be interpreted as the effect *per virion* on the viral load as one might have expected. It can be helpful to think about what you hope your coefficient will mean, and checking the math, before blindly running analyses.

Before moving on, what about the strange meaning of the intercept of the model? We can fix this easily by centering our data on some particular value (e.g., the mean or some useful point of reference). For instance, our lowest dose of exposure was $10^{1.4}$ pfu/mL, so we could subtract 1.4 from all of the doses so that the lowest dose now equates to zero.

```
summary( lm(Q ~ I(Dose-1.4), # <- surround by the "AsIs" function
            data=df3[df3$Died==1,])
)
Call:
lm(formula = Q ~ I(Dose - 1.4), data = df3[df3$Died == 1, ])

Residuals:
    Min      1Q  Median      3Q     Max
-3.2751 -1.4101 -0.0122  1.4921  3.4849

Coefficients:
              Estimate Std. Error t value Pr(>|t|)
(Intercept)     3.3394     0.6990   4.777 2.8e-05 ***
I(Dose - 1.4)   0.8057     0.2374   3.393 0.00166 **
---
Signif. Codes:  0 '***' 0.001 '**' 0.01 '*' 0.05 '.'
0.1 ' ' 1

Residual standard error: 1.677 on 37 degrees of freedom
Multiple R-squared:  0.2373,    Adjusted R-squared:  0.2167
F-statistic: 11.51 on 1 and 37 DF,  p-value: 0.001659
```

Note that the slope is unchanged, but the intercept has changed. Its meaning is still expected virus load in larvae when the predictor, dose, is zero, but now this occurs at the lowest dose (rather than no exposure). These models are mathematically equivalent, but this version is perhaps easier to interpret.

1.8.3   What to Do with Zeroes in Pathogen Load Data Set?

The second consideration when working with $\log_{10}$ quantities is what to do with the zeros. On the one hand, they may represent true negatives, and so they *should* be zero, but on the other hand, they might have come from a positive sample that simply did not test positive for one of a variety of reasons (e.g., they were below the limit of detection; assay failed). How can we discern which is which? The answer is that we cannot, at least not without outside information.

A more proximate, though related, problem is that $\log(0) = -\infty$. Values of negative infinity in our data tend to cause problems with our analyses; if we are lucky, it simply will not run, but if we are not, they might be excluded without us even knowing. So, what to do?

The answer largely depends on what you think these zeros represent, but there are no universally "right" answers. You might exclude the negative test results because you assume they represent real negatives, and you are only interested in values from infected animals. If this assumption is incorrect, then you will tend to bias your load estimates high. Alternatively, you might assume that they were simply low values that, because of the random nature of virus particles or gene targets getting into DNA extraction tubes, falsely tested negative, in which case you might be justified in setting those zeros to a low value, like the limit of detection. You also could add one to all values, since $\log_{10}(1) = 0$, which is a common approach to ensure that all values are ≥ 0. All of these options can be justified, but none are without consequence for your inference or estimates; even the value you decide to add to all samples can affect your results!

Alternatively, you can use zero-inflated or hurdle models. These models account for the probability that an individual is infected using the equivalent of a logistic model and then, given that an animal is infected, predicts the number of virions with another distribution, typically a Poisson or negative binomial, but continuous distributions are also possible to accommodate ranavirus titers. These models can also be applied to other surveillance data, such as the number of infected animals in a population, where there may be zeros because there is no infection in the population or because infected animals were missed during sampling. We direct the readers to Dohoo et al. (2003) and Zuur et al. (2012) for additional guidance on zero-inflated models.

1.9   Economics of Surveillance

The simplest way to reduce the cost of surveillance is not to do it. While it may seem drastic, there are situations where the outcomes of a surveillance program will not alter the management actions (see Value of Information analysis, Sect. 3.1).

However, there are scenarios where surveillance is necessary. In these cases, it's crucial to strike a balance between minimizing costs and maintaining the effectiveness of surveillance.

As discussed in Brunner et al. (2024), some hosts function as reservoirs for the virus and maintain subclinical infections resulting in low population prevalence, while other species serve as amplification hosts and initiate outbreaks. If funds are limited, a viable strategy would be to test amplification hosts, because these species tend to have lower resistance to ranavirus, and detection probabilities are therefore greater. Please see Halliday et al. (2007) for a discussion of how to select species as sentinels for community-wide risk.

A simple method for reducing project costs while maintaining effectiveness is sample pooling—also called batch sampling (Bilder 2019, Fig. 7). If n samples must be collected, pooling them into batches of size k means that if all batches are negative, only n/k batches need to be processed (Brunner 2020; Sabino-Pinto et al. 2019). If a batch is positive, the individual samples in that batch can be retested to find the positives. Effective batch sampling can reduce the number of diagnostics tests required, thus reducing the cost and time surveillance (Brunner 2020).

The economic efficiencies of batch sampling were first investigated by Dorfman (1943) and have since been used extensively to increase sample throughput and reduce costs, such as during the COVID-19 pandemic (Abdalhamid et al. 2020; Brunner 2020; Hogan et al. 2020). The expected savings from pooling samples can be calculated by comparing the cost of testing all individual samples to the cost of testing k pooled batches (Dorfman 1943). It is important to note that if a batch tests positive and data on individual samples is desired, the samples should be DNA

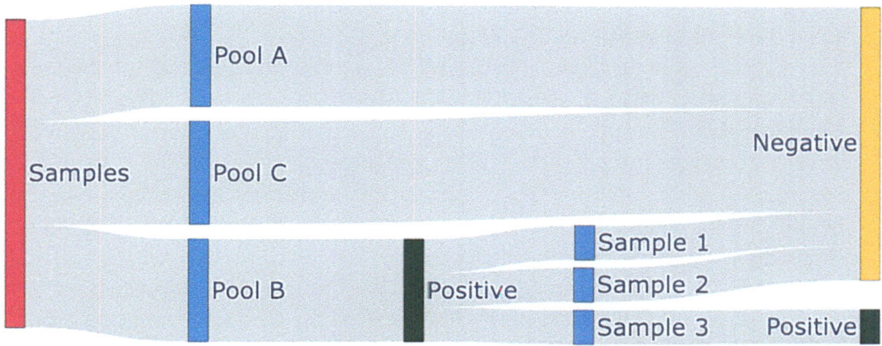

Fig. 7 Sankey diagram of the flow of samples through a two-level hierarchical pooled testing strategy. Starting with nine samples on the left, the samples are pooled into groups of three and tested. Pools A and C are negative, and so all the samples in those pools (six total) are recorded as negative (green). Pool B tests positive (red) and is deconvoluted into individual samples (1, 2, 3). The individual samples are tested again with the same assay, and sample 3 is found to be positive. By pooling the samples, only six out of nine potential tests were required (yellow boxes), resulting in a saving of 33% when using this testing strategy. Optimal pooling size should be determined for each project as savings will depend on factors such as prevalence, pool size, and assay sensitivity and specificity

extracted separately and only pooled during qPCR, which will increase the cost. Most diagnostic laboratories can pool samples and can provide comparative estimates (no pooling, qPCR pooling only, DNA extraction, and qPCR pooling).

Sample pooling has been utilized in ranaviral studies, although formal evaluation of pooling ranaviral samples is lacking (WOAH 2021). Brunner (2020) examined the utility of sample pooling and eDNA to facilitate pathogen surveillance in aquatic animal trade. While eDNA methods can be useful for screening large groups (e.g., in an aquatic trade setting where the goal is to find any infections), it cannot identify infected individuals or estimate prevalence precisely.

Sample pooling to determine individual infections is only effective when prevalence is low (e.g., <30%), although the optimal pooling strategy depends on many factors (Bilder 2019; Brunner 2020; Hitt et al. 2019). Hitt et al. (2019) provide an interactive shiny web app (https://bilder.shinyapps.io/PooledTesting) that can be used to determine optimal pooling size while considering factors such as testing algorithms, prevalence, and assay sensitivity and specificity. Natural groups (e.g., samples from the same locations) can be used to improve sample selection for pools. Positive samples are likely to be correlated due to local transmission chains; thus pooling related samples can reduce the number of pools that must be deconvoluted (Augenblick et al. 2022).

Pooling samples reduces the sensitivity of the assay, i.e., the more samples added to a pool, the more dilute the target will become (Brunner 2020). For this reason, assays utilizing pooled samples should be validated by comparing pooled results to individual samples. Usually, a maximum of three to five samples is recommended per batch.

Box 1 The Value of Simulating Data

It is difficult to provide broadly useful or appropriate guidance about the design and analysis of ranavirus studies. Most every study is different, and even those that follow a common pattern (e.g., experimental exposures, active surveillance among populations) have their idiosyncrasies. However, there is one piece of advice that we think is universal: simulating data from a study, ideally before conducting the study, is immensely helpful. See Electronic Supplement 3 for an example.

What do we mean by simulating data? We mean making up data as if it came from the planned study with samples from the treatments or across the gradient or among the populations you plan to use, assuming the effect sizes you presume will be at work and the sampling (and other) variability you expect. The process of simulating is often iterative. For instance, you might start by simply sketching graphs of the data you hope (or hope not) to see on a white board, then formalizing the relationship(s) you expect to see between the predictor(s) and response, and finally drawing values from distributions based on expected values (e.g., using *R*), revising and reconsidering your

(continued)

Box 1 (continued)

goals and approaches along the way. The further down this iterative process you go, the more you have to gain.

1. At a basic level, even sketching graphs can help you clarify your study, ensuring that your study is likely to answer the question(s) you hope to address. Many times, this process has helped us realize we needed to add treatment levels, use a stratified random sampling approach, or see similar improvements. Specifying a mathematical relationship between the predictor and response (e.g., linear, saturating) and choosing effect sizes (i.e., differences in means, slopes, etc.) then force us to consider whether we are looking for a strong or weak signal in what might be noisy data, whether we are comparing the right groups or levels (e.g., are our treatments distinct enough? Is there sufficient variation between wetlands in the predictor?), and whether our design might be able to distinguish between alternative explanations. We can then consider the variability we are likely to observe to give us a sense of whether the signal might be swamped by the noise in our data. Variability in the response variable comes, at a minimum, from sampling or measurement error. For instance, we do not expect repeated estimates of viral quantities would be identical; small (or large) differences in sample collection, DNA extraction, and qPCR assays will lead to at least some degree of sampling error. Add to this variability in underlying process (e.g., viral replication within a host), and we might reasonably expect estimates of viral quantities to vary by an order of magnitude or more between seemingly identical samples. We can simulate (e.g., with R or even Excel formulas) random observations from a distribution with the expected mean (from step 2) and this expected amount of variability to gain a sense of what our data might look like, even if everything is working as we expect. This can be especially helpful in settings where we lack a well-developed intuition, such as when working with binomial data. For instance, many would consider a sample size of $n = 10$ to be sufficient for estimating the probability of success (e.g., infection, death). However, with $p = 0.8$ observations from 5 of 10 to 10 of 10 are all quite likely to be seen from random chance alone.

2. If we are able to simulate data sets, then we are also able to apply our statistical analyses to those data. This is useful from the perspective of building a work flow—from the statistical models fit, statistics calculated, and graphs constructed—which is helpful to have ready for when the real data are available. However, this is even more useful in the sense that, unlike with real data, *you know the truth*. You know the slope or effect size used to simulate the data, so you can test whether your planned statistical analyses can consistently recover the true values of those parameters. With real data you will never know if everything is working right, but with simulated

(continued)

> **Box 1 (continued)**
>
> data you can! This gives us the opportunity to identify problems such as biased estimates or insufficient power to detect differences you know are there.
>
> In summary, the process of constructing and working with simulated data can help us do better science. It helps us ensure that we are embarking on experiments or observational studies that have the best chances of answering our questions. It also gives us confidence that we are doing our analyses well. While this approach places a lot of the hard work *before* the study, that makes the rest of the process much more tractable, powerful, and transparent.

2   Use of Dynamic Models

Dynamic models can be very useful in studying host-pathogen interactions. Within-host models can elucidate physiological mechanisms that lead to host infection and disease (e.g., Mideo et al. 2008, 2011; Woodhams et al. 2008; Mihaljevic et al. 2019). In comparison, between-host models focus on the fate of individuals and populations when a pathogen is introduced or circulating (Hastings 1997). In this section, we will focus on the latter because of their usefulness in predicting the effects of pathogens on populations. The main types of between-host models are Susceptible-Infected-Recovered (SIR) models based on differential equations, Individual-Based Models (IBMs) that track the fate of individuals, and population models that track the transitions between and abundances of life stages or age classes. Deciding which model to use will depend on the questions and the data available. SIR models are best if data relate to prevalence and questions are about the dominant mode of transmission or short-term disease dynamics during a single outbreak. IBMs are helpful when explicitly considering behavior or variation among individuals and can relate to various time scales. Population models are useful when questions relate to host population declines and population persistence, and data include survival rates of different life stages or age classes.

Each model type can provide different and unique insights. It is important to note that models may not necessarily answer all (or any of) your questions directly. However, used carefully, they can guide the next steps, highlight important data needs, and differentiate between plausible and improbable scenarios related to patterns seen in real systems. Model output is always reliant on the quality of information input to the model. Many times, we do not have data for every parameter necessary or even knowledge of when it is reasonable to make certain assumptions (DiRenzo and Grant 2019; Earl 2019). In this case, a useful technique is to consider a range of scenarios and a range of plausible parameter values, i.e., make educated guesses. If changing that parameter or assumption greatly changes model output, then that should guide future data collection.

In this section, we will discuss each type of model, examples that use ranavirus, and other novel examples from other pathogens that illustrate potential future directions for dynamic models of ranavirus-host interactions. We will also provide some limited guidance on model building and testing. Learning to build and test dynamic models requires a lot of time and effort and is beyond the scope of this chapter. However, we encourage laboratory and field scientists to collaborate with modelers. Having a better understanding of modeling fundamentals will help you with establishing collaborations and understanding papers that contain ranavirus dynamics modeling results.

2.1 SI/SIR Models: Transmission

SI/SIR models typically explore transmission dynamics using a system of ordinary differential equations that model and predict one of three outcomes: pathogen extinction, host extinction, or pathogen-host persistence (Allen 2006). In many simple cases, the total population of hosts is divided into three subpopulations: individuals susceptible to infection (S), infected individuals (I), and individuals that have recovered R from infection and cannot be re-infected or at least have temporary immunity. R can also be the individuals removed from the population (e.g., through mortality). A simpler version of the model is where individuals cannot become immune and do not die (i.e., they stay infected for their lifetime), the susceptible-infected (SI) model (Allen 2006). In another scenario, the SIS model, if individuals clear the infection, they become susceptible again. Here, we describe the basic SIR model.

In the simplest SIR model, the total population size (N) can be assumed constant:

$$N = S + I + R \tag{13}$$

where S, I, and R represent the number of individuals in each respective subpopulation (Hastings 1997). The rate of change of each subpopulation at time t can be modeled as

$$\frac{dS}{dt} = -\beta SI \tag{14a}$$

$$\frac{dI}{dt} = \beta SI - \gamma I \tag{14b}$$

$$\frac{dR}{dt} = \gamma I \tag{14c}$$

where β is the rate at which hosts contact and transmit the infection to each other and γ is the host recovery rate (or removal rate). Here, transmission is assumed to be density-dependent, as transmission is represented as βSI. Some evidence exists that transmission of ranavirus may be density-independent (Harp and Petranka 2006) and can be modeled as $\beta I/N$. McCallum et al. (2001) provide other forms of transmission functions, including nonlinear functions of host density. Because demography (birth, death, immigration, or emigration) is not included in this model, the only equilibrium occurs when all individuals are in the susceptible class (with $I = 0$). For an epidemic to occur, the number of infected individuals must increase, which corresponds with a basic reproductive number greater than one.

The basic reproductive number, R_0, is a key measure for estimating transmission potential and is fundamental in studying how diseases spread. R_0 is defined as the average number of secondary cases that arise from one primary case in a fully susceptible population. It is a measure of the spread of a disease at the beginning of an outbreak. For an epidemic to occur, the number of infected individuals must be initially increasing,

$$\left.\frac{dI}{dt}\right|_{t=0} > 0 \tag{15}$$

given a single infected cases $I(0) = 1$ in an otherwise susceptible population $S(0) = N - 1$. For large populations, we approximate $S(0) = N$ when we compute this initial spread of infection. If the model is simple enough, we can evaluate this initial spread directly from our equation. For example, consider the infection Eq. (14b), evaluating it at $t = 0$ assuming $I(0) = 1$ and approximating $S(0) = N$ yields

$$\left.\frac{dI}{dt}\right|_{t=0} = \beta N - \gamma. \tag{16}$$

The sign of this equation tells us whether the disease will spread or die out, and we formulate this as the basic reproductive number

$$R_0 = \frac{\beta N}{\gamma} \tag{17}$$

If $R_0 > 1$, the number of infected cases will increase, and there will be an outbreak. If $R_0 < 1$, the disease will die out. The expression for R_0 can be interpreted nicely into the product of biological components: the transmission rate β, the duration of infection $1/\gamma$, and the susceptible population density N. The expression for R_0 is model specific and will depend on model assumptions, for example, R_0 given in Eq. (17) is specific for model (14).

Many amphibian pathogens, like ranavirus or chytrid fungi, can be environmentally transmitted by virus particles or zoospores. It can be necessary to explicitly model the environmental pathogen pool with a differential equation coupled to the compartmental model of the host population. Letting P denote the concentration of pathogens in the environment, we can expand the basic SIR model (14) into the following SIRP four-dimensional system of equations:

$$\frac{dS}{dt} = -\beta_{\mathrm{d}}SI - \beta_{\mathrm{e}}SP \tag{18a}$$

$$\frac{dI}{dt} = \beta_{\mathrm{d}}SI + \beta_{\mathrm{e}}SP - \gamma I \tag{18b}$$

$$\frac{dR}{dt} = \gamma I \tag{18c}$$

$$\frac{dP}{dt} = \omega I - \delta P \tag{18d}$$

Here, β_{d} is the transmission rate via direct contact with infected individuals, and β_{e} is transmission rate via contact with environmental pathogens. Infected individuals shed pathogens at rate ω, which persist in the environment for duration δ^{-1}.

For the simple SIR model (14), we were able to consider whether the infected population is initially increasing to calculate R_0 (Eq. 17). This technique will not be possible for more complicated models; however there is rich mathematical theory using next-generation matrices (NGM) that can help (Diekmann et al. 1990; van den Driessche 2017). For disease transmission models, the NGM method is particularly useful when models have multiple infected compartments. For example, the classical SEIR model divides the infected population into two compartments: exposed (E) and infectious (I) individuals (see Blackwood and Childs 2018 for examples). Below, we show how to use the NGM to calculate the R_0 for model (18).

The NGM approach considers a sub-model of all the "disease" compartments. For model (18) the disease compartments are I and P. We consider a two-dimensional sub-model:

$$\frac{dI}{dt} = \beta_{\mathrm{d}}SI + \beta_{\mathrm{e}}SP - \gamma I \tag{19a}$$

$$\frac{dP}{dt} = \omega I - \delta P \tag{19b}$$

which can be written in a general vector form:

$$\frac{d\vec{x}}{dt} = \mathcal{F}(\vec{x}) - \mathcal{V}(\vec{x}) \tag{20}$$

where $\vec{x}$ is the vector of all disease compartments $[I, P]$. $\mathcal{F}(\vec{x})$ contains terms that yield new infections,[1] and $\mathcal{V}(\vec{x})$ contains all other terms or the transition terms into and out of these infected compartments.

---

[1] Here we assume that shedding is not a new infection and place the shedding term ωI into the transition vector $\mathcal{V}$. Alternatively, you can consider shed pathogen particles as new infection and get a different expression for the next-generation matrix, which should result in the same threshold for $R_0 = 1$, but may be harder to interpret biologically. See van den Driessche (2017) for details using a cholera model example.

$$\frac{d\vec{x}}{dt} = \begin{bmatrix} dI/dt \\ dP/dt \end{bmatrix} = \left(\begin{bmatrix} \beta_d SI + \beta_e SP \\ 0 \end{bmatrix} - \begin{bmatrix} \gamma I \\ -\omega I + \delta P \end{bmatrix} \right) = \mathcal{F}(\vec{x}) - \mathcal{V}(\vec{x}) \qquad (21)$$

Now, from vectors $\mathcal{F}(\vec{x})$ and $\mathcal{V}(\vec{x})$, we define matrices of partial derivatives evaluated at $(S, I, R, P) = (N, 0, 0, 0)$, the disease-free equilibrium:

$$F(\vec{x}) = \left[\frac{\partial \mathcal{F}_i}{\partial x_i} \right] = \begin{bmatrix} \dfrac{\partial \mathcal{F}_1}{\partial I} & \dfrac{\partial \mathcal{F}_1}{\partial P} \\ \dfrac{\partial \mathcal{F}_2}{\partial I} & \dfrac{\partial \mathcal{F}_2}{\partial P} \end{bmatrix} = \begin{bmatrix} \beta_d S & \beta_e S \\ 0 & 0 \end{bmatrix}\Bigg|_{(N,0,0,0)} = \begin{bmatrix} \beta_d N & \beta_e N \\ 0 & 0 \end{bmatrix} \qquad (22)$$

$$V(\vec{x}) = \left[\frac{\partial \mathcal{V}_i}{\partial x_i} \right] = \begin{bmatrix} \dfrac{\partial \mathcal{V}_1}{\partial I} & \dfrac{\partial \mathcal{V}_1}{\partial P} \\ \dfrac{\partial \mathcal{V}_2}{\partial I} & \dfrac{\partial \mathcal{V}_2}{\partial P} \end{bmatrix} = \begin{bmatrix} \gamma & 0 \\ -\omega & \delta \end{bmatrix}\Bigg|_{(N,0,0,0)} = \begin{bmatrix} \gamma & 0 \\ -\omega & \delta \end{bmatrix} \qquad (23)$$

These two matrices are used to define the next-generation matrix FV^{-1}. Here V^{-1} is the inverse of matrix V. R_0 is the spectral radius (dominant eigenvalue) of the next-generation matrix:

$$R_0 = \rho\left(FV^{-1} \right). \qquad (24)$$

For our example this expression takes the following form:

$$R_0 = \rho\left(\begin{bmatrix} \beta_d N & \beta_e N \\ 0 & 0 \end{bmatrix} \begin{bmatrix} \gamma & 0 \\ -\omega & \delta \end{bmatrix}^{-1} \right) = \left(\frac{\beta_d}{\gamma} + \frac{\beta_e \omega}{\gamma \delta} \right) N \qquad (25)$$

Our R_0 has two components that relate to the two routes of transmission in the model: direct transmission from contact with infected individuals $\left(\dfrac{\beta_d N}{\gamma} \right)$ and environmental transmission from contact with pathogen particles $\left(\dfrac{\beta_e \omega N}{\gamma \delta} \right)$. Each term in the R_0 can be nicely translated biologically:

$$\left[\underbrace{\underbrace{(\beta_d)}_{\text{trans. rate}} \underbrace{\left(\frac{1}{\gamma} \right)}_{\text{infection duration}}}_{\text{Direct transmission}} + \underbrace{\underbrace{(\beta_e)}_{\text{trans. rate}} \underbrace{\left(\frac{1}{\gamma} \right)}_{\text{infection duration}} \underbrace{(\omega)}_{\text{shed rate}} \underbrace{\left(\frac{1}{\delta} \right)}_{\text{pathogen persistence}}}_{\text{Environmental transmission}} \right] \underbrace{N}_{\text{Host density}}$$

For ranavirus and most natural populations, the basic SIR model (17) and SIRP model (18) are likely too simplistic, and many published SIR type models involving ranavirus expand on this model in different ways dependent on the host species and ranavirus strain. Duffus et al. (2019) used a discrete-time SI model to show that ranavirus could be maintained in a population of common frogs (*Rana temporaria*) in the UK with only transmission between adults. They also demonstrated that transmission between adults could maintain two syndromes of ranavirus (the ulcerative and hemorrhagic forms) in a single population. Farrell et al. (2018) modeled ranavirus transmission in tiger salamanders. They examined effects of the incidence function and the function describing life expectancy relative to time since infection on host persistence and population size after an epidemic.

Brunner and Yarber (2018) and Peace et al. (2019) modelled transmission in wood frogs (*Lithobates sylvaticus*) to determine the most likely route(s) of transmission. Both models compared the relative importance of environmental transmission, transmission from scavenging dead conspecifics (necrophagy), and from direct contact. Both models included terms to track the presence of carcasses available for necrophagy and the concentration of virions shed in the water available for environmental transmission. Peace et al. (2019) also included multiple stages of infection with different probabilities of mortality and virus shedding rates. Both models found that direct contact was the most important mode of transmission, but Peace et al. (2019) also found that environmental transmission could be more important with higher tadpole densities and later in an epidemic.

Other model expansions could be useful for predicting and/or gaining a better understanding of ranavirus dynamics in natural populations. For example, most ranavirus host species exist in communities where they are likely to interact with other susceptible species, possibly from different ectothermic vertebrate classes (Gray et al. 2009). Brenes et al. (2014a) demonstrated that interclass transmission of ranavirus through water was possible. He also showed that ranaviral disease outcomes depended on species composition in the amphibian community and which species was initially infected with ranavirus (Brenes 2013). These studies could serve as a starting point for determining transmission probabilities in aquatic communities with multiple species. In other disease systems, the addition of multiple species to transmission models had an effect on the focal host population, but depended on the host's competency as a reservoir and its dominance within the community (Keesing et al. 2006). The addition of multiple host species can make the analysis of SIR models challenging. To date, most models have included only two species and the pathogen (Keesing et al. 2006), which may be unrealistic for some ranavirus-host systems. Dobson (2004) dealt with the large number of parameters in multi-species models by scaling the parameters as allometric functions of host body size, although it is unclear such a relationship exists with transmission of ranavirus. Lélu et al. (2013) provide an example of a model including trophic transfer of a parasite (*Toxoplasma gondii*) from rats to cats and vertical transmission in cats. Similar complex interactions certainly occur among ranavirus hosts species, such as predation or necrophagy, and mechanical transmission by mosquitoes has been hypothesized (Allender et al. 2006; Johnson et al. 2007; Kimble et al. 2015). Despite the large number of possible interactions in a ranavirus-host system, several

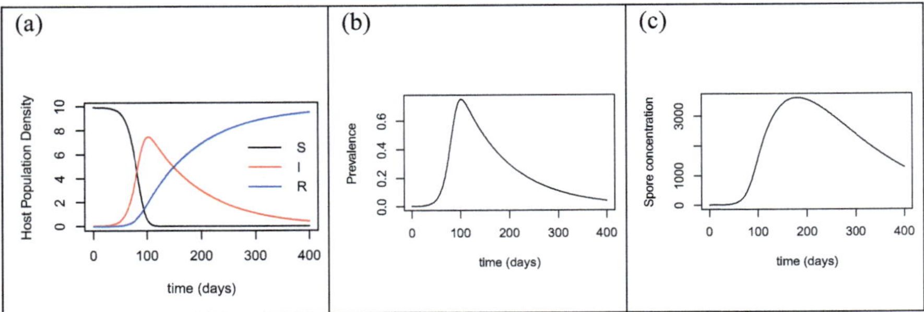

Fig. 8 Solutions for the base SIRP model (18) showing population densities (**a**), prevalence (**b**), and environmental spore loads (**c**) changing over time. Simulations were generated using ODE solvers in R's deSolve. *R* code for solving this model and generating these figures is at the end of Sect. 2.1

interactions are likely unimportant to its epidemiology. One strategy would be to create several competing models and fit them to data on dynamics in natural populations or in mesocosm studies to identify the most important mechanisms for transmission.

For researchers interested in using SIR models to examine ranavirus, we recommend Otto and Day's (2007) book, *A Biologist's Guide to Mathematical Modeling in Ecology and Evolution*, which reviews the mathematics and describes the process necessary for constructing and analyzing models primarily with ordinary differential equations. An understanding of computer programming and use of software (e.g., Matlab, Maple, Mathematica, *R*) will be necessary to construct models and perform simulations for most analyses. Sample *R* code for solving the base model (18) and producing Fig. 8 is below.

```
#R script to solve SIRP Model (18)

#required package for ODE solver:
install.packages("deSolve")
#load packages so we can use the ode function:
library(deSolve)

#Model Equations:
Model_SIRP <- function(t, variables, parameters)
{with( as.list(c(variables,parameters)),  #lets us access
variables and parameters by name
        { #System of ODES:
          dS = -betad*S*I-betae*S*P
          dI = betad*S*I+betae*S*P -gamma*I;
          dR = gamma*I;
          dP = omega*I-delta*P;
          #End of ODes
```

```r
        return(list(c(dS,dI,dR,dP)))
      })
} #End Model Equations

#Initialization: load parameter values:
betad=0.0003  #Direct transmission rate
betae=.0001  #Environmental transmission rate
gamma=0.01 #Recovery rate
omega=10  #spore shedding rate
delta=0.01 #Spore degradation rate
#set parameter list:
parameter_values<-c(betad=betad, betae=betae, gamma=gamma,
omega=omega, delta=delta)

#Initialization: Set time:
tend=400 #Days for simulation
time_values<-seq(0,tend)

#Initialization: Set initial conditions:
N0=10; #%Initial population density

initial_values <- c(
  S=N0*0.999,
  I=N0*0.001, #assume 0.1% of the population is initially
infected
  R=0,
  P=0
)

#Solve Model
Model_solutions <- ode(
  y = initial_values,
  times = time_values,
  func = Model_SIRP,
  parms = parameter_values
)

Model_solutions<-as.data.frame(Model_solutions) #convert output
to dataframe so we can use with()

#plotting
with(Model_solutions, {
  plot(time, S, type = "l", col = "black",
       xlab = "time (days)", ylab = "Host Population Density")
  lines(time, I, col = "red")
  lines(time, R, col = "blue")
```

```
legend("right", c("S", "I", "R"),
       col = c("black", "red", "blue"), lty = 1, bty = "n")
})


#Plot Prevalence
with(Model_solutions, {
  plot(time, I/(S+I+R), type = "l", col = "black",
       xlab = "time (days)", ylab = "Prevalence")
})


#Plot Pathogen load
with(Model_solutions, {
  plot(time, P, type = "l", col = "black",
       xlab = "time (days)", ylab = "Spore concentration")
})
```

2.2   *Individual-Based Models/Pattern-Oriented Modeling*

Individual-based models (IBMs), sometimes called agent-based models (ABMs), are also very useful for examining disease dynamics. IBMs are simulation-based, and during each time step, a set of rules or probabilistic events occur involving each individual. IBMs are often easier for biologists to construct than SIR models, because they do not require solving differential equations. However, IBMs can be complex and require computer programming skills. These models often operate on a set schedule of events that are implemented using sequential equations, a series of for-loops, and if-then statements that determine an individual's actions or fate. For disease IBMs, each individual's disease state is recorded, and their risk of infection can depend on their interaction with other individuals or the environment.

IBMs for ranavirus have been applied to Bosca's newt (Rosa et al. 2019) and the marbled newt (Rosa et al. 2022) in Portugal. The Bosca's newt example examined the length of time that outbreaks occurred and the effect of sex-biased mortality from ranavirus on population persistence. Researchers found that female-biased ranavirus mortality did not cause larger population declines but did slow down population recovery once outbreaks ceased (Rosa et al. 2019). In this model, mortality was applied as a percentage. In the Marbled Newt example, an individual-based ranavirus transmission model was linked to an individual-based population model and matched to existing population data covering 40 years (Rosa et al. 2022). Scenarios examined effects of ranavirus outbreaks and an invasive fish species separately and, when combined, found that both stressors were needed to generate patterns similar to those seen in the population data.

One attractive aspect of IBMs is that they can explicitly incorporate animal behavior. For ranavirus, researchers might be interested in how different behaviors, such as schooling or necrophagy, affect host populations and persistence with the pathogen. Further, many researchers use IBMs to better understand spatial dynamics by explicitly modeling animal movement. Incorporating animal movement can allow research to estimate the spread of pathogens to new areas. Animal movement can affect epidemiological patterns when space is heterogenous (Fofana and Hurford 2017).

There are also other types of IBMs that use differential equations. For example, Briggs et al. (2010) developed an IBM with differential equations that explicitly incorporated individual chytrid fungus load and examined how a pathogen reservoir and a long-lived tadpole stage affected whether frog populations could persist with chytrid infections or would experience local extinction. Similar models could be developed for ranavirus that include viral load and shedding to better understand how the virus might inter act with the host and factors that contribute to die-offs.

A useful technique for creating IBMs and determining plausible interactions is called pattern-oriented modeling (POM). In POM, data are used to determine several salient patterns seen in a natural system of interest that form the basis of model evaluation. Multiple possible forms of an IBM are created, representing different hypotheses about host-pathogen interactions. The different IBMs are evaluated based on their ability to recreate the patterns (Grimm et al. 2005; Grimm and Railsback 2012). When a model matches multiple patterns, it is more likely to be structurally realistic (Wiegand et al. 2003) and capable of producing testable predictions. In using POM, researchers can also contrast different hypotheses, determine a useful model structure, and reduce parameter uncertainty.

For researchers interested in developing IBMs, we recommend two books: Grimm and Railsback's (2005) *Individual Based Modeling and Ecology* and Railsback and Grimm's (2011) *Agent-Based and Individual-Based Modeling: A Practical Introduction.* Both titles describe a "best model practice" called object-oriented design and description (ODD), which is a standard format to describe various aspects of an IBM. The latter title goes through the process of building IBMs with examples and code for a relatively user-friendly and free program called NetLogo (http://ccl.northwestern.edu/netlogo/index.shtml). NetLogo includes a library of preconstructed models, including AIDS, Disease Solo, and Virus, which could form the basis for the development of models for ranavirus. Further, NetLogo's website includes a Modeling Commons, where NetLogo users can share their models to help others in their own model development. Another free software is Vortex (https://scti.tools/vortex/), which is an individual based population modelling framework. Vortex can be linked to Outbreak (https://scti.tools/outbreak/), which models transmission dynamics and can incorporate management scenarios, and Spatial (https://scti.tools/spatial/; tracks movements) through MetaModelManager (https://scti.tools/metamodelmanager/). All of the extensions are also free. Rosa et al. (2022) model ranavirus by linking Vortex and Outbreak through MetaModelManager. Other software, such as Matlab and *R*, can be used to develop and analyze IBMs.

2.3 Population Matrix Models

Population matrix models examine changes in population size and age structure over time. These models include parameters for the transition probability between each age class. To incorporate disease, the survival following exposure to ranavirus can be incorporated for each age class. Several population models have been developed for amphibian populations affected by ranavirus (Campbell et al. 2018; Earl and Gray 2014; Earl et al. 2016). These include either age- or stage-structured matrix models to predict the effects of ranavirus exposure during different life stages. These combine population models with data on mortality after ranavirus exposure (e.g., Haislip et al. 2011) to predict population outcomes. Earl and Gray (2014) and Earl et al. (2016) modeled populations of three different North American anuran species and demonstrated that local extinction of closed populations is possible. Earl et al. (2016) further assessed the impacts of immigration from nearby populations on population persistence. Campbell et al. (2018) modeled common frogs (*Rana temporaria*) and showed that ranavirus outbreaks could truncate age distributions, which made populations more susceptible to environmental stochasticity.

Population matrix models can also be combined with transmission models to more realistically model both dynamics simultaneously. For example, Briggs et al. (2005) merged a population model of yellow-legged frogs (*Rana muscosa*) and an SIR model of the infection dynamics of Bd based on the current knowledge of transmission and mortality rates. This model combined discrete-time between-year population dynamics with a continuous time transmission dynamics within each year. By running the model with different parameter values, Briggs et al. (2005) were able to determine which conditions resulted in extinction of the frog population, non-persistence of the pathogen, and persistence of the frog population and the pathogen.

Population models can also be scaled up to take into account metapopulation processes. A metapopulation is a set of spatially structured local populations that periodically interact via dispersal (Marsh and Trenham 2001; Smith and Green 2005). Several ranavirus host species are likely structured as metapopulations. Metapopulation models incorporate parameters for dispersal probability between local populations and demographic parameters in each local population. Metapopulation models are useful to understand the spatial spread of pathogens among populations and examine the effectiveness of disease intervention strategies (Hess 1996). For researchers interested in population matrix models, we recommend Caswell's (2000) *Matrix Population Models: Construction, Analysis, and Interpretation*. Hanski's (1999) *Metapopulation Ecology* will be useful for those interested in investigating ranavirus effects on metapopulation dynamics.

2.4 Modeling Disease Intervention Strategies

One goal of modeling host-pathogen dynamics is to identify intervention strategies that thwart disease outbreaks. Currently, there are few proposed control options for ranavirus, but vaccine development is possible in the future (Miller et al. 2011;

Chen et al. 2018). Other options include quarantining individuals or populations, culling, and creating captive populations for reinforcement or reintroduction if disease is likely to cause extremely high mortality to populations of conservation concern. Reinforcement efforts could easily be modeled with immigration parameters. Models also can be used to identify vulnerable points in the host-pathogen cycle that can be interrupted with intervention strategies. For example, if outbreaks are a consequence of density, emergent vegetation in wetlands can reduce the probability of transmission among amphibian larvae (Greer and Collins 2008; Malagon et al. 2020). If stressors in the aquatic environment (e.g., high nitrogen levels) are resulting in reoccurring outbreaks, strategies that improve those stressors (e.g., water quality) can be used. Rosa et al. (2022) modeled the removal of an invasive fish species that was affecting newt populations simultaneously with ranavirus outbreaks. The model showed that fish removal would likely prevent local extinction.

A thorough understanding of the factors responsible for outbreaks and the ranavirus-host system is essential to identifying plausible intervention strategies. In some cases, possible intervention strategies might be infeasible to implement, excessively costly, or undesirable in natural populations. However, if strategies are feasible, models can be used to determine when and how often the strategy should be employed for the best results. SIR models and their variants can be used to explore vaccination strategies (Hethcote 2000) and other control techniques such as culling (Lloyd-Smith et al. 2005). Cost of disease control can be incorporated into models to determine the best strategies given financial constraints (Fenichel et al. 2010). Woodhams et al. (2011) discussed possible intervention strategies for *Batrachochytrium dendrobatidis* and presented model results of their efficacy on individuals with and without an adaptive immunity. They also went on to show that reducing the host population size (i.e., decreasing transmission probability) could prevent extinction. For researchers interested in implementing optimal control models, we recommend Lenhart and Workman's (2007) *Optimal Control Applied to Biological Models*, which focuses on control of continuous ordinary differential equation models and includes sample code for the computer program Matlab. Optimal control can also be applied to IBMs, but effective techniques are still being developed (Federico et al. 2013).

2.5   Model Parameterization and "Evaludation"

There are a number of ways to parameterize models and integrate them with data. Frequently, modelers choose parameter values by searching the literature, but often not all parameter values are available. Another method is to construct a model and fit the output to an existing data sequence (e.g., Mihaljevic et al. 2019; Rosa et al. 2022), which unfortunately is not very common in amphibian population models (Earl 2019). In the case of ranavirus modeling, predictions could be fit to surveillance data that include abundances of infected and uninfected individuals or the magnitude and timing of a die-off. After the model is fit to the data, the parameter values that give the best fit or that match multiple patterns (as in POM) are then used.

If some parameters are known and researchers have a good idea of the possible range of other parameters, these ranges of values can be explored to determine how they change the model output. Assessing the effects of changes in parameter values is called sensitivity analysis (Cariboni et al. 2007). If the model is especially sensitive to a certain parameter, it suggests that better parameter estimation would be a valuable research direction (Biek et al. 2002; Cariboni et al. 2007), especially if the parameter estimate is not based on robust data (e.g., low sample sizes). Cariboni et al. (2007) suggest best practices for sensitivity analysis. An excellent review of parameter estimation for disease modeling of natural populations can be found in Cooch et al. (2012).

The aim of model evaluation is to determine if models typify natural systems well enough to represent the intended dynamics. This often involves determining whether or not they can be used to make accurate predictions. Frequently, the terms model evaluation, model validation, and model testing are used interchangeably. Because models are built on assumptions and simplifications, they are never truly "valid" or "correct." Augusiak et al. (2014) have suggested the term "evaludation" to represent the process of assessing the model's quality and reliability and discuss six elements for proper "evaludation" of a model:

- Assessing the quality of the data used to build the model
- Evaluating the simplifying assumptions structuring the model
- Verifying that the model is correctly implemented
- Verifying that the output matches the data used to design the model
- Exploring model sensitivity to changes in parameter values
- Assessing whether the model can fit an independent data set not used in original model formulation

It is recommended that model formation and "evaludation" follow a documentation procedure called TRACE (TRAnsparent and Comprehensive Ecological documentation) that is designed to ensure reliability of models and link the science to application (Grimm et al. 2014).

3 Risk Analysis for Introduction of Ranavirus into an Uninfected Area

The movement of animals inherently carries the risk of transmitting their parasites and diseases. In an ideal world, disease spread via animal trade and translocation would be prevented through testing of all animals for potential pathogens. However, real-world decisions must be made based on restricted information and finite resources. Risk analysis plays a crucial role in these decisions. This section delves into the multifaceted process of decision-making and risk analysis for the introduction of ranaviruses into uninfected areas.

3.1   Value of Information Analysis

There is an assumption that obtaining more information will always lead to better decision-making and management outcomes. Value of Information (VoI) analysis serves as a critical tool in the realm of decision-making (Schlai and Raiffa 1961). This methodology quantifies the potential benefit of obtaining additional information to enhance decision outcomes (Canessa et al. 2015). The key principle underlying VoI is that the acquisition of further data is deemed valuable only if it possesses the capacity to meaningfully influence or alter the preferred strategy currently considered by the decision-maker (Williams and Johnson 2015). When evaluating the merit of pursuing more information, one must account for the "cost" of this pursuit, which is typically characterized by the financial expenditure necessary to obtain the additional data. Thus, VoI analysis allows decision-makers to weigh the expected benefits of enhanced decision-making against these costs, facilitating more informed and effective decisions.

Consider the following scenario: Over time, a decline in a local frog population has been noted, and it is determined that it is necessary to bolster the population by translocating individuals from an external source population. Currently the population is free from ranaviral disease; however, there is a concern that ranaviruses could be introduced with the translocated frogs. Environmental DNA (eDNA)-based screening methods will be used to assess potential source populations for ranaviruses; however, there is limited funding available, and so unnecessary screening should be reduced where possible. Three potential source populations (*Sites 1*, *2*, and *3*) have been identified. The goal is to determine which, if any, of these three potential source populations should undergo eDNA screening for ranavirus given the desire to minimize screening expenses.

VoI analysis can be used to help decide which potential source populations should be prioritized for eDNA screening based on the potential impact of decision outcomes. Canessa et al. (2015) provide a comprehensive primer on VoI analysis, and here we adapt their methods to the scenario above.

The prior probability of ranavirus in each of the three populations is 0.1, 0.5, and 0.8 based on the ranaviral infection rate at surrounding sites. The eDNA screening method has a sensitivity (true-positive rate) and specificity (true negative rate) of 0.4 and 1, respectively. For simplicity, the population size is 100 frogs, and 35 will be introduced from a single source population. If ranavirus is introduced with the translocated frogs, then modeling predicts that the local population will be reduced to 55 frogs. If ranavirus is not introduced, no frogs will die leaving a population of 135 frogs. If no action is taken, the population remains stable (100 frogs total) regardless of ranaviral status at source sites. This information can be used to construct a consequence table (Table 1).

The expected value for translocating from each site with uncertainty, i.e., without testing (EVTU), represents the expected number of frogs if the decision is made to translocate regardless of the outcome. The EVTU (Table 2) can be calculated with the following:

Table 1 Consequence table of translocating or doing nothing under the hypothesis that ranavirus is present or not in the source population

	Ranavirus (rana+)	No ranavirus (rana−)
Translocate (T)	55	135
Do nothing (D)	100	100

Table 2 Components of EVPI calculations

Site	P(rana+)	EVTU	EVU	Decision uncertainty	EVC	EVPI	Testing priority
1	0.1	127	127	Translocate	131.5	4.5	3
2	0.5	95	100	Do nothing	117.5	17.5	1
3	0.8	71	100	Do nothing	107	7	2

Refer to Canessa et al. (2015) for equation details

$$V\left[T,\text{rana}+\right]*P\left(\text{rana}+\right)+V\left[T,\text{rana}\right]*P\left(\text{rana}-\right)$$

where $V[T, \text{rana}+]$ and $V[T, \text{rana}-]$ are the predicted population sizes if frogs are translocated under the hypothesis of ranavirus at the source site or not (55 and 135). $P(\text{rana}+)$ and $P(\text{rana}-)$ are the probabilities of ranavirus (0.1, 0.5, or 0.8) or no ranavirus (0.9, 0.5, or 0.2) at the source site. The EVTU represents the expected number of frogs if the decision is made to translocate regardless of the outcome.

Translocating is not the optimal decision under uncertainty at *Site 3* as the EVTU is 71 which is less than the expected value of doing nothing (100). The expected value (number of frogs) from making the optimal decision under uncertainty (EVU) is 100 in the case of *Site 3* as doing nothing has more expected value.

In each case, there is the potential to increase the expected value by gathering more information; however, this is dependent on whether the presence of ranavirus at the source site can be accurately determined. The expected value of making a decision under certainty (EVC) is the maximum value that could be gained by testing the potential source populations. By assuming a perfect screening method, i.e., no false positives or negatives EVC can be calculated as:

$$\left[\max\left(V\left[T,\text{rana}+\right],V\left[D,\text{rana}+\right]\right)*P\left(\text{rana}+\right)\right]$$
$$+\left[\max\left(V\left[T,\text{rana}-\right],V\left[D,\text{rana}-\right]\right)*P\left(\text{rana}-\right)\right],$$

where $V[D, \text{rana}+]$ and $V[D, \text{rana}-]$ are the predicted population sizes if no frogs are translocated (100 and 100). The expected value of perfect information (EVPI), that is, the expected value gained by eliminating uncertainty entirely, can be calculated by subtracting the expected values under certainty and uncertainty (Raiffa 1968).

Intuitively, a negative result from a perfect screening method at *Site 3* would indicate that this site is suitable as a source population. However, there is a slim chance of a negative result (as the prior suggests), and so the EVPI is low (Table 2). Similarly, at *Site 1* the prior (0.1) suggests a low chance of finding ranavirus. The

EVPI is low at *Site 1* because the results of a perfect test will likely be negative, so little value is gained by testing. At *Site 2*, the prior (0.5) indicates uncertainty around the ranavirus status. Screening the population at *Site 2* would relieve a lot of the uncertainty, and so the EVPI is high.

EVPI provides a useful measure of the maximum benefit from perfect information; however, like most real-world scenarios, the eDNA screening method is not perfect (sensitivity = 0.4). Bayesian pre-posterior analysis (Berger 1985) and the methods described in Canessa et al. (2015) can be used to calculate the expected value of sample information (EVSI), that is, the value expected to be gained by performing the imperfect eDNA screening at each site.

EVSI is calculated by considering how obtaining a positive (EV+) or negative (EV−) results would change beliefs (prior) and decisions (Decision+ and Decision−). Canessa et al. (2015) provide a useful spreadsheet as supplemental information that can be used to calculate the values in Table 3.

If the eDNA screening method is positive at any site, then it would not be chosen as source populations (Decision+) as the eDNA screening method does not produce false positives (100% specificity).

The probability of ranavirus at *Site 3* is high (0.8) and test sensitivity is low; therefore the negative predictive value (NPV) is also low. As a result of the low NPV, a negative screening of *Site 3* would likely be a false negative. Therefore, Site 3 can be definitively ruled out as a source population without any testing required.

In the case of *Site 1*, a lot of value is gained if the eDNA screening method is negative (higher NPV); however, the EVSI is low as there is a strong prior belief (0.1) that the test will be negative; therefore not much is gained by performing the test. A negative eDNA screening result at *Site 2* would change the decision from doing nothing (decision uncertainty, Table 2) to translocating (Decision−, Table 3), emphasizing the utility of VoI analysis to identify areas of focus.

In cases where the expected value of sample information is greater than zero, the cost of performing the eDNA screening must be weighed against the EVSI, i.e., 1.8 frogs at *Site 1* and 4 at *Site 2*. For example, if funds are only available for one site, then screening *Site 2* would give twice as much value. Following screening (assuming *Site 2* is negative), other factors could be used to select between *Site 1* and *2*, such as the absolute expected value, cost/difficulty of translocation, source population health, other pathogens, etc. (see Sect. 3.2).

When performing VoI analysis, it is important to keep in mind that Bayesian updating, EVPI, and EVSI calculations are just tools to inform decisions. They are based on estimates and models, which may not represent real-world complexities.

Table 3 Components of EVSI calculations

Site	P(rana+)	EV+	Decision+	EV−	Decision−	EVSI	Testing priority
1	0.1	100	Do nothing	130	Translocate	1.8	2
2	0.5	100	Do nothing	105	Translocate	4.0	1
3	0.8	100	Do nothing	100	Do nothing	0.0	No test

Refer to Canessa et al. (2015) for equation details

Therefore, while VoI analysis can provide useful insights, it should be used in conjunction with other analyses, knowledge, expertise, and judgment.

3.2 Risk Analysis

Risk analysis is a procedure that can be used to determine the threat of a pathogen entering a system. The consequences of pathogen introduction can be monitored directly (Sect. 1) or simulated using models (Sect. 2). Several standards exist for performing risk analysis of animal diseases, including the World Organization for Animal Health (WOAH, founded as OIE) risk analysis model and the OIE-IUCN Wildlife Disease Risk Analysis (DRA) guidelines (Brückner et al. 2010; Jakob-Hoff et al. 2014; Travis and Smith 2019). Both standards can be used to assess the potential threats of pathogens entering a system and the risks associated with their spread, although they differ in their focus, scope, and application (Fig. 9).

The OIE risk analysis model was primarily developed from a trade perspective as a tool for Import Risk Analysis (IRA) between two countries or regions. IRA is used to transparently assess the disease risk associated with the importation of animals or their products (Brückner et al. 2010). The OIE-IUCN Wildlife DRA guidelines were specifically developed with wildlife disease in mind. The OIE risk analysis model was adapted to encompass the special features associated with disease risk analysis as it is applied to wildlife and biodiversity conservation resulting

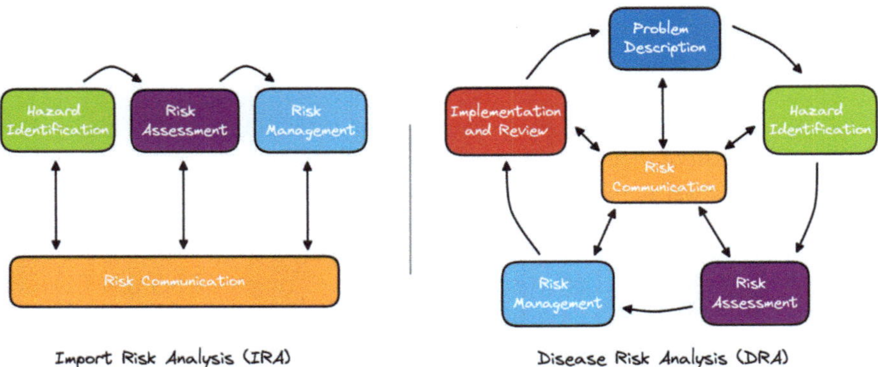

Fig. 9 Comparison of IRA and DRA frameworks. Boxes represent major components as identified by the respective guidelines. Both frameworks emphasize risk communication between interested stakeholders and experts throughout the analysis. The DRA framework identifies problem description (blue) and implementation and review (red) as distinct components (nested in other components in IRA). Starting and ending with the problem description, the circular nature of the DRA framework emphasizes the requirement of dynamic and ongoing analyses. The IRA is more linear and focused on producing an objective and defensible method of assessing disease import risks, ultimately resulting in sanitary measure recommendations

in the production of the Manual and Guidelines for Wildlife DRA (Jakob-Hoff et al. 2014).

In general IRA focuses on the potential infection of one species or several species within the same taxonomic class, whereas DRA can involve a broader range of species and ecosystems. As discussed in Duffus et al. (2015), many ranaviruses are multi-species pathogens that have the capability of infecting three vertebrate classes, which makes IRA more suited for the analysis of ranaviruses that infect a single class such as ENHV in fish. IRAs can be used to establish or revise trade or translocation guidelines for wildlife that could be subclinically infected with a pathogen (Smith et al. 2009). The WOAH lists ranaviruses that infect amphibians as notifiable pathogens, meaning that a subsample of amphibians that are involved in international trade should be verified ranavirus negative prior to shipment (Schloegel et al. 2010), although the degree to which these regulations are enforced globally is not clear (Kolby et al. 2014).

The procedures we outline below are based on principles and recommendations of the WOAH and IUNC (Brückner et al. 2010; Jakob-Hoff et al. 2014; Vose 2000); we provide examples of how they can be applied to parts of an IRA for the introduction of a ranavirus into an uninfected area. While this section focuses on IRA, many of the principles and recommendations can be applied to a wildlife DRA, which shares much of the terminology and concepts (Fig. 9).

3.2.1 Defining the Hazard

The first step in an IRA is defining an area of interest. The area could be a population of interest, such as one that contains an uncommon species that is susceptible to ranavirus, or it could be a geographic region or country (Rödder et al. 2009; OIE 2014). Generally, areas are defined based on artificial or natural barriers to animal movement or pathogen translocation (OIE 2014). For example, ranavirus virions can flow downstream in tributaries, and associated floodplains are often corridor for animal movement; thus, areas should be defined by watershed for lotic systems. In lentic systems, depressional wetlands or lakes containing possible ranavirus hosts could be defined as the area of interest, if it is hydrologically closed and surrounded by a terrestrial landscape. In zoological settings, the area of interest typically is the captive facility (OIE 2014).

The next step is determining the presence of ranavirus in the area of interest. Section 1 discussed surveillance studies, and additional guidelines are provided by OIE (2014). Minimum sample size to detect ranavirus depends on several factors (Sect. 1). Additionally, infrequent sampling can result in lack of detection. Todd-Thompson (2010) showed that ranavirus in Gourley Pond of the Great Smoky Mountains National Park appeared nonexistent except for a 3-week period in late spring when an outbreak occurred resulting in widespread mortality across multiple species. Thus, sampling sites every 2 weeks when hosts are present with a large sample size ($n > 30$) should result in a high detection probability. If resources are limited, sampling at least four periods per year while hosts are present may be sufficient. Using this sampling frequency, Hoverman et al. (2012) detected ranavirus at 33 of 40 sites. Given that ranavirus could have been present at all sites in this study, a ballpark estimate of detection probability was 82.5–100% with their sampling frequency. Sampling should be performed over several years to verify that a site is ranavirus negative. Value of information analysis (Sect. 3.1) may be useful for determining the cost benefit tradeoff of additional surveillance. For large areas of interest, multiple sites spaced no less than the average dispersal distance of hosts should be sampled, which for amphibians is about 1 km (Wells 2007). Thus, distinct populations should be sampled without leaving large gaps between them. Environmental DNA methods may also be useful for determining ranavirus status without extensive effort (Brunner 2020). If ranavirus is detected, there is no reason to conduct an IRA, unless there is concern of a foreign strain of ranavirus being introduced.

Although the primary interest in the introduction of ranavirus to an area typically is for a certain species of concern, it is important that all ranavirus hosts are considered in an IRA. Marschang et al. (2024) provide a list of known ranavirus hosts, and several challenge studies (e.g., Hoverman et al. 2011; Brenes et al. 2014b) can provide insight into relative difference in susceptibility between species.

3.2.2 Risk Assessment

Risk assessment involves three primary steps: identifying routes of introduction, identifying the consequence of introduction, and estimating risk. It is often useful to develop flow diagrams that illustrate each step of assessment (Figs. 9 and 10). To describe this process, below we provide an example of assessing risk to wild amphibians via import of aquacultured fish that are infected with ranavirus.

- *Routes of Introduction*

Routes of introduction could include dispersal paths of hosts or translocation of the virus on fomites attached to non-hosts (i.e., birds and mammals, Gray et al. 2009). Humans can play a large role in the possible introduction of ranavirus by moving between contaminated and uncontaminated sites. The environmental persistence of ranavirus in unsterile water and soil is probably at least 1 week (Nazir et al. 2012). Thus, recreationists that move among watersheds without decontaminating footwear or gear could be a major source of ranavirus introduction (Gray et al. 2009). Fish hatcheries are known sites of ranavirus outbreaks (Waltzek et al. 2014); thus, the release of clinically or subclinically infected fish or their effluent from the hatchery could be another major source of ranavirus introduction. For a particular area of interest, it is important to identify the most likely routes of introduction. It can be useful to divide routes of introduction into three stages: import, release, and exposure (Fig. 10). In the case of imported aquacultured fish, the following steps define the import stage:

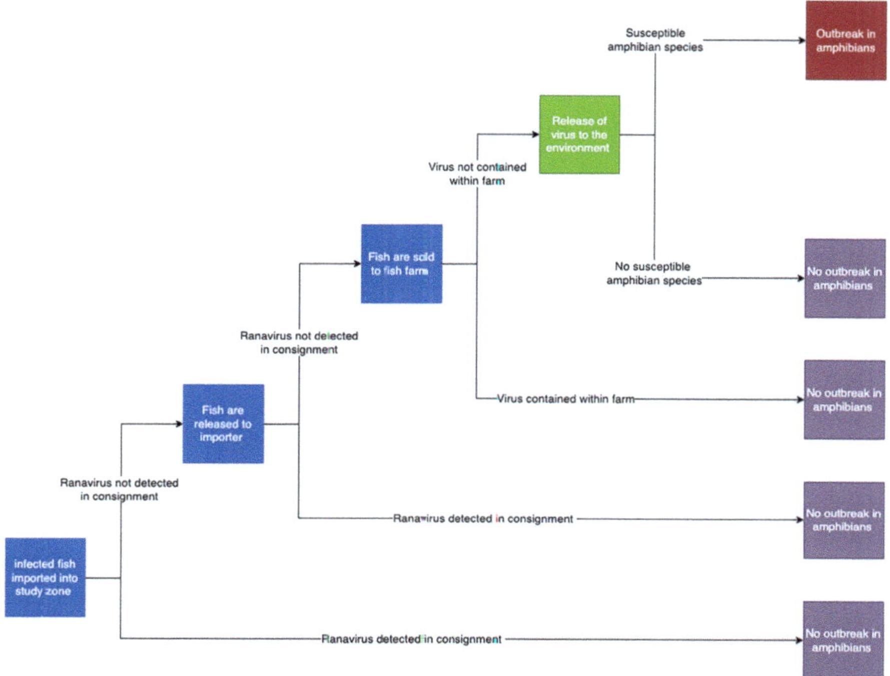

Fig. 10 Flow diagram for possible routes of transmission of ranavirus into a naïve susceptible population of amphibians in the wild. The routes of introduction are divided into three stages: import (blue), release (green), and exposure (red)

- Imported fish from an infected zone are infected with ranavirus.
- The infection passes undetected through border control.
- The infected fish are released to the retailer.
- The infected fish are sold to an aquaculture facility in the study zone.

Assuming that fish are contained in aquaculture ponds, ranavirus could be released into adjacent aquatic environments via several pathways:

- Virus contaminated effluent is released.
- Infected fish escape.
- Avian or mammalian predators could transport live or dead fish.
- Ranavirus hosts, such as amphibians or reptiles, could enter the pond, become infected, and disperse.
- Mechanical vectors, such as pets or humans, could transport the virus on fomites.

Finally, exposure to the virus could occur via several direct and indirect routes (Gray et al. 2009). Host species could be exposed to the virus in water, which is an efficient transmission medium, or the virus could be transmitted by direct contact or consumption of infected hosts (Miller et al. 2011). There is some evidence that ranavirus transmission can be density independent, which can increase extinction probabilities (Brunner et al. 2024).

- *Consequence Assessment*

The outcome of ranavirus infection in a species can be described qualitatively or quantitatively in terms of direct or indirect consequences. Direct consequences are the effect that ranavirus has on the species of interest, which typically includes estimating the likelihood of population declines and extinction (Sect. 2). Highly susceptible species that are rare have the greatest probability of extinction (Earl and Gray 2014), especially if these species co-occur with other ranavirus hosts. Indirect consequences are costs associated with pathogen surveillance (i.e., field and diagnostic expenses) and possible repatriation of populations following extinction.

- *Risk Estimation*

The assumption is that the virus will travel along the routes identified from an infected animal to a susceptible animal. In cases where it is determined that the consequence of ranavirus introduction is unacceptable, a series of critical control points (CCPs) should be established along the routes of introduction identified above, where the virus could be intercepted and the transmission terminated. The probability of the infection passing unnoticed through a CCP is estimated for each CCP by addressing several questions. This process can be summarized in a scenario tree, where each CCP has a "yes" and a "no" branch, and a likelihood of detection is assigned (Fig. 11). In Fig. 11, CCP 1 and 4 are predetermined for each border control post, while CCP 3 will depend on the training and experience of the inspectors. CCP 2 can be affected by viral load, water temperature, and animal health.

Detecting a pathogen in a laboratory test in CCP 5 is a function of two processes: sample size (Sect. 1) and performance of molecular tests (i.e., the sensitivity and specificity of PCR, Miller et al. 2024). The sensitivity and specificity of PCR for ranavirus are an ongoing research direction (Miller et al. 2024) and can be affected by sample type (i.e., lethal vs. nonlethal collection, Gray et al. 2012). In general, it is believed that the liver and kidney tissue provide the most reliable estimate of detection followed by tail, toe clips, and blood (Miller et al. 2024). Assuming perfect sensitivity and specificity of PCR, the probability of detecting ranavirus is approximately 95% (Sect. 1). Risk of not detecting ranavirus in an imported consignment is calculated as $-1 -$ the product of the detection probabilities at all CCPs (Fig. 11).

3.3   Risk Management and Communication

To manage the risk of ranavirus introduction, it is useful to perform a risk- consequence assessment. If risk is low but the consequence to the target species is high, risk management priority would be high. If, however, the risk of introducing ranavirus is high but the consequences are low, risk management priority would be low. If

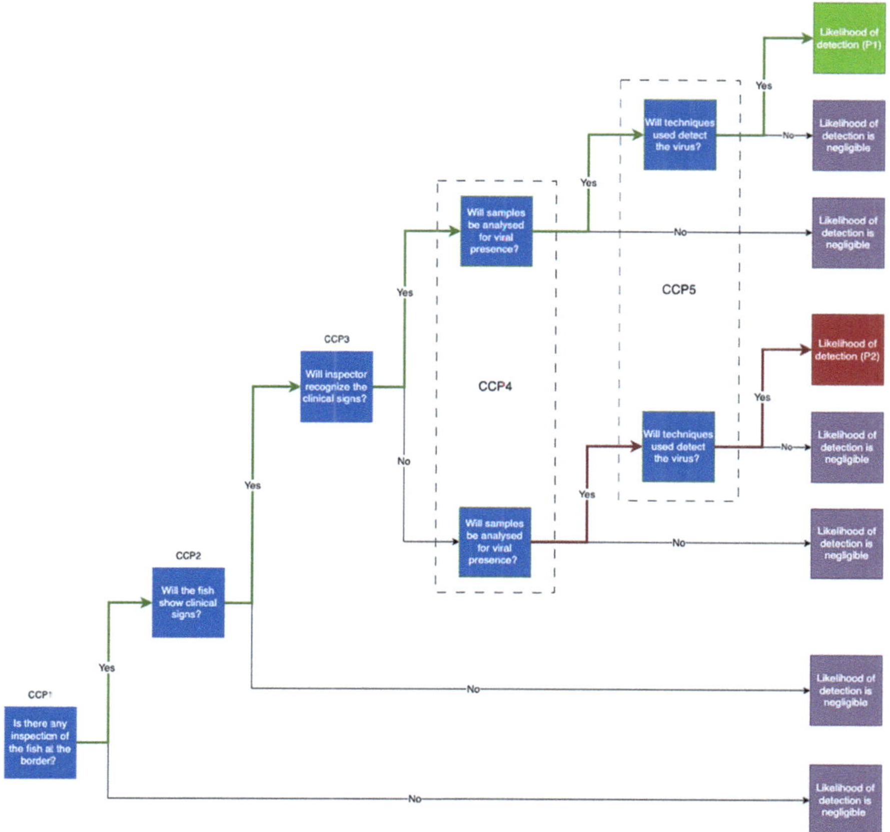

Fig. 11 Scenario tree for detection of ranavirus in a consignment of infected fish at the border inspection. Critical control points (CCP) 1–5 (blue) are opportunities identified where the virus could be detected and future transmission terminated. P1 (green) is the product of the "Yes" answer probabilities in the left branch of the tree. P2 (red) is the product of the two "yes" answer probabilities in the right branch of the scenario tree. The probability of ranavirus not being detected at the border is $-1 - (P1 + P2)$

the risk analysis indicates that the consequences are high, then the recommendations to management would focus on the CCPs and how to increase the likelihood of detecting and eliminating an infected consignment in a cost-effective manner.

Risk communication is an interactive and iterative process involving a two-way dialogue between decision-makers, experts, and stakeholders. Risk communication is used to gather information and opinions regarding hazards and risks and engage with relevant experts and stakeholders to maximize the quality of the analysis and the probability of recommendation implementation. It is essential for determining levels of acceptable risk. Risk communication is encouraged throughout risk analysis (Fig. 9) with stakeholders involved in the analysis from the outset. Jacobson (2009) is a useful resource of effective communication skills for conservation and resource management problems. Several tools and templates for effective risk communication are available in the Manual of Procedures for Wildlife Disease Risk Analysis (Jakob-Hoff et al. 2014).

Effective communication is required among stakeholders, both when collecting information to feed into risk analyses and in terms of informing end users of the findings, management options, and their implementation. Risk communication is often centered at government level, but individual organizations such as fish farmers or herpetological societies can investigate and implement their own quarantine and surveillance guidelines with qualified diagnostic support. Cooperation and awareness at all levels will reduce the risk of introducing ranavirus into an uninfected area.

Many of the facts needed to carry out a comprehensive risk analysis may already be available in the published scientific literature and should be used to substantiate the recommendation. It is important to consider the applicability and quality of the published literature before it is used in risk analyses. Published data might be from a different species, time of year, or continent. If published data do not exist for your species or region, a pilot study can be performed to generate data. Alternatively, obtaining expert opinion following the Delphi method can be an approach to secure preliminary estimates for use in the risk analysis (Helmer 1967; Vose 2000). We recommend that all organizations that are interested in performing an IRA (or DRA) consult experts that study ranaviruses. The GRC is a collection of scientists, veterinarians, and practitioners that can provide guidance with setting up risk analyses (https://www.ranavirus.org/). Each continent has a regional GRC representative that can assist or make necessary connections with experts in your region.

Acknowledgments We graciously thank Mark Wilber and Joe Mihaljevic for their numerous helpful comments on an earlier draft, especially MW's corrections to precision of prevalence in the small populations (Sect.). MJG was partially supported by NSF Division of Environmental Biology grants 1814520 and 2207922. JLB was partially supported by NSF DEB grant 1754474. We thank the following organizations for providing funds to support Open Access publishing of the second edition: University of Tennessee's (UT) Open Publishing Support Fund, UT School of Natural Resources, UT Center for Wildlife Health, Washington State University Libraries, University of Mississippi Medical Center, Gordon State College, Association of Reptile and Amphibian Veterinarians, and the Global Ranavirus Consortium.

References

Abdalhamid B, Bilder CR, McCutchen EL, Hinrichs SH, Koepsell SA, Iwen PC (2020) Assessment of specimen pooling to conserve SARS CoV-2 testing resources. Am J Clin Pathol 153(6):715–718

Allen LJS (2006) An introduction to mathematical biology. Pearson, New York

Allender MC, Fry MM, Irizarry AR et al (2006) Intracytoplasmic inclusions in circulating leukocytes from an eastern box turtle (*Terrapene carolina carolina*) with iridoviral infection. J Wildl Dis 42:677–684

Augenblick N, Kolstad J, Obermeyer Z, Wang A (2022) Pooled testing efficiency increases with test frequency. Proceedings of the National Academy of Sciences of the United States of America 119(2). https://doi.org/10.1073/pnas.2105180119

Augusiak J, Van Den Brink PJ, Grimm V (2014) Merging validation and evaluation of ecological models to "evaludation": a review of terminology and a practical approach. Ecol Model 280:117–128

Ben-Ami F, Ebert D, Regoes RR (2010) Pathogen dose infectivity curves as a method to analyze the distribution of host susceptibility: a quantitative assessment of maternal effects after food stress and pathogen exposure. Am Nat 175:106–115

Berger JO (1985) Statistical decision theory and Bayesian analysis. Springer-Verlag, New York

Biek R, Funk WC, Maxwell BA et al (2002) What is missing in amphibian decline research: insights from ecological sensitivity analysis. Conserv Biol 16:728–734

Bilder CR, Tebbs JM, McMahan CS (2019) Informative group testing for multiplex assays. Biometrics 75(1):278–288. https://doi.org/10.1111/biom.12988

Blackwood JC, Childs LM (2018) An introduction to compartmental modeling for the budding infectious disease modeler. Lett Biomath 5(1):195–221

Brenes R (2013) Mechanisms contributing to the emergence of ranavirus in ectothermic vertebrate communities. PhD dissertation, University of Tennessee

Brenes R, Gray MJ, Waltzek TB et al (2014a) Transmission of ranavirus between ectothermic vertebrate hosts. PLoS One 9:e92476

Brenes R, Miller DL, Waltzek TB et al (2014b) Susceptibility of fish and turtles to three ranaviruses isolated from different ectothermic vertebrate classes. J Aquat Anim Health 26(2):118–126

Briggs CJ, Vredenburg VT, Knapp RA et al (2005) Investigating the population-level effects of chytridiomycosis: an emerging infectious disease of amphibians. Ecology 86:3149–3159

Briggs CJ, Knapp RA, Vredenburg VT (2010) Enzootic and epizootic dynamics of the chytrid fungal pathogen of amphibians. Proc Natl Acad Sci USA 107:9695–9700

Brückner G, MacDiarmid SC, Murray N, Berthe F, Müller-Graf C, Sugiura K, Zepeda C, Kahn S, Mylrea G (2010) Handbook on import risk analysis for animal and animal products. In: Introduction and qualitative risk analysis, vol I, 2nd edn. World Organisation for Animal Health, Paris

Brunner JL (2020) Pooled samples and eDNA-based detection can facilitate the "clean trade" of aquatic animals. Sci Rep 10(1):10280. https://doi.org/10.1038/s41598-020-66280-7

Brunner JL, Yarber CM (2018) Evaluating the importance of environmental persistence for *Ranavirus* transmission and epidemiology. Adv Virus Res 101:129–148

Brunner JL, Richards K, Collins JP (2005) Dose and host characteristics influence virulence of ranavirus infections. Oecologia 144:399–406

Brunner JL et al (2024) Ranavirus ecology: from individual infections to population epidemiology to community impacts. In: Ranaviruses: Emerging pathogens of ectothermic vertebrates. Springer, Cham

Cameron AR, Baldock FC (1998) A new probability formula for surveys to substantiate freedom from disease. Prev Vet Med 34:1–17

Campbell LJ, Garner TWJ, Tessa G et al (2018) An emerging viral pathogen truncates population age structure in a European amphibian and may reduce population viability. PeerJ 6:e5949

Canessa S, Guillera-Arroita G, Lahoz-Monfort JJ, Southwell DM, Armstrong DP, Chadès I, Lacy RC, Converse SJ (2015) When do we need more data? A primer on calculating the value of information for applied ecologists. Methods Ecol Evol 6(10):1219–1228. https://doi.org/1 0.1111/2041-210X.12423

Cariboni J, Gatelli D, Liska R et al (2007) The role of sensitivity analysis in ecological modelling. Ecol Model 203:167–182

Caswell H (2000) Matrix population models: construction, analysis, and interpretation, 2nd edn. Sinauer, Sunderland

Chen Z, Li T, Gao X, Wang C, Zang Q (2018) Protective immunity induced by DNA vaccination against ranavirus infection in Chinese Giant salamander *Andrias davidianus*. Viruses 10(2):52

Cooch EG, Conn PB, Ellner SP et al (2012) Disease dynamics in wild populations: modeling and estimation: a review. J Ornithol 152(2):485–509

Diekmann O, Heesterbeek JA, Metz JA (1990) On the definition and the computation of the basic reproduction ratio R 0 in models for infectious diseases in heterogeneous populations. J Math Biol 28:365–382

DiRenzo GV, Grant EHC (2019) Overview of emerging amphibian pathogens and modeling advances for conservation-related decisions. Biol Conserv 236:474–483

Dobson A (2004) Population dynamics of pathogens with multiple host species. Am Nat 164:S64–S78

Dohoo I, Martin W, Stryhn H (2003) Veterinary epidemiologic research. AVC, Charlottetown

Dorfman R (1943) The detection of defective members of large populations. Ann Math Stat 14:436–440

Duffus ALJ, Waltzek TB, Stöhr AC, Allender MC, Gotesman M, Whittington RJ, Hick P, Hines MK, Marschang RE (2015) Distribution and host range of ranaviruses. In: Gray MJ, Chinchar VG (eds) Ranaviruses: lethal pathogens of ectothermic vertebrates. Springer, New York

Duffus ALJ, Garner TWJ, Nichols RA et al (2019) Modelling ranavirus transmission in populations of common frogs (*Rana temporaria*) in the United Kingdom. Viruses 11:556

Earl JE (2019) Evaluating the assumptions of population projection models used for conservation. Biol Conserv 237:145–154

Earl JE, Gray MJ (2014) Introduction of ranavirus to isolated wood frog population could cause local extinction. EcoHealth 11:581–592

Earl JE, Chaney JC, Sutton WB et al (2016) Ranavirus could facilitate local extinction of rare amphibian species. Oecologia 182:611–623

Farrell AP, Collins JP, Greer AL, Thieme HR (2018) Times from infection to disease-induced death and their influence on final population sizes after epidemic outbreaks. Bull Math Biol 80:1937–1961

Federico P, Gross LJ, Lenhart S et al (2013) Optimal control in individual-based models: implications from aggregated methods. Am Nat 181:64–77

Fenichel EP, Horan RD, Hickling GJ (2010) Management of infectious wildlife diseases: bridging conventional and bioeconomic approaches. Ecol Appl 20:903–914

Fofana AM, Hurford A (2017) Mechanistic movement models to understand epidemic spread. Philos Trans R Soc B 372:20160086

Gray MJ, Miller DL, Hoverman JT (2009) Ecology and pathology of amphibian ranaviruses. Dis Aquat Org 87:243–266

Gray MJ, Miller DL, Hoverman JT (2012) Reliability of non-lethal surveillance methods for detecting ranavirus infection. Dis Aquat Org 99:1–6

Gray MJ, Spatz JA, Carter ED, Yarber CM, Wilkes RP, Miller DL (2018) Poor biosecurity could lead to disease outbreaks in animal populations. PLoS One 13(3):e0193243

Greer AL, Collins JP (2008) Habitat fragmentation as a result of biotic and abiotic factors controls pathogen transmission throughout a host population. J Anim Ecol 77:364–369

Grimm V, Railsback SF (2005) Individual-based modeling and ecology. Princeton University Press, Princeton

Grimm V, Railsback SF (2012) Pattern-oriented modelling: a "mulit-scope" for predictive systems ecology. Philos Trans R Soc Lond Ser B Biol Sci 367:298–310

Grimm V, Revilla E, Berger U et al (2005) Pattern-oriented modeling of agent-based complex systems: lessons from ecology. Science 310:987–991

Grimm V, Augusiak J, Focks A et al (2014) Towards better modelling and decision support: documenting model development, testing, and analysis using TRACE. Ecol Model 280:129–139

Haislip NA, Gray MJ, Hoverman JT et al (2011) Development and disease: how susceptibility to an emerging pathogen changes through anuran development. PLoS One 6:e22307

Halliday JE, Meredith AL, Knobel DL, Shaw DJ, Bronsvoort BM, Cleaveland S (2007) A framework for evaluating animals as sentinels for infectious disease surveillance. J R Soc Interface 4:973–984

Hanski I (1999) Metapopulation ecology. Oxford University Press, New York

Harp EM, Petranka JW (2006) Ranavirus in wood frogs (*Rana sylvatica*): potential sources of transmission within and between ponds. J Wildl Dis 42:307–318

Hastings A (1997) Population biology: concepts and models. Springer, New York

Helmer O (1967) Analysis of the future: the Delphi method. RAND, Santa Monica

Hens N, Shkedy Z, Aerts M, Faes C, Van Damme P, Beutels P (2012) In: Hens N, Shkedy Z, Aerts M, Faes C, Van Damme P, Beutels P (eds) Estimating the force of infection from incidence and prevalence modeling infectious disease parameters based on serological and social contact data: a modern statistical perspective. Springer, New York

Hess G (1996) Disease in metapopulation models: implications for conservation. Ecology 77:1617–1632

Hethcote HW (2000) The mathematics of infectious disease. SIAM Rev 42:599–653

Hitt B, Bilder C, Tebbs J, McMahan C (2019) The objective function controversy for group testing: much ado about nothing? Stat Med 38:4912–4923

Hogan CA, Sahoo MK, Pinsky BA (2020) Sample pooling as a strategy to detect community transmission of SARS-CoV-2. JAMA 323(19):1967–1969

Hoverman JT, Gray MJ, Haislip NA et al (2011) Phylogeny, life history, and ecology contribute to differences in amphibian susceptibility to ranaviruses. EcoHealth 8:301–319

Hoverman JT, Gray MJ, Miller DL et al (2012) Widespread occurrence of ranavirus in pond-breeding amphibian populations. EcoHealth 9:36–48

Jacobson SK (2009) Communication skills for conservation professionals, 2nd Ed. Island Press, Washington, District of Columbia

Jakob-Hoff RM, MacDiarmid SC, Lees C, Miller PS, Travis D, Kock R (2014) Manual of Procedures for Wildlife Disease Risk Analysis. World Organisation for Animal Health, Paris, 160 pp. Published in association with the International Union for Conservation of Nature and the Species Survival Commission

Johnson AJ, Pessier AP, Jacobson ER (2007) Experimental transmission and induction of ranaviral disease in western ornate box turtles (*Terrapene ornata ornata*) and red-eared sliders (*Trachemys scripta elegans*). Vet Pathol 44:285–297

Keesing F, Holt RD, Ostfeld RS (2006) Effects of species diversity on disease risk. Ecol Lett 9:485–498

Kimble SJA, Karna AK, Johnson AJ, Hoverman JT, Williams RN (2015) Mosquitoes as a potential vector of ranavirus transmission in terrestrial turtles. EcoHealth 12:334–338

Kolby JE, Smith KM, Berger L et al (2014) First evidence of amphibian chytrid fungus (*Batrachochytrium dendrobatidis*) and ranavirus in Hong Kong amphibian trade. PLoS One 9:e90750

Lélu M, Langlais M, Poulle M et al (2013) When should a trophically and vertically transmitted parasite manipulate its intermediate host? The case of *Toxoplasma gondii*. Proc Biol Sci 280:20131143

Lenhart S, Workman JT (2007) Optimal control applied to biological models. Chapman & Hall/CRC, Boca Raton

Lloyd-Smith JO, Cross PC, Briggs CJ et al (2005) Should we expect population thresholds for wildlife disease? Trends Ecol Evol 20:511–519

Malagon DA, Melara LA, Prosper OF, Lenhart S, Carter ED, Fordyce JA, Peterson AC, Miller DA, Gray MJ (2020) Host density and habitat structure influence host contact rates and *Batrachochytrium salamandrivorans* transmission. Sci Rep 10:5584

Marschang RE et al (2024) Ranavirus distribution and host range. In: Ranaviruses: Emerging pathogens of ectothermic vertebrates. Springer, Cham

Marsh DM, Trenham PC (2001) Metapopulation dynamics and amphibian conservation. Conserv Biol 15:40–49

McCallum H (2008) Host–pathogen and host–parasite models. In: Population parameters: estimation for ecological models. John Wiley & Sons, New York, pp 284–312

McCallum H, Barlow N, Hone J (2001) How should pathogen transmission be modeled. Trends Ecol Evol 16:295–300

Mideo N, Barclay VC, Chan BHK et al (2008) Understanding and predicting strain-specific patterns of pathogenesis in the rodent malaria, *Plasmodium chabaudi*. Am Nat 172:E214–E238

Mideo N, Savill NJ, Chadwick W et al (2011) Causes of variation in malaria infection dynamics: insights from theory and data. Am Nat 178:174–188

Mijaljevic JR, Greer AL, Brunner JL (2019) Evaluating the within-host dynamics of *Ranavirus* infection with mechanistic disease models and experimental data. Viruses 11:396

Miller DL, Gray MJ, Storfer A (2011) Ecopathology of ranaviruses infecting amphibians. Viruses 3:2351–2373

Miller DL et al (2024) Pathology and diagnostics. In: Ranaviruses: Emerging pathogens of ectothermic vertebrates. Springer, Cham

Mosher BA, Brand AB, Wiewel ANM, Miller DAW, Gray MJ, Miller DL, Grant EHC (2019) Estimating occurrence, prevalence, and detection of amphibian pathogens: insights from occupancy models. J Wildl Dis 53:563–575

Nazir J, Spengler M, Marschang RE (2012) Environmental persistence of amphibian and reptilian ranaviruses. Dis Aquat Org 98:177–184

OIE (2014) Aquatic animal health code (online access). Office International des Epizooties, Paris. Accessed 21 May 2014

Otto SP, Day T (2007) A biologist's guide to mathematical modeling in ecology and evolution. Princeton University Press, Princeton

Peace A, O'Regan SM, Spatz JA et al (2019) A highly invasive chimeric ranavirus can decimate tadpole populations rapidly through multiple transmission pathways. Ecol Model 410:108777

Raiffa H (1968) Decision analysis: introductory lectures on choices under uncertainty. Addison-Wesley, Oxford, England

Railsback SF, Grimm V (2011) Agent-based and individual-based modeling: a practical introduction. Princeton University Press, Princeton

Rödder D, Kielgast J, Bielby J et al (2009) Global amphibian extinction risk assessment for the panzootic chytrid fungus. Diversity 1:52–66

Rosa GM, Bosch J, Martel A et al (2019) Sex-biased disease dynamics increase extinction risk by impairing population recovery. Anim Conserv 22:579–588

Rosa GM, Botto GA, Mitra AT et al (2022) Invasive host-pathogen dynamics leading to amphibian declines. Bio Conserv 276:109785

Sabino-Pinto J, Krause ET, Bletz MC, Martel A, Pasmans F, Steinfartz S, Vences M (2019) Detectability vs. Time and costs in pooled DNA extraction of cutaneous swabs: a study on the amphibian chytrid fungi. Amphib Reptil 40(1):29–39. https://doi.org/10.1163/15685381-20181011

Schlaifer R, Raiffa H (1961) Applied statistical decision theory. ClintonPress Inc, Boston

Schloegel LM, Ferreira CM, James TY et al (2010) The North American bullfrog as a reservoir for the spread of *Batrachochytrium dendrobatidis* in Brazil. Anim Conserv 14:53–61

Smith MA, Green DM (2005) Dispersal and the metapopulation paradigm in amphibian ecology and conservation: are all amphibian populations metapopulations? Ecography 28:110–128

Smith KF, Behrens M, Schoegel LM et al (2009) Reducing the risk of wildlife trade. Science 324:594–595

St-Amour V, Lesbarrères D (2007) Detecting ranavirus in toe clips: an alternative to lethal sampling methods. Conserv Genet 8:1247–1250

Todd-Thompson M (2010) Seasonality, variation in species prevalence, and localized disease for ranavirus in cades cove (Great Smoky Mountains National Park) amphibians. M.S., University of Tennessee. http://trace.tennessee.edu/utk_gradthes/665/. Accessed 21 May 2014

Travis DA, Smith K (2019) Risk analysis framework guidance for wildlife health professionals. In: Fowler's zoo and wild animal. Elsevier, St. Louis

Van den Driessche P (2017) Reproduction numbers of infectious disease models. Infect Dis Model 2:288–303

Vose D (2000) Risk analysis—a quantitative guide, 2nd edn. Wiley, Chichester

Waltzek TB, Miller DL, Gray MJ et al (2014) New disease records for hatchery-reared sturgeon. I. Expansion of frog virus 3 into *Scaphirhynchus albus*. Dis Aquat Org 111:219–227

Warne RW, Crespi EJ, Brunner JL (2011) Escape from the pond: stress and developmental responses to ranavirus infection in wood frog tadpoles. Funct Ecol 25:139–146

Wells KD (2007) The ecology and behavior of amphibians. University of Chicago Press, Chicago

Wiegand T, Jeltsch F, Hanski I et al (2003) Using pattern-oriented modeling for revealing hidden information: a key for reconciling ecological theory and application. Oikos 100:209–222

Williams BK, Johnson FA (2015) Value of information and natural resources decision-making. Wildl Soc Bull 39(3):488–496. https://doi.org/10.1002/wsb.575

Wilson EB (1927) Probable inference, the law of succession, and statistical inference. J Am Stat Assoc 22:209–212

WOAH (2021) Aquatic animal health code (online access). Office International des Epizooties, Paris. https://www.woah.org/en/what-we-do/standards/codes-and-manuals/aquatic-code-online-access

Wobeser GA (2006) Essentials of disease in wild animals. Blackwell, Ames

Woodhams DC, Alford RA, Briggs CJ et al (2008) Life-history trade-offs influence disease in changing climates: strategies of an amphibian pathogen. Ecology 89:1627–1639

Woodhams DC, Bosch J, Briggs CJ et al (2011) Mitigating amphibian disease: strategies to maintain wild populations and control chytridiomycosis. Front Zool 8:8

World Organisation for Animal Health (WOAH) (2021) Manual of diagnostic tests for aquatic animals. OIE, Paris. Chapter 2.1.2

Zuur AF, Savaliev AA, Ieno EN (2012) Zero inflated models and generalized linear mixed models with R. Highland Statistics, Newburgh